Designing Multi-Target Drugs

RSC Drug Discovery Series

Titles in the Series:

1: Metabolism, Pharmacokinetics and Toxicity of Functional Groups: Impact of Chemical Building Blocks on ADMET
2: Emerging Drugs and Targets for Alzheimer's Disease; Volume 1: Beta-Amyloid, Tau Protein and Glucose Metabolism
3: Emerging Drugs and Targets for Alzheimer's Disease; Volume 2: Neuronal Plasticity, Neuronal Protection and Other Miscellaneous Strategies
4: Accounts in Drug Discovery: Case Studies in Medicinal Chemistry
5: New Frontiers in Chemical Biology: Enabling Drug Discovery
6: Animal Models for Neurodegenerative Disease
7: Neurodegeneration: Metallostasis and Proteostasis
8: G Protein-Coupled Receptors: From Structure to Function
9: Pharmaceutical Process Development: Current Chemical and Engineering Challenges
10: Extracellular and Intracellular Signaling
11: New Synthetic Technologies in Medicinal Chemistry
12: New Horizons in Predictive Toxicology: Current Status and Application
13: Drug Design Strategies: Quantitative Approaches
14: Neglected Diseases and Drug Discovery
15: Biomedical Imaging: The Chemistry of Labels, Probes and Contrast Agents
16: Pharmaceutical Salts and Cocrystals
17: Polyamine Drug Discovery
18: Proteinases as Drug Targets
19: Kinase Drug Discovery
20: Drug Design Strategies: Computational Techniques and Applications
21: Designing Multi-Target Drugs

How to obtain future titles on publication:
A standing order plan is available for this series. A standing order will bring delivery of each new volume immediately on publication.

For further information please contact:
Book Sales Department, Royal Society of Chemistry, Thomas Graham House, Science Park, Milton Road, Cambridge, CB4 0WF, UK
Telephone: +44 (0)1223 420066, Fax: +44 (0)1223 420247, Email: books@rsc.org
Visit our website at http://www.rsc.org/Shop/Books/

Designing Multi-Target Drugs

Edited by

J. Richard Morphy
*Stirling, UK**

C. John Harris
Eynsford, Kent, UK

*Current address: Lilly Research Centre, Windlesham Research Centre, Surrey GU20 9PH, UK.

RSC Publishing

RSC Drug Discovery Series No. 21

ISBN: 978-1-84973-362-5
ISSN: 2041-3203

A catalogue record for this book is available from the British Library

Published by The Royal Society of Chemistry,
Thomas Graham House, Science Park, Milton Road,
Cambridge CB4 0WF, UK

Registered Charity Number 207890

For further information see our web site at www.rsc.org

Foreword

What is Multi-Targeted Drug Discovery (MTDD)?

This book consists of chapters concerned with a variety of aspects related to 'multi-targeted drug discovery' (MTDD). A good definition of MTDD is given by Metz and Hajduk[1] 'Multi-targeted drugs are promiscuous and exhibit activity against a wide range of molecular targets. In fact, it is now commonly accepted that the polypharmacology of these drugs (i.e. their ability to modulate the activity of multiple protein targets) is at least partly responsible for their efficacy'. MTDD is also described by the term 'designed multiple ligands' as described in multiple publications by Morphy.[2]

The terms 'promiscuous' or 'dirty drug' may have a pejorative aspect in that although multiple biological activities can be useful, leading to enhanced efficacy, they also may not be useful, leading to enhanced undesirable pharmacology (toxicity). Not to be confused with multi-targeting, the term 'promiscuous' has also been used in the realm of high throughput screening (HTS) to denote HTS assay biological activity related to usually undesirable chemical or physicochemical features. This phenomenon may be associated with covalent bond formation between ligand and target,[3] or undesirable *in vitro* biophysical effects such as colloidal particle aggregate formation.[4] In both cases, the observed promiscuity is not associated with useful biological activity. In contrast to the *in vitro* situation, it should be noted that colloidal aggregate formation *in vivo* in the gastrointestinal tract may be beneficial by enhancing oral absorption.[5] There is also a drug discovery viewpoint in favor of ligands with the potential to form covalent bonds between ligand and target.[6] However, in this author's opinion, these are minority viewpoints in dedicated drug discovery.

RSC Drug Discovery Series No. 21
Designing Multi-Target Drugs
Edited by J. Richard Morphy and C. John Harris

Published by the Royal Society of Chemistry, www.rsc.org

Why is there an Upsurge in Interest in MTDD?

The existence of polypharmacology, which provides the foundation for MTTD, has been known to medicinal chemists for decades. For example, the concept of privileged chemistry structures was first described by Evans *et al.* in 1988,[7] reviewed by Patchett in 2000,[8] and discussed in a drug discovery and library design context in 2010.[9] The Merck group's work on the use of the benzodiazepine scaffold originally found in anxiolytics provides a rich example of how relatively small structural changes to a scaffold can lead to a variety of unrelated biological activities. In their words, 'What is clear is that certain "privileged structures" are capable of providing useful ligands for more than one receptor and that judicious modification of such structures could be a viable alternative in the search for new receptor agonists and antagonists'. It is clear that in this early work that the concept of polypharmacology was well understood although it was uncertain if compounds with polypharmacology might be rare and difficult to find. We now know that privileged structures (i.e. promiscuous scaffolds) are much more numerous than previously supposed.[10]

This work on MTTD is being published in 2012. The reader may well ask what has changed over the last two decades to bring MTTD to greater attention. In this author's opinion, one major change is the gradual realization, especially in this last decade, that the superbly selective single drug with high affinity for a single biological target coupled with clinical efficacy is, charitably speaking, 'the exception'; more critically, some view this as a 'fundamentally flawed' approach to drug discovery.[11–14] Illustrating the charitable viewpoint, it is estimated that a complete single point pathway knockout results in a phenotypic response in only about 10–15% of cases. This low efficacy of the single-mechanism drug discovery approach is the explanation for the intense interest in target validation. How does one find the magic 10–15% of potential targets where the single-mechanistic approach has a chance of working? The low efficacy of the single-mechanism approach places HTS into context. HTS is only a tool and the HTS approach to drug discovery is critically dependent on target validation. Explaining the 'fundamentally flawed' viewpoint is the genomics-driven 'drug discovery factory' approach[15] of the early 1990s which wasted hundreds of millions of dollars and the efforts of many talented scientists.

A second major change is the realization that polypharmacology is the rule rather than the exception among clinically useful drugs.[16,17] Finally, the wealth of ligand to target database information in the current era allows the exploitation of a more chemo centric as opposed to molecular biology centric view of drug discovery. This change is well described in the following quote from the review by Shoichet:[18] 'What is new in the past few years is the quantitative restatement of classical ideas, allowing formal comparisons among targets and ligands at a scale not previously attempted. This has suggested unexpected relationships among receptors, identified targets active in phenotypic screens, and predicted off-targets and new disease indications for drugs.'

Chemical Space, Polypharmacology and MTDD

The distribution of biologically active compounds in chemistry space is critical to the concepts of polypharmacology and MTDD. If biologically active compounds are widely or uniformly distributed in chemical space then one might expect polypharmacology to be rare and MTDD would likely not work. Conversely, if biologically active compounds are clustered in chemistry space then polypharmacology should be common and MTTD should be tractable.

Chemical space is finite but exceedingly large. As discussed in a review by Reymond *et al.*[19] 'Is chemical space finite? Yes, if boundaries are defined. For small molecule drug discovery the natural limit is the molecular weight, which must be capped at 300–500 Da to ensure reasonable bioavailability. This chemical space of drug-like molecules has been estimated to be in excess of 10^{60} molecules.' The key medicinal chemistry question relevant to MTDD is whether biologically active compounds are evenly distributed in this incredibly large chemical space. In this and other authors' opinion the answer for synthetic compounds is a resounding 'no'. Multiple papers in the literature attest to the very uneven distribution of biologically active synthetic compounds in chemistry space.[20–22] Synthetically made biologically active compounds (as might be made by medicinal chemists) are most definitely not evenly distributed in chemical space. In fact, even without consideration of biological activity, the distribution of chemical structure scaffolds in the chemical literature is highly biased.[23] Screening truly diverse compounds is the worst way to discover a drug because the current evidence suggests that most of chemistry space is not populated by biologically active synthetic compounds.

An Issue of Timing: When is MTDD/Polypharmacology Undesirable?

Polypharmacology can be undesirable in a chemical biology context as opposed to a drug discovery context. Broadly speaking, chemical ligands can be tested in biology assays for two purposes: to discover drugs or to discover something about a biological process.[24] From a drug discovery perspective, polypharmacology is extremely useful. However, in a chemical biology context where one may be using a molecule as a tool or probe to learn something about a biological process,[25] perhaps to interrogate a step in a pathway or to discover a mechanism, selectivity is a key attribute and polypharmacology is a detriment. This is especially the case in phenotypic screening where the active chemical ligand becomes the tool or probe that is the starting point for the detective work to discover mechanism. Even when the stated screening goal is a chemical biology tool or probe, selectivity is difficult to achieve. For example, in a crowd sourcing evaluation of the 64 tools and probes resulting from the NIH roadmap HTS screening effort, about one-quarter were judged to be deficient with respect to selectivity.[26] The use of chemical biology probes with truly high selectivity can play a key role in understanding how to rationally design multi-targeted drugs, which is the key theme of this book.

Christopher Lipinski

References

1. J. T. Metz and P. J. Hajduk, *Curr. Opin. Chem. Biol.*, 2010, **14**, 498–504.
2. R. Morphy and Z. Rankovic, *Curr. Pharm. Des.*, 2009, **15**, 587–600.
3. J. B. Baell, *Future Med. Chem.*, 2010, **2**, 1529–1546.
4. A. Jadhav, R. S. Ferreira, C. Klumpp, B. T. Mott, C. P. Austin, J. Inglese, C. J. Thomas, D. J. Maloney, B. K. Shoichet and A. Simeonov, *J. Med. Chem.*, 2010, **53**, 37–51.
5. A. K. Doak, H. Wille, S. B. Prusiner and B. K. Shoichet, *J. Med. Chem.*, 2010, **53**, 4259–4265.
6. D. S. Johnson, E. Weerapana and B. F. Cravatt, *Future Med. Chem.*, 2010, **2**, 949–964.
7. B. E. Evans, K. E. Rittle, M. G. Bock, R. M. DiPardo, R. M. Freidinger, W. L. Whitter, G. F. Lundell, D. F. Veber and P. S. Anderson, *et al.*, *J. Med. Chem.*, 1988, **31**, 2235–2246.
8. A. A. Patchett and R. P. Nargund (Merck Research Laboratories, Rahway, NJ, USA), *Annu. Rep. Med. Chem.*, 2000, **35**, 289–298.
9. M. E. Welsch, S. A. Snyder and B. R. Stockwell, *Curr. Opin. Chem. Biol.*, 2010, **14**, 347–361.
10. Y. Hu and J. Bajorath, *J. Chem. Inf. Model.*, 2010, **50**, 500–510.
11. R. L. Ho and C. A. Lieu, *Drugs in R&D*, 2008, **9**, 203–216.
12. H. Kitano, *Nat. Rev. Drug Discovery*, 2007, **6**, 202–210.
13. D. Brown, *Drug Discovery Today*, 2007, **12**, 1007–1012.
14. F. Sams-Dodd, *Drug Discovery Today*, 2006, **11**, 465–472.
15. U. A. K. Betz, R. Farquhar and K. Ziegelbauer, *Curr. Opin. Chem. Biol.*, 2005, **9**, 387–391.
16. A. L. Hopkins, *Nat. Chem. Biol.*, 2008, **4**, 682–690.
17. Y. Hu and J. Bajorath, *J. Chem. Inf. Model.*, 2010, **50**, 2112–2118.
18. M. J. Keiser, J. J. Irwin and B. K. Shoichet, *Biochemistry*, 2010, **49**, 10267–10276.
19. J.-L. Reymond, R. van Deursen, L. C. Blum and L. Ruddigkeit, *MedChemComm*, 2010, **1**, 30–38.
20. J. Hert, J. J. Irwin, C. Laggner, M. J. Keiser and B. K. Shoichet, *Nat. Chem. Biol.*, 2009, **5**, 479–483.
21. P. Ertl, S. Jelfs, J. Muehlbacher, A. Schuffenhauer and P. Selzer, *J. Med. Chem.*, 2006, **49**, 4568–4573.
22. C. Lipinski and A. Hopkins, *Nature*, 2004, **432**, 855–861.

23. A. H. Lipkus, Q. Yuan, K. A. Lucas, S. A. Funk, W. F. Bartelt, III, R. J. Schenck and A. J. Trippe, *J. Org. Chem.*, 2008, **73**, 4443–4451.
24. T. Kodadek, *Nat. Chem. Biol.*, 2010, **6**, 162–165.
25. S. V. Frye, *Nat. Chem. Biol.*, 2010, **6**, 159–161.
26. T. I. Oprea, C. G. Bologa, S. Boyer, R. F. Curpan, R. C. Glen, A. L. Hopkins, C. A. Lipinski, G. R. Marshall, Y. C. Martin, L. Ostopovici-Halip, G. Rishton, O. Ursu, R. J. Vaz, C. Waller, H. Waldmann and L. A. Sklar, *Nat. Chem. Biol.*, 2009, **5**, 441–447.

Introduction

Why Design Multi-Target Drugs?

The promiscuity of a drug, that is, its tendency to bind to multiple drug targets is both a challenge and an opportunity for medicinal chemists. Drug designers attempt to reduce the 'off-target' toxicity liabilities of a compound by increasing the selectivity of a drug for one target over others. Analysis of the physicochemical properties of failed and successful drug candidates illustrate that drug target promiscuity, due to high lipophilicity, is linked to an increased risked of toxicity and failure in the clinic:[1,2] hence the traditional label of promiscuous compounds as 'dirty'.[3] Guiding the design principle of selectivity is not only a reduction in potential toxicity but also the assumption that the direct modulation of single proteins will produce clinical benefits: a reductionist paradigm summarised as 'one gene, one disease, one drug'.

Since the completion of the sequencing of the first draft of the human genome, evidence has been accumulating from functional genomics and the new field of network biology that biological systems are robust to perturbation of individual genes. The new insights are challenging the dominant assumption of single target-based drug discovery.[4–8] Insights into the robustness of phenotype to perturbation can be found from understanding the function of biological networks. The fundamental architecture of networks contributes to the robustness and redundancy of biological systems. Network analysis of biological pathways and interactions has revealed that much of the robustness of biological systems is derived from the structure of biological networks.[9,10] The scale-free nature of many biological networks produces systems that are resilient against random deletion of any one protein (node) but also critically dependent on a few highly connected hubs. Network biology analysis predicts that if, in most cases, deletion of individual nodes may have little effect on disease networks, modulating multiple proteins may be required to perturb

RSC Drug Discovery Series No. 21
Designing Multi-Target Drugs
Edited by J. Richard Morphy and C. John Harris

Published by the Royal Society of Chemistry, www.rsc.org

robust phenotypes.[6,9,11] A network approach to pharmacology[12] suggests two strategies: targeting highly connected hub proteins, which exhibit pleiotropic effects (such as the role of HSP90 in cancer) or the targeting of multiple proteins to increase the perturbation of a network.

The robustness of individual proteins to perturbation is also revealed by metabolic flux analysis where modulation of single components in a pathway rarely results in large changes in metabolic flux and therefore phenotype.[13–15] A greater degree of perturbation of phenotype is observed in systems where more than one gene product is modulated. The emergent phenotype that occurs from the perturbation of multiple proteins is demonstrated by the systematic experiments on synthetic lethality. Dual gene deletion experiments in model systems have shown that, whilst the deletion of any one of two genes by itself may show no effect, the deletion of both genes can lead to 'synthetic lethality' or 'synthetic sickness'.[16] When dual perturbations are introduced, by combining single gene genetic knock-outs with a second chemical perturbation, the number of essential genes in yeast is predicted to significantly expand the 18% of the genes for which singleton gene knockouts are lethal. A large-scale study by Hillenmeyer *et al.* demonstrates the extent of synthetic lethality when gene deletions are augmented by chemical interventions.[17] Under ideal conditions only 34% of single gene deletions result in lethality or sickness. When the whole genome panel of yeast single gene knock-outs was screened against a diverse, small molecular library and assayed against a wide range of environment conditions an additional 63% of gene knock-outs showed a growth phenotype.[17] Thus 97% of genes demonstrate a fitness defect when challenged with a small molecule under at least one environmental condition. The vast majority of genes may be redundant under any one environmental condition but there appears to be little redundancy across a spectrum of conditions when a genetic perturbation is combined with a chemical insult. Genes that may appear dormant and dispensable under one set of specific conditions may prove essential under other stresses.[18,19]

The fundamental property of inherent robustness of biological networks has profound implications for drug discovery; instead of searching for a single 'disease-modifying' gene, network biology suggests that the strategy should be to perturb the disease network.[20,21] Hellerstein has argued that the true targets of drugs are not individual proteins but functionally important biochemical pathways embedded in larger biological networks.[22]

The intellectual foundations of network pharmacology challenge deep assumptions behind target selection and validation. Responding to the new biological insights into the complexity, robustness and redundancy in disease phenotype is helping to drive the emergence of a new approach to drug discovery, that of polypharmacology or multi-target drug discovery (MTDD).[3–6,11,12,23–30] Therefore, understanding the polypharmacology of a drug and its effect on biological networks and phenotype is essential if we wish to improve efficacy and understand toxicity.

Over the past decade the assumption of the desirability of single drug target mechanisms has begun to be questioned.[6,12,31] In certain circumstances, it may be advantageous for a drug to act on multiple drug targets, deliberately and specifically, rather than be too selective. The chapters written for this book gather together in one volume the state-of-the-art of the emerging new field of MTDD. Successes in rational MTDD have already been reported, such as the approval of lapatinib, discussed in Chapter 8. Moreover, new tools are now emerging to aid the medicinal chemists to discover multi-target drugs. We hope this book serves as a record of the achievements of the field to date and provides inspiration for the development of rational MTDD as the next paradigm in drug discovery.

Andrew L. Hopkins
Division of Biological Chemistry and Drug Discovery
College of Life Sciences, University of Dundee,
D11 5EH, UK

References

1. P. D. Leeson and B. Springthorpe, *Nat. Rev. Drug Discovery*, 2007, **6**, 881–890.
2. J. D. Hughes, *et al.*, *Bioorg. Med. Chem. Lett.*, 2008, **18**, 4872–4875.
3. S. Frantz, *Nature*, 2005, **437**, 942–943.
4. B. L. Roth, D. J. Sheffler and W. K. Kroeze, *Nat. Rev. Drug Discovery*, 2004, **3**, 353–359.
5. C. G. Wermuth, *Drug Discovery Today*, 2004, **9**, 826–827.
6. P. Csermely, V. Agoston and S. Pongor, *Trends Pharm. Sci.*, 2005, **26**, 178–182.
7. C. T. Keith, A. A. Borisy and B. R. Stockwel, *Nat. Rev. Drug Discovery*, 2005, **4**, 71–78.
8. A. Petrelli and S. Giordano, *Curr. Med. Chem.*, 2008, **15**(5), 422.
9. A. L. Barabási and Z. N. Oltvai, *Nat. Rev. Genet.*, 2004, **5**, 101–113.
10. R. Albert, H. Jeong and A. L. Barabasi, *Nature*, 2000, **406**, 378–382.
11. T. Korcsmáros, M. S. Szalay, C. Böde, I. Kovács and P. Csermely, Expert Opin, *Drug Discovery*, 2007, **2**, 1–10.
12. A. L. Hopkins, *Nat. Chem. Biol.*, 2008, **4**, 682–690.
13. J. E. Bailey, *Nat. Biotechnol.*, 1999, **17**, 616–618.
14. J. E. Bailey, *Metab. Eng.*, 2001, **3**, 111–114.
15. M. K. Hellerstein, *J. Pharmacol. Exp. Ther.*, 2008, **325**, 1–9.
16. S. L. Ooi, *et al.*, *Trends Genet.*, 2006, **22**, 55–63.
17. M. E. Hillenmeyer, *et al.*, *Science*, 2008, **320**, 362–365.
18. L. M. Blank, L. Kuepfer and U. Sauer, *Genome Biol.*, 2005, **6**, R49.
19. R. Harrison, B. Papp, C. Pál, S. G. Oliver and D. Delneri, *Proc. Natl. Acad. Sci. USA*, 2007, **104**, 2307–2312.
20. Y. Chen, *et al.*, *Nature*, 2008, **452**, 429–435.
21. E. Schadt, *Mol. Nat.*, 2009, **461**, 218–223.
22. M. K. Hellerstein, *Metab. Eng.*, 2008, **10**, 1–9.

23. S. K. Mencher and L. G. Wang, *BMC Clin. Pharma.*, 2005, **5**, 3.
24. A. L. Hopkins, J. S. Mason and J. P. Overington, *Curr. Opin. Struct. Biol.*, 2006, **16**, 127–136.
25. C. S. Flordellis, A. S. Manolis, H. Paris and A. Karabinis, *Curr. Top. Med. Chem.*, 2006, **6**, 1791–1798.
26. N. Dessalew and M. Workalemahu, *Curr. Comput.-Aided Drug Design*, 2008, **4**, 76–90.
27. O. Keskin, A. Gursoy, B. Ma and R. Nussinov, *Curr. Topics Med. Chem.*, 2007, **7**, 943–951.
28. G. R. Zimmermann, J. Lehár and C. T. Keith, *Drug Discovery Today*, 2007, **12**, 34–42.
29. A. L. Hopkins, *Nat. Biotech.*, 2007, **25**, 1110–1111.
30. A. Schrattenholz and V. Soskic, *Curr. Med. Chem.*, 2008, **15**, 1520–1528.
31. R. Morphy, C. Kay and Z. Rankovic, *Drug Discovery Today*, 2004, **9**, 641–651.

Preface

Forming part of the Royal Society of Chemistry Drug Discovery Series, this book is intended to provide an integrated and comprehensive overview of modern approaches to multi-target drug discovery (MTDD) and the state of our knowledge in the over-arching field of polypharmacology. Given the intense current interest in this field, we hope that this book will be of significant interest to medicinal and computational chemists in the commercial sector and in academia, as well as the wider drug discovery community.

Many readers will already be aware of the serendipitous nature of the discovery of many existing multi-target drugs. In this book we intend to focus on the rational and practical execution of MTDD. The chapters are written by widely recognized experts and opinion leaders in the field. The first challenge of MTDD is to identify biologically validated combinations of targets relevant to a disease state. However, it is equally important that these disease-relevant combinations are chemically tractable from a medicinal chemistry perspective. The book thus follows a natural thread from target identification and validation, through lead generation and lead optimization, and finally to clinical development. A key feature of the book is a collection of seminal case studies chosen to illustrate the challenges and opportunities of MTDD. These include compounds at various stages of development from preclinical to marketed drugs.

In the forward, two eminent proponents of MTDD, Andrew Hopkins and Chris Lipinski, discuss the motivation and background rationale for designing multi-target drugs.

Although MTDD has been applied in the context of many diseases, it is particularly relevant to complex polygenic diseases such as cancer, inflammation and central nervous system (CNS) disorders. In Chapters 1 and 2, Jorrit Hornberg and Mo Shahid discuss the pathophysiological context of these diseases that makes the development of multi-target drugs such an attractive proposition.

RSC Drug Discovery Series No. 21
Designing Multi-Target Drugs
Edited by J. Richard Morphy and C. John Harris

Published by the Royal Society of Chemistry, www.rsc.org

A growing area of interest for target identification is network pharmacology. In Chapter 3, Malcolm Young, Steven Zimmer and Alan Whitmore discuss the properties of biological networks and how a judicious selection of certain combinations of targets (or nodes) using network simulation algorithms can have a disproportionate effect on the robustness of a disease network, thereby enhancing therapeutic effect.

The chemical tractability of a given target combination has been assessed using an increasing range of cheminformatics tools, as described by Michael Keiser and Elisabet Gregori-Puigjané in Chapter 4. These predictive polypharmacology methods typically rely upon various measures of chemical similarity between known ligands for each target and can reveal links between the binding sites of distant proteins, unrelated by overall sequence or structure or function. Potentially the same methods can also identify undesired targets associated with toxicity.

Whilst *in silico* predictions of polypharmacology are increasingly useful, currently the most reliable method for exploring the polypharmacology of compounds is *in vitro* panel screening using a broad and diverse set of targets. In Chapter 5, Jonathan Mason explains how these direct readouts of promiscuity and selectivity can be used to select tool compounds for target validation, the best starting compounds for further optimization and to provide differentiation amongst drug candidates.

The last quarter century has witnessed a seismic change from a world of drug discovery based upon phenotypic screening in whole organisms to the reductionist target-centric approach of today. Given the high clinical attrition seen in recent years, there are signs that this trend is reversing and in Chapter 6, Chris Lipinski describes how phenotypic screening is highly relevant to the MTDD field and how many active compounds identified in such screens are likely to display polypharmacology. Whilst it is feasible to optimize and clinically develop compounds without knowledge of mechanism, if such knowledge were to be available, then optimization of potency would be more straightforward. In Chapter 7, Kilian Huber, Giulio Superti-Furga and Georg Winter describe a range of methods for deconvoluting mechanism of action for hits from phenotypic screening.

Many different lead generation approaches have been used for MTDD projects and, given the added complexity of optimizing against multiple targets, finding high quality starting compounds is critical. Both rational design and screening approaches are described by Richard Morphy in Chapter 8, and *in silico* screening is described in more detail by Yuzong Chen in Chapter 9.

Lead optimization is a pivotal stage in the history of any drug discovery project and all the more so in the MTDD area, where the simultaneous optimization of multiple parameters must be addressed. In Chapter 10, Richard Morphy discusses the importance of optimizing the activity ratio, wider selectivity, physicochemical properties and the pharmacokinetic profile.

There are two possible scenarios for multi-target therapy, drug combinations and multi-target drugs, which are sometimes seen as competing world views but which are actually complementary in their value and applicability. In

Chapter 11, Janet Dancey and Jose Monzon discuss the challenges of clinical development of drug combinations in the cancer area and compare and contrast such combinations with multi-target agents such as the multi-kinase inhibitors like sunitinib.

The storyline of the book then moves to a series of case studies. Continuing the oncology theme from Chapter 11, Karen Lackey describes how the dual EGFR/erbB2 kinase inhibitor, lapatinib, was discovered and clinically developed to provide a marketed drug for treating breast cancer. The following three case studies are also oncology-focussed, illustrating the extent of current interest in MTDD for this highly complex and heterogeneous disease. Carlos García-Echeverría and Andreas Karlsson describe dual PI3K/mTOR inhibitors, Xiong Cai and Changgeng Qian discuss multi-target HDAC/kinase inhibitors, and Michael Wendt tells the story of how dual inhibitors of Bcl-2 and Bcl-xL were discovered.

The next three case studies describe the discovery of compounds with multiple targets within the CNS. First, John Lowe, one of the inventors of the anti-schizophrenia drug, ziprasidone, explains how the compound was rationally designed starting from the structures of dopamine and a serotonin receptor antagonist. Next, Robert Weikert writes about triple blockers of the serotonin, noradrenaline and dopamine transporters. Then in Chapter 18, Maria Laura Bolognesi, Carlo Melchiorre, Cornelis Van der Schyf and Moussa Youdim discuss a range of multi-target approaches for neurodegenerative diseases, such as the dual monoamine oxidase/acetylcholinesterase inhibitor, Ladostigil, for Alzheimer's disease and Parkinson's disease.

The book concludes with two final case studies from the cardiovascular and anti-infective disease areas. The first from Natesan Murugesan describes dual angiotensin-1/endothelin-A receptor antagonists for treating hypertension and in the second, Stephen East, Lloyd Czaplewski and David Haydon discuss dual inhibitors of DNA gyrase/topoisomerase IV as broad-spectrum antibacterial agents.

This is the first time that a single volume has gathered together in one place the current state-of-the-art, the achievements and the challenges of the MTDD field and, importantly, the lessons learned by researchers to date. Through an intentional focus on the rational and practical execution of MTDD, we hope that this book will play a significant role in facilitating the future development of the field.

Richard Morphy and John Harris

Contents

RSC Drug Discovery Series No. 21
Designing Multi-Target Drugs
Edited by J. Richard Morphy and C. John Harris

Published by the Royal Society of Chemistry, www.rsc.org

CHAPTER 1

Simple Drugs Do Not Cure Complex Diseases: The Need for Multi-Targeted Drugs

JORRIT J. HORNBERG

H Lundbeck A/S, Ottiliavej 9, 2500 Valby, Denmark
Email: JJH@Lundbeck.com

1.1 Introduction

The contents of this book provide a comprehensive overview of the field of polypharmacology and modern approaches to identify drugs that hit multiple targets, including a number of case studies. But why do we actually need such multi-targeted drugs? This introductory chapter aims to answer that question. First, there is clearly a need for better and safer drugs in the clinic and also to improve output (productivity) of drug discovery and development in general. Second, many diseases with unmet medical needs are in essence complex and multi-factorial. I will discuss two disease areas, cancer and rheumatoid arthritis, to exemplify this complexity. Systems biology and network control analysis have shown that the systems underlying complex diseases are robust against perturbations and are always controlled by more than one biochemical process. Therefore, aiming to hit multiple targets is a better strategy than to hit a single target. Finally, though polypharmacology is naturally associated with toxicology and off-target side effects, it can be argued that multi-targeted drugs,

RSC Drug Discovery Series No. 21
Designing Multi-Target Drugs
Edited by J. Richard Morphy and C. John Harris

Published by the Royal Society of Chemistry, www.rsc.org

when rationally designed, can actually have a larger therapeutic window than those hitting a single target and thus prove to be safer drugs.

1.2 The Need for Better and Safer Drugs

There are two main reasons why we need better and safer drugs.

Firstly, there is the unmet medical need in the clinic. Patients need safe cures and complex diseases are difficult to cure. Cancer survival rates, for example, are still lower than desired, roughly 50–65% in Europe and the US, making it a leading cause of death, responsible for almost 25% of all deaths in the US.[1,2] Other examples include autoimmune disorders, some cardiovascular diseases, diabetes and neurodegenerative diseases. The incidence of some of these diseases is expected to increase with the increasingly ageing population. Dementia, for instance, affects almost 1% of those at 60–64 years of age and that number doubles for each subsequent 5-year cohort to 25–33% of those ≥85 years of age.[3] Furthermore, the incidence of serious adverse drug reactions in hospitalized patients is so high that it ranks as the 4th–6th leading cause of death.[4]

Secondly, the low success rate of drug development calls for better and safer drugs. In the past few years, an average of only ~20 new drugs were approved annually.[5,6] This is the result of the high attrition rates in clinical development: about 90% of all new drugs fail after first-in-human testing, varying from 80% for cardiovascular diseases to 95% for cancer.[7] The main underlying causes were identified to be lack of efficacy and poor safety (toxicology and clinical safety), each accounting for ~30% of all failures.[7]

One may therefore argue that we need to hit better targets. However, going after novel targets has, in itself, not proven to be a particularly successful strategy for drug development. Attrition of candidates with a novel mechanism of action is higher than average.[8] In addition, many diseases with unmet medical needs are complex and multi-factorial. Therefore, an approach to hit multiple targets may be more successful. The next two sections of this chapter discuss the complexity of cancer and rheumatoid arthritis.

1.3 Cancer

Two decades ago, Fearon and Vogelstein proposed their genetic model for colorectal tumorigenesis.[9] From comparing cells from multiple stages of colorectal cancer, it became apparent that, at each stage, cells had acquired at least one additional mutation, compared to the previous stage. The fact that carcinogenesis is a multi-step process requiring multiple sequential mutations has since been confirmed many times, e.g. by similar models for other types of cancer and by the artificial creation of tumour cells by introducing defined genetic alterations.[10–13] Besides mutations that change the structure and function of a gene product, (epi)genetic alterations which influence gene expression also contribute to carcinogenesis, such as gene amplification, changes in DNA methylation and histone acetylation, and the functioning of micro-RNAs.[14–17] The list of genes which are causally implicated in cancer via genetic alteration currently contains 436 genes.[18,19] Though some genes are frequently mutated in many cancer types, it is rather a combination of low frequency mutations that drive the cancer phenotype and they differ per cancer type.[20] This (epi)genetic heterogeneity of cancer presents a major challenge.[21] Disturbances of the signal transduction pathways in which most of these cancer genes function leads to the so-called 'hallmarks of cancer', including evasion of apoptosis and growth control, self-sufficiency in growth signals, induction of angiogenesis, the ability to metastasize, evade immune surveillance and, indirectly, stress phenotypes.[20,22–24] It is important to recognize that cancer is a multi-factorial disease already at its origin (the genetic level) and in its essence (the hallmarks). But this is merely where the complexity starts.

The information flow through signalling pathways that ultimately constitute these hallmarks is highly complex. The many sequential steps in a pathway encompass many different biochemical processes, such as protein binding (e.g. recruitment of pathway components by scaffolds, binding of ligand to receptor, receptor dimerization, transcription factor binding) and enzymatic reactions (e.g. phosphorylation, methylation, acetylation, ubiquitination). The output of a signalling pathway does not follow the input in a linear fashion. In part, this is caused by non-linear kinetics of biochemical reactions and different expression levels of pathway components. Signalling is also subject to spatio-temporal control and macromolecular crowding, since pathway components can be physically separated or highly concentrated locally.[25–27] Moreover, about a decade ago, the concept that signal transduction pathways have a linear architecture had to be abandoned. First, the topology of regulatory circuitry, such as negative feedback loops, appeared to be a recurrent theme in signalling networks.[28] Second, pathways are so highly interconnected by direct interactions (e.g. via phosphorylation), indirect regulation (e.g. via gene expression) or by sharing pathway components, that they form complex signalling networks rather than linear pathways.[29–32] This network structure itself can give rise to new 'emergent properties' or 'systems behaviour'.[33–36]

Beyond the multitude of contributing factors on the genetic level and in signalling networks, the complexity extends on the supra-cellular level. The population of cancer cells in a tumour is heterogenic, e.g. with respect to their state of differentiation.[37] Cells communicate with each other either by direct interaction or by stimulating each other in paracrine growth factor loops. Targeting those loops is a strategy for treatment of certain cancers.[38,39] And within the microenvironment, complex interactions between tumour cells and stromal cells, like paracrine growth factor networks, stimulate tumour progression.[40,41] Then, there is the interaction with the host immune system. On the one hand, tumour cells can secrete cytokines which trigger macrophages to differentiate to a subtype that actually promotes tumour progression.[42] On the other hand, immune cells can clean up tumour cells, which makes the therapeutic strategy to stimulate the host immune system to launch an immunological attack on the tumour very promising.[43] A further example of an important supra-cellular interaction is the activation of angiogenesis.[44] Under hypoxic conditions, tumour cells secrete pro-angiogenic growth factors, thereby stimulating the proliferation of endothelial cells. This leads to new blood vessel formation and provides the tumour cells with access to more oxygen. If tumours do not trigger this 'angiogenic switch', they will not be able to grow and will remain dormant.[45]

In summary, it can be concluded that a combination of multiple genetic alterations that affect multiple processes give rise to the development of cancer.

1.4 Rheumatoid Arthritis

A similar conclusion can be drawn from the biology of many autoimmune disorders, such as rheumatoid arthritis (RA). This systemic disease is characterized by joint inflammation and subsequent cartilage and bone destruction. The development of the disease occurs roughly in three steps: (i) the onset of a modest local inflammation in the synovium by the innate immune system; (ii) triggering of an adaptive immune response by the recruitment and activation of immune cells, leading to full-blown inflammation; and (iii) the invasion of the cartilage and increased bone resorption leading to bone destruction.[46]

Many different cell types are involved during this process. Mast cells produce pro-inflammatory mediators, such as cytokines, and also tissue destructive proteases. They contribute to the initiation of the adaptive immune response by recruiting T-cells, influencing T-cell skewing and antigen presentation.[47] Also dendritic cells, in their function as antigen-presenting cells, are thought to play a role in initiation of the adaptive immune response.[48] Macrophages produce a large variety of cytokines and chemokines. Their abundance and activation in the synovium correlates with the severity of RA.[49] Abnormal lymphocyte function is thought to be a central part of RA pathology. T-cells mediate both the recruitment and the activation of other immune cells by direct cell–cell contact and cytokine production.[50] B-cells produce self-reactive antibodies, function as antigen-presenting cells and secrete cytokines.[51] Fibroblast-like synoviocytes in the synovial intimal lining also play a key role by producing cytokines that perpetuate inflammation and proteases that contribute to cartilage destruction. They develop a unique aggressive phenotype that increases invasiveness into the extracellular matrix and further exacerbates joint damage.[52] Another important cell type for bone destruction are osteoclasts.[53] Joint degeneration is ultimately mediated by a disturbance in bone homeostasis, which is normally kept by balancing bone formation (by osteoblasts) and bone resorption (by osteoclasts). In RA, bone loss is due to excess bone resorption by osteoclasts, which are activated by inflammatory factors, and simultaneously, bone formation by osteoblasts is also impaired at erosion sites.

These many different cell types (and their subtypes) all contribute to RA and they communicate with each other by direct cellular contact and by an extensive extracellular network of cytokines and chemokines.[46] Factors like TNFα, IL-1β, IL-2, IL-6, IL-7, IL-8, IL-15, IL-17, IL-18, IL-21, IL-22, IL-23,

GM-CSF, MMPs, RANKL and RANTES function in autocrine and paracrine networks that mediate the onset and propagation of the inflammation, as well as bone destruction.[54] The central role of cytokines in RA pathology is clearly evidenced by the relatively successful application of biologics that block cytokine function to treat the disease.[55]

Zooming in on the individual cell types presents a further level of complexity: the structure and function of signal transduction pathways mediating the inflammatory response and regulating cell proliferation, survival and differentiation, and expression of aforementioned extracellular factors. The activation of lymphocytes often requires two signals.[56] T-cell activation occurs after engagement of the T-cell receptor (TCR) with its cognate peptide–major histocompatibility complex (signal 1) and subsequent engagement of co-stimulatory molecules (signal 2). This 'second signal' contributes to T-cell activation by promoting proliferation, survival and effector function. A multitude of factors are involved in processing the signal from the TCR and activating downstream signalling pathways.[57,58] Similarly, B-cell activation upon stimulation of the B-cell receptor (BCR) requires co-stimulation for the development of complete effector function, and a complex cascade is involved in propagation of the signal.[56,59] The key downstream signalling pathways in inflammation are similar to, overlap with or are often even the same as the signalling pathways involved in cancer, e.g. JAK-STAT, NF-κB, MAP kinase pathways.[60–62] As discussed above in Section 1.3, these pathways are highly complex with respect to the large number of signalling molecules, the different nature of the reactions they are involved in, the structure of the pathways, with cross-talk and feedback loops, etc. Many kinases in these pathways are currently being pursued as drug targets for treatment of RA.[60,63–66]

In conclusion, RA is a complex disease, which depends on the combined action of many cell types and multiple factors at the supra- and intra-cellular levels.

1.5 Control of Complex Biological Systems

As complex diseases, such as cancer and rheumatoid arthritis, are indeed dependent on so many factors, it is a major challenge to produce successful treatment strategies. Systems biology is needed to integrate the available knowledge and develop comprehensive understanding of how systems behaviour depends on the components of the system and their interactions.[67–69] Substantial advances in understanding (and prediction) have been achieved by integrating computational modelling with quantitative experimental data on molecular and cellular networks, in particular in the field of cancer systems biology.[70,71] On the basis of network models, it has been proposed that partial inhibition of multiple drug targets is more effective than full inhibition of a single target.[72] This can be explained by the intrinsic robustness (the ability to maintain homeostasis) of living systems.[73] Biological networks are built up with a high degree of redundancy and a small fraction of essential components.[74] In order to modulate the phenotype of a complex and robust disease system, several perturbations of non-essential components must be combined. Network analysis can then further aid to identify successful combinations of drug targets.[71,75–79] It is essential to recognize that the robust nature of complex diseases is caused by their network properties and it is therefore the network that should be targeted, rather than isolated parts.[79,80]

A similar lesson can be drawn from the application of metabolic control analysis (MCA). MCA quantifies the extent to which individual reactions or network parts control the entire reaction network, such as the control of a kinase on downstream phosphorylation, the control of a signalling pathway on cell proliferation, or the control of a particular cell type on disease pathology. In that way, MCA aids in the selection of drug targets, based on the magnitude of their control.[81,82] Applying MCA to computational models of signal transduction indeed allowed for rank-ordering individual reactions: some reactions exert more control than others. Perhaps more surprisingly, it was also found that control tends to be distributed over more than a single reaction.[83–85] In other words, there was no single rate-limiting step. One can thus argue that multi-targeted drugs will be more effective than mono-targeted drugs.

1.6 Safety of Multi-Targeted Drugs

As mentioned above, where ~90% of all drug development projects fail after first-in-human testing, ~30% do so because of toxicology or clinical safety criteria.[7] Since many animal toxicity studies are done in pre-clinical development, the attrition related to safety/toxicology issues of development candidates produced by drug discovery research is even higher. Serious adverse drug reactions leading to attrition and drug withdrawal manifest themselves in a variety of different ways and organs, in particular in cardiovascular side effects and in liver toxicity.[86] Because of its association with toxicity (often referred to as side effects), polypharmacology has long been considered undesired. However, there are some arguments that polypharmacology in itself does not necessarily have to be associated with toxicity and that multi-targeted drugs can actually have a better efficacy/safety ratio than mono-targeted drugs. The underlying causes for safety/toxicology-related issues can roughly be grouped into three categories.

1.6.1 Target-Related Toxicity

Target-related toxicity (also often referred to as mechanism-based toxicity or as exaggerated primary pharmacology), is caused by hitting the intended pharmacological target.

An example of target-related toxicity is the proarrhythmic effect of class III antiarrhythmic drugs, such as dofetilide. These drugs modulate the activity of cardiac potassium channels, such as hERG, which leads to prolongation of the refractory period in the myocardium, and reduction of arrhythmias.

But paradoxically, this prolongation of the so-called QT interval also increases the risk for proarrythmia in predisposed individuals, in particular torsade de pointes, and sudden death.[87–89] This makes the risk-benefit assessment of this drug class very important.[90]

Intuitively one could argue that hitting more targets implies a higher risk for target-related toxicity. However, polypharmacology can also be used to reduce target-related toxicity. By rationally selecting multiple drug targets, such that the therapeutic effects of those targets overlap, but the detrimental effects do not overlap, one automatically designs in a therapeutic window based on target choice.

When analysing signalling pathways with MCA, it was found that activation of one of the pathway components, for example one which occurs in the event of an oncogenic mutation, reduces the control exerted by that component.[91] It is a given that total control of all pathway components (i.e. the sum of the quantified control for each individual pathway component) on signal transduction always sums to a constant value.[92,93] Therefore, a change in control by one pathway component will always be compensated for by a change in control of other pathway components. As a result, the distribution of control will be different in diseased versus healthy cells.[76,91] It should therefore be possible to hit multiple targets which, in combination, exert more control on the diseased system than on the healthy system. This would provide an inherent therapeutic window, and a smaller risk for target-related toxicity.

1.6.2 Off-Target Toxicity

Off-target toxicity is caused by hitting one or more targets other than the intended pharmacological target.

Polypharmacology has in particular been associated with toxicity or side effects caused by hitting off-targets. Though there are many, a good example is again the inhibition of hERG, which via prolongation of the QT interval leads to cardiac side effects in many drugs that exert their therapeutic effect via other targets.[88] However, such type of polypharmacology-related toxicity is caused by the compound hitting *unintended* targets, and such targets were often not identified until later compound profiling efforts. In contrast, a modern polypharmacology approach focuses on the design of drugs aimed at multiple *intended* targets. Moreover, polypharmacology can nowadays be predicted using *in silico* approaches and monitored carefully by *in vitro* profiling on a large set of targets in early stages of drug discovery (discussed elsewhere in this book). It is perhaps noteworthy that unintended targets can also contribute to efficacy. Sorafenib, for example, was originally developed as a Raf kinase inhibitor to treat malignant Raf-dependent melanoma, but it turned out to be a multi-kinase inhibitor that, because of its effect on receptor tyrosine kinases such as the VEGF receptor, is now approved for the treatment of renal cell carcinoma and hepatocellular carcinoma.[94,95]

1.6.3 Chemistry-Related Toxicity

Chemistry-related toxicity is caused by the chemical properties of the compound.

Examples of compound chemistry-related toxicity include oxidative damage, where reactive oxygen species cause oxidation of DNA, proteins or lipid membranes, and covalent binding to macromolecules by reactive metabolites.[96–98] The occurrence of those, often idiosyncratic, toxicities is clearly related to the daily dose: drugs dosed at 10 mg/day or less rarely cause toxicity.[96,97] Recently, this was further refined with the introduction of a zone classification system, which captures the risk for idiosyncratic toxicity in varying zones, based on daily dose and reactivity in human hepatocytes.[99] As mentioned in Section 1.5, it has been proposed that if multiple targets are hit simultaneously, partial inhibition of those targets may be sufficient in order to perturb the robust disease system, and will be more effective than complete inhibition of a single target.[72] That implies, with the assumption that target-binding kinetics are comparable, that the free drug concentration required to obtain a therapeutic effect is lower for a multi-targeted drug than for a mono-targeted drug. At lower dose levels, the risk for idiosyncratic chemistry-related toxicity is reduced.

In summary, toxicity has long been considered inherent to drugs that hit more than a single target. However, it is important to recognize that modern polypharmacology is actually an opportunity to produce safer drugs, by combining rational selection of target combinations with rational design of compounds that hit those targets.

1.7 Concluding Remarks

In clinical practice, combination therapy is a well-established concept. Numerous examples exist in which multi-factorial diseases can be treated more successfully either with a mixture of drugs or with a multi-targeted drug, including the treatment of cancer with multi-targeted kinase inhibitors and RA with methotrexate and other agents.[100–106] The approach to hit multiple targets has, however, not always been part of the initial strategy for their discovery (one gene, one drug, one disease). If we appreciate the underlying complexity of multi-factorial diseases, rational design of multi-targeted drugs becomes essential: Simple drugs do not cure complex diseases.

Many diseases with high medical need are multi-factorial. Here, I have highlighted that cancer and RA are diseases that originate and develop because of the combined action of multiple factors. A single mutation does not cause cancer; a single immune cell type does not sustain inflammation in RA. In both disease areas, highly complex genetic and signal transduction networks are affected; many different cell types are involved. Systems biology and network analysis have shown that the multi-factorial nature of complex diseases and the intrinsic robustness of the complex disease networks require perturbation of multiple targets simultaneously. Control of such networks is always dispersed over multiple steps. Multi-targeted drugs may not only be more efficacious, they may also be safer and reduce toxicity-related attrition. It is precisely these drugs that we need in the pharmaceutical pipelines and that patients need in the clinic.

Acknowledgements

I thank Ellinor Hornberg, Morten Laursen and Tomas Mow for their valuable input to the manuscript. I am indebted to Hans V. Westerhoff for introducing me to the systems biology field.

References

1. American Cancer Society, *Cancer Facts & Figures 2010*, Atlanta; American Cancer Society, 2010.
2. A. Verdecchia, S. Francisci, H. Brenner, G. Gatta, A. Micheli, L. Mangone and I. Kunkler, *Lancet Oncol.*, 2007, **8**, 784.
3. C. P. Ferri, M. Prince, C. Brayne, H. Brodaty, L. Fratiglioni, M. Ganguli, K. Hall, K. Hasegawa, H. Hendrie, Y. Huang, A. Jorm, C. Mathers, P. R. Menezes, E. Rimmer and M. Scazufca, *Lancet*, 2005, **366**, 2112.
4. J. Lazarou, B. H. Pomeranz and P. N. Corey, *JAMA*, 1998, **279**, 1200.
5. K. I. Kaitin and J. A. Dimasi, *Clin. Pharmacol. Ther.*, 2011, **89**, 183.
6. K. Traynor, *Am. J. Health Syst. Pharm.*, 2011, **68**, 186.
7. I. Kola and J. Landis, *Nat. Rev. Drug Discovery*, 2004, **3**, 711.
8. P. Ma and R. Zemmel, *Nat. Rev. Drug Discovery*, 2002, **1**, 571.
9. E. R. Fearon and B. Vogelstein, *Cell*, 1990, **61**, 759.
10. L. L. McCluskey and L. Dubeau, *Curr. Opin. Oncol.*, 1997, **9**, 465.
11. C. A. Reznikoff, C. Kao, E. M. Messing, M. Newton and S. Swaminathan, *Semin. Cancer Biol.*, 1993, **4**, 143.
12. G. J. Walker, J. M. Palmer, M. K. Walters and N. K. Hayward, *Genes Chromosomes Cancer*, 1995, **12**, 134.
13. W. C. Hahn, C. M. Counter, A. S. Lundberg, R. L. Beijersbergen, M. W. Brooks and R. A. Weinberg, *Nature*, 1999, **400**, 464.
14. D. P. Cahill, K. W. Kinzler, B. Vogelstein and C. Lengauer, *Trends Cell. Biol.*, 1999, **9**, M57.
15. S. Uribe-Lewis, K. Woodfine, L. Stojic and A. Murrell, *Expert Rev. Mol. Med.*, 2011, **13**, e2.
16. S. Valastyan and R. A. Weinberg, *Cell Cycle*, 2009, **8**, 3506.
17. J. M. Wagner, B. Hackanson, M. Lubbert and M. Jung, *Clin. Epigenetics*, 2010, **1**, 117.
18. www.sanger.ac.uk/genetics/CGP/Census/
19. P. A. Futreal, L. Coin, M. Marshall, T. Down, T. Hubbard, R. Wooster, N. Rahman and M. R. Stratton, *Nat. Rev. Cancer*, 2004, **4**, 177.
20. J. Luo, N. L. Solimini and S. J. Elledge, *Cell*, 2009, **136**, 823.

21. H. H. Heng, G. Liu, J. B. Stevens, S. W. Bremer, K. J. Ye and C. J. Ye, *Curr. Drug Targets*, 2010, **11**, 1304.
22. D. Hanahan and R. A. Weinberg, *Cell*, 2000, **100**, 57.
23. G. Kroemer and J. Pouyssegur, *Cancer Cell*, 2008, **13**, 472.
24. B. Vogelstein and K. W. Kinzler, *Nat. Med.*, 2004, **10**, 789.
25. B. N. Kholodenko and H. M. Sauro, *SEB Exp. Biol. Ser.*, 2008, **61**, 141.
26. J. Pouyssegur and P. Lenormand, *Eur. J. Biochem.*, 2003, **270**, 3291.
27. R. J. Ellis and A. P. Minton, *Biol. Chem.*, 2006, **387**, 485.
28. R. Milo, S. Shen-Orr, S. Itzkovitz, N. Kashtan, D. Chklovskii and U. Alon, *Science*, 2002, **298**, 824.
29. S. L. Campbell, R. Khosravi-Far, K. L. Rossman, G. J. Clark and C. J. Der, *Oncogene*, 1998, **17**, 1395.
30. Y. J. Huang, D. Hang, L. J. Lu, L. Tong, M. B. Gerstein and G. T. Montelione, *Mol. Cell. Proteomics*, 2008, **7**, 2048.
31. W. Kolch, *Biochem. J.*, 2000, **351**(Pt 2), 289.
32. G. Weng, U. S. Bhalla and R. Iyengar, *Science*, 1999, **284**, 92.
33. U. S. Bhalla and R. Iyengar, *Science*, 1999, **283**, 381.
34. J. J. Hornberg, M. R. Tijssen and J. Lankelma, *Eur. J. Biochem.*, 2004, **271**, 3905.
35. A. J. Ninfa and A. E. Mayo, *Sci. STKE*, 2004, **2004**, e20.
36. T. Pawson and N. Warner, *Oncogene*, 2007, **26**, 1268.
37. P. B. Gupta, C. L. Chaffer and R. A. Weinberg, *Nat. Med.*, 2009, **15**, 1010.
38. T. M. Browder, C. E. Dunbar and A. W. Nienhuis, *Cancer Cells*, 1989, **1**, 9.
39. X. Liu, W. Yao, R. C. Newton and P. A. Scherle, *Expert Opin. Investig. Drugs*, 2008, **17**, 997.
40. S. S. McAllister and R. A. Weinberg, *J. Clin. Oncol.*, 2010, **28**, 4022.
41. M. P. Shekhar, J. Werdell, S. J. Santner, R. J. Pauley and L. Tait, *Cancer Res.*, 2001, **61**, 1320.
42. A. Mantovani, S. Sozzani, M. Locati, P. Allavena and A. Sica, *Trends Immunol.*, 2002, **23**, 549.
43. R. Abes and J. L. Teillaud, *Cancer Metastasis Rev.*, 2011, **30**, 111.
44. D. Hanahan and J. Folkman, *Cell*, 1996, **86**, 353.
45. G. N. Naumov, J. Folkman and O. Straume, *Clin. Exp. Metastasis*, 2009, **26**, 51.
46. G. S. Firestein, *Nature*, 2003, **423**, 356.
47. K. K. Eklund, *Immunol. Rev.*, 2007, **217**, 38.
48. S. Khan, J. D. Greenberg and N. Bhardwaj, *Nat. Rev. Rheumatol.*, 2009, **5**, 566.
49. R. W. Kinne, R. Brauer, B. Stuhlmuller, E. Palombo-Kinne and G. R. Burmester, *Arthritis Res.*, 2000, **2**, 189.
50. A. P. Cope, H. Schulze-Koops and M. Aringer, *Clin. Exp. Rheumatol.*, 2007, **25**, S4.
51. M. J. Townsend, J. G. Monroe and A. C. Chan, *Immunol. Rev.*, 2010, **237**, 264.

52. B. Bartok and G. S. Firestein, *Immunol. Rev.*, 2010, **233**, 233.
53. N. C. Walsh and E. M. Gravallese, *Immunol. Rev.*, 2010, **233**, 301.
54. F. M. Brennan and I. B. McInnes, *J. Clin. Invest.*, 2008, **118**, 3537.
55. M. Kukar, O. Petryna and P. Efthimiou, *Biologics*, 2009, **3**, 443.
56. N. J. Felix, A. Suri, L. Salter-Cid, S. G. Nadler, S. Gujrathi, M. Corbo and R. Aranda, *Autoimmunity*, 2010, **43**, 514.
57. D. R. Fooksman, S. Vardhana, G. Vasiliver-Shamis, J. Liese, D. A. Blair, J. Waite, C. Sacristan, G. D. Victora, A. Zanin-Zhorov and M. L. Dustin, *Annu. Rev. Immunol.*, 2010, **28**, 79.
58. D. Qian and A. Weiss, *Curr. Opin. Cell Biol.*, 1997, **9**, 205.
59. T. Kurosaki, *Immunol. Rev.*, 2010, **237**, 5.
60. D. Hammaker, S. Sweeney and G. S. Firestein, *Ann. Rheum. Dis.*, 2003, **62 Suppl 2**, ii86.
61. G. He and M. Karin, *Cell Res.*, 2011, **21**, 159.
62. S. M. Steward-Tharp, Y. J. Song, R. M. Siegel and J. J. O'Shea, *Ann. NY Acad. Sci.*, 2010, **1183**, 123.
63. A. G. Tristano, *Int. Immunopharmacol.*, 2009, **9**, 1.
64. S. Cohen and R. Fleischmann, *Curr. Opin. Rheumatol.*, 2010, **22**, 330.
65. P. Cohen, *Curr. Opin. Cell Biol.*, 2009, **21**, 317.
66. A. O. Vassilev, H. E. Tibbles, D. DuMez, T. K. Venkatachalam and F. M. Uckun, *Curr. Drug Targets*, 2006, **7**, 327.
67. J. J. Hornberg, F. J. Bruggeman, H. V. Westerhoff and J. Lankelma, *Biosystems*, 2006, **83**, 81.
68. I. G. Khalil and C. Hill, *Curr. Opin. Oncol.*, 2005, **17**, 44.
69. F. J. Bruggeman, J. J. Hornberg, F. C. Boogerd and H. V. Westerhoff, *EXS*, 2007, **97**, 1.
70. P. K. Kreeger and D. A. Lauffenburger, *Carcinogenesis*, 2010, **31**, 2.
71. A. D. Boran and R. Iyengar, *Curr. Opin. Drug Discovery Devel.*, 2010, **13**, 297.
72. P. Csermely, V. Agoston and S. Pongor, *Trends Pharmacol. Sci.*, 2005, **26**, 178.
73. H. Kitano, *Nat. Rev. Drug Discovery*, 2007, **6**, 202.
74. A. L. Barabasi and Z. N. Oltvai, *Nat. Rev. Genet.*, 2004, **5**, 101.
75. A. S. Azmi, Z. Wang, P. A. Philip, R. M. Mohammad and F. H. Sarkar, *Mol. Cancer Ther.*, 2010, **9**, 3137.
76. E. Oxelmark and J. J. Hornberg, *Drug Discovery Today: Technol.*, 2008, **4**, e91.
77. W. Alkema, T. Rullmann and A. van Elsas, *Expert Opin. Ther. Targets*, 2006, **10**, 635.
78. J. K. Morrow, L. Tian and S. Zhang, *Crit. Rev. Biomed. Eng.*, 2010, **38**, 143.
79. A. L. Hopkins, *Nat. Chem. Biol.*, 2008, **4**, 682.
80. T. Pawson and R. Linding, *FEBS Lett.*, 2008, **582**, 1266.
81. M. Cascante, L. G. Boros, B. Comin-Anduix, P. de Atauri, J. J. Centelles and P. W. Lee, *Nat. Biotechnol.*, 2002, **20**, 243.
82. J. J. Hornberg, F. J. Bruggeman, B. M. Bakker and H. V. Westerhoff, *Prog. Drug Res.*, 2007, **64**, 171, 173.

83. J. J. Hornberg, B. Binder, F. J. Bruggeman, B. Schoeberl, R. Heinrich and H. V. Westerhoff, *Oncogene*, 2005, **24**, 5533.
84. A. E. Ihekwaba, S. J. Wilkinson, D. Waithe, D. S. Broomhead, P. Li, R. L. Grimley and N. Benson, *FEBS J.*, 2007, **274**, 1678.
85. B. Schoeberl, E. A. Pace, J. B. Fitzgerald, B. D. Harms, L. Xu, L. Nie, B. Linggi, A. Kalra, V. Paragas, R. Bukhalid, V. Grantcharova, N. Kohli, K. A. West, M. Leszczyniecka, M. J. Feldhaus, A. J. Kudla and U. B. Nielsen, *Sci. Signal.*, 2009, **2**, ra31.
86. W. S. Redfern, L. Ewart, T. G. Hammond, R. Bialecki, L. Kinter, S. Lindgren, C. E. Pollard, R. Roberts, M. G. Rolf and J.-P. Valentin, *The Toxicologist*, 2010, **114[S1]**, 1081.
87. N. K. Jurkiewicz and M. C. Sanguinetti, *Circ. Res.*, 1993, **72**, 75.
88. W. S. Redfern, L. Carlsson, A. S. Davis, W. G. Lynch, I. MacKenzie, S. Palethorpe, P. K. Siegl, I. Strang, A. T. Sullivan, R. Wallis, A. J. Camm and T. G. Hammond, *Cardiovasc. Res.*, 2003, **58**, 32.
89. M. C. Sanguinetti, *Hypertension*, 1992, **19**, 228.
90. B. Brendorp, O. Pedersen, C. Torp-Pedersen, N. Sahebzadah and L. Kober, *Drug Saf.*, 2002, **25**, 847.
91. J. J. Hornberg and H. V. Westerhoff, *Mol. Biotechnol.*, 2006, **34**, 109.
92. J. J. Hornberg, F. J. Bruggeman, B. Binder, C. R. Geest, A. J. de Vaate, J. Lankelma, R. Heinrich and H. V. Westerhoff, *FEBS J.*, 2005, **272**, 244.
93. D. Kahn and H. V. Westerhoff, *J. Theor. Biol.*, 1991, **153**, 255.
94. S. M. Wilhelm, C. Carter, L. Tang, D. Wilkie, A. McNabola, H. Rong, C. Chen, X. Zhang, P. Vincent, M. McHugh, Y. Cao, J. Shujath, S. Gawlak, D. Eveleigh, B. Rowley, L. Liu, L. Adnane, M. Lynch, D. Auclair, I. Taylor, R. Gedrich, A. Voznesensky, B. Riedl, L. E. Post, G. Bollag and P. A. Trail, *Cancer Res.*, 2004, **64**, 7099.
95. R. Iyer, G. Fetterly, A. Lugade and Y. Thanavala, *Expert Opin. Pharmacother.*, 2010, **11**, 1943.
96. J. L. Walgren, M. D. Mitchell and D. C. Thompson, *Crit. Rev. Toxicol.*, 2005, **35**, 325.
97. J. Uetrecht, *Curr. Opin. Drug Discovery Devel.*, 2001, **4**, 55.
98. R. Kohen and A. Nyska, *Toxicol. Pathol.*, 2002, **30**, 620.
99. S. Nakayama, R. Atsumi, H. Takakusa, Y. Kobayashi, A. Kurihara, Y. Nagai, D. Nakai and O. Okazaki, *Drug Metab. Dispos.*, 2009, **37**, 1970.
100. T. Rath and A. Rubbert, *Clin. Exp. Rheumatol.*, 2010, **28**, S52.
101. J. S. Smolen, R. Landewe, F. C. Breedveld, M. Dougados, P. Emery, C. Gaujoux-Viala, S. Gorter, R. Knevel, J. Nam, M. Schoels, D. Aletaha, M. Buch, L. Gossec, T. Huizinga, J. W. Bijlsma, G. Burmester, B. Combe, M. Cutolo, C. Gabay, J. Gomez-Reino, M. Kouloumas, T. K. Kvien, E. Martin-Mola, I. McInnes, K. Pavelka, R. P. van, M. Scholte, D. L. Scott, T. Sokka, G. Valesini, R. F. van Vollenhoven, K. L. Winthrop, J. Wong, A. Zink and D. van der Heijde, *Ann. Rheum. Dis.*, 2010, **69**, 964.
102. V. Abramson and C. L. Arteaga, *Clin. Cancer Res.*, 2011, **17**, 952.

103. A. Sartore-Bianchi, K. Bencardino, A. Cassingena, F. Venturini, C. Funaioli, T. Cipani, A. Amatu, L. Pietrogiovanna, R. Schiavo, N. F. Di, S. Artale, A. Bardelli and S. Siena, *Cancer Treat. Rev.*, 2010, **36(Suppl 3)**, S1.
104. M. Shabbir and R. Stuart, *Expert Opin. Investig. Drugs*, 2010, **19**, 427.
105. T. G. Hopkins, M. Marples and D. Stark, *Eur. J. Surg. Oncol.*, 2008, **34**, 844.
106. M. W. Karaman, S. Herrgard, D. K. Treiber, P. Gallant, C. E. Atteridge, B. T. Campbell, K. W. Chan, P. Ciceri, M. I. Davis, P. T. Edeen, R. Faraoni, M. Floyd, J. P. Hunt, D. J. Lockhart, Z. V. Milanov, M. J. Morrison, G. Pallares, H. K. Patel, S. Pritchard, L. M. Wodicka and P. P. Zarrinkar, *Nat. Biotechnol.*, 2008, **26**, 127.

CHAPTER 2

Clinical Need and Rationale for Multi-Target Drugs in Psychiatry

MOHAMMED SHAHID

Preclin-2D Ltd, Glasgow, UK
Email: mo.shahid@preclin-2d.com

2.1 Introduction

Productivity, as judged by the number of annual new drug approvals, is continuing to pose a major performance issue for the pharmaceutical industry.[1,2] Indeed this, combined with the fast approaching 'patent cliff', leading to loss of exclusivity on high revenue generating blockbuster drugs,[3] has been a primary driver for some of the recent dramatic consolidation and strategic re-organisation observed within leading pharma companies. Greater focus on commercial opportunities and market access for existing products in emerging economies, such as China and India, has been now expanded to include unprecedented levels of outsourcing of R&D activities to these sectors. The yearly total of new innovative medicines has been gradually developing a downward trend for the past two decades whilst R&D costs have rocketed to an all time high.[4,5] There is considerable variation in the level of success over therapeutic categories which has prompted a shift in R&D focus for some companies. Cancer and central nervous system (CNS) related drug discovery, in particular, represent domains of relatively high attrition,[6] fuelling reduction of or exit from investment in research in

RSC Drug Discovery Series No. 21
Designing Multi-Target Drugs
Edited by J. Richard Morphy and C. John Harris

Published by the Royal Society of Chemistry, www.rsc.org

these therapeutic areas for some companies. And this is despite the strong unmet medical need and compelling justification for improved drug treatment options.

Within the CNS field, psychiatric drug discovery is undoubtedly one of the more challenging areas. With limited insight on molecular pathology and high failure rate in development, it is perhaps not surprising that many leading pharma companies have de-emphasised effort in this area. Multiple factors, operating in concert, are likely to underlie the poor rate of success in psychiatric drug research and have been discussed extensively in recent reviews.[2,7–9] Given the likely complex multi-genic, and possibly developmental, as well as multi-factorial environmental basis for these diseases, one area of debate has been the continued focus and priority on developing highly selective compounds working through a single molecular drug target – the so-called single target agents (STAs). Indeed the experience so far, in general, for psychiatric therapeutics has been disappointing; a notable number of examples have now accumulated demonstrating lack of success with STA-based monotherapy approaches.[9,11] It is conceivable, however, with increased insight on molecular pathology and disease mechanisms that this situation may improve in the future. In this respect it will be interesting to track the rate of success in molecular pathology-driven drug discovery in neurodegenerative disorders such as Alzheimer's and Parkinson's disease where relatively better progress has been achieved towards delineating genes and cellular pathways contributing to disease development.[12,13] In the meantime, however, for new psychiatric drugs refinement of successful multi-target agents (MTAs) or perhaps more interestingly development of novel tailored MTAs still represents a promising avenue for exploration. Indeed the introduction of successive generations of MTAs for schizophrenia and mood disorders over the past five decades or so has provided new treatment options for severely ill patients whose therapeutic needs were not being adequately met by available drugs. Furthermore, consideration of recent successful development programmes and approvals shows that most of these have been drugs with an MTA profile.

Even with a better understanding of pathology and the advent of disease mechanism-based therapeutics, it is doubtful if an STA-based approach will be adequate in effectively treating the diverse range of symptoms associated with multi-dimensional illnesses such as schizophrenia, bipolar disorder and major depression. A multi-target approach could, in principle, be achieved through polypharmacy with the combination of several STAs. An interesting analogy, in this respect, is the development of highly selective inhibitors of oncogenic protein kinases as anti-cancer agents. It seems that, when tested alone, a number of these agents only provide limited and/or transient efficacy[14,15] whilst, at least based on preclinical work, simultaneous targeting of multiple kinases may be a more effective strategy. Indeed the rationale is so compelling that it led to Merck (AKT inhibitor MK-2206) and AstraZeneca (MEK inhibitor AZD6244) to share proprietary compounds to investigate the concept at a surprisingly early stage of development.[73] Polypharmacy involving combination of pharmacological agents, however, also has significant limitations and challenges such as ensuring pharmacokinetic and dose compatibility as well as drug–drug interaction issues to mention a few. It is clear that the alternative option of developing tailored MTAs is a more attractive proposition in this respect. Therefore, not surprisingly, interest has grown in trying to identify compounds that are dual kinase inhibitors.[14,15] Certainly in psychiatry some of the most efficacious and useful as well as successful therapeutic agents have been MTAs. Quetiapine is perhaps a profound illustration in support of the latter point. It started off as an antipsychotic agent but is now approved for bipolar mania, bipolar depression, treatment resistant depression as well as showing evidence for treatment of major depression and anxiety.[16] However, the issue with classical MTAs has been aggressive receptor promiscuity causing significant side effect issues. Historically, MTAs have evolved from chemical templates with multi-receptor activity which was then further optimised to dial-in or dial-out particular receptor properties – the so-called 'chop down' approach. This, however, still left considerable undesirable receptor activities which led to side effects. It is clear that a different medicinal chemistry strategy needs to be developed that will enable building tailored MTAs perhaps from knowledge gained from the development of STAs.[11,17]

The present chapter is aimed at outlining the relative merits of multi-target agents in terms of addressing clinical need and scientific rationale for their utility in psychiatric diseases such as schizophrenia, bipolar disorder and major depression. In addition, a brief update on recently introduced new drugs and compounds in clinical development for schizophrenia and mood disorders is provided.

2.2 Clinical Need

Given the large pharmacological armamentarium available for the therapeutic management of schizophrenia and mood disorders, why is there still a need for new drugs for these disorders? Well, whilst historically, considerable success has been achieved in introducing a wide array of useful therapeutic agents for these psychiatric diseases, they unfortunately still do not fully meet the treatment needs of all patients suffering from these disabling conditions. Some of the main reasons for the continued significant medical need in these brain disorders include the following.

- limited response and high rates of relapse – e.g. in depression only $<50\%$ respond to first treatment and $\sim 30\%$ entered remission[18]
- treatment resistance or refractoriness[19] – failure of response despite target engagement
- high discontinuation rates and adherence problems – a high proportion of patients are unsatisfied with initial treatment and switch to another option within 1 year[20]
- lack of *a priori* prediction of response to a particular treatment option – treatment algorithms are based on an iterative process dependent on physician experience rather than neurobiological measures of pathology or reliable predictors of response which makes it difficult to personalise treatment to individual patient needs[21]

- restricted symptom control such as poor relief of negative symptoms, cognitive impairment and affective dysfunction. Poor tolerability due to adverse events including motor, somnolescense, sexual dysfunction
- long-term safety issues such as cardiovascular, metabolic dysregulation and weight gain
- need for treatment-related improvement in patient functional capacity

The genomics revolution, which started in the 1980s, has enabled considerable progress towards elucidation of predisposing genes and genetic factors in many disease areas such as cancer and neurodegenerative diseases. This has allowed the investigation of single target-based therapeutic interventions with some, albeit limited, success particularly in the field of anti-cancer drug discovery. Unfortunately, despite intense effort over the past 10 years, similar progress has not been achieved in psychiatric disease genetics. Nevertheless, it is clear that further genetics research is essential to delineate cellular pathology and identify novel molecular drug targets or cellular pathways that can stimulate drug discovery based on disease mechanism(s). Limited progress has been achieved with identification of some reproducible genetic markers.[22,23] Technological improvements coupled with better experimental strategies (e.g. genetic neuropathology using patient brain tissue[24]) as well as more realistic expectations provide a sense of optimism in this area for more informative progress over the next ten years. An important dynamic in this highly challenging endeavour will be the need for a stronger and more effective alliance between industry and basic research leaders in academic and government supported research.[10] Given the almost routine collection of patient DNA samples in most industry-sponsored studies, the potential to access high-quality material from well-powered studies offers an invaluable resource and opportunity to aid drug discovery. Development of personalised medicine, tailoring therapy to individual patient need, is recognised and will be an important beneficiary from progress in pharmacogenetic and pharmacogenomic analysis of psychiatric disorders. Future effort in this area needs to involve focus on specifically designed, well-controlled prospective studies to determine the genetic basis of why some patients respond to a particular therapeutic intervention while other subjects fail to benefit. To be of informative use for a physician, genetic biomarkers should show high sensitivity and reproducibility in terms of predictiveness as well as a clear rationale for clinical utility.[25–27] Although psychiatric genetics will continue to be one of the toughest areas for dramatic progress, the prospect, at some point in the future, of diagnosing and more accurately defining patient subpopulations on the basis of biological factors is a strong motivator to continue the effort. Comparative evaluation of the relatively more successful research strategy pursued in this regard, say for instance in cancer, may provide insight on further refinement of genetics research for psychiatric disorders. A more definitive understanding will undoubtedly also reinvigorate industrial interest in psychiatric drug discovery as it has done it other areas, for example in neurodegenerative disorders. An aspirational goal would be to establish the trend and output achieved in neurodegenerative disorders such as Alzheimer's and Parkinson's disease. Advances in understanding some of the genetic aspects of the molecular pathology has also opened up the tantalising prospect of modifying disease progression through pharmacological intervention. With reliable genetic markers one can then expect more targeted investigations in pre-symptomatic or prodromal patients to monitor and improve long-term prognosis.

The continued need for more effective, better tolerated and safer therapeutic agents make schizophrenia and mood disorders medically important and commercially attractive areas in neuroscience related drug discovery. But recent developments in the pharma industry indicate a trend towards a decline of investment in these therapeutic categories. Exact causes for this development are difficult to identify but some of the potential factors detailed elsewhere[2,7,9,10] include:

- insufficient knowledge of the molecular basis for disease pathology
- poor target validation
- lack of translation of preclinical data – poor reliability of disease-based animal models
- high failure rate of novel approaches
- complex heterogeneous patient population with varying degrees of symptom presentation
- drug discovery being driven by mode of action of pharmacological disease mimicking agents and/or current drugs, leading to step rather than quantum change in treatment options
- clinical trial issues – growing placebo response, difficulties in replication

Repeated failure to establish proof of concept with unprecedented approaches suggests a major issue in quality of target validation. It is clear that without a direct link to disease pathology, selection of highly novel molecular targets carries perhaps too high a risk with strong likelihood of failure. More robust and critical evaluation of target validation is necessary before embarking on expensive clinical trials. Traditional behavioural pharmacology alone is no longer a viable strategy in this respect and more disease construct driven multi-dimensional animal models are needed. In recognition of this need, animal models based on disease factors (genetic, environmental, developmental) but also engaging multiple variables (e.g. neurophysiology, neurochemistry and behaviour) are increasingly being developed which should assist future drug discovery effort (e.g. see refs. 28–33). Moreover, it is critical that these models as much as possible must have aspects that are amenable to translation to clinical testing. Greater use of quantitative pharmacology involving preclinical-clinical pharmacokinetic/pharmacodynamic (PK/PD) modelling, not only in informing about dose selection for efficacy but also potential for side effect or safety issues, will help to reduce suboptimal compounds entering clinical development.[34] This can also be an effective strategy for molecular targets that lack tools for assessing receptor engagement. The recent study by Bursi *et al.*[35] is a good illustration, demonstrating the value of preclinical/clinical

PK/PD modelling to assist clinical dose determination for the AMPA receptor potentiator Org 26576.

Taking these significant challenges into consideration it would seem prudent to attempt to develop compounds with a mixture of precedented and unprecedented molecular target activity. Hopefully these approaches will yield drugs that combine improvements in efficacy (broader symptom coverage) as well as side effect profile and enable progress towards more effective agents.

2.3 Rationale For Multi-Target Agents: Multifunctional Pharmacology and Multi-Therapeutic Application

In view of the considerable technical challenge involved in early drug discovery it is appropriate to examine the quality and strength of the rationale for the utility of MTAs in psychiatric disease treatment. Several lines of argument provide a strong justification for the continued development of MTAs for the treatment of schizophrenia and mood disorders.[9–11,15,17,36]

Firstly, what is known about disease phenotype and characteristics? Schizophrenia, bipolar disorder and major depression are complex and multidimensional disorders with a heterogeneous patient population showing a range of symptoms with varying intensity. Although each disorder manifests a broad range of symptoms and co-morbidity[11] with considerable overlap, there is a core set of symptoms rather specific to each disorder.

Although current insight into disease pathology is clearly inadequate, there is consensus that a combination of genetic, environmental and developmental factors appears to play a significant role in disease development.[22,23,37] Brain imaging studies, albeit in small patient studies, indicate the presence of abnormalities in brain structure and metabolism for all three indications.[38–40] For example, cortical thinning and diminution of frontal cortex volume seems to be a reproducible finding in schizophrenia patients.[38] Whilst caution should be exercised in extrapolating from these findings in terms of applicability to the general patient population, if these findings are an accurate hallmark of the pathology they represent a powerful indicator of likelihood of dysregulation in multiple pathways and neurotransmitter systems as precursor to symptom generation. In addition, the opinion that multiple genes of low effect are likely to be involved in the pathology of these disorders is also somewhat in tune with the idea of multi system dysregulation. So it would not seem too speculative to suggest that the molecular pathology of schizophrenia and mood disorders is likely to involve multiple genes that trigger malfunction in multiple neuronal pathways, multiple neurotransmitter systems and consequently multiple receptors. Based on this it seems, at least intuitively, discordant to suggest that a highly selective STA could be an effective monotherapy providing multi-symptom relief.

All currently used pharmacological treatments for schizophrenia, bipolar disorder and major depression are MTAs.[9] This is most clearly represented by the class of atypical antipsychotics (e.g. clozapine, olanzapine and risperidone) but also antidepressants (e.g. mirtazapine, venlafaxine, agomelatine) and mood stabilisers (e.g. quetiapine). SSRIs can be considered as STAs, since they primarily work by inhibiting a single molecular target – the neuronal serotonin transporter. However, these drugs could also be viewed as indirect MTAs since they work by increasing synaptic 5-HT levels, causing stimulation of a range of 5-HT receptors. Similarly, non-pharmacological treatment options such as electroconvulsive therapy, deep brain stimulation or cognitive behavioural therapy are also likely to engage multiple neurotransmitter systems.

Several attempts to investigate the therapeutic utility of STAs for major psychiatric disorders have met with failure in clinical development, as summarised by Wong *et al.*[9] For some projects this was at the very costly stage of clinical development (e.g. for the 5-HT_{2A} selective antagonist MI100907.[41] These failures have highlighted the need to improve the quality of target validation as well as identify reliable translatable biomarkers for efficacy. Until this situation improves the appetite for further testing of STAs will remain subdued. In sharp contrast, several programmes with MTAs have been successful in achieving regulatory approval over the past five years (e.g. agomelatine, paliperidone, asenapine, iloperidone, lurasidone, vilazodone).

Multi-target agents can provide a multi-functional pharmacology through synergistic interaction between complementary pharmacological components in a single molecule. This is critical with regards to multi-symptom efficacy and there should be a clear scientific rationale validating engagement of multiple therapeutically promising receptors at clinically relevant drug exposure levels. Furthermore, a multi-functional pharmacology opens up the prospect of multi-indication utility. The gradual broadening of the use of atypical antipsychotics from schizophrenia to other indications such as bipolar disorder[42] and treatment-resistant depression[43] is a powerful illustration in this respect. The almost panacea-like use of quetiapine in several indications is a reflection of its considerable multi-functional pharmacology, including the significant contribution of its metabolite, desalkyl-quetiapine which adds noradrenaline transporter inhibition to its pharmacology.[16,44,45] Given the widespread use of quetiapine in mood disorders it raises the question whether it should still be labelled as an atypical antipsychotic. Indeed this group of drugs is now composed of a heterogeneous group of molecules with diverse chemistry, pharmacology and therapeutic utility. Perhaps it is opportune to reconsider and redefine the classification or at least the terminology used to describe these therapeutic agents. It has been suggested that it should perhaps be based on the drug receptor pharmacology, which can now help to explain the efficacy and side effect profile of each drug.[46]

For a balanced appraisal, it is also important to outline some of the issues associated with MTAs. The combination of a limited number (e.g. 2 to 3) receptor activities in a drug may not pose a significant disadvantage with respect to side effect potential when compared to an STA. In fact it could be argued that dialling in additional activity can have beneficial effects with respect to side effect issues (e.g. combination of D_2 and 5-HT_{2A} reduces liability for motor side effects). However, a highly multi-receptorial promiscuous profile, as in the case of some atypical antipsychotics (e.g. clozapine), raises the potential for a diverse range of side effects as well as safety issues. Establishing evidence-based proof of concept, in accordance with modern drug discovery practice, will perhaps be more challenging for an MTA when compared to an STA. This will be particularly true if the targets are highly novel and multiple biomarkers for target engagement as well as efficacy have to be established. A tendency for an unclear dose-response for efficacy may also be more of an issue with MTAs. Clearly, attempts to optimise activity and selectivity for multiple receptors are going to be considerably more challenging and resource intensive. This difficulty can, however, be managed by limiting the number of receptors to two or at the most three. Indeed with this approach it is feasible to generate hybrid compounds from substantially different receptor families. For example, Millan[11] has provided evidence to show that serotonin transporter inhibition can be combined with a range of receptors (e.g. H_3, α_2-adrenergic, NK_1 receptors, acetycholinesterase).

2.4 New Introductions

Despite the challenges and high attrition in psychiatric drug discovery it is, however, still possible to achieve success with MTAs. There have been a number of drugs achieving market authorisation for the treatment of schizophrenia and mood disorders since 2006. A brief summary of the pharmacology of two antidepressants (agomelatine, vilazodone) and three antipsychotics (asenapine, lurasidone, iloperidone) is presented here (Figure 2.1). In relation to mood stabilisers it seems that the trend to borrow

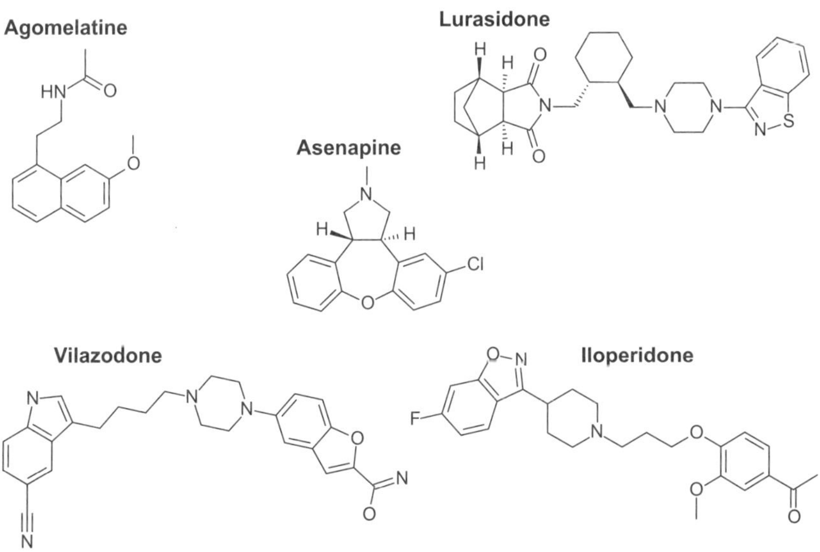

Figure 2.1 Examples of recent drug approvals for psychiatric diseases.

therapies from other therapeutic categories, particularly antipsychotics, is continuing. Given the unique characteristics of bipolar disorder, it is clear that greater effort needs to be devoted towards drug discovery specifically dedicated to this indication. Perhaps agomelatine and asenapine are the closest example of tailored MTAs. These two agents also seem to be the most well characterised, in particular the human receptor pharmacology of asenapine was compared to a range of other antipsychotics in the same study.[52] Although asenapine has a heavily multi-receptorial profile, it does introduce a new chemotype in the antipsychotic class of compounds. Regulatory approval is clearly a very important milestone; however, the true success of these drugs will be determined by how much extra value they bring towards meeting patient treatment needs and improving functional capacity. One good thing, nevertheless, is beyond doubt, which is that the availability of the agents will provide new therapeutic options to patients whose treatment needs are not being effectively addressed by other historical drugs.

2.4.1 Agomelatine

Agomelatine is an innovative MTA that combines melatonin (MT_1 and MT_2) receptor agonism with somewhat modest 5-HT_{2C} receptor antagonist properties.[47] Following an extensive clinical development programme, it gained marketing authorisation, albeit in Europe only, in 2009 for the treatment of major depression. The two distinct pharmacologies, through complementary action, provide a rationale for treatment of circadian rhythm as well as mood dysregulation in depressed patients. Agomelatine is an interesting illustration as it shows that it is possible to generate a successful MTA which is a hybrid of two very distinct receptor subtypes but also different functional activity at these receptors.

2.4.2 Vilazodone

Vilazodone is essentially a dual MTA with serotonin transporter inhibition activity combined with partial agonism at the 5-HT_{1A} receptor.[48,49] It was approved in the US for the treatment of major depression in 2011. The tagging on of 5-HT_{1A} activity is hypothesised to confer improved antidepressant properties (e.g. faster onset) as well as better side effect profile (e.g. less disruption of sexual function). Clinical data provide some support for this profile.

2.4.3 Asenapine

Asenapine is the first tetracyclic antipsychotic approved in the US for the treatment of schizophrenia and bipolar mania; it was approved in 2009.[50,51] The unique chemical template of asenapine, in the antipsychotic class of compounds, is associated with a very distinctive multi-receptor binding profile.[52] Asenapine is an interesting MTA that was simultaneously developed in two indications and the development programme included dedicated trials to examine effects on negative symptoms of schizophrenia.

It is a multi-functional drug as it has at least three distinct therapeutically promising pharmacologies connected together in the one molecule (Figure 2.2) to provide a scientific rational for efficacy beyond controlling psychosis and mania. First, it is a potent and selective antagonist at 5-HT_{2C}> 5-HT_{2A}> 5-HT_7> 5 H-T_6 receptors over other monoamine receptors. Based largely on preclinical and some clinical observations this profile is interesting with respect to its potential for the amelioration of affective and negative symptoms as well as cognitive impairment.[52,53] For example, the 5HT_7 receptor antagonism has been suggested as a putative mechanism for fast-onset anti-depressant action.[54] Second, the presence of α_2-adrenergic receptor antagonism may deliver a synergistic influence to further boost these properties.[11] Secondary *post hoc* analysis of data from bipolar trials in acute manic or mixed episode bipolar patients suggests efficacy on depressive symptoms with asenapine.[55] Combination of 5-HT_{2A}/D_2 antagonism provides the driver for anti-psychotic and antimanic activity with low extrapyramidal symptom liability. Whilst the latter has been clearly established, the greater potential of asenapine in terms of broader efficacy still remains to be confirmed by further clinical evaluation. Asenapine also has modest H_1 and α_1-adrenergic antagonist properties, which raise an alert with respect to side effects such as sedation and hypotension, respectively. The latter, however, may be offset by the potent α_2-adrenergic antagonist properties of asenapine. Finally, despite strong affinity for the 5-HT_{2C} receptor and modest activity at the H_1 receptor, asenapine does not have the same propensity to elevate weight gain as olanzapine.[56,57] Lack of anitmuscarinic activity may also be relevant with regards to potential for metabolic and cognitive dysfunction.

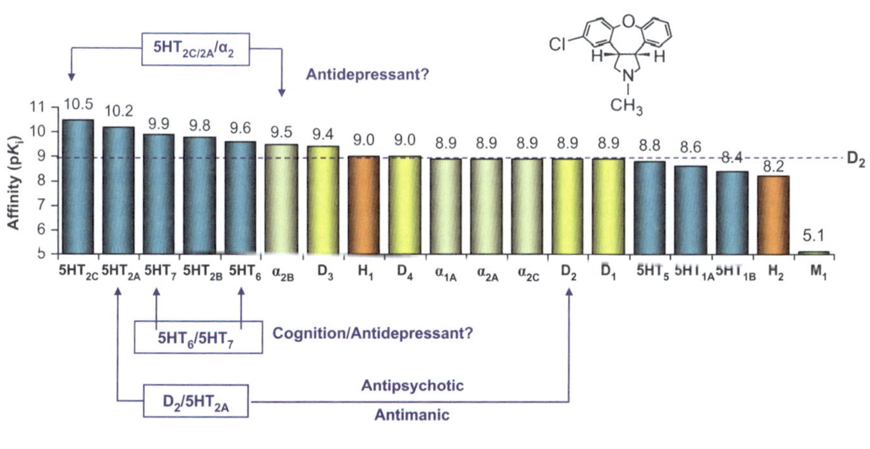

Figure 2.2 Multifunctional pharmacology of asenapine: cloned human receptor binding profile of asenapine (adapted from Shahid *et al.*, ref. 52).

2.4.4 Lurasidone

An azapirone class derivative, lurasidone gained FDA approval in 2010 for the treatment of schizophrenia.[58] As well as D_2 (K_i: 1.68 nM) and $5\text{-}HT_{2A}$ (K_i: 2.03 nM) receptor blockade, it has $5\text{-}HT_7$ (K_i: 0.49 nM) and α_{2C}-adrenergic (K_i: 10.8 nM) antagonism as interesting ancillary receptor activities[59] suggesting potential for antidepressant and pro-cognitive properties. Indeed there are interesting preclinical data to support antidepressant activity.[59] Furthermore, lurasidone showed superior activity in a impaired learning and memory task in rats when compared to risperidone, clozapine, aripiprazole and haloperidol.[60] Selective $5\text{-}HT_7$ and $5\text{-}HT_{1A}$ antagonists mimicked the effects of lurasidone suggesting the involvement of these receptors in the cognition-improving properties of lurasidone.[61] From a side effect liability perspective lurasidone looks promising as it has no appreciable activity at α_1-adrenergic, H_1 or muscarinic receptors. Studies in bipolar depression are currently in progress with lurasidone and should provide insight on its multi-indicational potential.

2.4.5 Iloperidone

Iloperidone is a piperdinyl-benzisoxazole derivative, structurally somewhat similar to risperidone, and was approved in the US for the treatment of schizophrenia in 2009.[62,63] Unlike asenapine a comprehensive picture of the human receptor binding profile of iloperidone is not available in the published literature. However, like other atypical antipsychotics it binds to $5\text{-}HT_{2A}$ (K_i: 5.6 nM) and D_2 (K_i: 6.3 nM) receptors although unusually the affinity is similar for both receptors.[64] Iloperidone shows highest affinity for the α_1-adrenergic receptor (K_i: 0.4 nM), which raises an alert with respect of cardiovascular side effect liability.

2.5 Emerging Promising Compounds in Development

With the drop of interest in psychiatric drug discovery by a number of companies it has become uncertain which compounds are still under active investigation in clinical trials. However, a few with potential to make it to phase III development or regulatory submission are summarised in Figure 2.3. It is encouraging that this group represents a mixed portfolio of three validated and three non-validated approaches. Projects in earlier stages of drug discovery have been covered in recent reviews.[9]

2.5.1 Cariprazine

Cariprazine (RGH-188) is in phase III development for both schizophrenia and bipolar disorder.[65] Data from phase II trials has demonstrated antipsychotic and antimanic efficacy in schizophrenia and bipolar patients, respectively. Its receptor pharmacology seems to be similar to the benzamide

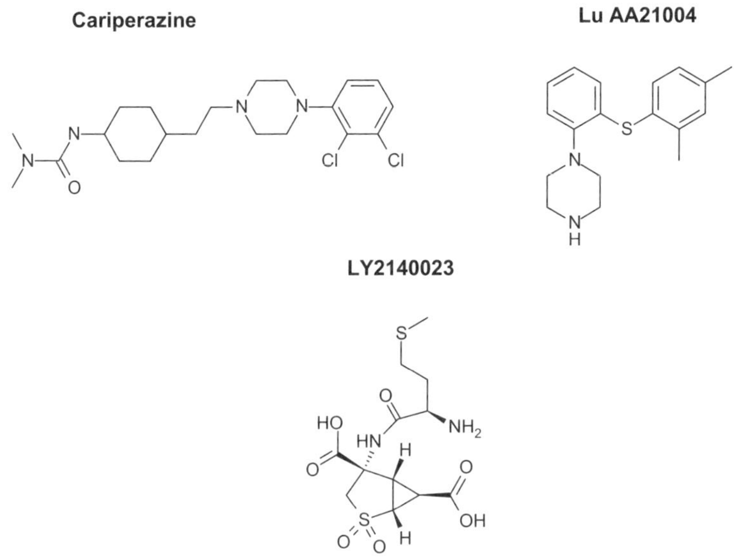

Figure 2.3 Examples of promising compounds in development for psychiatric diseases.

class of antipsychotics, showing rather selective affinity and blocker properties at the D_2 (K_i: 0.69 nM), D_3 (K_i: 0.19 nM) and 5-HT_{1A} (K_i: 2.6 nM) receptors.[66] Partial agonism at these receptors has been argued to be a mechanistic differentiator with potential to confer additional therapeutic properties (e.g. improvement of depressive symptoms). There is some animal data showing pro-cognitive activity in animal models.[67] Consistent with its core receptor pharmacology, cariprazine has shown efficacy as an antipsychotic and antimanic in short-term trials in schizophrenia and bipolar patients.[65]

This molecule, clearly, has much more restricted receptor activity – it could in fact, based on current information, be considered as a dual MTA. Thus it will be interesting to see how its overall clinical profile compares, both in terms of efficacy and side effects, to other antipsychotics, particularly those with a stronger serotonergic and multi-receptor profile.

2.5.2 Lu AA21004 and Zicronapine

Lundbeck continues to show strong commitment to psychiatric research, including the investigation of MTAs. Lu AA21004 and zicronapine (Lu 31-130) are two promising MTAs in its development pipeline.

Lu AA21004 possesses serotonin transporter (SERT) inhibition (K_i: 1.6 nM) blended with 5-HT_{1A} agonism (K_i: 15 nM), $5HT_3$ (K_i: 3.7 nM) antagonism[68] and 5-HT_7 (K_i: 19 nM) antagonism, albeit considerably below the human

SERT affinity. The 5-HT_{1A} agonism and $5HT_3$ antagonism are hypothesised to synergistically boost antidepressant action whilst improving side effect liability (e.g. nausea and sexual dysfunction). The outcome from a double-blind placebo-controlled trial with venlafaxine as an active drug has been published, indicating that Lu AA21004 showed antidepressant efficacy at both 5 or 10 mg.[69] Incidence of sexual side effects was comparable to placebo and less than that observed with venlafaxine. Further data are required in order to judge the benefits of Lu AA21004 over existing antidepressant therapies.

Currently there are few published data on zicronapine but Lundbeck issued a press release in January 2011[74] that a phase III programme in schizophrenia had been initiated following positive data from two phase II exploratory studies. Four doses of zicronapine (3, 5, 7 and 10 mg/day) were examined in one study with 7 and 10 mg/day doses showing significant separation from placebo. In the second study a flexible dose of zicronapine (5–7 mg/day) and olanzapine (10–15 mg/day) were evaluated. The first phase III study is examining the efficacy and metabolic profile of zicronapine (7.5 mg/day) following 6 months of treatment. Risperidone (5 mg/day) will be included as an active control. Additional short-term studies will also be conducted.

In addition to D_2 receptor block, zicronapine has D_1 receptor antagonist properties. This is postulated to confer pro-cognitive properties. The suggestion is that zicronapine may have a more restricted receptor profile, perhaps a dual MTA; however, further details are required to confirm this and assess how it differentiates from other drugs particularly since a number of compounds (e.g. asenapine) also have this activity.

2.5.3 LY2140023

A glutamatergic mechanism-based compound that caused high excitement in the schizophrenia field is the mGluR2/3 agonist LY2140023 which is a prodrug of LY404039.[70] The current development status of this interesting molecule, a dual MTA, remains uncertain as the initial ground-breaking positive result from an acute schizophrenia trial demonstrating monotherapy antipsychotic efficacy[71] has not been reproduced.[72] However, the latter study was a failed trial as the active control, olanzapine, also failed to separate from placebo. The trial, unlike the initial proof of concept study, involved multiple centres and was complicated by a large placebo effect. This issue, which has historically been more apparent in depression studies, is now also increasingly observed in schizophrenia and is posing a major challenge in the design of clinical trials for new compounds in development. Anyhow, if the initial data obtained with LY2140023 are replicated it would provide a new class of antipsychotic agent with a non-dopaminergic mode of action. This approach, although free of the typical motor side effects associated with D_2 receptor block, has potential for other side effects such as cognitive impairment and disruption of affect as well as safety concerns (e.g. proconvulsant activity).

2.6 Summary and Future Perspectives

Schizophrenia, bipolar disorder and major depression are complex multi-dimensional psychiatric disorders that require multi-functional pharmacology for effective multi-symptom control. This can either be achieved through polypharmacy, involving combination of several drugs, or polypharmacology in a single molecule. Historically, polypharmacologic MTAs have provided useful therapies whilst STAs have generally been rather unsuccessful. This trend has continued as a number of new MTAs have been approved recently which provide incremental benefits as well as important additional options for patients who do not benefit from existing agents. Furthermore, there are a number of MTA and STA compounds in clinical development, some of which will hopefully translate into improved therapies with better efficacy and tolerability. From a future discovery perspective, however, the MTA strategy has to evolve from being serendipity driven towards rational design approaches. In order to progress away from the promiscuity of the historical MTA templates, new chemical space and chemotypes need to be defined. Random library screening and STAs may be useful in providing suitable chemical starting points in this respect. The introduction of hypothesis-based synergistic hybrid pharmacophores that combine a validated target with an unprecedented molecular target component(s) provide the promise for differentiation and improved therapy. To maximise technical feasibility, the number of pharmacophores could be limited to two to three receptors. Additionally, it is important to avoid receptor mechanisms that are now known to be problematic in terms of side effects such as sedation (e.g. H_1, α_1-adrenergic antagonism), cardiovascular (e.g. α_1-adrenergic antagonism), metabolic (e.g. H_1 and muscarinic antagonism) and cognitive disruption (e.g. muscarinic antagonism).

From a pharmacology perspective, the translation of a MTA receptor signature to multi-functional pharmacology is critical. This has to be done in a systematic fashion so that a well-integrated picture is provided, starting with the demonstration of *in vitro* receptor properties in whole animals using receptor occupancy or other read outs for molecular target engagement. Application of quantitative PK/PD analysis should then be engaged to drive dose selection in disease-relevant animal models as well as side effect assays. Ideally this preclinical data should be amenable to clinical translation to aid dose selection in proof of concept studies. Alignment of preclinical and clinical work will be important in facilitating the validation or invalidation of scientific concepts. Furthermore through reverse translation, opportunities for improvement of molecules and/or disease models may also be identified.

The quality of the molecular target is a critical factor in determining the likelihood of success; this is of course challenging given the lack of insight into molecular pathology. Identification and progress in the understanding of genes and molecular targets as well as neurochemical systems involved in psychiatric disease pathology should assist in this regard. In the meantime, investigations on the mode of action of existing drugs and experimental agents (e.g. ketamine, scopolamine) as well as non-pharmacological treatments (e.g. electroconvulsive therapy, sleep deprivation) will continue to provide new therapeutic targets for consideration. However, this needs to be associated with sound brain function hypotheses on how the modulation of a particular target may lead to symptom relief. It is informative that even after decades of use of dopamine receptor blockers and serotonin transporter inhibitors as therapies we still do not properly understand the cellular, neurochemical and neurophysiological changes that eventually lead to their therapeutic effects. However, new ideas are emerging, such as systems involved in modulating neuroplasticity, that will continue to drive knowledge and insight as well as targets for drug discovery. Mechanistic studies on disease mimic agents (e.g. phencyclidine) provide an important complementary source of molecular targets.

Whether from genetics or mechanistic studies, identification and in particular validation of novel disease pathways will take considerable effort and represents part of the long-term strategy. Stronger and more effective collaboration between government, academia and industry will be important in driving this effort. This includes an emphasis on intra-industry sharing of technological developments (e.g. biomarkers, disease models) and experience (e.g. publication of negative or failed trials) without compromising proprietary assets. Although psychiatric drug discovery is currently under considerable scrutiny and pressure, it is hoped that more risk-managed projects and rationally designed novel MTAs containing synergistic pharmacophores will provide multi-functional drugs that will more effectively address patient's needs.

Acknowledgements

The author is grateful to Dr R. Morphy for providing help with obtaining structures for some of the drugs described in this article.

References

1. S. M. Paul, D. S. Mytelka, C. T. Dunwiddie, C. C. Persinger, B. H. Munos, S. R. Lindborg and A. L. Schacht, *Nat. Rev. Drug Discovery*, 2010, **9**, 203.
2. M. J. Williams, *Pharmacol. Exp. Ther.*, 2011, **336**, 3.
3. C. Harrison, *Nat. Rev. Drug Discovery*, 2011, **10**, 12.
4. B. Munos, *Nat. Rev. Drug Discovery*, 2009, **8**, 959.
5. J. Koenig, *Drug Discovery Today*, 2011, **16**, 377.
6. I. Kola, *Clin. Pharmacol. Ther.*, 2008, **83**, 227.
7. Y. Agid, G. Buzsáki, D. M. Diamond, R. Frackowiak, J. Giedd, J. A. Girault, A. Grace, J. J. Lambert, H. Manji, H. Mayberg, M. Popoli, A. Prochiantz, G. Richter-Levin, P. Somogyi, M. Spedding, P. Svenningsson and D. Weinberger, *Nat. Rev. Drug Discovery*, 2007, **6**, 189.
8. S. J. Enna and M. J. Williams, *Pharmacol. Exp. Ther.*, 2009, **329**, 404.
9. E. H. Wong, F. I. Tarazi and M. Shahid, *Pharmacol. Ther.*, 2010, **126**, 173.

10. E. H. Wong, F. Yocca, M. A. Smith and C. M. Lee, *Int. J. Neuropsychopharmacol.*, 2010, **13**, 1269.
11. M. J. Millan, *Neurotherapeutics*, 2009, **6**, 53.
12. L. Bertram, *Curr. Neurol. Neurosci. Rep.*, 2011, **11**, 246.
13. J. M. Shulman, P. L. De Jager and M. B. Feany, *Annu. Rev. Pathol.*, 2011, **28**, 193.
14. Z. A. Knight, H. Lin and K. M. Shokat, *Nat. Rev. Cancer*, 2010, **10**, 130.
15. R. Morphy, C. Kay and Z. Rankovic, *Drug Discovery Today*, 2004, **9**, 641.
16. R. S. McIntyre, J. K. Soczynska, H. O. Woldeyohannes, M. Alsuwaidan and J. Z. Konarski, *Expert Opin. Pharmacother.*, 2007, **8**, 1211.
17. E. H. Wong, S. S. Nikam and M. Shahid, *Curr. Opin. Investig. Drugs*, 2008, **9**, 28.
18. A. J. Rush, *Am. J. Psychiatry*, 2007, **164**, 201.
19. S. P. Van Sant and P. F. Buckley, *Expert Opin. Pharmacother.*, 2011, **12**, 411.
20. J. A. Lieberman, *J. Clin. Psychiatry*, 2007, **68**, e04.
21. A. H. Kemp, E. Gordon, A. J. Rush and L. M. Williams, *CNS Spectr.*, 2008, **13**, 1066.
22. A. Foster, del D. Miller and P. Buckley, *Clin. Lab. Med.*, 2010, **30**, 975.
23. T. Goltser-Dubner, E. Galili-Weisstub, R. H. Segman and J. Isr, *Psychiatry Relat Sci.*, 2010, **47**, 72.
24. J. E. Kleinman, A. J. Law, B. K. Lipska, T. M. Hyde, J. K. Ellis, P. J. Harrison and D. R. Weinberger, *Biol. Psychiatry*, 2011, **69**, 140.
25. P. B. Mitchell, B. Meiser, A. Wilde, J. Fullerton, J. Donald, K. Wilhelm and P. R. Schofield, *Clin. Lab. Med.*, 2010, **30**, 829.
26. J. P. Zhang and A. K. Malhotra, *Expert Opin. Drug Metab. Toxicol.*, 2011, **7**, 9.
27. S. Porcelli, A. Drago, C. Fabbri, S. Gibiino, R. Calati and A. J. Serretti, *Psychiatry Neurosci.*, 2011, **36**, 87.
28. A. Markou, C. Chiamulera, M. A. Geyer, M. Tricklebank and T. Steckler, *Neuropsychopharmacology*, 2009, **34**, 74.
29. O. Malkesman, D. R. Austin, G. Chen and H. K. Manji, *Dis. Model. Mech.*, 2009, **2**, 238.
30. J. C. Neill, S. Barnes, S. Cook, B. Grayson, N. F. Idris, S. L. McLean, S. Snigdha, L. Rajagopal and M. K. Harte, *Pharmacol. Ther.*, 2010, **128**, 419.
31. E. J. Nestler and S. E. Hyman, *Nat. Neurosci.*, 2010, **13**, 1161.
32. C. M. O'Tuathaigh and J. L. Waddington, *Schizophr. Bull.*, 2010, **36**, 243.
33. M. Möller, J. L. Du Preez, R. Emsley and B. H. Harvey, *Eur. Neuropsychopharmacol.*, 2011, **21**, 471.
34. J. Gabrielsson, A. R. Green and P. H. Van der Graaf, *J. Pharmacol. Toxicol. Methods*, 2010, **61**, 146.
35. R. Bursi, G. Erdemli, R. Campbell, M. M. Hutmacher, T. Kerbusch, D. Spanswick and R. Jeggo, K. R. Nations, P. Dogterom, J. Schipper and M. Shahid, *Psychopharmacology (Berl.)*, 2011, Dec, **218**, 713.
36. B. L. Roth, D. J. Sheffler and W. K. Kroeze, *Nat. Rev. Drug Discovery*, 2004, **3**, 353.

37. P. McGuffin, N. Perroud, R. Uher, A. Butler, K. J. Aitchison, I. Craig, C. Lewis and A. Farmer, *Eur. Psychiatry*, 2010, **25**, 275.
38. J. J. Levitt, L. Bobrow, D. Lucia and P. Srinivasan, *Curr. Top. Behav. Neurosci.*, 2010, **4:2**, 43.
39. R. Rigucci, G. Serafini, M. Pompili, G. D. Kotzalidis and R. Tatarelli, *World J. Biol. Psychiatry*, 2010, **11**(2 Pt 2), 165.
40. D. A. Kupferschmidt and K. K. Zakzanis, *Psychiatry Res.*, 2011, **193**, 71.
41. T. de Paulis, *Curr. Opin. Investig. Drugs*, 2001, **2**, 123.
42. S. S. Liauw and R. S. McIntyre, *Expert Opin. Pharmacother.*, 2010, **11**, 2827.
43. C. DeBattista and J. Hawkins, *CNS Drugs*, 2009, **23**, 369.
44. N. H. Jensen, R. M. Rodriguiz, M. G. Caron, W. C. Wetsel, R. B. Rothman and B. L. Roth, *Neuropsychopharmacology*, 2008, **33**, 2303.
45. A. V. Ravindran, A. Al-Subaie and G. Abraham, *Expert Opin. Investig. Drugs*, 2010, **19**, 1187.
46. R. D. Jindal and M. S. Keshavan, *CNS Drugs*, 2008, **22**, 1047.
47. C. de Bodinat, B. Guardiola-Lemaitre, E. Mocaër, P. Renard, C. Muñoz and M. J. Millan, *Nat. Rev. Drug Discovery*, 2010, **9**, 628.
48. Z. A. Hughes, K. R. Starr, C. J. Langmead, M. Hill, G. D. Bartoszyk, J. J. Hagan, D. N. Middlemiss and L. A. Dawson, *Eur. J. Pharmacol.*, 2005, **510**, 49.
49. A. Khan, *Expert Opin. Investig. Drugs*, 2009, **18**, 1753.
50. L. Citrome, *Int. J. Clin. Pract.*, 2009, **63**, 1762.
51. F. I. Tarazi and M. Shahid, *Drugs Today (Barc).*, 2009, **45**, 865.
52. M. Shahid, G. B. Walker, S. H. Zorn and E. H. Wong, *J. Psychopharmacol.*, 2009, **23**, 65.
53. H. Y. Meltzer and B. W. Massey, *Curr. Opin. Pharmacol.*, 2011, **11**, 59.
54. O. Mnie-Filali, C. Faure, L. Lambás-Señas, M. El Mansari, H. Belblidia, E. Gondard, A. Etiévant, H. Scarna, A. Didier, A. Berod, P. Blier and N. Haddjeri, *Neuropsychopharmacology*, 2011, **36**, 1275.
55. A. Szegedi, J. Zhao and A. van Willigenburg, K. R. Nations, M. Mackle and J. Panagides, *BMC Psychiat.*, 2011, **11**, 101.
56. J. Schoemaker, D. Naber, P. Vrijland, J. Panagides and R. Emsley, *Pharmacopsychiatry*, 2010, **43**, 138.
57. J. M. Gonzalez, P. M. Thompson and T. A. Moore, *Patient Prefer., Adherence*, 2011, **5**, 333.
58. L. Citrome, *Int. J. Clin. Pract.*, 2011, **65**, 189.
59. T. Ishibashi, T. Horisawa, K. Tokuda, T. Ishiyama, M. Ogasa, R. Tagashira, K. Matsumoto, H. Nishikawa, Y. Ueda, S. Toma, H. Oki, N. Tanno, I. Saji, A. Ito, Y. Ohno and M. Nakamura, *J. Pharmacol. Exp. Ther.*, 2010, **334**, 171.
60. T. Enomoto, T. Ishibashi, K. Tokuda, T. Ishiyama, S. Toma and A. Ito, *Behav. Brain Res.*, 2008, **186**, 197.
61. T. Horisawa, T. Ishibashi, N. Nishikawa, T. Enomoto, S. Toma, T. Ishiyama and M. Taiji, *Behav. Brain Res.*, 2011, **220**, 83.
62. B. L. Crabtree and J. Montgomery, *Clin. Ther.*, 2011, **33**, 330.

63. L. Citrome, *Expert Opin. Drug Metab. Toxicol.*, 2010, **6**, 1551.
64. H. O. Kalkman, N. Subramanian and D. Hoyer, *Neuropsychopharmacology*, 2001, **25**, 904.
65. G. Grunder, *Curr. Opin. Investig. Drugs*, 2010, **11**, 823.
66. B. Kiss, A. Horváth, Z. Némethy, E. Schmidt, I. Laszlovszky, G. Bugovics, K. Fazekas, K. Hornok, S. Orosz, I. Gyertyán, E. Agai-Csongor, G. Domány, K. Tihanyi, N. Adham and Z. J. Szombathelyi, *Pharmacol. Exp. Ther.*, 2010, **333**, 328.
67. I. Gyertyán, B. Kiss, K. Sághy, J. Laszy, G. Szabó, T. Szabados, L. I. Gémesi, G. Pásztor, M. Zájer-Balázs, M. Kapás, E. A. Csongor, G. Domány, K. Tihanyi and Z. Szombathelyi, *Neurochem. Int.*, 2011, Nov, **59**(6), 925.
68. B. Bang-Andersen, T. Ruhland, M. Jørgensen, G. Smith, K. Frederiksen, K. G. Jensen, H. Zhong, S. M. Nielsen, S. Hogg, A. Mørk and T. B. Stensbøl, *J. Med. Chem.*, 2011, **54**, 3206.
69. E. Alvarez, V. Perez, M. Dragheim, H. Loft and F. Artigas, *Int. J. Neuropsychopharmacol.*, 2011, **18**, 1.
70. M. Mezler, H. Geneste, L. Gault and G. J. Marek, *Curr. Opin. Investig. Drugs*, 2010, **11**, 833.
71. S. T. Patil, L. Zhang, F. Martenyi, S. L. Lowe, J. A. Jackson, B. V. Andreev, A. S. Avedisova, L. M. Bardenstein, I. Y. Gurovich, M. A. Morozova, S. N. Mosolov, N. G. Neznanov, A. M. Reznik, A. B. Smulevich, V. A. Tochilov, B. G. Johnson, J. A. Monn and D. D. Schoepp, *Nat. Med.*, 2007, **13**, 1102. [Epub 2007 Sep 2]. Erratum in *Nat. Med.*, 2007, **13**, 1264.
72. B. J. Kinon, L. Zhang, B. A. Millen, O. O. Osuntokun, J. E. Williams, S. Kollack-Walker, K. Jackson, L. Kryzhanovskaya, N. Jarkova and the HBBI Study Group, *J. Clin. Psychopharmacol.*, 2011, **31**, 349.
73. www.merck.com/licensing/partnership success/astrazeneca.html/
74. http://investor.lundbeck.com/releasedetail.cfm?ReleaseID=608573

CHAPTER 3

Drug Molecules and Biology: Network and Systems Aspects

MALCOLM P. YOUNG,[a,b,*] STEVEN ZIMMER[a] AND ALAN V. WHITMORE[a,c]

[a] e-Therapeutics plc, Holland Park, Holland Drive, Newcastle NE2 4LZ, UK; [b] Institute of Neuroscience, Newcastle University NE2 4HH, UK; [c] School of Pharmacy, University of London, Brunswick Square, London WC1N 1AX, UK
*Email: Malcolm.Young@etherapeutics.co.uk

3.1 Biological Robustness and Therapeutic Discovery

Therapeutic discovery is the process of finding beneficial ways to affect biology. The means by which therapy may be accomplished are very varied – small bioactive molecules, large ones, antibodies, vaccines, nucleic acids, radioligands, surgery and so on – but the task of finding new ways to affect biology beneficially with drugs has evidently proven challenging.[1–4]

One reason for this challenge is that biology is (thankfully) extraordinarily robust to almost every kind of insult or assault, since evolution has selected biological function for this robustness.[5–11] Whole organisms, and the important functional processes within them, have to be robust against variability in a wide range of parameters, including temperature, radiation, salinity, pH, oxygen partial pressure, loss or partial failure of their components, against invasion by parasites, and against changes in the chemical environment.[12] To accomplish this, functional systems in biology are composed of a complex set of pathways, cycles, feedback loops, signalling sequences, homeostatic balancing mechanisms, redundancies, sub-networks

RSC Drug Discovery Series No. 21
Designing Multi-Target Drugs
Edited by J. Richard Morphy and C. John Harris

Published by the Royal Society of Chemistry, www.rsc.org

and clusters, competitive interactions, long- and short-distance interfacing systems and higher-order topological motifs.[2,13–16,57] These network features intervene between adaptive, robust biological function and the components of these networks, such as proteins, nucleic acids, lipids, second messengers and metabolites, which tend to be labile, transient, prone to damage and generally much less robust.[7,8] An analogy might be that of military organisation, such as the organisation of soldiers into platoons, regiments and battalions, which groupings intervene between the individual soldier, who is prone to injury and incapacity during a battle, and a robust performance of the army.

Where biological systems malfunction, or where perfectly functioning biology is nonetheless undesirable for humans (for example in disease caused by pathogens), therapeutic discovery has to confront the complexity and robustness of biological systems head on.[4,17,18] Success in doing so results in efficacious and safe new medicines.[2,17] Failure to do so results in weak efficacy, development of therapeutic resistance, poor side effect and tolerance profiles, drug candidate failures in the clinic more often than not, and unsustainably low productivity (see e.g. ref. 1).

The best way to approach therapeutic discovery is more in dispute now than at any time in the last many decades, not least through a slow but increasingly general acknowledgement of the astonishingly poor productivity of conventional discovery programmes,[3,17,19–21] and the slow realisation of the ramifications of advances in network biology and network mathematics for discovery.[2,17,22–24]

To gain increased traction in discovery, do we need to consider and address network properties rather than the properties of single components, such as protein 'targets'? Is inhibition of one protein 'target' ever a scientifically plausible explanation of therapeutic function, or indeed a scientifically plausible motivation for systematic drug discovery? Is one 'target' ever enough?

3.2 Biological Networks and Their Properties

An encouraging development in network science has been the discovery that networks of biological complexity are strikingly robust to deletion or interruption of their components (e.g. refs. 15, 25–27). These results are encouraging because they are in keeping with the expectation that it is these network structures and topologies that bridge between unreliable components and reliable systems-level function in biological systems, and so they reduce the mystery of how biology is able to generate such robustness from such unreliable components.[23,28]

A further interesting feature of the analysis of biological networks is that individual components of them (nodes) differ greatly in importance for the network (see the glossary at the end of this chapter for a summary of network-related nomenclature). For example, in many biological networks, 'degree' – the extent to which a node is connected to other nodes – varies in a way that approximates a power law, so that a very small number of 'hubs' have an unusually great number of connections.[26,29] This network property has been related in several datasets to the biological essentiality of network components; and to impact on network integrity, the greater the degree of the node to some extent reflecting the greater impact on the network from its deletion.[15,30,58]

The gross connectedness of nodes, though, is only one of many ways in which nodes can have differential significance in a network of biological complexity. Nodes that are entirely unremarkable in terms of degree can nonetheless be highly significant in the structure of the network because they bear other structurally important properties. These other structurally important properties are numerous, and include, for example, 'betweenness', the extent to which a node connects otherwise largely unconnected clusters, and 'redundancy', the extent there are pathways through a network other than through a specific node.[31–34] In addition, it is known that nodes with high values in these properties are associated statistically with biological essentiality in well-studied systems (e.g. refs. 17, 35, 36). Hence, network components can vary greatly in their significance for network structure and function; this significance can derive from many different network properties; and across many of these network properties, there is statistical evidence of association with biological importance (e.g. refs. 30, 35). Biological network structure is illustrated in Figure 3.1, which exemplifies some of the network properties that have proven informative.

The association of these network properties with biologically important features suggests that networks should be susceptible to attack or intervention that is guided by knowledge of the network properties of nodes. And indeed this is what has been observed in 'deletion experiments', in which the effects on a measure of network integrity are recorded when cumulative deletions are made in the network.[15,30,58] Comparisons between cumulative deletions prioritised by several different network properties and deletions prioritised at random show that networks are highly robust to random deletions, but much more susceptible to targeted ones (see Figure 3.2). This result is helpful for the prospects of systematic drug discovery, since it suggests ways of prioritising approaches to therapeutic intervention, at least to the extent of identifying 'targets' in a scientifically rational way.[2]

These results, though, offer a direct scientific challenge to the central precept of conventional single-target drug discovery (STDD) activities. For STDD to be a plausible approach to discovering new medicines, a single targeted intervention should be sufficient to affect biological network integrity significantly. However, to date, no known network property, no plausible measure of the impact of an intervention on network integrity, and no biological network that has been analysed has shown that a single deletion can deliver significant change to network integrity and so to function.[4,17,28,37] This is, in one way, not at all a surprising result: returning to our military analogy, there is no single soldier we can shoot whose demise would significantly affect the performance of the army. Even shooting the General would only affect the army significantly in the very special

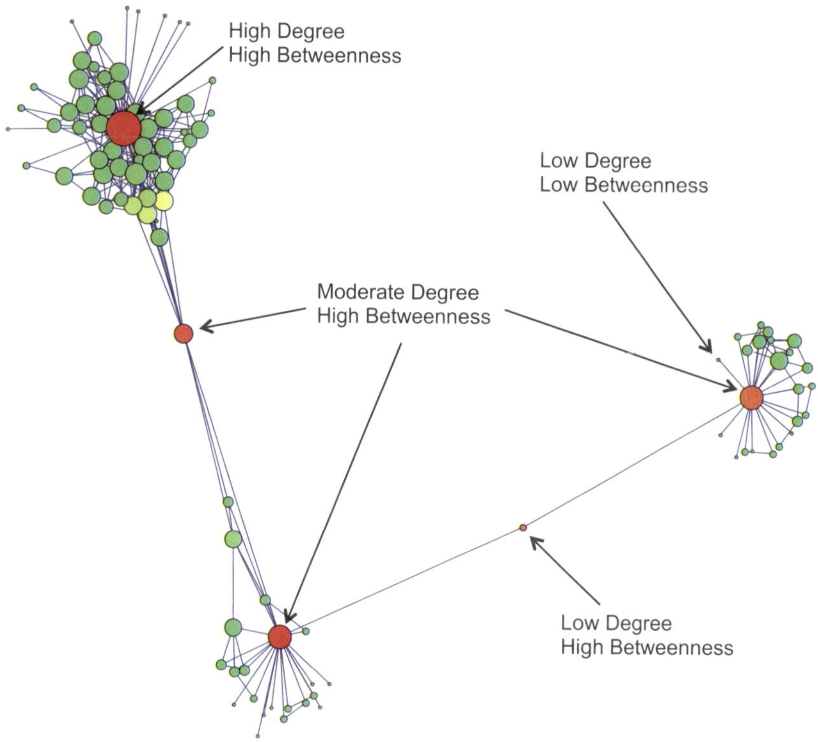

Figure 3.1 Measures of node significance in a network: an example of a sub-component of a real biological network (protein–protein interactions around the microtubule associated protein Tau). The degree of each protein node (the number of neighbours to which it is connected) is represented by the size of the circles (low degree-small). The parameter betweenness (see Glossary) is represented by colour: green (low), yellow (intermediate) and red (high). One or two nodes with high degree also have high betweenness, but a significant number have only low or moderate degree but have high betweenness and are critical for connecting sub-clusters of this network (bright red).

circumstance that his functions cannot be replaced by his general staff.[33,34] These results are illustrated in Figure 3.2, which shows the effects of one and many deletions, prioritised by a variety of different network properties, on two commonly used measures of network integrity.

With only one known important exception (see below), multiple targeted interventions in biological networks always produce greater network impact than a single intervention. Viewed from this perspective, a strategy for developing drug candidates with modest or insufficient efficacy would be to base one's discovery efforts on STDD. Conversely, a reasonable framework for developing efficacious ones would be to try to target multiple network vulnerabilities in disease-related cells, informed by network analysis, and hence to follow a multi-target drug discovery (MTDD) strategy.

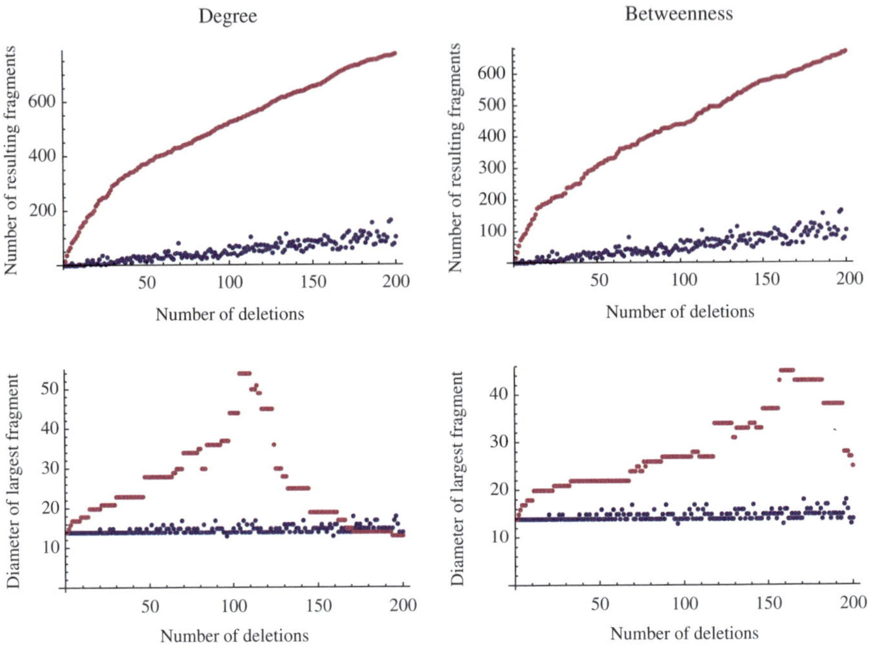

Figure 3.2 Multiple targeted interventions are required to perturb biological networks: the effects on integrity of a network of biological complexity (the giant component of the yeast interactome downloaded from http://interactome.dfci.harvard.edu/S_cerevisiae/) when progressively increasing numbers of random (blue squares) or targeted (mauve squares) node deletions are made. Two measures of node importance were used for targeting: 'degree' and 'betweenness' (see Glossary). Two measures of network integrity are illustrated for each type of targeted intervention – the number of disconnected 'fragments' into which the network has been broken by the interventions (upper panel), and 'diameter', which is the longest shortest path between any pair of nodes in the network (lower panel). Deletions prioritised by 'degree' and 'betweenness' have a much greater impact than random ones (which have almost no effect unless a very large proportion of nodes are removed). Even with prioritisation, multiple interventions are still necessary to perturb the network significantly.

3.3 Multiple Interventions: Expect the Unexpected

There is one known and potentially very important exception to the observation that multiple interventions always yield greater network impact than single interventions. This exception relates to the possibility of negative synergies between multiple interventions, in which one intervention negates or diminishes the effect of another. These effects are 'paradoxical' in that function pre-dating the interventions may be restored by further interventions in a system affected by the first intervention. For example, in disease, the original dysfunctional state may be inadvertently restored if one intervention, intended to improve the condition, is accompanied by another intervention made in an unfortunate place in the network. Such paradoxical effects are well known and characterised in, for example, neurology, where they are known as 'paradoxical restorations of function' (e.g. refs. 15, 38). There is ample opportunity for these effects to manifest in any biological network where there is both sub-network structure and competitive interaction between sub-networks, as appears to be the case in all biological networks of relevance to drug discovery.[15,38,39]

The simplest case in which these negative synergetic effects manifest occurs when two simultaneous interventions in the network are made.[15] In this case, one intervention might negate or diminish the other, by, for example, restoring balanced competition between competing subsystems when such balanced competition is an important feature of system function (e.g. refs. 15, 38). In the context of drug discovery, these effects might manifest as lack of experimental efficacy despite one's drug candidate certainly diminishing the function of an important 'target' protein – because, unbeknownst to the drug designer, the molecule also intervenes on a protein in a different sub-network that has competitive interactions with the sub-network containing the primary 'target', restoring function based on balanced competition. Note that studying only the local pathway in which the primary target is believed to participate would not be sufficient to anticipate these paradoxical synergies, which are network-based, systems-level effects. Note also that without checking for these types of synergetic effects, efficacy is not reliably predictable from potency at a single 'target' in the general case.[2,40]

If the presence of a bioactive molecule in or near a cell involves any pleiotropy (effects on proteins or other signalling molecules downstream in the network from a directly affected protein), promiscuity (direct effects on more proteins than the designed 'target'), or polypharmacy (more species of molecule present than the bioactive molecule itself – such as the molecule's long-lived metabolites, each of which can exhibit both pleiotropy and promiscuity), then multiple interventions in the network of higher order than two will be present. We examine whether and how drug molecules intervene on multiple proteins in the next section. However, determining the net effect of more complex sets of interventions in biological networks than two is a combinatorial problem that is very difficult to address experimentally, since large numbers of permutations cannot effectively be explored in that way. The need to determine the net effects of these synergies as early as possible in drug development, in order to de-risk candidates prior to expensive clinical trials, means that there is an irreducible computational step in discovery and development programmes that do not wish to rely on luck.[2,33,34,40]

3.4 Effective Drugs and Multiple Interventions in Networks

The likely presence of negative synergetic effects from making multiple interventions in networks of biological complexity might suggest that MTDD is a riskier strategy than STDD, since intervening at only a single 'target' would not engage these factors. This would only be true, however, if drug molecules motivated by STDD did in fact only intervene at the designed 'target'. Hence, an important question is: do any drugs affect only one protein 'target'?

We noted above some theoretical reasons for doubting that any efficacious drug affects only a single 'target'. Chemical biology also suggests that the assertion that efficacy derives from a single 'target' intervention is a very implausible scientific claim, perhaps one so dubious as to require special evidential support (e.g. refs. 17, 40). A molecule designed to have very high affinity at one protein is likely also to have affinities at other proteins, since proteins are configurationally similar. This promiscuity has been discussed extensively recently,[32,41–44] and is illustrated by the spectra of affinities shown across screens for binding affinity with other proteins.[41] The binding spectra so far explored in the literature relate only to a very small proportion of the protein *dramatis personae* in human and pathogenic cells, typically being fewer than 100 proteins (e.g. ref. 41). However, while the number of proteins in human and other disease-related cells is not known definitively, it is likely that it exceeds many hundreds of thousands.[45] Hence, there is great scope for drug–protein promiscuity in medicine, only a small proportion of which is typically addressed in discovery programmes. It cannot safely be assumed that these off-target interactions will not have consequences for efficacy (see Section 3.3) or safety, and we hypothesise that these unpredicted effects are one reason for the prevalence of clinical stage attrition of drugs motivated by STDD, since clinical trials test expensively for the global effect of the presence of the molecule in patients, including these currently poorly attended off-target interventions.[40]

The presence of a molecule in the body can affect the functional contribution of proteins for which the molecule does not itself have high affinity. For example, the presence of a molecule can pleiotropically affect the expression and abundance of proteins to which it does not bind directly. A celebrated example of this is the effect of CGK1026 on the expression of telomerase.[46] CGK1026 forms a complex with HDAC2 and E2F, which has the effect of de-repressing TERT (Telomerase Reverse Transcriptase).[46] Hence, CGK1026 profoundly affects the expression of telomerase, with potentially significant effects on cell biology, without binding telomerase directly. Similarly, the presence of a molecule can pleiotropically affect the phosphorylation or other functional state of proteins to which it does not bind (e.g. refs. 47, 48). Indeed, it can be argued that the importance of kinases as therapeutic 'targets' derives largely from the pleiotropic effects they exert on many other proteins.[49] There is evidence for even less direct pleiotropic effects in networks of biological complexity: deletion of nodes can readily affect other unusually significant nodes through network-mediated effects via these nodes' network associates.[2,15,46,60]

Binding promiscuity and these several varieties of pleiotropy can also attend each downstream metabolite of a bioactive molecule that is sufficiently long-lived to interact biologically. Hence, the presence of a molecule in the body can entail a pattern of interventions in networks far beyond its designed primary 'target'. These wide-ranging effects can be mediated through wider promiscuous binding; through the pleiotropic downstream effects of each of these interactions; and its metabolites, if sufficiently long-lived to be active, can each also entail wide-ranging effects mediated through their own promiscuity and pleiotropy. Hence, most, and perhaps all, bioactive molecules cause multiple interventions in networks of biological complexity, and so will engage the potentially complex consequences explored above.[2,40]

In regard to defining the best way to approach therapeutic discovery – following the recent advances in network science – drugs motivated by STDD principles are hence no better placed to avoid the problem of the paradoxical effects of multiple interventions in networks, since these drugs are almost certainly themselves multiple interveners. Motivating discovery by MTDD principles is better supported theoretically, and also perhaps psychologically, since MTDD programmes might be disposed to be more alive to a potentially wide range of interventions caused by their candidate molecules. In the cases of both STDD and MTDD strategies, though, network analysis suggests that there may be some further hills to climb before drug discovery and development will generate noticeably higher productivity.

3.5 Drug Discovery Problems in Light of Network Science

Consideration of impact graphs, such as those in Figure 3.2, suggests that greater impact attends greater numbers of network interventions. If network impact is a good proxy for efficacy, then, in general, the more 'multi' a multiply intervening drug is, the more effective it should be, all else being equal. Furthermore, it has been suggested, on the basis of research on the transcriptional regulatory networks of bacteria and yeast, that such interventions can be quite weak and yet remain very effective in combination.[37] This implies that it would be reasonable to try to design for many more than two 'targets'.[41,42,50,51] The technical difficulty of doing this is addressed elsewhere in this volume. However, we have seen that there are potentially important negative synergies that can be engaged should some of our molecule's interventions be at unfortunate places in the network, the effect of which could be to greatly diminish efficacy, even when a number of interventions in biological networks are made. We have also seen that drug molecules tend to make rather many interventions in biological networks, in a number of different ways.

These considerations suggest some scientific elements that may underlie low productivity in drug discovery. Official statistics suggest that only around 1 in 10 000 lead-optimised molecules eventually becomes a marketed drug.[52] Further observations are that discovery and development across the industry have turned exponentially improving inputs (e.g. bases sequenced per hour; compounds synthesised and screened per hour) into exponentially decreasing output (e.g. new drugs gaining market approval divided by R&D spend) over recent decades; and that the Net Present Value of small molecule-oriented

R&D is now negative when considered across the industry as a whole.[1,21] Before becoming overly enthused by new approaches to drug discovery, it would perhaps be prudent to consider some scientific explanation of what has gone wrong with the conventional approach, so that the same or similar difficulties can be avoided in the future.

A consequence of the features of promiscuity, pleiotropy and metabolite polypharmacy (Section 3.4) may be that the great majority of the interventions in biological networks that are caused by the presence of a drug or candidate are not the one (or two) that motivated its design. Even when a molecule binds with high affinity to its designed 'target', it may affect many other biological system components in an undesigned way. Since unfortunate undesigned interventions can diminish both efficacy and safety, and fortunate undesigned interventions can occasionally improve both efficacy and safety, a poorly controlled factor would appear to be present in drug discovery programmes that do not take account of these features. From first principles, we expect that this uncontrolled factor would manifest as great difficulty in predicting efficacy and safety reliably in patients from preclinical information. This does indeed seem to be a feature of current problems in drug discovery and development.[2,3,17,40,52]

Very many more candidate drug molecules are developed than go on to become effective drugs,[52] with very many failing for efficacy or safety reasons in clinical evaluation with human volunteers or patients. The large number of failures and the small number of successes provides a very strong opportunity for *selection* to operate. Observations suggest that across very many discovery programmes, there is an occasional success, usually offering an incremental benefit in efficacy or safety, framed by many candidates failing in the clinic.[1] It is tempting to believe that the successes are derived by rational design. Confidence in rational design might be reinforced by survivor bias, that is, podium presentations being biased toward projects that delivered clinical results broadly those being looked for, and biased against those that did not.[1] However, an equally plausible explanation, taking account of the large number of failures in 'targeted' development programmes, is that the successes are the result of *clinical selection* operating over plausible chemical diversity, and not principally the result of design. Each candidate will doubtless possess high affinity for its designed target, and it will also likely cause a variety of undesigned interventions that mean that its efficacy and safety proposition in human beings will be highly unpredictable from its affinity for its primary 'target'.

Clinical evaluation then tests for the global net effect on efficacy and safety of the presence of the molecule in patients, including all the 'off-target' interventions arising from promiscuity, pleiotropy and metabolite polypharmacy. Most combinations of designed and undesigned interventions would be expected to be useless or unsafe (and often both), since biological networks are robust to extensive random interruptions. Clinical evaluation will fail these candidates very often, at great cumulative expense. But occasionally the combination of designed and undesigned interventions would be expected to be beneficial, simply by chance. The net effect, evaluated by the clinic, would support these rare candidates, and they have opportunity to proceed through development to become successful medicines.

This model of clinical selection over plausible chemical diversity – as opposed to medicinal chemistry design – suffers from the great expense of clinical trials, and the great number of them that are required for even a small number of successful new drugs to emerge.[19,53] The model is sustainable only when the revenues from the occasional success are sufficient to support the large number of clinical failures each success requires. If the bar is raised in any way, however, through, for example, it being more difficult to charge a premium for new drugs that are only marginally superior, or more aggressive generic erosion, or increased regulatory distaste for safety risk, or areas of remaining unmet medical need being harder to gain traction by using conventional methods, then it will generally fail, no matter what the scale of resources that are dedicated to it.

Even if it remains difficult at present to design molecules with specific multi-intervention profiles, there may be ways to increase the efficiency of a selection-based model of drug development. One way would be to pay much greater attention to the potentially wide-ranging interventions caused by candidate drug molecules, to pay much greater attention to the network impacts of these interventions, and to reject a far greater proportion of drug candidates on this basis before they reach the clinic.

3.6 Network Pharmacology: Exploiting Advances in Chemical Biology and Network Science

In what ways can improved understanding of chemical biology and network science be leveraged to improve upon the unsustainably low productivity of conventional, typically single 'target', drug discovery? Historically, heuristics that were insufficiently productive to be sustainable have often been replaced by approaches based on the application of engineering principles (e.g. ref. 54), and this is the intent of network pharmacology.[2,40] In the case of improving the productivity of drug discovery and development, the pressing needs are to obviate failure during the process, and to deliver high efficacy safely in areas of unmet medical need.

Since all bioactive molecules intervene on multiple proteins (see Sections 3.3 and 3.4 above), we consider particularly the case of the network impact of multiple interventions in complex biological networks. There are very many different combinations of, for the sake of the argument, five proteins that can be removed simultaneously from any network of full biological complexity without affecting network integrity in any significant way. This reflects evolutionary selection in favour of biological systems that are robust to random damage and chemical perturbation, through redundancy, amongst other organisational adaptations. However, a small number of combinations of the same five proteins can have a very great impact on network integrity in most biological networks. Figure 3.3A–C show how

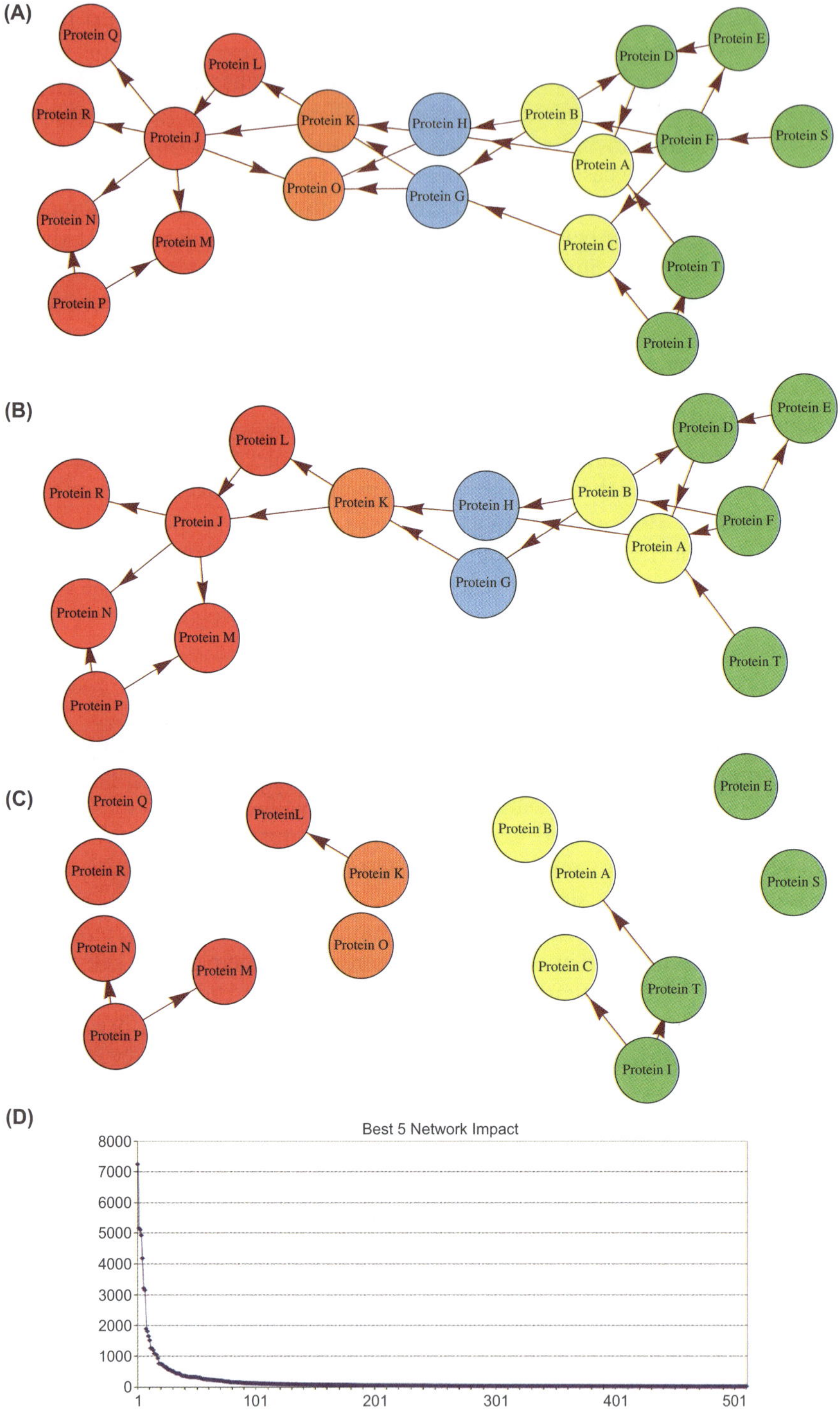
(A)
Protein Q
Protein R
Protein J
Protein L
Protein K
Protein H
Protein B
Protein D
Protein E
Protein F
Protein S
Protein A
Protein O
Protein G
Protein N
Protein M
Protein P
Protein C
Protein T
Protein I
(B)
Protein L
Protein R
Protein J
Protein K
Protein H
Protein B
Protein D
Protein E
Protein F
Protein A
Protein G
Protein N
Protein M
Protein P
Protein T
(C)
Protein Q
Protein R
Protein N
Protein M
Protein P
ProteinL
Protein K
Protein O
Protein B
Protein A
Protein C
Protein T
Protein I
Protein E
Protein S
(D)
Best 5 Network Impact
8000
7000
6000
5000
4000
3000
2000
1000
0
1
101
201
301
401
501

multiple interventions can be optimised by combinatorial impact analysis using a simplified artificial network for a hypothetical pathogenic bacterium. The objective is to find a way to disrupt the building of the cell wall using a maximum of five interventions. The network shows considerable redundancy in that there are a number of alternative possible paths between proteins responsible for component synthesis (green circles) and for the final assembly process (red circles). Random deletion of five nodes has no effect on the bacterium's ability to assemble a cell wall, but when the highest degree nodes and the two nodes with highest betweenness are deleted, the network disintegrates such that the bacterium cannot make a cell wall. In fact in this example it might be sufficient just to delete the two blue nodes to disconnect the 'parts department' from the 'assembly line', although some components might still reach the assembly complex by simple diffusion if this were not also disabled. Figure 3.3D shows how the distribution of combinatorial impacts on network integrity for a real biological network tends to be exponential, showing a small number of highly impactive combinations and a long tail of irrelevant ones. Hence, the strategy of network pharmacology-based drug discovery is to find the very few highly impactive sets of proteins that should be addressed simultaneously, and to develop molecules that deliver these optimally impactive sets of interactions, in order to yield high efficacy.

Combinations of simultaneous interventions that yield very high network impact represent only a tiny proportion of all the possible combinations

---

Figure 3.3 Multiple interventions can be optimised by combinatorial impact analysis. (A) Simplified artificial bacterial network for cell wall building. Green circles are proteins responsible for post-translational modification of newly synthesised cell wall components. The cell wall components (yellow) must then interact with the carrier proteins (blue) to form a complex (orange) that can be assembled by the wall assembly proteins (red). For a viable systems-level function to be present there must be a continuous path between each type of protein. The objective is to overcome the redundancy (there are a number of alternative possible paths from green to red) using a maximum of five interventions. (B) Deletion of five nodes at random has no material effect since there remains a viable path from green to red. (C) Targeted deletions of the three highest degree nodes and the two highest betweenness nodes have been made. The redundancies have been overcome and the bacterium can no longer make a cell wall. (D) Example distribution of combinatorial impacts on network integrity for a real biological network. Most ways of removing five nodes at random yield a very minor impact on network integrity. All single node deletions would be placed far to the right in the distribution, yielding negligible impact on network integrity, and most combinations of a designed single deletion with four other deletions associated with the single deletion semi-randomly would have negligible effect on network integrity. However, there are a small number of combinations of five node deletions that have a profound effect on network integrity (far left). The goal of network pharmacology-guided drug discovery is to identify such impactful combinations and to develop molecules that intervene on them appropriately to deliver high efficacy.

of interventions. Such high-impact combinations are unlikely to be found unless they are specifically optimised or searched for. These combinations yield many-fold greater impact than those achieved through single designed interventions and also from single designed interventions accompanied by a number of undesigned interventions. From first principles, therefore, where network impact is a good predictor of efficacy in target cells, much greater efficacy would be expected from optimising for network impact through selecting specific combinations of interventions, than could be derived from intervening on one (or two) 'target' proteins, with or without associated undesigned interventions. Bearing in mind that interventions at specific proteins can be weak in terms of binding affinity, and indeed pleiotropic, yet be highly effective in combination (e.g. ref. 37), greatly enhanced efficacy should be obtainable by optimising for network impact, rather than nanomolar binding potency at a single protein 'target'. From this perspective, network pharmacology optimises a completely different feature of chemical biology than conventional discovery, and so represents a significant, scientifically motivated, departure from conventional approaches. Similarly, where conventional discovery processes are industrialised around optimising for nanomolar potency at single 'targets', as they appear often to be, this may in large part explain the exponential decline in conventional productivity as the outcome of industrialising design for a largely irrelevant feature (see also ref. 55).

Employing network pharmacology in drug discovery requires the optimisation step of combinatorial network impact analysis to be embedded in a number of other processes. Network pharmacology is relatively new, but most discovery projects motivated in this way have begun by identifying a specific medical need that is poorly served by existing therapies, which would represent a sufficient market to justify the expense of development, and which can be framed in an affordable and implementable development and regulatory pathway. Once the specific medical need is identified, the pathophysiology of the disease is explored, the optimal medical intervention point in the pathophysiological development of the disease determined and systems-level functions operating at this point in the pathophysiology mapped in detail. The genomic and proteomic networks that are implicated in these functions are then curated, and then combinatorial impact analysis is undertaken to derive optimally impactive sets of proteins to intervene upon simultaneously in order to drive the network(s) in a therapeutically desirable direction.

Once optimally impactive sets of proteins have been derived for the disease, molecules are sought whose presence in the human body is associated with an appropriate pattern of interaction with the proteins in the optimal set. These interactions can be exerted through direct, allosteric and promiscuous binding affinity; or pleiotropically, through effects on protein expression, phosphorylation, or post-translational modification; or through indirect network-mediated effects, typically by affecting a near topological neighbour of a protein in the optimally impactive set. This process requires significant chemoinformatics resources, but the inclusion of non-binding interventions in the mapping between an optimally impactive protein set and molecules that could deliver it has the effect of opening the bandwidth of molecular selection beyond that available from direct high-affinity binding alone. Molecules that deliver an optimally impactive set of interventions are likely to be efficacious in delivering the specific therapeutic functionality sought in the design of the discovery project.

The principle of combinatorial network impact, employed to focus discovery on high efficacy above, can also be applied very early in development to de-risk candidate molecules for safety and tolerance. Conventional *in vitro* and *in vivo* preclinical toxicological studies plainly do not de-risk molecules for safety in human patients nearly well enough, evidenced by the high number of safety failures in clinical stages (e.g. refs. 19, 59). However, insight into the effect on a wide variety of normal human cell types of the presence of a candidate molecule in the body can be derived by calculating the combinatorial impact of those proteins affected by the molecule in each of hundreds of normal human cell types. These cell types represent, as far as current bioinformatics data are able, the full range of tissue types. This process requires significant chemo- and bioinformatics resources, and, unlike targeting for efficacy, is undesirably open-ended as a scientific problem, since there are very many different cell types in patients, and data on the networks within them are far from complete. Notwithstanding the data limitations, which are lessening as knowledge of normal function increases, it is certainly feasible to triage a proportion of molecules with the potential for significant impact in these normal cell types, and these can then be discarded.

This workflow embeds the process of combinatorial network impact analysis and the promiscuity and pleiotropy of bioactive molecules to derive molecules that are probably both efficacious in the specific indication sought, and safe. The workflow is designed to predict accurately what risks a candidate has as cheaply and early as possible, and does not rely on expensive clinical trials as the principal means of de-risking molecules. In terms of the effect of this approach on discovery and development productivity, conventional processes cannot predict either clinical efficacy or safety accurately at present, and so most drug candidates fail after substantial sums are spent on them, since expensive tests and trials are the only way to evaluate these risks (e.g. ref. 56). This means that at any point the vast majority of discovery and development resource is tied up in drug candidates that are going to fail, and productivity is unsustainably low. But if early stage predictions could be improved, a lower proportion of candidates that are predicted to be worth developing will fail, and expensive trials will usually evaluate the risks as low. In this scenario, at any one time, the majority of discovery and development resource will be invested in drug candidates that will probably not fail because technical risks manifest, and productivity will be much higher. Commercial risk for individual companies would not diminish in this scenario. Indeed it will rise if more good drugs come to the market, but that would perhaps be a better problem to have, especially from the perspective of patients.

3.7 Prospects for Multi-Target Drug Discovery in Light of Network Science

The advances in network science that we have touched on in this chapter hold out great promise for enabling systematic multi-target drug discovery into the future.

First, the robustness of biological systems, which confronts head-on all attempts to change biological systems for the therapeutic better, may be mediated by network features that we can now target specifically, through network analysis.

Second, network science advances throw a harshly critical light on some of the fundamental principles of most drug discovery activity, such as the single 'target' idea, suggesting that resources could be diverted toward activities more likely to be productive, such as MTDD strategies. There may be a need to influence the education of scientists in parallel with any paradigm shift in the industry.

Third, some of the pitfalls of making multiple interventions in networks of biological complexity, such as paradoxical synergies between interventions, are becoming clear in a form that permits empirical testing and navigation around them.

Fourth, the exponential distribution of combinatorial impact in networks, together with the mediocre positioning of both single 'target' and poorly targeted multi-'target' drugs in these distributions, suggest that much greater efficacy than seen hitherto could be engineered into drug molecules using this type of design principle.

Fifth, the many different ways in which bioactive molecules can cause interventions in biological networks are becoming clearer, enabling the prospect first of rational selection of molecules from their chemoproteomic signatures for specific medical tasks, and, perhaps in future, the design of molecules on these principles.

Sixth, diseases that have proven highly resistant to conventional therapeutic discovery, which tend to be those diseases in which a new drug can still be priced at a premium, are also those that seem unusually suited to therapeutic targeting by approaches based on these principles of network pharmacology.

Glossary

Betweenness The extent to which a node (or connection) in a network appears on the paths between other nodes.

Cluster, fragment or sub-network Community structure within a complex network: in the simplest case a group of nodes that share more connections with each other than with other nodes in the network; or in functional terms an interconnected group of nodes whose interactions underlie an identifiable, emergent, systems-level function.

Combinatorial impact Denotes the impact on the structural integrity of a complex network of removing a set of nodes and/or connections.

Degree The number of connections that a node has with other nodes in a network.

Essentiality In biological networks, this denotes the importance of a node (such as a gene or protein) to the cell's viability or to the ability to sustain a systems-level function.

Heuristic An experience-based strategy or tactic that, using available information, is intended to yield the desired outcome.

Hub A highly connected network node; a node that has high degree.

Impact (network) A measure of the importance of an intervention on the integrity or functional state of a complex network. Metrics include the histogram of the shortest path; or a plot of the magnitude of the longest (or average) shortest path in a network versus different types of intervention(s).

Integrity (network) The structural state of a network in which the network is able to function normally to generate its emergent, systems-level function.

Metabolite polypharmacy In this context is used to represent the situation where not only a compound but also its (possibly multiple) metabolites are pharmacologically active and therefore have additional (and often unpredicted) network consequences.

Negative synergy An interaction between two or more interventions within a complex network whose net impact on the emergent function of that network process is either neutral (because they cancel each other out) or has an effect opposite to that intended.

Network analysis Denotes the analysis of the structure and nature of relationships between nodes and edges in a network.

Pleiotropy In this context describes potentially multiple consequences immediately downstream of the effect of a bioactive molecule at a particular protein, which may change the phosphorylation state, abundance, methylation or post-translational modification of multiple other proteins.

Promiscuity Denotes the ability of a molecule to interact with many different proteins either through the 'active sites' of those proteins or through allosteric binding.

Redundancy This denotes a mechanism through which complex networks generate robustness to loss of components or to other insults. The degree of fault tolerance in these networks is related to the presence of network redundancies (e.g. alternative information flow pathways or clusters with redundant sub-functions). In biology, redundancy is a fail-safe mechanism for systems-level networks, permitting organisms to survive mutations or environmental insults with minimal or no consequence to the systems-level function controlled by the network.

Robustness Usually refers to the ability of a complex system or network to maintain its functional characteristics in the face of perturbations. In biology, it can mean the ability of an organism to maintain a particular systems-level function or phenotype in the presence of mutations or external stressors. In network science, it denotes the ability of a network to withstand attack or component failures.

References

1. J. W. Scannell and A. Blanckley, *The Long View: Pharmaceutical R&D productivity – after six decades of failed cures, isn't it time for a diagnosis?*, Bernstein Research, Sanford C Bernstein Ltd, 2010.
2. S. Zimmer and M. P. Young, *Innov. Pharm. Technol.*, 2009, **18**, 38–41.
3. G. Duyk, *Science*, 2003, **302**, 603–605.
4. M. Hellerstein, *J. Pharmacol. Exp. Ther.*, 2008, **325**, 1–9.
5. H. Kitano, *Nat. Rev. Genet.*, 2004, **5**, 826–837.
6. H. Kitano, *Mol. Systems Biol.*, 2007, **3**, 1–7.
7. A. Wagner and J. Wright, *Biosystems*, 2007, **88**, 163–172.
8. A. Wagner, *Proc. R. Soc. B*, 2007, **275**, 91–100.
9. J. Masel and M. Siegal, *Trends Genet.*, 2009, **25**, 395–403.
10. J. Masel and M. Trotter, *Trends Genet.*, 2010, **26**, 406–414.
11. K. Raman and A. Wagner, *Mol. Biosystems*, 2011, **7**, 1081–1092.
12. T. Piersma and J. Van Gils, *The Flexible Phenotype*, Oxford University Press, 2011.
13. M. P. Young, *Nature*, 1992, **358**, 152–155.
14. M. P. Young, *Proc. R. Soc. Biol. Sci.*, 1993, **252**, 13–18.
15. M. P. Young, C. C. Hilgetag and J. W. Scannell, *Philos. Trans. R. Soc. Biol. Sci.*, 2000, **355**, 147–161.
16. D. Krakauer, *Complex Systems Sci., Biomed.*, 2006, **2**, 183–205.
17. A. Hopkins, *Nature Chem. Biol.*, 2008, **4**, 682–690.
18. T. Hase, T. Hiroshi, Y. Suzuki, S. Nakagawa and H. Kitano, *PloS Comp. Biol.*, 2009, **5**, 1–9.
19. I. Kola and J. Landis, *Nat. Rev. Drug Discovery*, 2004, **3**, 711–716.
20. I. Kola, *Clin. Pharmacol. Ther.*, 2008, **83**, 227–230.
21. J. W. Scannell, T. Anderson, J. Redenius and C. L. Bedell, *The Long View: Pharmaceutical R&D productivity follows Moore's Law backwards. Does anyone know why?*, Bernstein Research, Sanford C Bernstein Ltd, 2011.
22. P. Csermely, V. Agoston and S. Pongor, *Trends Pharmacol. Sci.*, 2005, **26**, 178–182.
23. A. Pujol, R. Mosca, J. Farres and P. Aloy, *Trends Pharmacol. Sci.*, 2010, **31**, 115–123.
24. M. P. Young and A. V. Whitmore, *Eur. Biopharm. Rev.*, 2010, **132**, 2692.
25. R. Albert, H. Jeong and A.-L. Barabasi, *Nature*, 2000, **406**, 378–382.
26. A.-L. Barabasi and Z. Oltvai, *Nat. Rev. Genet.*, 2004, **5**, 101–113.
27. R. J. Albert, *Cell Sci.*, 2005, **118**, 4947–4957.
28. A. Wagner, *Nat. Genet.*, 2000, **24**, 355–361.
29. A.-L. Barabasi and R. Albert, *Science*, 1999, **286**, 509–512.
30. H. Jeong, S. Mason, A.-L. Barabasi and Z. Oltvai, *Nature*, 2001, **411**, 41–42.
31. M. Kaiser and C. C. Hilgetag, *Biol. Cybern.*, 2004, **90**, 311–317.
32. G. V. Paolini, R. H. Shapland, W. P. van Hoorn, J. Mason and A. L. Hopkins, *Nature Biotech.*, 2006, **24**, 805–815.

33. M. P. Young, P. Andras and M. A. O'Neill, *European Patent Office*, 2002, EP1556819.
34. M. P. Young, P. Andras and M. A. O'Neill, *US Patent Office*, 2002, US7466663 B2.
35. O. Idowu, S. Lynden, M. Young and P. Andras, *Proc. IEEE Computational Systems Bioinformatics Conference, Stanford, USA*, California, 2004, pp. 623–625.
36. O. Idowu and P. Andras, *Proceedings of Computational Intelligence in Medicine and Healthcare (CIMED)*, 2005, pp. 330–333.
37. V. Agoston, P. Csermely and S. Pongor, *Phys. Rev. E*, 2005, **71**, 051909.
38. C. C. Hilgetag, R. Kotter and M. P. Young, *Prog. Brain Res.*, 1999, **121**, 121–141.
39. V. Spirin and L. Mirny, *Proc. Natl. Acad. Sci. USA*, 2003, **100**, 12123–12128.
40. M. P. Young, *Drug Discovery World*, 2008, **3**, 9–12.
41. A. L. Hopkins, J. Mason and J. P. Overington, *Curr. Opin. Struct. Biol.*, 2006, **16**, 127–136.
42. R. Morphy, C. Kay and Z. Rankovic, *Drug Discovery Today*, 2004, **9**, 641–651.
43. B. L. Roth, D. J. Sheffler and W. K. Kroeze, *Nature Rev. Drug Discovery*, 2004, **3**, 353–359.
44. G. Weitz-Schmidt, *et al.*, *Nature Med.*, 2001, **7**, 687–692.
45. O. N. Jensen, *Curr. Opin. Chem. Biol.*, 2004, **8**, 33–41.
46. J. Won, S. Chang, S. Oh and T. T. Kim, *Proc. Natl. Acad. Sci. USA*, 2004, **101**, 11328–11333.
47. J. Liao and U. Laufs, *Annu. Rev. Pharmacol. Toxicol.*, 2005, **45**, 89–118.
48. J. Greenwood, L. Steinman and S. Zamvil, *Nat. Rev. Immunol.*, 2006, **6**, 358–370.
49. R. Morphy, *J. Med. Chem.*, 2010, **53**, 1413–1437.
50. C. Wermuth, *Drug Discovery Today*, 2004, **9**, 826–827.
51. J. T. Metz and P. J. Hajduk, *Curr. Opin. Chem. Biol.*, 2010, **14**, 498–504.
52. European Commission, *Innovative Medicines Initiative: Better Tools for Better Medicines*, Office for Official Publications of the European Communities, Luxembourg, 2008, p. 5.
53. J. A. Di Masi, R. W. Hansen and H. G. Grabowski, *J. Health Econom.*, 2003, **22**, 151–185.
54. R. E. Barlow, *Engineering Reliability*, ASA and SIAM, New York, 1998.
55. D. C. Swinney and J. Anthony, *Nat. Rev. Drug Discovery*, 2011, **10**, 507–519.
56. C. Adams and V. Brantner, *Health Aff (Millwood)*, 2006, **25**, 420–428.
57. J. W. Scannell and M. P. Young, *Curr. Biol.*, 1993, **3**, 191–200.
58. M. Kaiser, R. Martin, P. Andras and M. P. Young, *Eur. J. Neurosci.*, 2007, **25**, 3185–3192.
59. J. A. Di Masi, *Clin. Pharmacol. Ther.*, 2001, **69**, 297–307.
60. M. P. Young, P. Andras and M. A. O'Neill, *US Patent Office*, 2011, US79990878 B2.

CHAPTER 4

Chemoinformatic Approaches to Target Identification

ELISABET GREGORI-PUIGJANÉ[a] AND MICHAEL J. KEISER[a,b,*]

[a] UCSF Department of Pharmaceutical Chemistry, UCSF MC 2550, 1700 4th St, San Francisco, CA 94158, USA; [b] SeaChange Pharmaceuticals Inc., QB3 MC 2522, 1700 4th St Ste 214, San Francisco, CA 94158-2330, USA
*Email: mkeiser@seachangepharma.com

4.1 Introduction

Our understanding of pharmacodynamic drug action has become increasingly complex and network based. We now know that many drugs once thought of as being target specific are active at therapeutic doses against multiple targets. Such unintended 'off-target' drug activity can have consequences for ill – but also for good. Whereas off-target activities account for some undesired side effects, they can also contribute to increased efficacy (e.g., by modulating several targets in a single pathway) and to new indications for old drugs.[1]

Drug discovery efforts now attempt prediction of full pharmacological compound profiles and of their interaction consequences; improvements in experimental and computational capabilities together drive this shift. Not only has *in vitro* testing of isolated compounds increased in throughput, but more complex phenotypic and high-content screens are also becoming increasingly available. Simultaneously, growth in chemoinformatics capacities has expanded the reach of virtual screening from single-target analysis to rapid profiling of millions of compounds at thousands of targets. In an effort to integrate and

RSC Drug Discovery Series No. 21
Designing Multi-Target Drugs
Edited by J. Richard Morphy and C. John Harris

Published by the Royal Society of Chemistry, www.rsc.org

interpret the rising tide of complementary experimental and computational data, network pharmacology methods have emerged to combine the strengths of both.

4.2 Approaches

4.2.1 Representing Ligands for Similarity Calculations

A small molecule carries a wealth of disparate information, whose import may vary for a medicinal chemist, a pharmacologist, a molecular biologist – or for a computer. Lacking a central dogma of information flow such as in biology, '*from DNA to RNA to protein*,' we have no canonical representation of small molecules. Instead, we often collapse the molecule's structure into a tractable linear fingerprint. This collapse is a *loss-based* one – much as a digital photograph's grid of pixilated colors alone cannot later perfectly recreate the scene from which it was once derived. What is worse, no single theoretical basis exists to inform our choice of metric for *comparing* these already loss-based fingerprints. The following subsections discuss the chemoinformatic response to these challenges.

4.2.1.1 Chemical Fingerprints

In chemoinformatics, we encode each molecule into a computer-readable *fingerprint* or *descriptor*, which is a limited representation of the molecule's information, often derived from direct physical properties. Fingerprints such as Daylight[2] or Scitegic's 'extended connectivity fingerprint' (ECFP)[3] focus on the two-dimensional structure of a small molecule – e.g., atom types and the bond connectivity among them – whereas others such as Chemically Advanced Template Search (CATS) descriptors[4,5] encode physical binding property types such as cations, anions, and hydrogen bond donors and acceptors. Other work has expanded to 'affinity fingerprints,'[6,7] which represent a molecule not by its direct physical properties, but instead by its responses to a high-throughput screen, or by other indirect observed properties. Ultimately, most fingerprints are a fixed-length sequence of bits, whose pattern of '1's and '0's is a nearly unique signature for a single small molecule. Some fingerprints stretch this definition, such as those encoding molecules by their three-dimensional structures or surfaces; these include ROCS,[8,9] FEPOPS,[10] and morphological similarity.[11]

4.2.1.2 Limitations

A fingerprint is an imperfect stand-in for a small molecule. Fingerprints do not guarantee uniqueness and may gloss over important information because they collapse a ligand's multidimensional information space into a single sequence of bits, in return for speed. When comparing two small molecules that are encoded by a *knowledge-based* fingerprint such as MDL keys,[12] each bit denotes a specific chemical pattern. Often the *similarity metric* to compare the bits is ignorant of any given bit's meaning, and must treat them all equally, although the contribution of different chemical patterns might vary greatly. The responsibility for imparting this meaning falls squarely on the shoulders of the fingerprint designer. Knowledge-based fingerprints, for example, contain several hundred bits whose meaning is chosen to cover the space of patterns corresponding with human chemical intuition. Nonetheless, the performance of knowledge-based fingerprints often falls short of *information-theoretic* ones,[13] whose bits represent bond-connectivity or atom-neighborhood patterns derived from computational algorithms.

4.2.1.3 Similarity Metrics

Leaving aside fingerprint encoding schemes, we turn here to the means of comparing any two fingerprints of the same type; this is the *similarity metric*. A similarity metric is a score for how well the patterns from two sequences of bits match each other. In the most common approach, a *Tanimoto coefficient*[14,15] (Tc) compares the number of matching ON bits between two fingerprints to all the ON bits that could have been matched between them. Developed in 1957,[16] the Tanimoto coefficient extends the *Jaccard coefficient*, once used to compare similarity and diversity of sample sets in alpine flower populations in 1901.[17] The Tc measures overall similarity between two molecules and is symmetric, e.g., for fingerprints 'fp_a' and 'fp_b', then $Tc(fp_a,fp_b) = Tc(fp_b,fp_a)$.

By comparison, the *Tversky index*[18] (Ti) asks whether one fingerprint is a subset of the other – thus, if one molecule is a perfect substructure of a larger one, it would achieve a perfect score, even when additional moieties on the larger molecule remain unmatched. Given that a substructure can score perfectly, the Ti metric is asymmetric;[19] e.g., if fp_b contains a substructure perfectly matching the entire molecule encoded as fp_a, then $Ti(fp_a,fp_b) = 1.0$, whereas $Ti(fp_b,fp_a) < 1.0$. Any method using Tversky indices must consider the directionality of the comparison; one solution is to always calculate the Ti in both directions, and take the best score. But should we also consider the 'poorer' score of the two? Again, specifics of the task at hand – and not an established theory or acknowledged best approach – often inform this choice.

4.2.2 Organizing Biological Targets by their Ligands

4.2.2.1 Network Pharmacology

Computational methods augment classical pharmacology, using small molecule interactions to quantify and infer protein-to-protein relationships. Such inferred relationships are chemo-centric links revealing new patterns from domain knowledge visualized on a broad scale, as when Hopkins *et al.*[20] demonstrated cross-target *polypharmacology* within a large dataset by asking which receptors shared identical ligands. Similarly, Vidal and colleagues[21] have analyzed connectivity patterns within drug-receptor networks, and Mestres *et al.*[22] have expanded such networks with additional publicly available information relating ligands and proteins. In these network representations (Figure 4.1A), the *nodes*

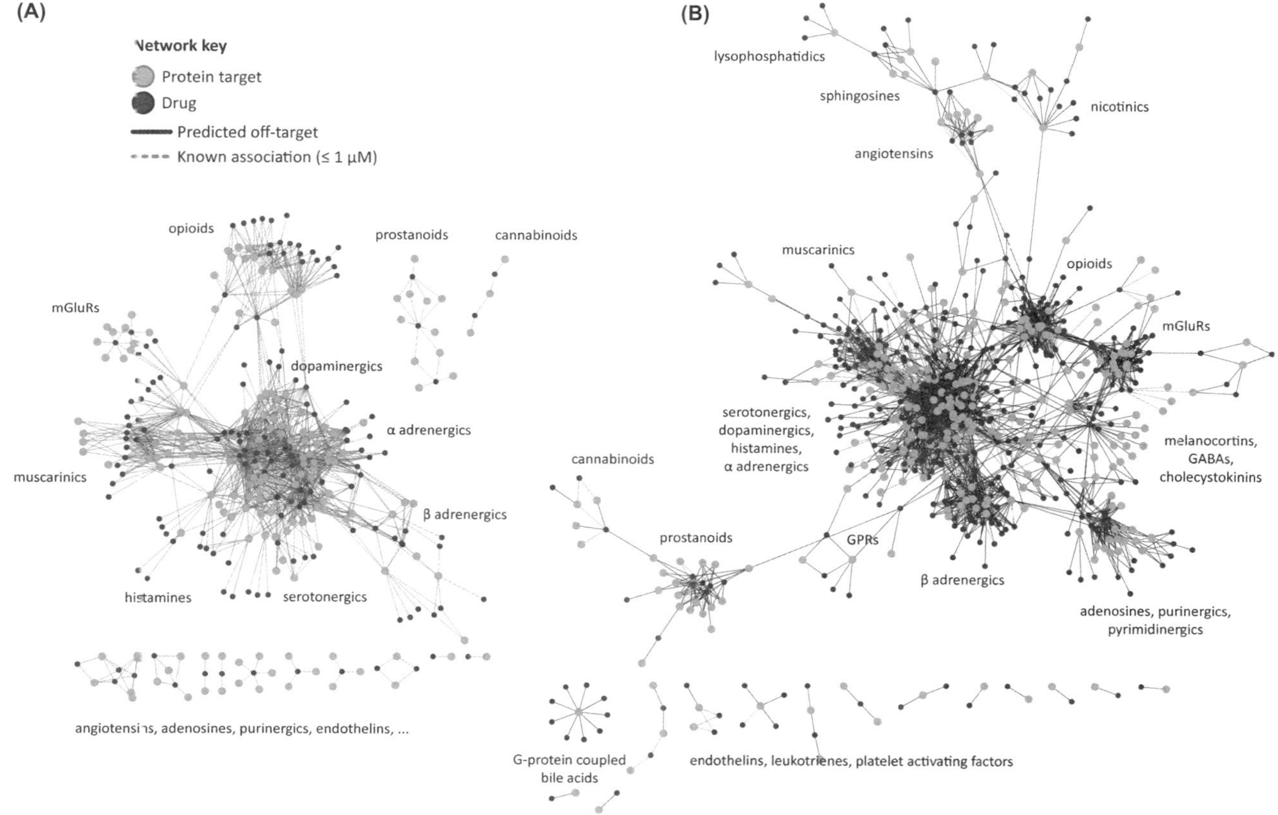

Figure 4.1 Predicting drug-target networks using chemical informatics. (A) Drugs (dark gray) are linked via dotted edges to their known targets (light gray). (B) Prediction network expanding on panel (A), where additional solid edges link drugs to targets predicted for them by SEA (http://sea.bkslab.org).[83] In both networks, 1216 drugs were derived from the US EPA DSSTox database and annotated or predicted against 344 GPCRs.

are proteins, or a mixture of proteins and their ligands. The *edges* may link proteins that share ligands, or here, link small molecules to proteins they are known to bind. The overall structures of these chemo-centric networks often evince connectivity patterns similar to those observed in biological networks, such as scale-free and small-world properties.[21]

In this era of molecular biology, we seek to understand drug off-target binding – be it desired or undesired – through the receptors with which they interact. Thus we consider two proteins similar when their sequences or structures are similar, and it is to these bioinformatic criteria that we may first turn when considering the feasibility of possible multi-target profiles. But pharmacological networks reshuffle the protein landscape, quantifying those relationships first known to traditional pharmacology, wherein investigators once began with the small molecule and inferred the targets. Indeed, proteins distant in sequence, such as the folate recognition enzymes dihydrofolate reductase and thymidylate synthase or the metabotropic 5-HT_3 and ionotropic 5-HT_4 receptors, become neighbors in ligand space. Where two targets are pharmacological neighbors, we might expect greater success in targeting both simultaneously, although the mechanics of such an inference are yet to be fully worked out. Nonetheless, the design of small molecules with intentional polypharmacology plays a growing role in cancer,[23] depression, and neurological disorders. Traditional medicinal chemistry studies have designed several such,[24] but computational methods also now contribute to early-stage rational polypharmacology design. This is discussed in Section 4.2.3.1.

4.2.2.2 Predicting Polypharmacology

Extending on this notion, some chemical similarity approaches now use *set-wise* ligand chemical similarity as a proxy for the pharmacological similarities of their protein targets.[20,25–28] This reorganizes protein space not only by known but also by statistically likely inferred relationships (Figure 4.1B). The idea exploits the internal similarity of most ligands for a particular target[29] and the observation that similar ligands often have similar protein binding patterns.[30,31] Whereas this hypothesis may be violated in specific cases, chemical similarity on the whole is a good guide to the biological action[32] or medicinal chemistry design[33] of an organic molecule, and this is the guiding principle of chemoinformatics.

Consequently, an extensive chemoinformatic literature explores methods to compare ligand pairs for similarity[34] and predict pharmacological profiles.[7,35–38] Leveraging pair-wise chemical similarity to divine relationships among receptors, Izrailev and Farnum link ligand sets representing receptors, by focusing on the most similar molecules between them.[25] Unlike prior metrics that relied on overall similarity between ligand sets, this focus on the 'average nearest neighbor'[25] detected similarity arising from small sub-groups of ligands that would otherwise be drowned out by the majority. Likewise, Shoichet *et al.* introduced a similarity ensemble approach (SEA)[27,39] to link receptors, based on the statistical significance of similarity among high-scoring ligand pairs across receptor sets. The statistics were motivated by BLAST,[3,40] with ligands standing in for BLAST's unordered 'words,' and leverage extreme value distributions (EVD). The EVD type used for BLAST and SEA was identified in 1927 by Maurice Fréchet,[41] and has since expanded to a variety of uses, such as wind speed forecasts, wireless communications fading, and survival analysis.

These networks expand on known relationships by filling in receptor–drug gaps left by incomplete testing or lack of publicly accessible data. While chemocentric networks do not benefit from the evolutionary theory and deep understanding that bioinformatics networks enjoy, they may nonetheless encode consistent information reflective of other underlying principles. Intriguingly, the structures of chemical similarity prediction networks are stable in the face of varied ligand representations, and like the known polypharmacology networks discussed above, share the scale-free properties found also in their bioinformatics analogs.[3]

4.2.3 Profiling

Late attrition is a major pharmaceutical concern. By the time a drug candidate reaches clinical testing, it has cost years of research and tens of millions of development dollars. Nevertheless, 89% of drug candidates that enter clinical trials do not reach market.[42] Reasons for this late failure vary over time; the main causes in the 1990s were related to pharmacokinetic issues,[43] engendering models to predict these properties early in the pipeline. Lack of efficacy and safety have since stolen the limelight, each accounting for 30% of drug candidate failures in clinical study phases.[43] Whereas efforts in early safety prediction and mechanism-based drug design have increased, FDA safety regulations have also become stricter. In this context, computational methods are especially time- and cost-effective tools.[44,45]

4.2.3.1 Multi-Target Profiling

The search for 'magic bullet' compounds that bind specifically to a single predetermined target has proven effective in many projects. However, recent studies show that, even for drugs initially designed to be target selective, the pharmacological profile is more complex. In a recent study, Yildirim *et al.*[21] applied network analysis to public drug-target information from DrugBank.[46] Although these data were not comprehensive, the authors observed an interconnected network, instead of the isolated 'islands' of bipartite nodes that would be expected of drugs acting selectively on single targets. In a following study, Mestres *et al.*[47] showed that when extending the available experimental information with virtual target predictions, the drug-target network becomes even denser.

Many systems chemical biology studies, which model entire biological network structures,[48] predict that one must modulate multiple proteins simultaneously to modify phenotype. Pathways are often in dynamic equilibrium, and biological systems find alternative compensatory routes to single point perturbations.[49] Targeting a single protein alone can be harmful, leading to rapid

Table 4.1 Selection of chemoinformatics virtual profiling methods and their features.

Method	*Descriptor*	*Similarity method*	*Statistical layer*
Similarity Ensemble Approach[27]	2D molecular descriptor (Daylight/ECFP_4)	Tanimoto coefficient	Extreme value distribution (E-value)
Hopkins *et al.*[20]	2D molecular descriptor (FCFP_6)	Tanimoto coefficient	Laplacian-corrected Bayesian classifier
Bayes affinity fingerprints[7]	Target affinity fingerprints	Pearson correlation coefficient	Bayes theorem
Mestres *et al.*[66]	2D molecular descriptor (SHED)	Euclidean distance	None
Muresan *et al.*[60]	9 different molecular descriptors: Daylight Unity AlFi[61] Hologram™ CATS TRUST[62] Molprint 2D[63,64] ChemGPS[65] ALOGP	Tanimoto coefficient	None

onset of resistance in human immunodeficiency virus type 1[50] or cancer[51] therapy. Correspondingly, complex diseases such as depression,[52] inflammation,[53] and cancer[54] respond more efficiently to simultaneous multi-targeting of pathological proteins.[55] The anti-cancer drug quercetin exhibits favorable polypharmacology at multiple levels, regulating cell signaling, cell cycle, and apoptosis.[56] Such multi-target pharmacological profiling has become crucial for efficacy against complex diseases.[57]

Ligand-based virtual profiling computational approaches aid polypharmacological drug design.[58] These methods are intended to profile large numbers of molecules for ca. 1000 protein targets at a time. They exhibit varying degrees of mathematical complexity (Table 4.1), ranging from traditional similarity based approaches (see Section 4.2.1) to statistical-based machine learning methods (Figure 4.2). Similarity based virtual profiling methods (Figure 4.2A) rely on the assumption that similar compounds will have similar pharmacological profiles (see Section 4.2.2.2, and ref. 59). Descriptors ensuring this ‘neighborhood behavior’ hypothesis enable transfer of target annotations from a compound to its closest neighbors; this is discussed in Section 4.2.1.1. The other factor differentiating similarity based virtual profiling methods is the similarity metric they use (Section 4.2.1.2). For instance, Muresan and colleagues[60] use nine different fingerprints (Daylight, Unity, AlFi,[61] Hologram™, CATS, TRUST,[62] Molprint 2D,[63,64] ChemGPS,[65] and ALOGP) with the Tanimoto similarity coefficient. They propose that combining fingerprints compensates for individual weaknesses. By contrast, the SHED-based approach, developed by Mestres and colleagues,[66] does not rely on group fusion[13,67] or substructure matching, but rather on the relative distribution of pharmacophoric features in the whole molecule. As SHED descriptors are not binary fingerprints, the approach uses Euclidean distance to compare molecules instead of the common Tanimoto coefficient.

A virtual profiling method halfway between pair-wise similarity methods and the statistical model-building methods is the Similarity Ensemble Approach (SEA).[27] SEA uses extended connectivity fingerprints[68] and Tanimoto coefficients to determine the pair-wise distances among all compounds between any two sets. These distances are then compared to a statistical model for random set similarity that yields BLAST-like expectation values (Figure 4.2B). This additional statistical consideration allows the method to

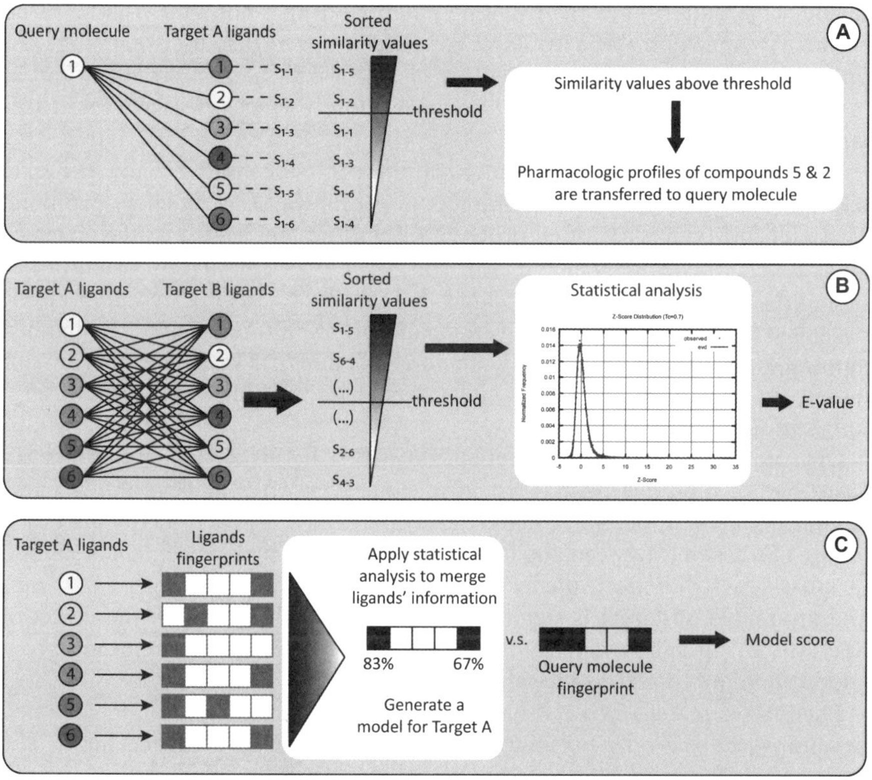

Figure 4.2 Diagram of several chemoinformatics approaches. Diagram (A) summarizes a traditional approach, where pair-wise similarity comparisons are considered relevant if they fall above a predetermined threshold. In diagram (B) we see a mixed similarity-/statistics-based method, in which all similarities are calculated and, instead of a predetermined threshold, a statistical analysis is performed to assess if the similarity is statistically relevant. Diagram (C) shows a statistics-based method, which generates a model for each protein based on fingerprints of all its known ligands.

determine whether the observed similarity is significant with respect to the entire set of known ligands for a certain target.

Machine-learning approaches are a subset of the statistical virtual profiling methods (Figure 4.2C). These approaches do not compare molecules by their direct pair-wise similarity, but instead build a statistical model for each target, which implicitly encodes binding motifs that may be responsible for activity. One example is the Bayes affinity fingerprint,[7] which leverages Bayes' theorem as a statistical correction.

4.2.3.2 Predicting Side Effects

Miniaturization and parallel screening enable *in vitro* compound profiling against a wide range of targets, but logistics as yet limit this approach's truly high throughput application. Such multi-target physical profiling is best suited to late-stage leads, to prioritize among potential drug candidates, as little or no chemistry capacity may be available for corrections at this discovery stage. At this scale, virtual profiling protocols may assist lead safety assessment and in prioritizing among lead series, by favoring those predicted to have fewer off-target interactions.[42]

Furthermore, hit and lead optimization protocols traditionally evaluate selectivity for only a handful of targets in preclinical *in vitro* safety profiling. These few off-targets are often proteins within the same family as the target of interest or proteins for which a clear link with a certain adverse reaction has been established (i.e., hERG-related K^+ channel, 5-HT_{2B} receptor, PXR nuclear receptor).[69] Consequently, virtually profiling a chemical series against more than a thousand structurally unrelated proteins can reveal unexpected off-targets, or suggest *in vitro* selectivity targets to add to the preclinical *in vitro* safety profiling panel.

One of the distinctive characteristics of SEA, its ability to compare entire ensembles of molecules, can assist in preclinical *in vitro* safety profiling panel design. Using SEA, one can compare known ligand sets for targets against each other. The E-value between any two ligand ensembles may reflect the likelihood of cross-activity between them. If the E-value between the target of interest and a possible off-target is significant, molecules with activity at the target of interest may also show activity for the off-target. Such an off-target may be a good candidate to add to the selectivity panel.

Phenotypic screens and cell-based high content screening (HCS) are a step beyond standard *in vitro* profiling for safety and toxicity evaluation, and will be discussed in Section 4.3.1.

4.3 Applications

The chemical organization of biological information yields predictions amenable to testing by direct assay of the small molecules that articulate them. Whereas a conserved fold across two receptor structures may present a similar ligand-binding site in each, it also may not. Bioinformatic measures are not always predictive of a receptor's pharmacological profile. Conversely, a chemocentric prediction of similarity between a receptor's ligands and an 'off-target' small molecule drug is often directly testable, and hence falsifiable. Some off-targets bolster drug on-target action, whereas others are consistent with drug side effects.

4.3.1 Target Identification

Physiological and pathological processes are complex, but failure to understand them is a liability in drug discovery. This has led, on one hand, to polypharmacological target profiling in certain drug discovery processes (see Section 4.2.3) and, on the other, to the reappearance of phenotypic screens in the early drug discovery processes.

4.3.1.1 Targets of Phenotypic and High-Throughput Screening

Phenotypic screens are information-rich processes, providing valuable data on *in vivo* compound efficacy and toxicity. As phenotypic screening applies both to whole organisms and to cell cultures, we differentiate whole organism screens here as full 'phenotypic screens',[70] and refer to high-throughput cell-based screens as 'high content screens' (HCS).[71] Both techniques link with computational target profiling to predict the protein (or proteins) involved with the *in vivo* effect of the compound as a hypothesis for its mechanism of action.[72]

Central nervous system (CNS) therapeutics is among the most profitable sectors in the pharmaceutical market. Most approved atypical antipsychotics have complex pharmacological profiles, with significant affinities for a variety of aminergic GPCRs.[73] The discovery and design of these polypharmacological drugs is challenging because classical *in vitro* medicinal chemistry approaches such as HTS or *in vitro* assays at individual purified protein targets are not sufficiently accurate or do not scale. Phenotypic screens may bridge this gap. In work by Peterson and colleagues[70] the zebrafish serves as a model organism to identify compounds modulating CNS behavior; in this case, the phenotype is the zebrafish's reaction to a light pulse stimulus. After quantifying and clustering the observed behavioral phenotypes, Peterson *et al.* used SEA virtual profiling to predict targets modulating each response. These predictions were confirmed *in vitro* via purified protein assays, leading to new mode of action hypotheses.

Assays at the comparatively simpler whole cell level allow for higher-throughput analysis. Historically, cellular toxicity screening relied on single parameter readouts for toxicity markers such as cell proliferation, mitochondrial activity, or membrane permeability. Although useful to an extent, single-parameter predictability for compound toxicity *in vivo* is poor.[74] In contrast, HCS enables multiplex analysis, wherein two or more discrete responses may be measured in a single assay, all within a cellular context.[75] By incorporating a compound's simultaneous effects on many measured parameters, HCS assays may achieve a high level of cytotoxicity predictability.[74]

HCS encodes complex phenotypes as unbiased morphological descriptors of cellular structure.[76] Each screening compound acquires a signature, based on its observed effects on a particular cell type. As reviewed by Davies and colleagues,[76] several studies have examined small molecule activity via HCS – however, they primarily detect cell integrity and toxicity only. HCS lacks a path to specific target hypotheses to account for the welter of data a particular compound may evoke. Virtual screening tools, however, may assist in winnowing potential mechanism hypotheses. Mitchison and colleagues[77,78] showed that drugs with common targets clustered together based solely on observed phenotypes, using only hypothesis-free HCS data and unsupervised clustering algorithms. In an other study, Feng and colleagues derived mechanism of action inferences from a HCS of fluorescent stains of multiple cell cycle markers.[72] They derived a phenotypic stain profile for each compound that, when clustered, elucidated structure–activity relationships consistent with structural patterns and known activities.[72] They then computationally predicted a common target, α-tubulin, for three groups of phenotypically similar yet structurally distinct molecules (e.g., colchicine, quinoline, and pseudolarix acid B) – and confirmed this hypothesis via micrograph.[72]

4.3.1.2 Drug Repositioning

An off-target can be an opportunity to repurpose a drug for diseases unrelated to its initial indication. Di Bernardo *et al.* showed potential to use Fasudil, a Rho-kinase inhibitor and vasodilator, in cancer and in some neurodegenerative diseases.[79] Using mode of action by network analysis (MANTRA), the authors grouped drugs into 'communities' by similarities in their connectivity map[80] profiles of induced transcriptional responses. By identifying drugs similar to 2-deoxy-D-glucose, a known inducer of autophagy, they predicted and demonstrated activation of autophagic degradation by Fasudil in human fibroblasts and HeLa cells.[79] Fasudil has a good safety profile, which may prove useful in Alzheimer's treatment,[81] consistent with cellular autophagy's role in disorders thought to arise from protein misfolding.

Likewise, Bork and colleagues noted that the acetylcholinesterase inhibitor donepezil binds to the serotonin reuptake transporter and may thus find use against depression, depending on the therapeutic profile. Distefano's group demonstrated that the antifungals Monistat and Spectazole bound protein farnesyltransferase and that Monistat disrupts H-ras oncogene localization in cells, consistent with prediction.[82] While these examples illustrate relatively weak off-targets, they can also be potent; Shoichet *et al.* showed that the antihypertensive Doralese bound to the dopamine D4 receptor a log order more tightly (18 nM) than it does to its canonical α-adrenergic 'on-target' (200–600 nM).[83]

4.3.2 Safety and Target-Specific Toxicity

Unintended drug off-targets call to mind the specter of adverse drug reactions. Correspondingly, some off-targets indeed have proven consistent with known drug side effects.[27,83] Bork and colleagues developed a predictive method that *uses* known side effect information, organizing drugs into networks by similarities among the side effect profiles listed on their package inserts.[84] From these networks, they predicted and experimentally confirmed thirteen cases of novel drug off-target activity. In one, they identified a sub-network in which the CNS drugs pergolide, Paxil, Prozac, and zolmitriptan were clustered around the antiulcer drug rabeprazole, a proton pump inhibitor. This led them to discover that rabeprazole also bound two CNS targets known for these drugs, the dopamine D3 (1.6 μM) and 5-HT_{1D} (7.6 μM) receptors. As rabeprazole concentrations reach these levels in plasma, this may suggest that it should also be investigated for the side effects already associated with these nervous system targets.[84] Thus, analyzing hitherto unappreciated similarities in drug side effect profiles may also reveal new side effects, which may be fed back into the method in a self-boosting process.

One need not start with side effects to find side effects, however. Leveraging instead the statistical patterns of atom and bond topology among small molecules, Shoichet *et al.* found novel off-targets consistent with drug side effects.[27,83] The amebicide emetine, whose side effects include hypotension, tachycardia, and congestive heart failure, also bound the α_2-adrenergic receptor;[27] methadone's side effects were consistent with its novel muscarinic M3 binding;[27] and Motilium, imported by nursing mothers to stimulate lactation despite a ban by the FDA due to cardiac arrest, bound α_{1A}-adrenergic receptors at 71 nM.[83] These off-targets were consistent with clinical concentrations. The SSRIs Prozac and Paxil bound β_1 adrenergic receptors, consistent with SSRI discontinuation syndrome and the sexual dysfunction induced by these antidepressants.[83] A pilot study has since correlated a human β_1 adrenergic gene single nucleotide polymorphism with these observations.[85]

Others have demonstrated chemoinformatics methods whose predictions across multiple therapeutic areas were successfully confirmed upon deep literature review, revealing side effect consistent off-targets known in the literature but entirely unknown to the datasets used to predict them.[42,69,86–89]

4.3.3 Applicability

Chemoinformatics approaches leverage large numbers of individually imperfect comparisons to arrive at novel and statistically sound conclusions. The pairwise similarity scores underlying these comparisons are rapid but limited in scope, often focusing on only a handful of key properties at a time. Furthermore, chemoinformatics' central guiding principle, that similar molecules often have similar properties, limits its ability to predict completely novel chemotypes or binding to poorly characterized targets.

But chemoinformatics' sole reliance on similarity is also its strength. By automating a limited yet highly scalable view of classical pharmacology, chemoinformatics inherits its wide applicability. These methods can predict ligands for receptors whose crystal structures are not yet solved, or find molecules exhibiting a desired phenotype without requiring detailed domain knowledge. As data sources grow nonlinearly in size, detail, and quality, the scope of the methods that operate on these data also grow. While any single similarity score used in a chemoinformatics method may, like a firefly, cast only a small point of light, even fireflies are bright – when collected in the billions.

References

1. J. A. Allen and B. L. Roth, *Annu. Rev. Pharmacol. Toxicol.*, 2011, **51**, 117–144.
2. C. James, D. Weininger and J. Delany, *Daylight Theory Manual*, Daylight Chemical Information Systems Inc., Mission Viejo, CA, 1992–2005.
3. J. Hert, M. J. Keiser, J. J. Irwin, T. I. Oprea and B. K. Shoichet, *J. Chem. Inf. Model.*, 2008, **48**, 755–765.
4. U. Fechner, L. Franke, S. Renner, P. Schneider and G. Schneider, *J. Comput.-Aided Mol. Des.*, 2003, **17**, 687–698.
5. G. Schneider, W. Neidhart, T. Giller and G. Schmid, *Angew. Chem., Int. Ed. Engl.*, 1999, **38**, 2894–2896.
6. L. M. Kauvar, D. L. Higgins, H. O. Villar, J. R. Sportsman, A. Engqvist-Goldstein, R. Bukar, K. E. Bauer, H. Dilley and D. M. Rocke, *Chem. Biol.*, 1995, **2**, 107–118.
7. A. Bender, J. L. Jenkins, M. Glick, Z. Deng, J. H. Nettles and J. W. Davies, *J. Chem. Inf. Model.*, 2006, **46**, 2445–2456.
8. T. S. Rush, 3rd, J. A. Grant, L. Mosyak and A. Nicholls, *J. Med. Chem.*, 2005, **48**, 1489–1495.
9. OpenEye Scientific Software, Santa Fe, New Mexico, 2004.
10. J. L. Jenkins, M. Glick and J. W. Davies, *J. Med. Chem.*, 2004, **47**, 6144–6159.
11. A. N. Jain, *J. Comput.-Aided. Mol. Des.*, 2000, **14**, 199–213.
12. J. L. Durant, B. A. Leland, D. R. Henry and J. G. Nourse, *J. Chem. Inf. Comput. Sci.*, 2002, **42**, 1273–1280.
13. J. Hert, P. Willett, D. J. Wilton, P. Acklin, K. Azzaoui, E. Jacoby and A. Schuffenhauer, *J. Chem. Inf. Comput. Sci.*, 2004, **44**, 1177–1185.
14. P. Willett, *Similarity and Clustering in Chemical Information Systems*, Research Studies Press, Wiley, Letchworth, Hertfordshire, England; New York, 1987.
15. R. D. Brown and Y. C. Martin, *J. Chem. Inf. Comput. Sci.*, 1996, **36**, 572–584.
16. T. Tanimoto, *IBM Internal Report, 17th Nov. 1957*, 1957.
17. P. Jaccard, *Bulletin de la Société Vaudoise des Sciences Naturelles*, 1901, **37**, 547–579.
18. A. Tversky, *Psychol. Rev.*, 1977, **84**, 327–352.
19. A. Tversky and I. Gati, *Psychol. Rev.*, 1982, **89**, 123–154.
20. G. V. Paolini, R. H. B. Shapland, W. P. van Hoorn, J. S. Mason and A. L. Hopkins, *Nat. Biotechnol.*, 2006, **24**, 805–815.
21. M. A. Yildirim, K. I. Goh, M. E. Cusick, A. L. Barabasi and M. Vidal, *Nat. Biotechnol.*, 2007, **25**, 1119–1126.

22. J. Mestres, E. Gregori-Puigjane, S. Valverde and R. V. Sole, *Mol. BioSystems*, 2009, **5**, 1051–1057.
23. Z. A. Knight, H. Lin and K. M. Shokat, *Nat. Rev. Cancer*, 2010, **10**, 130–137.
24. B. Apsel, J. A. Blair, B. Gonzalez, T. M. Nazif, M. E. Feldman, B. Aizenstein, R. Hoffman, R. L. Williams, K. M. Shokat and Z. A. Knight, *Nat. Chem. Biol.*, 2008, **4**, 691–699.
25. S. Izrailev and M. A. Farnum, *Proteins*, 2004, **57**, 711–724.
26. M. Vieth, R. E. Higgs, D. H. Robertson, M. Shapiro, E. A. Gragg and H. Hemmerle, *Biochim. Biophys. Acta*, 2004, **1697**, 243–257.
27. M. J. Keiser, B. L. Roth, B. N. Armbruster, P. Ernsberger, J. J. Irwin and B. K. Shoichet, *Nat. Biotech.*, 2007, **25**, 197–206.
28. S. L. Schreiber, *Nat. Chem. Biol.*, 2005, **1**, 64–66.
29. M. A. Johnson and G. M. Maggiora, *Concepts and Applications of Molecular Similarity*, Wiley, New York, 1990.
30. S. V. Frye, *Chem. Biol.*, 1999, **6**, R3–7.
31. M. Bredel and E. Jacoby, *Nat. Rev. Genet.*, 2004, **5**, 262–275.
32. H. Matter, *J. Med. Chem.*, 1997, **40**, 1219–1229.
33. M. Whittle, V. J. Gillet, P. Willett, A. Alex and J. Loesel, *J. Chem. Inf. Comput. Sci.*, 2004, **44**, 1840–1848.
34. P. Willett, *J. Med. Chem.*, 2005, **48**, 4183–4199.
35. M. Nidhi, J. W. Glick and J. L. Davies, *J. Chem. Inf. Model.*, 2006, **46**, 1124–1133.
36. T. M. Steindl, D. Schuster, C. Laggner and T. Langer, *J. Chem. Inf. Model.*, 2006, **46**, 2146–2157.
37. A. Schuffenhauer, P. Floersheim, P. Acklin and E. Jacoby, *J. Chem. Inf. Comput. Sci.*, 2003, **43**, 391–405.
38. D. Horvath and C. Jeandenans, *J. Chem. Inf. Comput. Sci.*, 2003, **43**, 680–690.
39. M. J. Keiser and J. Hert, *Methods Mol. Biol. (Clifton, NJ)*, 2009, **575**, 195–205.
40. S. F. Altschul, W. Gish, W. Miller, E. W. Myers and D. J. Lipman, *J. Mol. Biol.*, 1990, **215**, 403–410.
41. M. Fréchet, *Ann. Soc. Polon. Math.*, 1927, **6**, 93–116.
42. A. Bender, J. Scheiber, M. Glick, J. W. Davies, K. Azzaoui, J. Hamon, L. Urban, S. Whitebread and J. L. Jenkins, *ChemMedChem*, 2007, **2**, 861–873.
43. D. C. Liebler and F. P. Guengerich, *Nat. Rev. Drug Discovery*, 2005, **4**, 410–420.
44. G. Klebe, *Drug Discovery Today*, 2006, **11**, 580–594.
45. E. Gregori-Puigjané and J. Mestres, *Curr. Opin. Chem. Biol.*, 2008, **12**, 359–365.
46. D. S. Wishart, C. Knox, A. C. Guo, D. Cheng, S. Shrivastava, D. Tzur, B. Gautam and M. Hassanli, *Nucl. Acids Res.*, 2008, **36**, D901–D906.
47. J. Mestres, E. Gregori-Puigjane, S. Valverde and R. V. Sole, *Nat. Biotechnol.*, 2008, **26**, 983–984.
48. T. I. Oprea, A. Tropsha, J.-L. Faulon and M. D. Rintoul, *Nat. Chem. Biol.*, 2007, **3**, 447–450.

49. A. L. Hopkins, J. S. Mason and J. P. Overington, *Curr. Opin. Struct. Biol.*, 2006, **16**, 127–136.
50. S. M. Hammer, M. S. Saag, M. Schechter, J. S. G. Montaner, R. T. Schooley, D. M. Jacobsen, M. A. Thompson, C. C. J. Carpenter, M. A. Fischl, B. G. Gazzard, J. M. Gatell, M. S. Hirsch, D. A. Katzenstein, D. D. Richman, S. Vella, P. G. Yeni and P. A. Volberding, *JAMA*, 2006, **296**, 827–843.
51. S. K. Mencher and L. G. Wang, *BMC Clin. Pharmacol.*, 2005, **5**, 3.
52. M. J. Millan, *Eur. J. Pharmacol.*, 2004, **500**, 371–384.
53. C. Charlier and C. Michaux, *Eur. J. Med. Chem.*, 2003, **38**, 645–659.
54. A. Jimeno and M. Hidalgo, *Crit. Rev. Oncol. Hematol.*, 2006, **59**, 150–158.
55. T. Klabunde, *Br. J. Pharmacol.*, 2007, **152**, 5–7.
56. S. C. Janga and A. Tzakos, *Mol. BioSystems*, 2009, **5**, 1536–1548.
57. H. Kitano, *Nat. Rev. Drug Discovery*, 2007, **6**, 202–210.
58. R. Morphy and Z. Rankovic, *J. Med. Chem.*, 2005, **48**, 6523–6543.
59. D. E. Patterson, R. D. Cramer, A. M. Ferguson, R. D. Clark and L. E. Weinberger, *J. Med. Chem.*, 1996, **39**, 3049–3059.
60. T. Kogej, O. Engkvist, N. Blomberg and S. Muresan, *J. Chem. Inf. Model.*, 2006, **46**, 1201–1213.
61. D. Cosgrove, AIFi - an alternative to Daylight fingerprints, AstraZeneca Internal Software Document.
62. E. Jacoby, A. Schuffenhauer, M. Popov, K. Azzaoui, B. Havill, U. Schopfer, C. Engeloch, J. Stanek, P. Ackin, P. Rigollier, F. Stoll, G. Koch, P. Meier, D. Orain, R. Giger, J. Hinrichs, K. Malagu, J. Zimmermann and H.-J. Roth, *Curr. Top. Med. Chem.*, 2005, **5**, 397–411.
63. A. Bender, H. Y. Mussa, R. C. Glen and S. Reiling, *J. Chem. Inf. Comput. Sci.*, 2003, **44**, 170–178.
64. A. Bender, H. Y. Mussa, R. C. Glen and S. Reiling, *J. Chem. Inf. Comput. Sci.*, 2004, **44**, 1708–1718.
65. T. I. Oprea and J. Gottfries, *J. Comb. Chem.*, 2001, **3**, 157–166.
66. E. Gregori-Puigjané and J. Mestres, *Comb. Chem. High Throughput Screening*, 2008, **11**, 669–676.
67. J. Hert, P. Willett, D. J. Wilton, P. Acklin, K. Azzaoui, E. Jacoby and A. Schuffenhauer, *J. Chem. Inf. Model.*, 2006, **46**, 462–470.
68. D. Rogers and M. Hahn, *J. Chem. Inf. Model.*, 2010, **50**, 742–754.
69. J. Scheiber, B. Chen, M. Milik, S. C. Sukuru, A. Bender, D. Mikhailov, S. Whitebread, J. Hamon, K. Azzaoui, L. Urban, M. Glick, J. W. Davies and J. L. Jenkins, *J. Chem. Inf. Model.*, 2009, **49**, 308–317.
70. D. Kokel, J. Bryan, C. Laggner, R. White, C. Y. J. Cheung, R. Mateus, D. Healey, S. Kim, A. A. Werdich, S. J. Haggarty, C. A. MacRae, B. K. Shoichet and R. T. Peterson, *Nat. Chem. Biol.*, 2010, **6**, 231–237.
71. C. Liptrot, *Drug Discovery Today*, 2001, **6**, 832–834.
72. D. W. Young, A. Bender, J. Hoyt, E. McWhinnie, G.-W. Chirn, C. Y. Tao, J. A. Tallarico, M. Labow, J. L. Jenkins, T. J. Mitchison and Y. Feng, *Nat. Chem. Biol.*, 2007, **4**, 59–68.

73. B. L. Roth, D. J. Sheffler and W. K. Kroeze, *Nat. Rev. Drug Discovery*, 2004, **3**, 353–359.
74. O. Rausch, *Curr. Opin. Chem. Biol.*, 2006, **10**, 316–320.
75. K. Korn and E. Krausz, *Curr. Opin. Chem. Biol.*, 2007, **11**, 503–510.
76. A. Bender, D. W. Young, J. L. Jenkins, M. Serrano, D. Mikhailov, P. A. Clemons and J. W. Davies, *Comb. Chem. High Throughput Screening*, 2007, **10**, 719–731.
77. T. J. Mitchison, *ChemBioChem*, 2005, **6**, 33–39.
78. Z. E. Perlman, M. D. Slack, Y. Feng, T. J. Mitchison, L. F. Wu and S. J. Altschuler, *Science*, 2004, **306**, 1194–1198.
79. F. Iorio, R. Bosotti, E. Scacheri, V. Belcastro, P. Mithbaokar, R. Ferriero, L. Murino, R. Tagliaferri, N. Brunetti-Pierri, A. Isacchi and D. di Bernardo, *Proc. Natl. Acad. Sci. USA*, 2010, **107**, 14621–14626.
80. J. Lamb, *Nat. Rev.*, 2007, **7**, 54–60.
81. M. J. Huentelman, D. A. Stephan, J. Talboom, J. J. Corneveaux, D. M. Reiman, J. D. Gerber, C. A. Barnes, G. E. Alexander, E. M. Reiman and H. A. Bimonte-Nelson, *Behav. Neurosci.*, 2009, **123**, 218–223.
82. A. J. DeGraw, M. J. Keiser, J. D. Ochocki, B. K. Shoichet and M. D. Distefano, *J. Med. Chem.*, 2010, **53**, 2464–2471.
83. M. J. Keiser, V. Setola, J. J. Irwin, C. Laggner, A. I. Abbas, S. J. Hufeisen, N. H. Jensen, M. B. Kuijer, R. C. Matos, T. B. Tran, R. Whaley, R. A. Glennon, J. Hert, K. L. Thomas, D. D. Edwards, B. K. Shoichet and B. L. Roth, *Nature*, 2009, **462**, 175–181.
84. M. Campillos, M. Kuhn, A. C. Gavin, L. J. Jensen and P. Bork, *Science (New York, NY)*, 2008, **321**, 263–266.
85. K. L. Thomas, V. L. Ellingrod, J. R. Bishop and M. J. Keiser, *Psychopharmacol. Bull.*, 2010, **43**, 11–22.
86. J. Scheiber and J. L. Jenkins, *Methods Mol. Biol. (Clifton, NJ)*, 2009, **575**, 207–223.
87. J. Scheiber, J. L. Jenkins, S. C. Sukuru, A. Bender, D. Mikhailov, M. Milik, K. Azzaoui, S. Whitebread, J. Hamon, L. Urban, M. Glick and J. W. Davies, *J. Med. Chem.*, 2009, **52**, 3103–3107.
88. K. Azzaoui, J. Hamon, B. Faller, S. Whitebread, E. Jacoby, A. Bender, J. L. Jenkins and L. Urban, *ChemMedChem*, 2007, **2**, 874–880.
89. D. Rognan, *Mol. Inf.*, 2010, **29**, 176–187.

CHAPTER 5

Designing Multi-Target Drugs: In Vitro *Panel Screening – Biological Fingerprinting*

JONATHAN S. MASON

Heptares Therapeutics Ltd, BioPark, Broadwater Road, Welwyn Garden City, AL7 3AX, UK & Lundbeck Research, Ottiliavej 9, Valby, DK-2500, Denmark
Email: jonathan.mason@heptares.com

5.1 Introduction: Biological Fingerprints – A Biological View of Compounds

In vitro panel screening, also known as biological fingerprinting,[1–4] in addition to providing direct information on the polypharmacology of compounds, enables a biologically relevant description of molecules based on the way they bind to a broad and diverse set of relevant targets. This approach is quite different to the significant efforts made over the years to characterise molecules by fingerprints based on their chemical structure. Such structurally defined fingerprints are often based on the 2D structure (e.g. substructures, atom paths and circular connectivity), sometimes using the more relevant 3D structure-based descriptors such as pharmacophores or molecular interaction fields (MIFs) that much better represent how the protein binding sites would 'see' a molecule. The 2D structure-based approaches can only profile the underlying structure that gives rise to the properties recognised by a biological target, whereas the more relevant 3D structure-based approaches have the problem

RSC Drug Discovery Series No. 21
Designing Multi-Target Drugs
Edited by J. Richard Morphy and C. John Harris

Published by the Royal Society of Chemistry, www.rsc.org

that the 'bioactive' conformation for different targets or sites may not be known for all (or any) targets, and may indeed be different for different target binding sites. Thus, ideally, an ensemble of conformations needs to be used, and even this ensemble may or may not include the bioactive conformation(s), and it may contain a lot of 'noise' from biologically irrelevant conformations.

Generation of a fingerprint based on experimental binding affinities for a diverse range of pharmacologically relevant targets means that such issues and limitations are inherently avoided, and provides such fingerprints with a unique description of how biological targets 'see' a molecule. By using a fingerprint of a compound's multi-target/polypharmacology, it can be positioned and clustered with other compounds in biological space. Thus in the design of molecules with a desired multi-target pharmacology, the other 'off-target' activities can be used to differentiate compounds and highlight those with the 'best' or most differentiated profile. In the serotonin/noradrenaline reuptake inhibitor (SNRI) example described below, the only hit series suitable for progression to a clinical candidate was highlighted from the beginning and key selectivity assays identified.

Many names have been given to this type of experimental biological description of molecules. In addition to 'biological fingerprints' and 'biological profiles',[1–5] 'biological spectra/biospectra',[6–8] 'bioactivity spectra'[9] and 'affinity fingerprints',[10,11] 'chemical genomic profiles'[12] and 'chemical-genetic fingerprints'[13] have been recorded. *In silico* approaches to calculate such fingerprints are discussed elsewhere (see Chapters 4 and 9), but so far these have had mixed success; only the experimentally derived *in vitro* fingerprints will be discussed in this chapter. Reliable bioactivity models are not available for many targets, particularly those for which there is limited activity data (including inactives), or for the profiling of new 'chemotypes' that are outside the predictive 'space'/capability of the model. It is hoped, with improved descriptors (e.g. the use of 3D pharmacophoric or molecular interaction field descriptors where there is less chemical structural dependence) and methods, that this will improve over time. Large-scale biological data generation and integration enables the development of *in silico* models for a subset of relevant targets[14] but only a 'partial' biological fingerprint can be produced at the moment, which may miss a key off-target activity or, when over-predicting, miss a new selectivity. In the design of compounds with multi-target pharmacology, when predictive models for the desired activities and (where possible) key undesirable activities can be developed, a computational multi-dimensional optimisation approach can be applied, continuously updated and improved by the incorporation of new experimental profiling data.

The largest effort in the area of *in vitro* panel screening has been the Cerep BioPrint® initiative,[15–20] involving several major large pharmaceutical companies. A large amount of new, and internally consistent, information on the polypharmacology of drugs, attrited compounds and medicinal chemistry project compounds has been generated with BioPrint®. Lessons learnt from these analyses will be a focus in this chapter, being the first hand experience of the author.

5.2 The Cerep Bioprint® Database

There have been several initiatives to generate biological fingerprints systematically, with the largest and most significant in terms of broad biological fingerprints being BioPrint®[15–20] from Cerep, who provide broad profiling services for many companies. Other smaller initiatives focusing on subsets of targets such as kinases have also been undertaken.[21] This major and costly undertaking, systematically producing 'full-matrix' data for a large and reasonably diverse number of targets (>90–150 pharmacological and 30 ADME-related assays), with dose–response data for all compounds with >30% inhibition at 10 μM, has been supported by several major pharmaceutical companies who helped select 'interesting' targets to include, particularly from an attrition prediction perspective. BioPrint® is a very well-established project, started in 1997 with Bristol Myers Squibb as the first partner, with Pfizer and then others such as Astra-Zeneca becoming involved later. The BioPrint® package consists of a large database of measured *in vitro* data and curated *in vivo* data, together with a set of tools to access both the data and models generated from the data. The core of the database is *in vitro*, *in vivo* and structural data on most marketed pharmaceuticals and a variety of other reference compounds. The general concept of BioPrint® and its utility is illustrated in Figure 5.1.

The assays were selected primarily for their scientific interest, but consideration was also given to the robustness and the consistency and quality of the data from the assay, coverage of relevant therapeutic areas, phylogenetic analysis, the concept of the 'druggable' proteome[22] and various technical and other constraints. The highest proportion is receptors, with GPCRs being the most represented, followed by enzymes, ion channels, transporters and nuclear receptors. One of the strengths of the BioPrint® initiative is that many experienced medicinal chemists, research and safety assessment biologists, computational chemists and bioinformaticians in the pharmaceutical industry have been involved in guiding its development.

A key application of BioPrint® is in the differentiation of structures (and their underlying 'chemotypes') and it became clear from the initial phase of the project, where most marketed drugs and some reference compounds were

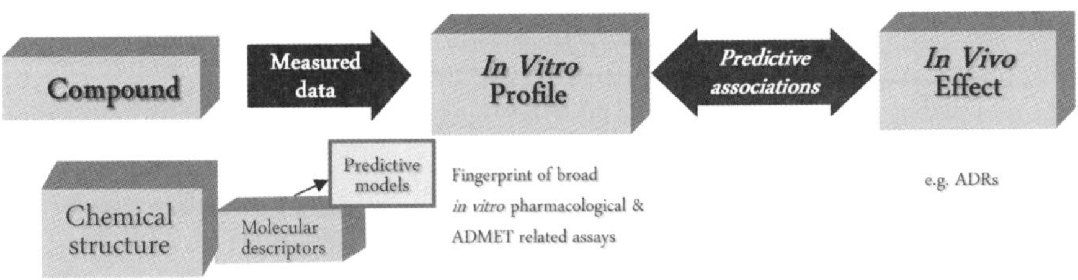

Figure 5.1 The Cerep BioPrint® approach. The full-matrix dataset (*in vitro* profile) for drugs etc. is all measured in a consistent manner with full dose response for any activity >30% at 10μM. Curated *in vivo* data is assembled from available data sources for the compounds.

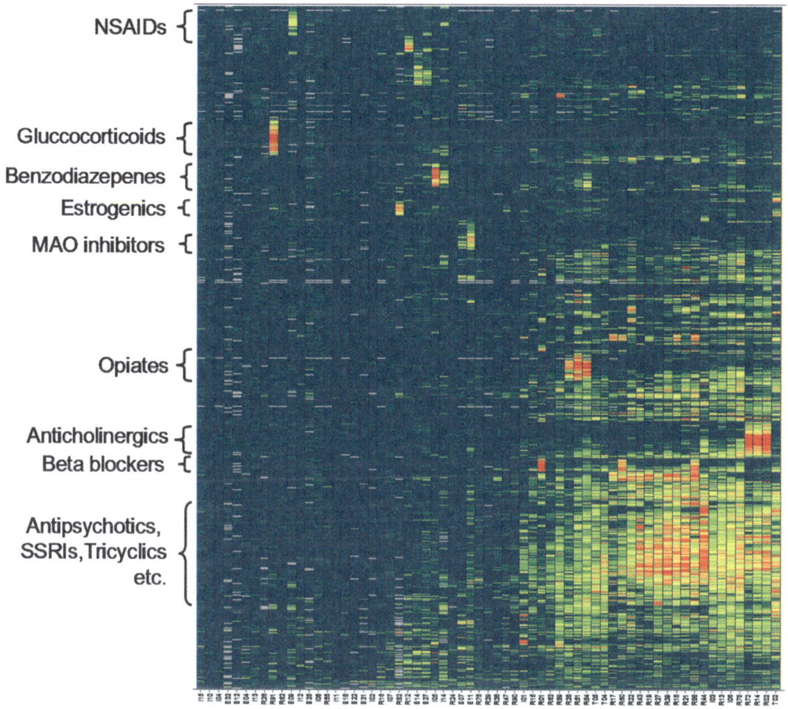

Figure 5.2 A heatmap of compounds (~2000 drugs and related compounds as of year 2000) versus assays (70 pharmacological from the BioPrint® database; pIC_{50}). The rows contains the biological fingerprint of a compound as a heatmap of the biological assay data (*x*-axis). Hierarchical clustering has been performed on both axes: compounds by their fingerprint of biological activities and targets by the fingerprint of the activities of the same set of compounds for each target. The activities are coded from red (most active) through yellow to blue-green (inactive). Some therapeutic areas that the drug compounds tended to cluster into are indicated on the left.

profiled (~1500 compounds), that multi-target/polypharmacology was the norm rather than being unusual, and that at varying levels compared to the nominated primary target(s) most drug-like compounds bind to other so-called 'off-targets'. Figure 5.2 shows a 'heatmap' of these drugs and related compounds assessed in a subset of the BioPrint assays; this is described further in Section 5.4.

To use *in vitro* biological fingerprints most effectively as a way to describe molecules and utilise polypharmacology, a measurement from a dose–response study is needed, otherwise much important detail of differential activity is masked. Figure 5.3 illustrates this for Clozapine, with the partial BioPrint® biological fingerprint showing only assays with a % inhibition >90% at 10 μM on the upper heatmap (all dark grey). The importance of using more precise information is shown in the lower heatmap, with the related IC_{50} values

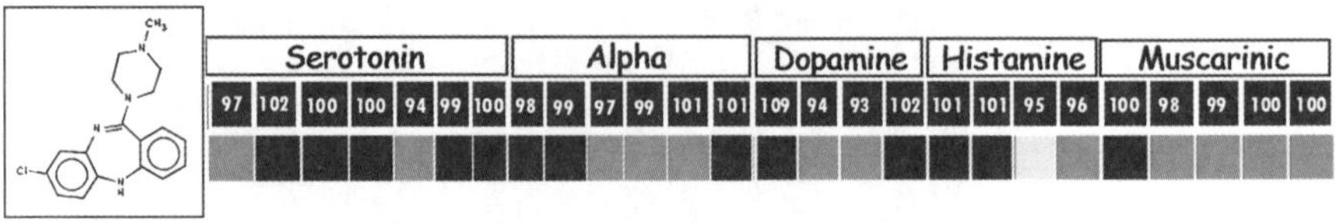

Figure 5.3 Biological fingerprint (BioPrint® broad panel screen) for Clozapine, showing the results for assays with a % inhibition > 90% at 10 μM (upper line, dark grey). The IC_{50} values are shown coded on the lower line (dark grey < 100 nM, grey < 1 μM, light grey < 5 μM).

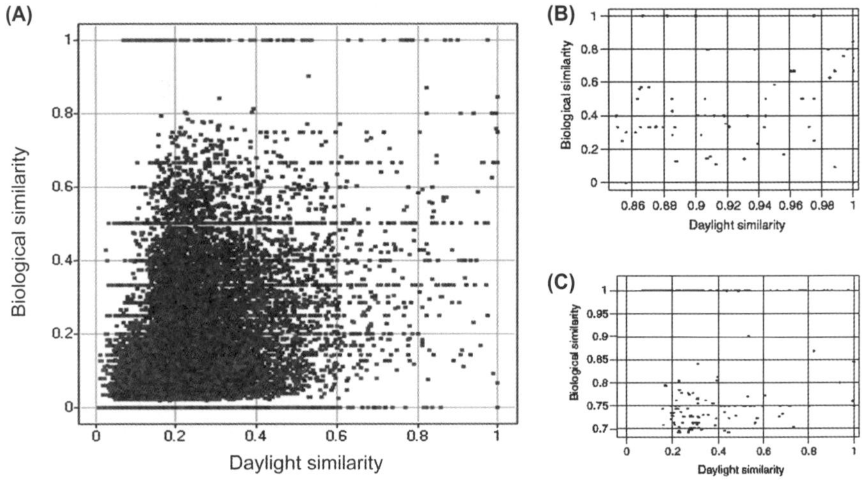

Figure 5.4 Plots showing the general lack of similarity in broad biological space versus similarity in chemical structure space for 347 drugs from the BioPrint® dataset. (A) Pairwise Tanimoto distances (0–1, 1 = identical) from Daylight structural fingerprints for structural similarity on *x*-axis and from BioPrint® biological activity fingerprints (154 assays) where active is defined as an IC_{50} < 100 μM) on *y*-axis. (B) An enlarged view of the region where structural similarity is high (Daylight fingerprint similarity > 0.85). (C) An enlarged view of the region where biological activity similarity is high (> 0.7).

colour coded with dark grey most active (< 100 nM), where differential binding to the various targets is clear (but is masked in the % inhibition data).

Whilst most 'off-target' activities were found at levels less than that for the nominated primary target(s), some were at similar levels, with others only quite weak. Also of great interest, and not expected, is that these off-target activities are often quite different for relatively chemically similar molecules. Using only a binary coding of activity (> 30% at 10 μM) for the fingerprints, no correlation was found of similarity in broad biological space with that in chemical space (as defined using Daylight 2D structural keys, see Figure 5.4). Using a similarity metric that takes the activity level into account, more correlation was found for very similar compounds with, for example kinase targets, but this failed more generally, emphasising that quite similar compounds can still have very different biological fingerprints. As 'similar' compounds are often ones in which the key pharmacophoric elements for a desired activity are retained and other substituents are varied in medicinal chemistry studies, their activities are thus likely to be similar, leading to the observation that 'similar compounds have similar activity' but a potential bias in many published analyses. With 'random' variations, as seen in drugs for different targets, such a structure–activity (multi-target pharmacology) relationship is much less evident.

An issue with such biological fingerprints is that there may be redundancy in the assay data, particularly as data from several receptor subtypes are included. This was investigated using the KEM approach,[23,24] developed by Ariana[25] (a systematic rule-based method that identifies all relations of the type A→B, A & B→C & not D etc.). An analysis of a subset of 80 assays and 1600 compounds was performed to verify that there were no obvious correlations. Even when using very broad activity bins, the analysis showed that no such non-contradicted relations where found, showing that there are valid signals in all the assay data, including that from subtype assays.

5.3 Profiling Concepts and Practice

Even a broad pharmacological/biological profile can only describe a limited number of biological targets. Each of these targets may have direct relevance, and be part of a polypharmacological design, but an important concept is to use broad pharmacological profiles, and the associated assays, as surrogates for a far larger set of targets. This principle was used in the affinity fingerprint method[10,11] (see Stanton and Cao[3] for discussion), in which a small 'diverse' and 'orthogonal' set of protein targets and compounds were used to model the activity of a new protein.

Another way of looking at the broad biological profiles is as 'biological spectra' of compounds. Using all % inhibition data gives a continuous numerical value, so avoiding 'missing data', as even with a 30% inhibition cut-off for IC_{50} determination many compounds will have no associated IC_{50} value. Fliri *et al.* have published several interesting papers on the use of such BioPrint data.[6–8] In this study. early safety issues were addressed (e.g. to find a new series to faster and more effectively avoid muscle toxicity issues) as well as to prioritise compounds to avoid certain adverse drug reactions (ADRs). However, the use of continuous % inhibition data can lack the resolution seen in conventional IC_{50} based dose–response analysis.

Both continuous and IC_{50}-based approaches have yielded useful results; in this chapter the focus is on using biological fingerprints based on dose–response data (i.e. binding IC_{50} values from Cerep BioPrint® for any compounds with >30–50% inhibition at 10 μM). These have been used extensively by the author as a tool to aid decision making in drug discovery projects. Rather than using the entire assay set, most of the cited examples use a more economic subset of data from 70–100 of the assays which give the strongest signal, and from which unambiguous decisions for the prioritisation of future work could normally be made. These abbreviated but nevertheless information-rich biological fingerprints can save time and money; however, for the analysis of key tool and reference compounds a full BioPrint® profile was used and this is highly recommended for finding unexpected off-target activities that cannot be predicted.

5.4 Profiling of Drugs: The Multi-Target/ Polypharmacology of Drugs

It is clear from the profiles shown in Figure 5.2 that most drug-like compounds do not bind to a single target. This multi-target/polypharmacology of many drugs, both expected and unexpected, was a key early finding from the BioPrint® initiative. Whilst many activities are less potent than the 'primary'/ desired one(s), the sheer diversity of 'off-target' binding of many drugs (even between 'similar' ones, from a 2D structural perspective) is an interesting insight into their potential pharmacological activities. Many of the 'off-target' activities are not for targets with any phylogenetic similarity, and sometimes the levels can be close to the 'primary' activity.

Analyses of the BioPrint dataset[4] showed that compounds that are active at <1 μM on more than 10 targets are generally more lipophilic, with a clogP >3 (see Figure 5.5). BioPrint data for drugs (*y*-axis) against assays (*x*-axis) is shown in Figure 5.2 for a subset of targets and drugs (as in the database ca. 2000). The data has been clustered hierarchically both by

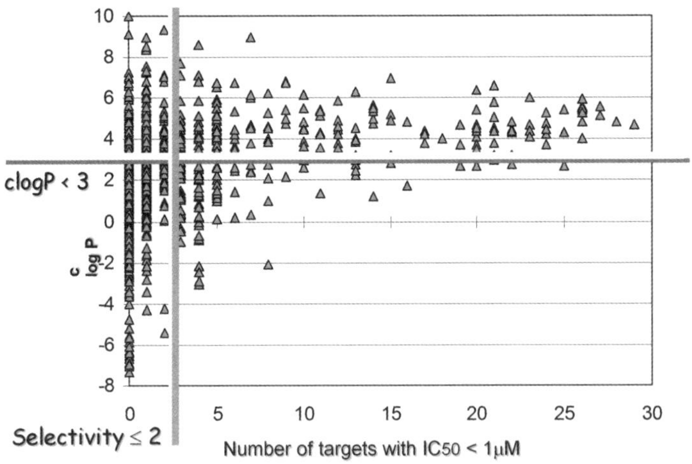

Figure 5.5 A view on 'promiscuity' or lack of selectivity as defined by the number targets hit (*x*-axis) for 1098 drugs profiled in the BioPrint® assay panel (with active defined as an $IC_{50} < 1$ μM) versus clogP (hydrophobicity, *y*-axis).

compounds' biological fingerprints and by target using the nominated activities of the drug compounds, with colour-coding of red indicating high binding affinity, yellow medium and green-blue low. It can be noted from the clustered data that many drugs of a particular therapeutic class tend to cluster together; however, through use of *in vitro* panel screening of different hit series it was found that compounds with a desired profile of single or multi-pharmacology could often be identified that clustered into a different part of space, opening up opportunities to get candidates with just the desired major (multi-target) pharmacology, with quite different patterns of weaker activity against other targets. These findings may enable differentiation in a manner more relevant to biological systems than using 'chemotype' similarities and differences (see Section 5.5.1).

By applying this systematic *in vitro* broad profiling to project candidate compounds from newer target classes such as kinases, these were often shown to have unrelated multi-target/polypharmacology, for example with potent activities at unrelated targets such as the aminergic GPCRs, highlighting that selectivity outside of the target class can be as important as that for 'related' targets. The knowledge provided by early *in vitro* pharmacological profiling enables such issues to be addressed at an earlier stage in the next generation of compounds. Multi-target/polypharmacology is a two-edged sword: it can be important for the efficacy of some drugs, but equally it can be associated with undesired side effects which are often due to other therapeutically unnecessary activities. Thus knowledge of the broad polypharmacology of a compound is very important in all cases.

5.5 Profiling of Project Compounds

The differentiation and prioritisation of compounds is key at all stages in the drug discovery process. At the target validation stage, it is very important to exclude tool compounds having 'off-target' activities that could affect the biological response. At the hit/lead identification stage, significant time can be saved by selecting the 'best' starting point and being aware early of key selectivity targets. Indeed, the 'best' series can be missed if all hits are not followed up, and at candidate selection it can be that the 'best' compound from a sub-optimal series is selected instead. At this stage, it is important to select a compound that will be differentiated (both in terms of attrition risk and commercial attractiveness) from compounds already in development or from existing competitor compounds. *In vitro* profiling and the use of biological fingerprints provide a more relevant approach for decision making than one based on the 2D structure and 'chemotype'. The examples described below illustrate these various scenarios. Differentiation (from existing compounds and between new compounds) by a broad pharmacological profile is a useful approach for selecting one or multiple clinical candidates.

An unexpected finding from profiling sets of hit/lead compounds across a very broad range of targets was that one compound would normally stand out as being the best 'clean' starting point, having mainly the desired (multi-target) pharmacology, in contrast to other 'similar' compounds. As 'similar' compounds can have different biological fingerprints then the recommendation is to profile at least two compounds from a hit or lead series to be sure of undesirable activities before a 'negative' decision is made, but as noted above the prioritisation of resources could often be made at least initially from single compounds as one compound would be a clear 'winner' in terms of off-target activities.

5.5.1 Choosing the Best Hit or Lead Compound and Differentiation

Choosing the best hit or lead compound to develop further is a core task in medicinal chemistry, and significant time/cost savings can be made if the best choice can be made upfront rather than wasting resources on chasing many less promising leads. The early identification of the best leads is thus critical, as it can be very difficult at later stages to make major changes to the lead series chemistry.

As well as achieving the desired (multi-target) pharmacological profile, selectivity and physicochemical/ADMET properties, the selection of development candidates needs to address the major challenge in the drug discovery process, that of attrition. The candidate compound should have the best chance of survival (safety, efficacy etc.), and where there are multiple candidates, or other compounds in development, the attrition risk should be orthogonalised as much as possible, to avoid multiple compounds attriting for the same unexpected cause. This is where *in vitro* panel screening has multiple roles: to facilitate the prioritisation of the leads of most interest that have the best chance of becoming a suitable development candidate and possess the most desired profile of biological and ADME related properties, and by using the broad profile, allowing the inclusion of compounds with weak/different 'off-target' activities, thus providing differentiation in 'biological space' from other compounds (be they in-house or competitor).

The power of this approach was validated in an early pilot study at Pfizer,[26,27] in which a parallel approach was pursued. Four potent hit/lead series from a SNRI project were identified from HTS; these were analysed and clustered using their BioPrint® profiles but also all were pursued by medicinal chemistry teams. The key structural features of the hit compounds and how they cluster in both biological space, using the BioPrint® fingerprints, and in chemical space, using Daylight structural fingerprints,[28] is illustrated in Figure 5.6, together with a reference compound in clinical development at the time, Duloxetine (see Figure 5.7). The desired primary multi-target pharmacology (in this case dual target activities) was shown by all the compounds, but from the BioPrint® analysis it became clear that they had quite different off-target activities in their overall biological profiles. A key selectivity target was also identified that gave the project team the advantage of being able

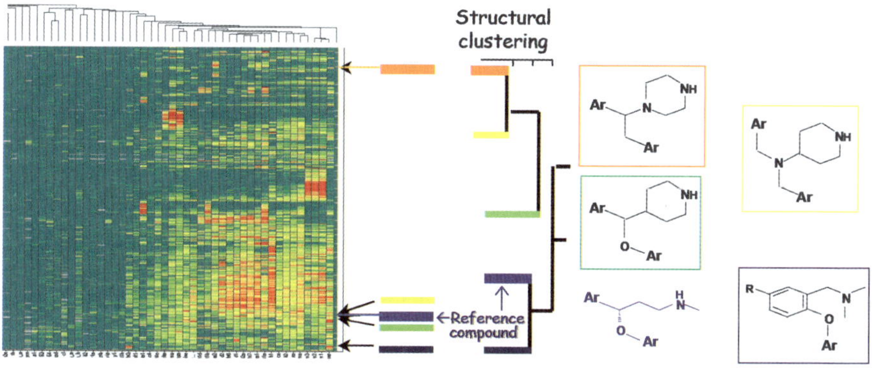

Figure 5.6 Clustering of four hit/lead compounds based on their biological fingerprint (BioPrint®) together with the BioPrint® drug compounds (left). A clinical reference compound Duloxetine is included; all compounds had similar potent 'primary' activities. The rows show as a heatmap the activity for a compound (subset of BioPrint assays on the *x*-axis). Activity is colour-coded from red (very active) through yellow to blue-green (inactive). In the centre for comparison is shown a structure-based clustering using Daylight fingerprints, highlighting the very different similarities in structural versus biological space.

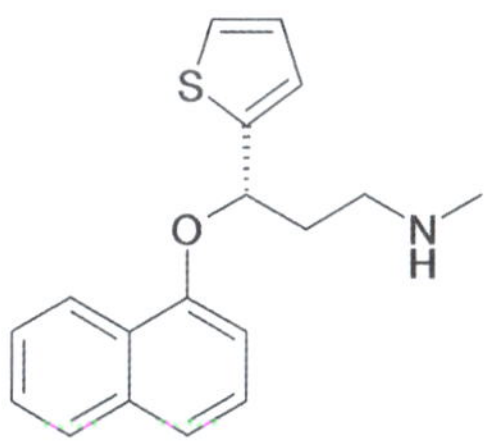

Figure 5.7 Structure of Duloxetine.

to screen for this from the beginning. Many of the 'off-target' activities were significant and interestingly only the compound that was highlighted by the BioPrint® profile clustering as being a much cleaner and differentiated starting point could be optimised to a clinical candidate (the piperazine in Figure 5.6). The activity at the troublesome selectivity target that affected the other series could not removed whilst retaining other desired activities/properties. A simple structural analysis would not lead to the choice of this compound as preferred, nor do predictive models based on large-scale data integration[29] predict this differentiation. Indeed the opposite can be seen, with many false positives and negatives.

A 'chemotype'-based analysis would not obviously lead to the best starting point being highlighted, with the structures being relatively similar in terms of key features (basic and aromatic groups) and the most interesting compound, the piperazine (highlighted with orange), is not obvious from a structural viewpoint. A nitrogen has effectively moved one bond along, which would affect pharmacophoric distances, but as noted above predictive models tended to show inverse profiles for the number and strength of off-target activities. This small nitrogen atom shift between the amino piperidine (yellow) and piperazine (orange) compounds produces the most differentiated compound in biological space, moving it into a drug space more occupied by opiates. Similar non-selectivity to the reference compound duloxetine (blue) was found in the amino piperidine compound (yellow), yet a typical structural clustering, illustrated in the centre of Figure 5.6 with the 2D Daylight fingerprints, shows it as quite different. Such 2D structural clustering is commonly used to select representatives for further analysis and, for example, to reduce compound lists, but it can be quite ineffective and misleading in 'biological space' Selecting the amino piperidine compound (yellow) to represent this part of structural 'diversity', instead of the piperazine, would lead to a compound with similar undesirable off-target issues to the reference compound being pursued, and the most interesting piperazine compound (orange) missed; the other non-amino piperidine (green) has similar undesirable polypharmacology issues. Interestingly the aromatic ether compound (dark blue) clusters in 2D with the reference compound, but this compound actually has a somewhat differentiated biological profile; note that the yellow and green piperidine compounds have similar biological profiles to the reference compound even though they cluster differently based on 2D structure-derived Daylight fingerprints. The use of other structure-based fingerprints can give different results; in this set of structures a change of descriptor to the Scitegic FCFP6 circular fingerprints enabled more differentiation of the 'clean' compound, but unfortunately this is not a universal solution.

As, in this early example, all the lead series were followed up, armed with the knowledge of key selectivity assays, this selection approach could be 'validated', in that only the orange piperazine highlighted by the BioPrint® clustering was moved into clinical development, its profile remaining relatively 'clean'. The selectivity issues could not be resolved for the other compounds.

This study clearly illustrates the power of early biological broad *in vitro* profiling and that a 'better' starting point can be critical to project success. Following this success a larger and successful initiative was started with Cerep to use BioPrint® profiling for hit and lead compounds from all therapeutic areas. As a result of the systematic application of BioPrint® profiling, many examples were found where decisions could be clearly made from the biological profile (using the 70–90 assays with the highest hit rate) that were not evident from a chemical structure-based analysis. Many insights into unexpected activities (from unrelated proteins at the sequence level and outside of the target class) were obtained. Indeed, examples were found where the off-target activity difference for compounds of the same 'chemotype' could be quite dramatic (clean versus promiscuous), which could prove to be very important if the compound were to be used for *in vivo* studies. An example for a pair of compounds with 2–4 nM activity, each with a distinctive bicyclic polyheteroaromatic core ('chemotype') linked to a cyclic base and two substituents,

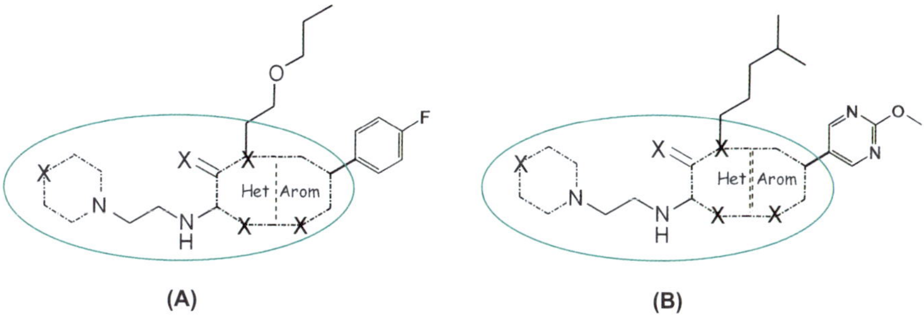

Figure 5.8 Two compounds with similar structure/'chemotype' and potent primary activities (IC_{50}: 4 and 2 nM). Compound A (clogP = 5) is promiscuous, with sub μM activity on 31 off-targets, whereas compound B (clogP = 2), with the same core features, is a much cleaner with sub μM activities for only 2 off-targets + CYP3A4. Thus two 'similar' compounds could behave quite differently *in vivo* because of the polypharmacology, in this case undesired.

is shown in Figure 5.8. Compound A binds at <1 μM to 31 pharmacological off-targets, whereas the other (B) is relatively clean, binding to only 2 pharmacological off-targets and CYP3A4 at <1 μM.[27] The substituent change of alkoxy to alkyl and pyrimidine to phenyl causes this dramatic effect, which may be associated with both a pharmacophoric change and a clogP increase (see Section 5.4 and Leeson and Springthorpe[30]). However, the situation is more complex than a simple lipophilicity/clogP difference, as an analogue in which one of the nitrogens is moved from the pyrimidine substituent to the core bicyclic ring, giving a compound with a similar clogP (2.2), has increased pharmacological promiscuity (2→10 <1 μM pharmacological off-targets) and ADME issues (1→3 <1 μM CYP inhibition + Pgp efflux). Thus a biological fingerprint is quite critical in choosing a suitable compound from this 'chemotype' for further evaluation.

5.5.2 Profiling of Tool Compounds: Target Validation

Obtaining broad *in vitro* pharmacological profiling data, beyond any designed single or multi-target activities, is a critical step in the selection of a 'tool' compound to be used *in vivo* to investigate if a desired biological effect is obtained through a hypothesised mechanism. The examination of BioPrint profiles of many reported 'tool' compounds has shown that the term 'selective' is a function of the limited range of the related assays that are often used and that these 'tools' can have significant activity on other targets. Such activities could indeed be responsible for the desired biological effect, and thus a wrong and wasteful decision to pursue a target for a particular indication could be made.

An example of this is shown in Figure 5.9, where full BioPrint® profiling of several published $5HT_7$ antagonists revealed that one of them, SB-269970, was

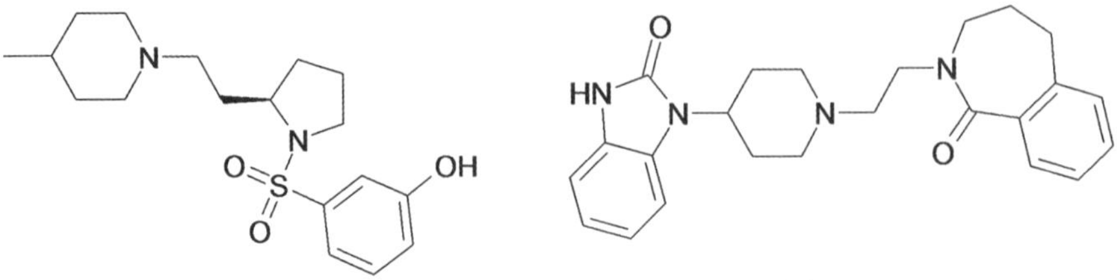

(A) (B)

Figure 5.9 Two compounds reported to be selective $5HT_7$ inhibitors: Using the Cerep BioPrint® profiling (A) (SB-269970) is shown to be selective and the best reference compound, versus (B) (SB-691673) that is <3× selective against 5 targets and could thus give misleading *in vivo* results.

more than 80 fold selective in all the BioPrint® assays and would indeed be a good *in vivo* tool compound to evaluate a mechanistic hypothesis, whereas others were less selective, such as SB-691673 that is less than three fold selective against five targets. In this case it would indeed have been a poor choice to evaluate the hypothesis, as one of the 'off-target' activities is known to cause the desired *in vivo* effect, thus a wrong target validation could have been made, the 'clean' compound turned out not to be active in the *in vivo* assay, thus further time was not wasted pursuing that hypothesis.

5.5.3 Selectivity and the Use of the Broad *In Vitro* Biological Profile to Predict *In Vivo* Effects and Safety Issues

In addition to the main focus of this chapter on how biological fingerprints provide a powerful, medicinal chemistry-relevant, way of describing and differentiating compounds, other uses of biological profiling data have been reported, focusing more on safety issues and compound promiscuity. Leeson[30] has published a very informative paper using a newer version of the BioPrint® dataset to look for promiscuity–property relations, finding relationships similar to those illustrated in Figure 5.5, together with many other interesting analyses of structure–activity data. Bamborough *et al.*[31] at GSK have published on kinome space and selectivity amongst kinase targets, showing that compounds often exhibited various off-target kinase activities that could not be predicted from similarity in binding site amino acids.

The similarity of the BioPrint® profile of individual hits to known compounds was used by Migeon and co-workers[1,18,19] to look for potential adverse drug reaction (ADR) liabilities. They have found biological profiling to be particularly useful in placing new drug candidates in the context of known drugs and related compounds, where much *in vivo* data is available. They also analyse the binding activities within the profile to assess for potential ADR liabilities as an extensive collection of ADR associations exists within BioPrint®. Pharmacokinetic data is also used to confirm that the strength of the *in vitro* hit is consistent with *in vivo* exposure levels. Groups at Novartis have also reported, in several interesting and informative papers,[32–35] on the analysis of pharmacology data and the prediction of adverse drug reactions and off-target effects, both from biological profiles and from chemical structure similarity.

5.5.4 Multi-Target/Polypharmacology of Attrited Compounds

The multi-target/polypharmacology of a diverse set of 130 attrited compounds from Pfizer pre-2003 was investigated using the BioPrint® panel screens. The results were very interesting, highlighting 15 assays that were hit multiple times by these compounds, compared to results for the same assays for the general drug set.[1] There was no obvious pattern, with multiple but different combinations of these assays often, but not always, hit. It was also clear that differentiation by chemical structure fails to separate 'clean' and failed compounds and that structurally dissimilar compounds may actually have similar off-target effects, for example due to similar 'decoration' on a different 'scaffold'. Figure 5.10 shows the partial BioPrint profile of eight compounds developed for activity against a serotonin receptor. There are clearly large differences in the activity profiles on both other serotonin receptors and transporters and the broad panel of assays, with a richness of

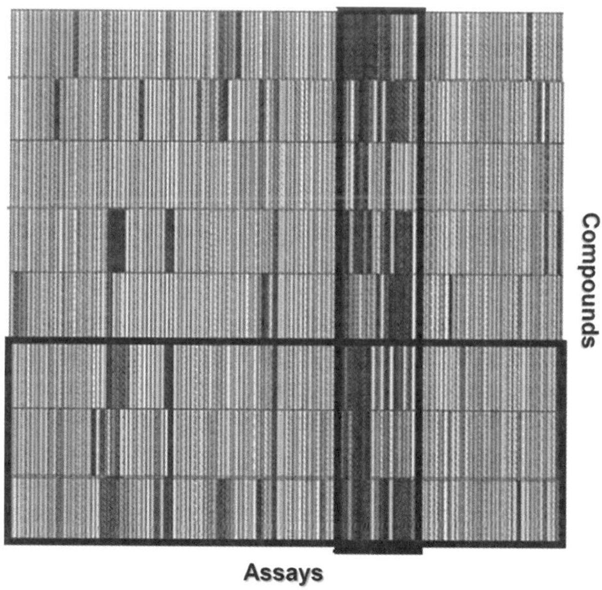

Figure 5.10 The partial BioPrint® profile of eight compounds developed for activity against a serotonin receptor. The three compounds that attrited for some type of toxicity issue are shown at the bottom in a box. The serotonin receptors and transporters are highlighted with a vertical box. The dark grey bands are for the highest activities (< 100 nM).

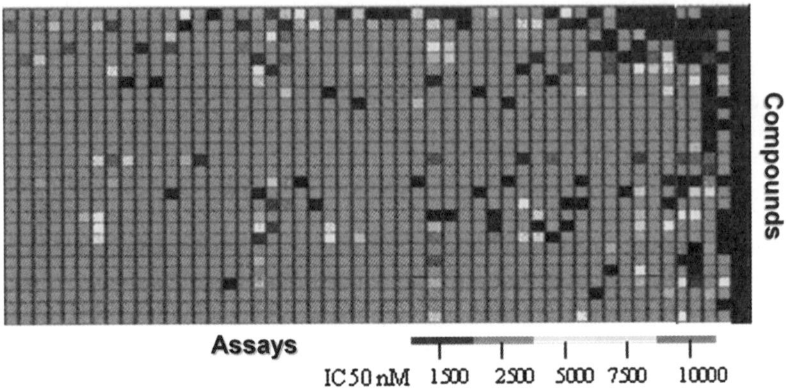

Figure 5.11 BioPrint® (partial) data for a set of compounds developed against the same enzyme (PDE) target, illustrating the very different multi-pharmacology/selectivity possible.

polypharmacology probably beyond the desired multi-target profile; three compounds that attrited for the same type of toxicity issue are shown at the bottom in a box. Figure 5.11 shows a partial BioPrint® analysis of a set compounds developed for the same phosphodiesterase (PDE) enzyme target, with the primary activity (part of the profile) shown in the last column. The diversity of broad pharmacological profiles show that potent yet broad pharmacologically 'clean' compounds beyond the desired PDE target(s) could be developed, but also that many of the compounds had significant off-target binding on non-PDE targets, much of which was not expected.

5.6 Profiling and Clustering of Compounds: *In Silico* Descriptors and Similarity Issues

Drug molecules are very often grouped by 'scaffold', based on the 2D structures and derived fingerprints etc., which is not relevant to how a protein target sees a molecule. The biological fingerprints from *in vitro* profiling provide a powerful 'biological view'. In terms of *in silico* representations, a more pharmacophoric description (hydrogen bond acceptor/donor, lipophilic etc.) is better, particularly if calculated from a 3D structure rather than simple connectivity. Fuzzy pharmacophoric descriptors were found to be best for matching the neighbourhood behaviour of biological fingerprints[36,37] but, in practice, identifying nearest neighbours or predicting a broad biological profile by *in silico* approaches still produces many false positives and negatives. As data sets get larger and descriptors improve the results should improve, and useful results are already being obtained.[14,29] Molecular interaction fields (MIFs), such as those generated by the well-established program GRID,[38–40] provide a powerful *in silico* descriptor in which the properties of the molecule are projected into 'receptor space' and analysed in this 'protein space', for example by the FLAP method.[41–43] The ligand conformation problem remains a challenge for approaches based on 3D structures, as the bioactive conformation is not known for all (or any) targets, and is possibly different for different targets. Failing to sample a bioactive conformation means potentially missing key descriptors, and noise is added by sampling conformations not relevant to binding. Thus similarity approaches from simpler 2D-based descriptors may have higher enrichment rates than those from 3D descriptors, but 3D methods will often reveal very interesting 'scaffold hopping' style compounds not found by the 2D structure-based methods.

A recent study from Steffen *et al.*[44] at AstraZeneca using a more recent and larger version of BioPrint® with 146 assays, confirmed that fingerprint methods which describe global features of a molecule such as pharmacophore patterns and physicochemical properties are likely to be better suited to describe similarity of biological activity profiles than purely structural fingerprint methods. The authors suggest that the usage of these integrated fingerprint methods could increase the probability of finding molecules with a similar biological activity profile but a different chemical structure and this has been the experience of the author with 4-point pharmacophore fingerprints.[45–48] Nevertheless, conformational space uncertainties can still cause poorer results with the more specific 3- and 4-point pharmacophore fingerprints.

The issue of structural versus biological similarity remains much debated. As chemists tend to make 'analogue' compounds where some key pharmacophoric components are kept constant, 2D-similar compounds will tend to have similar primary activities, biasing many analyses that seek to illustrate a basic concept in medicinal chemistry, that similar compounds have similar activities. This general concept can be misleading, being based on biased datasets. The much quoted claim that compounds with a 2D (Daylight) fingerprint Tanimoto similarity >0.85 will most likely have similar activities has more recently[49] been modified to 'only a 30% chance'. When the bias in datasets is reduced and broad activity is considered, results such as the poor correlation ($R^2 = 0.13$) of Figure 5.4 are obtained, and large differences between 'similar' compounds can be found (e.g. as illustrated in Figure 5.8).

In particular when designing multi-target compounds the 'similar structure, similar activity' medicinal chemistry concept should not be assumed, and experimental results should be obtained, backed up by the use of appropriate *in silico* models. The poor correlation between structural similarity and broad biological profile similarity (using a binary fingerprint of broad biological activity) (Figure 5.4,) is potentially quite enabling, suggesting that optimisation from quite similar starting points, including fragments, can lead to compounds with very differentiated biological profiles. Bender *et al.*[50] have used 'Bayes affinity fingerprints' to improve retrieval rates in virtual screening and to define orthogonal bioactivity space, discussing when are multi-target drugs a feasible concept. Sutherland *et al.* have published recently on the use of chemical fragments for understanding target space and activity prediction.[51]

5.7 *In Vitro* Panel Screening: The Future

Biological fingerprints from *in vitro* profiling will continue to have a key role in quantifying and understanding the role of multi-target/polypharmacology, as both an exploitable but also potentially undesirable property of a compound. The ever increasing knowledge of activities beyond the desired primary pharmacology of drugs, attrited and project compounds, emerging from broad *in vitro* profiling and large-scale integration of published data, enables better analyses and better predictive models. However, the consistency of the data from diverse sources remains a challenge and the value of having *in vitro* data generated in a consistent fashion year after year is one of the key advantages of key initiatives like BioPrint®. The potential of such profiles to characterise a medicinal chemistry compound in biological space will increase, especially as functional screening approaches emerge with the same speed and cost as binding assays. *In silico* approaches will become more effective for both the prediction of the *in vitro* profiles, and in their use for the prediction of *in vivo* effects, including ADRs of compounds. *In vitro* pharmacological fingerprints are just the start of our increasing knowledge of compounds from a biological perspective, and other initiatives such the IMI eTox initiative[52] are bringing together more *in vivo* data generated for compounds, both in humans and animals. Thus more powerful analyses to seek structural and *in vitro*/*in vivo* associations will be possible in the near future. The work discussed in this chapter involved the use of binding assays only, but it is now possible, where relevant, to use functional assays for profiling, providing key differentiation of agonist and antagonist effects. In terms of designing multi-target compounds, knowledge of the broad pharmacological profile at all stages is very important. Together with increased understanding of the associations of certain activities with adverse effects, it will be possible to improve the design and selection of candidates that only contain the desired or low risk multi-target/polypharmacological activities, leading hopefully to reduced attrition in preclinical and clinical development.

References

1. J. S. Mason, J. Migeon, P. Dupuis and A. Otto-Bruc, *Antitargets: Prediction and Prevention of Drug Side Effects*, Vol. 38 of *Methods and Principles in Medicinal Chemistry*, Wiley Online Library, 2008, p. 23.
2. J. S. Mason, in *Burger's Medicinal Chemistry, Drug Discovery, and Development*, 7th edn., ed. D. J. Abraham and D. P. Rotella, Wiley, 2010.
3. R. Stanton and Q. Cao, in Vol. 4 *of Comprehensive Medicinal Chemistry II*, ed. J. S. Mason, J. B. Taylor and D. J. Triggle, Elsevier, Oxford, 2007, p. 807.
4. A. L. Hopkins, J. S. Mason and J. P. Overington, *Curr. Opin. Struct. Biol.*, 2006, **16**, 127.
5. A. Bender, D. W. Young, J. L. Jenkins, M. Serrano, D. Mikhailov, P. A. Clemons and J. W. Davies, *Comb. Chem. High Throughput Screening*, 2007, **10**, 719.

6. A. F. Fliri, W. T. Loging, P. Thadeio and R. A. Volkmann, *Proc. Natl. Acad. Sci., USA*, 2005, **102**, 261.
7. A. F. Fliri, W. T. Loging, P. Thadeio and R. A. Volkmann, *Nat. Chem. Biol.*, 2005, **1**, 389.
8. A. F. Fliri, W. T. Loging, P. Thadeio and R. A. Volkmann, *J. Med. Chem.*, 2005, **48**, 6918.
9. D. A. Filimonov, W. Poroikov and E. I. Karaicheva, *Exp. Clin. Pharmacol. (Rus.)*, 1995, **58**, 56.
10. L. M. Kauvar, D. L. Higgins, H. O. Villar, J. R. Sportsman, A. Engqvist-Goldstein, R. Bukar, K. E. Bauer, H. Dilley and D. M. Rocke, *Chem. Biol.*, 1995, **2**, 107.
11. S. L. Dixon and H. O. Villar, *J. Chem. Inf. Comput. Sci.*, 1998, **38**, 1192.
12. S. J. Haggarty, P. A. Clemons and S. L. Schreiber, *J. Am. Chem. Soc.*, 2003, **125**, 10543.
13. Y. K. Kim, M. A. Arai, T. Arai, J. O. Tamenzo, E. F. Dean, 3rd, N. Patterson, P. A. Clemons and S. L. Schreiber, *J. Am. Chem. Soc.*, 2004, **126**, 14740.
14. G. V. Paolini, R. H. B. Shapland, W. W. P. van Hoorn, J. S. Mason and A. H. Hopkins, *Nat. Biotech.*, 2006, **24**, 805.
15. T. Jean and B. Chapelain, *Method of Identification of Leads or Active Compounds*, CEREP, 1999, International publication number WO-09915894.
16. C. M. Krejsa, D. Horvath, S. L. Rogalski, J. E. Penzotti, B. Mao, F. Barbosa and J. C. Migeon, *Curr. Opin. Drug Discovery Devel.*, 2003, **6**, 471.
17. N. Froloff, V. Hamon, P. Dupuis, A. Otto-Bruc, B. Mao, S. Merrick and J. Migeon, in *Chemogenomics: Knowledge-based Approaches to Drug Discovery*, ed. E. Jacoby, Imperial College Press, London, 2006, p. 175.
18. C. M. Krejsa, D. Horvath, S. L. Rogalski, J. E. Penzotti, B. Mao, F. Barbosa and J. C. Migeon, *Curr. Opin. Drug Discovery Devel.*, 2003, **6**, 471.
19. D. Armstrong, J. Migeon, M. G. Rolf, J. Bowes, M. Crawford and J.-P. Valentin, in *Preclinical Development Handbook: Toxicology*, Wiley, 2008, p. 581.
20. http://www.cerep.fr/cerep/users/pages/productsservices/bioprintservices.asp and http://www.chemaxon.com/forum/vberenz_cerep_2006_v2-download1572.pdf
21. http://www.millipore.com/drug_discovery
22. A. L. Hopkins and C. R. Groom, *Nat. Rev. Drug Discovery*, 2002, **1**, 727.
23. J. Sallantin, C. Dartnell and M. Afshar, in *Discovery Science*, Vol. 4265 of *Lecture Notes in Computer Science*, ed. N. Lavrac, L. Todrovski and J. P. Jantke, Springer Verlag, Berlin, Heidelberg, 2006, p. 231.
24. M. Afshar, A. Lanoue and J. Sallantin, in Vol. 4 of *Comprehensive Medicinal Chemistry II*, ed. J. S. Mason, J. B. Taylor and D. J. Triggle, Elsevier, Oxford, 2007, p. 767.
25. Ariana Pharmaceuticals, Paris: www.arianapharma.com
26. J. S. Mason, J. E. Mills, C. Barker, J. Loesel, K. Yeap and M. Snarey, in *Abstracts of Papers*, 225th ACS National Meeting, New Orleans, LA, United States, March 23–27, 2003, COMP-343.

27. J. S. Mason, in *First European Conference on Chemistry for Life Sciences: Understanding the Chemical Mechanisms of Life*, Rimini (Italy), October 4–8, 2005, Proceeding/Abstracts book.
28. Daylight Fingerprints. Daylight Chemical Information Systems, Inc. www.daylight.com: Irvine, CA.
29. M. Cases and J. Mestres, *Drug Discovery Today*, 2009, **14**, 479.
30. P.-D. Leeson and B. Springthorpe, *Nat. Rev. Drug Discovery*, 2007, **6**, 881.
31. P. Bamborough, D. Drewry, G. Harper, G. K. Smith and K. J. Schneider, *Med. Chem.*, 2008, **51**, 7898.
32. J. Scheiber, J. L. Jenkins, S. C. K. Sukuru, A. Bender, D. Mikhailov, M. Milik, K. Azzaoui, S. Whitebread, J. Hamon, L. Urban, M. Glick and J. W. Davies, *J. Med. Chem.*, 2009, **52**, 3103.
33. A. Bender, J. Scheiber, M. Glick, J. W. Davies, K. Azzaoui, J. Hamon, L. Urban, S. Whitebread and J. L. Jenkins, *ChemMedChem*, 2007, **2**, 861.
34. S. Whitebread, J. Hamon, D. Bojanic and L. Urban, *Drug Discovery Today*, 2005, **10**, 1421.
35. B. Faller, J. Wang, A. Zimmerlin, L. Bell, J. Hamon, S. Whitebread, K. Azzaoui, D. Bojanic and L. Urban, *Expert Opin. Drug Metab. Toxicol.*, 2006, **2**, 823.
36. D. Horvath and C. Jeandenans, *J. Chem. Inf. Comput. Sci.*, 2003, **43**, 680.
37. D. Horvath and C. Jeandenans, *J. Chem. Inf. Comput. Sci.*, 2003, **43**, 691.
38. P. J. Goodford, *J. Med. Chem.*, 1985, **28**, 849.
39. *Molecular Interaction Fields: Applications in Drug Discovery and ADME Prediction*, Vol. 27 of *Methods and Principles in Medicinal Chemistry*, Wiley Online Library, 2006.
40. T. Fox, in *Molecular Interaction Fields: Applications in Drug Discovery and ADME Prediction*, Vol. 27 of *Methods and Principles in Medicinal Chemistry, Wiley Online Library*, 2006, p. 45.
41. M. Baroni, G. Cruciani, S. Sciabola, F. Perruccio and J. S. Mason, *J. Chem. Inf. Model.*, 2007, **47**, 279.
42. F. Perruccio, J. S. Mason, S. Sciabola and M. Baroni, in *Molecular Interaction Fields: Applications in Drug Discovery and ADME Prediction*, Vol. 27 of *Methods and Principles in Medicinal Chemistry, Wiley Online Library*, 2006, p. 83.
43. S. Sciabola, R. V. Stanton, J. E. Mills, M. M. Flocco, M. Baroni, G. Cruciani, F. Perruccio and J. S. Mason, *J. Chem. Inf. Model.*, 2010, **50**, 155.
44. A. Steffen, T. Kogej, C. Tyrchan and O. Engkvist, *J. Chem. Inf. Model.*, 2009, **49**, 338.
45. J. S. Mason, I. Morize, P. R. Menard, D. L. Cheney, C. Hulme and R. F. Labaudiniere, *J. Med. Chem.*, 1999, **42**, 3251.
46. A. C. Good, J. S. Mason and S. D. Pickett, in *Virtual Screening for Bioactive Molecules*, Vol. 10 in *Methods and Principles in Medicinal Chemistry*, 2000, p. 131.
47. J. S. Mason and D. L. Cheney, in *Pacific Symposium on Biocomputing '99*, ed. R. B. Altman, World Scientific Publishing Company, 1999.

48. J. S. Mason and D. L. Cheney, in *Pacific Symposium on Biocomputing 2000*, ed. R. B. Altman, K. Lauderdale and T. E. Klein, World Scientific Publishing Company, 1999.
49. Y. C. Martin, J. L. Kofron and L. M. Traphagen, *J. Med. Chem.*, 2002, **45**, 4350.
50. A. Bender, J. L. Jenkins, M. Glick, Z. Deng, J. H. Nettles and J. W. Davies, *J. Chem. Inf. Model.*, 2006, **46**, 2445.
51. J. J. Sutherland, R. E. Higgs, I. Watson and M. Vieth, *J. Med. Chem.*, 2008, **51**, 2689.
52. http://www.etoxproject.eu/

CHAPTER 6

Phenotypic and In Vivo *Screening: Lead Discovery and Drug Repurposing*

CHRISTOPHER A. LIPINSKI

Scientific Advisor, Melior Discovery, Waterford, CT, USA
Email: clipinski@meliordiscovery.com

6.1 Changes in Screening Philosophy

Over the past 40 years, drug discovery has moved from a mechanism agnostic and chemocentric approach based on phenotypic screening in whole animals to a reductionist mechanism-based screening approach which is largely molecular biology centric.[1] In this transition, the major type of drug discovery screening became the search for a single compound that was superbly selective for a single target with a known mechanism. The focus on high throughput target mechanism-based discovery strategies led to a transition away from the 1970s animal-based phenotypic drug discovery. This trend accelerated in the 1990s and reached its zenith in the early 2000s, concomitant with the deciphering of the human genome. Pharmaceutical companies launched the exploration of new genomic-based targets, giving rise to the new science of chemical genomics.[2] Large collaborations were established to mine the new genomic targets and massive HTS campaigns were started to discover ligands for these new genomic targets. Large companies had assays for 500 different targets with a million data points and a wish to screen 100 000 compounds per day in a drug discovery factory[3] and a wish to 'make a drug for each target before we knew

RSC Drug Discovery Series No. 21
Designing Multi-Target Drugs
Edited by J. Richard Morphy and C. John Harris

Published by the Royal Society of Chemistry, www.rsc.org

we were interested in the target'.[4] As history attests, many of these reductionist genomic target-based efforts, much facilitated by automation, whilst being scientifically fascinating, have proved economically unsustainable,[5] difficult to implement,[6] and have failed to enhance or improve drug discovery productivity.[7,8]

Certain drug discovery cultural beliefs solidified coincident with this era in drug discovery. Among these beliefs were the assertion that the only way to discover a drug was to run an HTS using a diverse library against a single target whose biochemical mechanism was known. Success in this process would be the discovery of a lead that could be optimized in chemistry so that, as ascertained from *in vitro* assays, a superbly selective low nanomolar inhibitor of the target would result. This paradigm was taught, directly or indirectly, so thoroughly that even today it is possible to find academic biologists who believe that the single target single mechanism approach is the only way to discover a drug. Along with this molecular biology centric reductionist approach was the belief that it was essential to know mechanism and that it was impossible to progress a compound into the clinic unless the mechanism was known. In practice, knowing mechanism became a more easily manageable type of surrogate for knowing about safety and many younger researchers still believe that knowing mechanism is a regulatory requirement for approval to enter clinical studies. Knowing mechanism is not a regulatory requirement; knowing about safety is a regulatory requirement.

Not fully appreciated until into the latter part of this decade is the robustness of network signaling in organisms generally and humans in particular.[9] Most major diseases involve multiple pathways; from experiments in yeast it is estimated that 85–90% of targets or pathways that are completely blocked by a small molecule result in no observable phenotypic response.[10] Other things being equal, the complete pathway block in the majority of cases will not result in a robust phenotypic drug response. Moreover, as literature on signaling pathway perturbation accumulated, it became clear that several linked pathway interventions more modest than a complete block were likely to be more efficacious in eliciting a robust phenotypic block than a single point complete block. Thus was born the pathway biology (systems biology) basis for multi-target drug discovery (MTDD).

The trick in drug discovery then became to discover the 10–15% of targets where a single mechanism selective pathway block might be efficacious in human medicine, since, from experience, it was known that some drugs with this profile were in fact highly effective in human therapy. From this came the recognition of the extreme importance of biological target validation.[11] The predominant existing mode of drug discovery was doomed to failure unless the target was validated. In theory if the nascent science of systems biology was robust enough, that science could have been used to pick the validated target. In the real world, despite huge progress, the new science fell short of current needs and something else had to fill the gap. The gap filler in target validation was to rely on academic biology experts and to set up collaborations between the pharmaceutical company developers and the experts in the biology realm.

The direction of the collaborations made sense because in terms of sheer numbers the overwhelming supply of basic research biologists exists in academia with sometimes decades of experience in a particular area of biology. The validated target was most likely to be found in those biology areas with in-depth knowledge and decades of researcher experience.

As the decade of the 2000s progressed it became distressingly clear that the number of drugs progressing to clinical use was not increasing and that, of all the causes for clinical failure, it was lack of efficacy that was the most problematic.[12] For example, a currently quoted statistic is that current pre-clinical data has only 10% predictive value for clinical-stage efficacy compared to nearly 50% predictivity for toxicity and 60% for chemistry-related issues. Although in most cases the cause of clinical attrition could not be directly linked to the method of pre-clinical screening, the concordance of the clinical efficacy-related attrition problem with the documented limitations of the mechanism-based screening added to the discomfort level with the pre-clinical approach and thus made the industry more receptive in considering complementary approaches to the predominant mechanism-based screening approach. Thus was reborn the renewed industry interest in phenotypic screening, *in vivo* screening, in MTDD, and in drug repurposing.

Here it is instructive to draw an analogy between the chemistry and biology realms. In medicinal chemistry, rules and filters to optimize the probability of achieving oral activity appeared in the 1990s; the rule of five (RO5) was first published in 1997. In that early time period nobody would have contemplated that simple physical properties like MWt and lipophilicity (log *P*) would be related to something as complex as clinical attrition. Yet, subsequent to the RO5 publication, multiple strikingly consistent papers reported the association of simple physicochemical properties to clinical attrition. In particular, as the properties in the clinical candidate moved out of the favorable drug-like pre-clinical range the probability of clinical failure increased. Biology is much more complex than chemistry. Nevertheless, it is tempting to wonder whether a similar analogy exists. Is there likely to be a greater degree of clinical attrition for those drug candidates whose pre-clinical biology discovery occurred in the least physiologically relevant screens? Certainly this hypothesis is consistent with industry-wide investments in high content screening, an increasing focus on more physiologically relevant cell-based assays[13] at the expense of biochemical assays, as well as a new focus on drugs that display polypharmacology (read MTDD).

6.2 Phenotypic Screening: Advantages, Disadvantages, Ligand Matching and MTDD

Phenotypic screening was the norm in the 1970s and as previously discussed was subsequently largely replaced by mechanistic screening. Many drugs useful in human medicine were discovered in the 1970s. Because the clinical data for a direct comparison are unavailable, we do not know whether phenotypic screening is advantageous in terms of ultimate clinical success over mechanistic screening. After all, one could argue that today's unmet medical need directs us to address inherently different, more complex, diseases than 30–40 years ago. However, a retrospective analysis of marketed drugs over the time period 1999–2008 of drug discovery against arguably more complex diseases exits.[14] This study comparing marketed drugs developed by phenotypic versus mechanistic screening suggests that phenotypic screening is more effective at discovering first-in-class agents. Mechanistic screening is more effective at discovering follow-on agents once the breakthrough compound has been discovered.

Phenotypic screening carries great advantages in terms of target opportunity space. When one screens successfully against a mechanistic target, one only discovers a ligand binding to that particular target. Most often this is an antagonist. More rarely an agonist or one of the more recently described classes of ligand, e.g. inverse agonist, inverse antagonist, etc. is discovered. In a phenotypic mechanistically unbiased screen one can discover any single or multiple mechanisms consistent with the experimentally observed phenotype. For example, a search in the ThomsonReuters (formerly Prous) Integrity Database for anti-obesity drugs results in over 13 000 compounds annotated to 349 different molecular mechanisms, including the category of unspecified mechanism. Thus a phenotypic screen for an orally active compound that reduces weight gain in mice fed a high fat diet could result in an active compound with hundreds of possible molecular mechanisms. In this sense a mechanistically unbiased phenotypic screen is broad in target opportunity space because a compound could be active through any one or more of a very large number of mechanisms. Phenotypic screening in mechanistically unbiased screens is therefore one possible method of detecting a multi-targeted compound. Because of the breadth in target opportunity space, a mechanism-unbiased phenotypic screen will in general require far fewer compounds to be screened to have a decent probability of detecting an active one than would be the case in a mechanistic screen.

As we have seen, screens can be categorized as to breadth of target opportunity, with phenotypic screens having much the broader target opportunity space. In a similar manner, ligands can be categorized as to broadness in chemistry space. The small ligands as used in fragment screening are the broadest in chemistry space followed by the typical synthetic medicinal compound. Narrowest in chemistry space are the complex natural products whose complicated topology and frequent sp^3 centers result in a narrow specificity of binding. The idea in screening is to optimally match the type of screen to the type of ligands being screened. Natural products make ideal partners for mechanism-unbiased phenotypic screens because the narrowness of the chemistry space coverage of natural products is offset by the target opportunity breadth of the phenotypic screen. It is no accident that this type of pairing is frequently found in academic basic research. The student relies on the specificity of the natural product to detect a hit in a mechanism-unknown phenotypic screen and then spends the rest of his/her PhD studies doing the challenging mechanistic detective work. This pattern works well in academia but as one might suspect works not well at all in industry because of time pressures. How does this all fit in with MTDD? On first principles one would anticipate the probability of detecting multi-target activity would be highest in fragments, then next in typical medicinal chemistry compounds, and least in complex natural products. With respect to types of screening one would predict that detection of multi-target activity would be higher in phenotypic screens than in pair-wise mechanistic screens.

Of course, phenotypic screening carries a number of disadvantages. The issue of uncertainty in mechanism has been mentioned. In early pre-clinical discovery a frequent problem is how to deal with a toxic response in a compound. Is the toxicity on or off target? If the untoward effect is off target then simply changing to a different chemical structure with unchanged on target mechanism will most often alleviate the problem. There is no doubt that not knowing a primary mechanism greatly complicates the assessment of a compound's safety profile. However, it is important to remember that safety is the paramount consideration and not mechanism. Another disadvantage is the negative reaction of medicinal chemists when asked to optimize activity in a 'black box' screen. Most of the chemists, having learned their trade in the era of screening against mechanistic targets, are unenthusiastic on working on a project without precise mechanistic screening data. To some degree the medicinal chemists are correct; there is very little precedence from the earlier 1970s era of phenotypic screening for achieving a 1000-fold improvement in potency as in the common mechanistic screening situation of moving from micromolar to nanomolar activity. However, in a phenotypic screen the potency of the starting point is likely to be better especially if the phenotypic screen is *in vivo*. Potency improvements in the range 10- to 30-fold were routinely achieved in the 1970s and, as the productivity figures show, were more than adequate for clinical success.

Whilst it is arguable that optimization of a hit compound is perfectly possible without knowing its mechanism of action, the enthusiasm of medicinal chemists for phenotypic primary screening would likely be enhanced if more robust methods for deconvoluting mechanism were available.[12] This would allow a more rational approach to optimization using, for example, structural biology and biophysical measurements. Much progress is being made in deconvoluting multiple targets, using such techniques as gene knock-outs,[15] RNAi,[16,17] and 'fishing' in cellular systems using sophisticated ligand affinity-based chemical proteomic techniques,[18,19] and it is likely that medicinal chemists will embrace phenotypic screening more thoroughly as these technologies progress. These methods are described in more detail in the following chapter.

6.3 Drug Repurposing: Leveraging Signaling Network Activities

Drug repurposing is finding a new use for an old drug, sometimes with a change in formulation.[20,21] At its heart, drug repurposing capitalizes on the robustness of signaling in biological networks.[22] The signaling robustness, which is a limitation in mechanistic screening and which limits genomic-based drug discovery, is an advantage in drug repurposing. Numerous observations show that a clinically useful drug never (or almost never) has just a single biological action.[23] Almost by definition the clinically useful drug is able to perturb at least one biological signaling pathway. By luck or by design the drug that elicits a clinical response is the exception to the more common failure of the mechanistic screening approach. The probability that this compound will have multiple useful clinical activities is very high. In the past 'off-label' effects for drugs were discovered sometimes very late in clinical studies[24] and, even more distressingly for the innovator company, showed up post marketing approval. This late appreciation of beneficial clinical activity has very negative financial consequences to the sponsor because in the USA off-label activity cannot be legally promoted and so a burgeoning industry sprung up to detect useful unanticipated biological activity early in the clinical pipeline while there was still time to take advantage of the finding.

Thirty percent of drug-like compounds exhibit otherwise unpredicted biological activity when they are run through a panel chosen from 45 mechanistically unbiased phenotypic *in vivo* rodent screens.[25] When a new use is found for a drug-like compound, is the new activity 'on target' or 'off target'? This is a very important question because if the activity is 'on target' then the compound is likely already optimized for the particular mechanism and no medicinal chemistry optimization will be needed. Experience from the group at Melior Discovery on over 200 compounds suggests that the vast majority (90%) of new activity is 'on target'.[25] However, there is a caveat to this observation in that if biological target/mechanism were fully explored the 'on target' category would likely be lower and the 'off target' would most likely be higher. Operationally, the 'on target' category is determined in the following manner. The mechanism originally assigned to the compound is matched with the new biology observation. The newly observed drug action is labeled as 'on target' if the biology literature supports a plausible connection between the original compound mechanism and the new biology observation. This observation points out that despite the large number of publically available disease target databases[26,27] there are large gaps in biology knowledge underlying the inability to predict useful new activity attributable to a known mechanism. At this point in time only about 14% of clinically used drugs are mechanistically selective as defined by the NIH's tool and probe criteria[23] so it is likely that for any new compound in clinical study there could well be new biology related to unexpected mechanisms.

A reduced emphasis on molecular pharmacology and integrative pharmacology concomitant with an increase in reductionist thinking has been cited as a contributor to the decline in new drug approvals.[28] To this author, a striking component in this decline has been the reduction of *in vivo* rodent screening. Rodent-based disease screens represent an attractive strategy for rapidly and efficiently identifying new therapeutic indications for existing compounds. Animal testing continues to offer the best predictive value for most disease areas;[29] ensures chemistry, biology, and pharmacokinetics are mutually optimized;[30,31] and provides a rich data output to support go/no-go candidate evaluation as well as elucidation of target activity to inform next-generation drugs. Some recent *in vivo* surrogate technologies, such as the use of small genetically modified vertebrates such as the zebrafish, are able to address some of the well-known cost and scale issues of using traditional small mammals in animal screening.[32,33]

References

1. M. J. Keiser, J. J. Irwin and B. K. Shoichet, *Biochemistry*, 2010, **49**, 10267–10276.
2. A. Sehgal, *Curr. Opin. Drug Discovery Dev.*, 2002, **5**, 526–531.
3. R. Archer, *Nat. Biotechnol.*, 1999, **17**, 834.
4. P. N. Goodfellow, *EMBO Rep.*, 2004, **5**, 843–846.
5. U. A. K. Betz, *Drug Discovery Today*, 2005, **10**, 1057–1063.
6. U. A. K. Betz, R. Farquhar and K. Ziegelbauer, *Curr. Opin. Chem. Biol.*, 2005, **9**, 387–391.
7. M. K. Hellerstein, *Metab. Eng.*, 2008, **10**, 1–9.
8. S. M. Paul, D. S. Mytelka, C. T. Dunwiddie, C. C. Persinger, B. H. Munos, S. R. Lindborg and A. L. Schacht, *Nat. Rev. Drug Discovery*, 2010, **9**, 203–214.
9. M. K. Hellerstein, *J. Pharmacol. Exp. Ther.*, 2008, **325**, 1–9.
10. A. L. Hopkins, *Nat. Chem. Biol.*, 2008, **4**, 682–690.
11. F. Sams-Dodd, *Drug Discovery Today*, 2005, **10**, 139–147.
12. A. Bakker, A. Caricasole, G. Gaviraghi, G. Pollio, G. Robertson, G. C. Terstappen, M. Salerno and P. Tunici, *ChemMedChem*, 2009, **4**, 923–933.
13. R. M. Eglen, A. Gilchrist and T. Reisine, *Comb. Chem. High Throughput Screening*, 2008, **11**, 566–572.
14. D. C. Swinney and J. Anthony, *Nat. Rev. Drug Discovery*, 2011, **10**, 507–519.
15. B. P. Zambrowicz and A. T. Sands, *Drug Discovery Today*, 2004, **3**, 198–207.
16. S. Morgan-Lappe, K. W. Woods, Q. Li, M. G. Anderson, M. E. Schurdak, Y. Luo, V. L. Giranda, S. W. Fesik and J. D. Leverson, *Oncogene*, 2006, **25**, 1340–1348.
17. Z. A. Knight, H. Lin and K. M. Shokat, *Nat. Rev. Cancer*, 2010, **10**, 130–137.
18. G. C. Terstappen, C. Schlüpen, R. Raggiaschi and G. Gaviraghi, *Nat. Rev. Drug Discovery*, 2007, **6**, 891–903.
19. D. C. Schriemer, D. Kemmer and M. Roberge, *Comb. Chem. High Throughput Screening*, 2008, **11**, 610–616.
20. D. Cavalla, *Nat. Rev. Drug Discovery*, 2009, **8**, 849–853.
21. D. Grau and G. Serbedzija, *Drug Discovery Dev.*, 2005, http://www.dddmag.com/innovative-strategies-for-drug.aspx/

22. J. A. Lowe, P. Jones and D. M. Wilson, *Curr. Opin. Drug Discovery Dev.*, 2010, **13**, 524–526.
23. R. Garcia-Serna and J. Mestres, *Drug Discovery Today*, 2010, **16**, 99–106.
24. J. von Eichborn Joachim, M. S. Murgueitio, M. Dunkel, S. Koerner, P. E. Bourne and R. Preissner, *Nucleic Acids Res.*, 2011, **39** (Database issue), D1060–1066.
25. P. F. Dimond, *Gen. Eng. News*, 2010, **30**, http://www.genengnews.com/gen-articles/drug-repositioning-gains-in-popularity/3263/?page=1/
26. L. Harland and A. Gaulton, *Expert Opin. Drug Discovery*, 2009, **4**, 857–872.
27. G. Hu and P. Agarwal, *PLoS One*, 2009, **4**, http://www.plosone.org/article/fetchObjectAttachment.action?uri=info%3Adoi%2F10.1371%2Fjournal.pone.0006536&representation=PDF/
28. S. J. Enna and M. Williams, *Adv. Pharmacol. (San Diego)*, 2009, **57**, 1–30.
29. W. Pawlik, *Folia Medica Cracoviensia*, 1998, **39**, 175–182.
30. T. J. H. Bueters, J. Hoogstraate and S. A. G. Visser, *Drug Discovery Today*, 2009, **14**, 89–94.
31. B. M. Amore, J. P. Gibbs and M. G. Emery, *Comb. Chem. High Throughput Screening*, 2010, **13**, 207–218.
32. T. C. Tran, B. Sneed, J. Haider, D. Blavo, A. White, T. Aiyejorun, T. C. Baranowski, A. L. Rubinstein, T. N. Doan, R. Dingledine and E. M. Sandberg, *Cancer Res.*, 2007, **67**, 11386–11392.
33. C. A. McRae, *Exp. Opin. Drug Discovery*, 2010, **5**, 619–632.

CHAPTER 7

Target/s Identification Approaches – Experimental Biological Approaches

GIULIO SUPERTI-FURGA,* KILIAN HUBER AND GEORG WINTER

CeMM – Research Centre for Molecular Medicine of the Austrian Academy of Sciences, Lazarettgasse 14/AKH BT 25.3, A-1090 Vienna, Austria
*Email: GSuperti-Furga@cemm.oeaw.ac.at

7.1 Introduction

Interest in both cell-based and whole organism phenotypic screening as a powerful tool for multi-target drug discovery is increasing (also see Chapter 4, Section 4.3.1 and Chapter 6). For instance, high-content screening assays with complex (cellular) readouts like protein localization, cell motility or gene signatures have become increasingly amenable and popular for high-throughput campaigns. However, in such settings, the cellular targets, inhibition of which elicits the phenotype of interest, usually remain elusive, creating a necessity for subsequent target deconvolution approaches.

Moreover, a considerable proportion of drugs that are used in the clinic today were developed without *a priori* knowledge of the actual target that causes the beneficial clinical effect. Dependent on the indication, a given therapeutic agent might as well have multiple, very different, targets that contribute in a context-specific manner to the observed phenotypes;

RSC Drug Discovery Series No. 21
Designing Multi-Target Drugs
Edited by J. Richard Morphy and C. John Harris

Published by the Royal Society of Chemistry, www.rsc.org

dihydrofolate reductase (DHFR) is for instance the well-established cellular target for methotrexate in the context of cancer therapy, while the mechanism of action of methotrexate in other therapeutic indications such as rheumatoid arthritis still remains to be elucidated.[1] Another layer of complexity is represented by small molecules that simultaneously target, due to their polypharmacological properties and promiscuity, a number of gene products. This promiscuity can, dependent on the target spectra, either cause unwanted off-target effects or be key to the efficacy of a therapeutic agent. In this regard, target deconvolution approaches can facilitate the prediction of side effects or explain unanticipated clinical responses.[2]

In order to unravel functionalities of drugs, or small molecules in general (such as hits from cell-based or whole organism phenotypic screening campaigns), scientists can choose between a broad variety of genomic and proteomic approaches. In the course of the following chapter, a variety of different approaches will be introduced and exemplified.

7.2 Yeast Genomic Assays

Both from a genomic and proteomic perspective, *S. cerevisiae* is an extremely well-characterized model organism to study biology. A lot of -omics-based knowledge has been assembled in the last years and sophisticated tools like genome-wide barcoded deletion strains are available allowing parallel assessment of gene and protein function in truly 'systems views' of the processes under investigation.[3] Additional beneficial features, especially in the light of high-throughput techniques, are the facile and inexpensive cultivation conditions as well as the rapid generation time.

Trying to understand the mechanism of action of drugs for purposes relevant to human pharmacology and physiology using yeast appears paradoxical and inefficient, given the enormous evolutionary distance of the two organisms and the various fundamental differences in organismal organization (unicellularism, cell wall, etc.) Nevertheless, for the sake of gaining basic insights into central and evolutionarily conserved pathways, yeast is a model organism with outstanding properties and has been widely used to understand the molecular mode of action of small molecules.[4] In fact, 45% of yeast genes have human correspondants.[5]

7.2.1 Drug-Induced Haplo-Insufficiency Profiling (HIP)

The availability of genome-wide barcoded deletion strains mentioned above forms the basis of drug-induced haplo-insufficiency profiling (HIP).[3] The major advantage of the molecular barcode strategy is the ability to perform screens using pools of genetically distinct cells. Those allow the assessment of growth advantages or disadvantages of a distinct strain (or a selected group of strains) in response to a selection pressure in an environment made competitive by the simultaneous presence of the other barcoded strains. The strains deficient for those genes that are important for cellular growth under the chosen experimental conditions will be depleted over time. Relative abundance of every strain in a pool can be assessed by extraction of genomic DNA followed by PCR amplification of the barcodes and subsequent hybridization of the amplified barcodes to DNA microarrays (Figure 7.1).

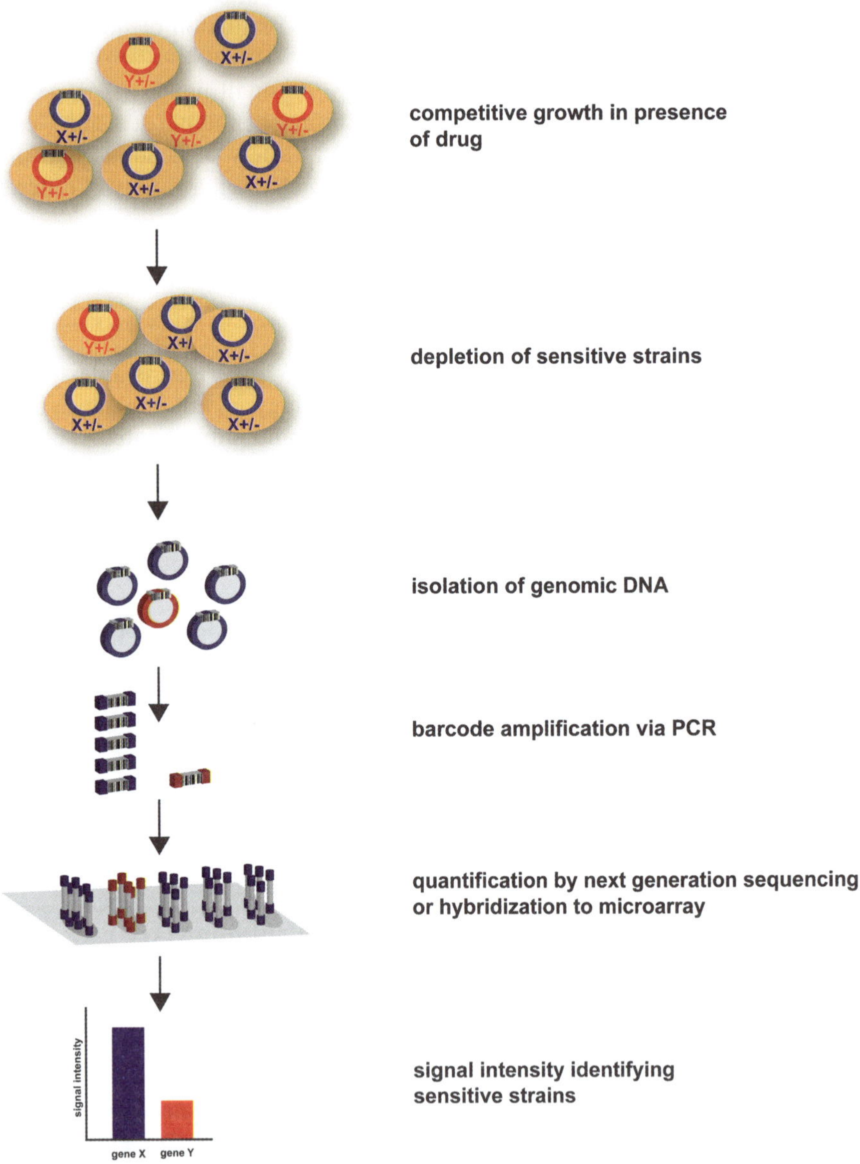

Figure 7.1 Drug-induced haplo-insufficiency profiling.

This parallel growth strategy was used in the context of HIP using initially 233, and later on genome wide, deletion strains to broadly test heterozygous deletion strain for their specific vulnerability to small molecules. If the small molecule targets the gene product of this heterozygous locus, it creates a growth difference, termed 'hypersensitivity'.[6] Moreover, strategies have been developed for cases where lowering the gene dosage by 50% is not sufficient for observing a phenotype and full deletions are necessary.[7] General limitations of this kind of approach include that the action of the gene-drug is indirect. Thus, the data do not imply evidence of a physical interaction of the small molecule and the gene product. Furthermore, drug targets that do not impair the growth abilities or are redundant, as well as targets that lack a counterpart in *S. cerevisiae*, will not be picked up by HIP.[8]

7.2.2 Homozygous Profiling (HOP)/Haploid Deletion Chemical Genetic Profiling

The concept and design of homozygous profiling is similar to HIP profiling, with the major difference that the used yeast strains are completely deficient of individual (non-essential) genes. Testing all viable mutants for hypersensitivity to a compound of interest reveals a chemical-genetic interaction profile that can, when intersected with genetic interaction profiles, provide information of the targets and pathways affected by the small molecule.[9] This approach has proven to be efficient especially for bioactive molecules that, like alkylating or DNA-damaging agents in general, lack a distinct protein target.[10] However, as the barcoded strains contain complete deletions of non-essential genes, HOP assays do not directly lead to drug target identification.

7.3 Genomic Assays in Mammalian Cells

7.3.1 Comparative Gene Expression Profiling

Based on the findings that gene expression compendia can be used for assigning the impact of uncharacterized perturbations in *S. cerevisiae*,[11] and the fact that gene expression profiles, by their complexity, produce robust patterns, Lamb and colleagues set out to create the first reference assortment of gene expression profiles resulting from treating human cultured cell lines with bioactive small molecules.[12] This approach, named 'Connectivity map' (also known as cmap), was successfully applied first by Todd Golub and colleagues in order to formulate hypotheses on the molecular nature of two natural compounds, celastrol and gedunin. These compounds had been identified in a gene expression high-throughput screen for inhibition of androgen receptor signalling.[12] Comparing gene expression signatures caused by treatments with the two compounds, at that point still uncharacterized agents, with the Connectivity map database revealed high similarity with signatures of known HSP90 inhibitors, suggesting the possibility that the two compounds were acting by inhibiting HSP90 as well.[13] Further biochemical characterization validated this hypothesis and furthermore revealed a novel binding mode independent of the chaperone's ATP binding pocket. The Connectivity map approach is widely used. As the community can contribute profiles (http://www.broadinstitute.org/cmap/) to the database maintained at the Broad Institute, the accuracy of prediction is expected to grow with time.

7.3.2 RNA Interference-Based Screens

Large-scale screens based on transient or stable RNA interference have been used extensively, not only to find the direct target of a drug, but primarily to spot modulator genes that, upon knock-down, either cause resistance to drug treatment or further enhance the effect of a drug of interest.[14] Notably, the topology of identified genes and gene products on maps of molecular networks and pathways could also be used to generate hypotheses on the mechanism of action and for drug target identification.[15] Furthermore, RNA interference-based screens have been combined with proteomics approaches, which are introduced later on, for target deconvolution approaches.[16]

7.4 Proteomic Approaches

7.4.1 Compound-Centred Chemical Proteomics (CCCP)

Even though genomics-based target deconvolution approaches such as the Connectivity map (see Section 7.3.1) can provide broad, i.e. genome wide, valuable information on global cellular changes following drug action, the interpretation and translation of such results into a functional understanding of cellular signalling is far from straightforward. Unfortunately, in many cases there is no interpretable relationship between gene activation and drug-induced phenotype. This is not surprising as, although the genome conserves all required information for cellular maintenance, the functional effectors are the encoded proteins, their entirety being referred to as the proteome. Hence hardly any drugs interact with DNA directly, instead they are usually targeted against the products of transcriptional activity. Moreover, cells have numerous ways to modify protein activity at various post-transcriptional stages, e.g. alternative splicing and phosphorylation, which are beyond the scope of chemical genomics let alone biochemical kinetic assays. Considering these facts, an investigation of potential interactors within the proteome in its natural environment seems more straightforward. In order to identify (most hopefully) all possible interactors of a bioactive small molecule (the so-called 'interactome'), various proteomic approaches have been suggested. One strategy is to use the small molecule drug of interest as a bait to 'pull out' interaction partners (Figure 7.3).[17] A matrix, for instance beads, provides the sort of 'fishing rod' which allows subsequent separation and purification of the captured protein targets. In comparison to *in vitro* assays, this procedure exposes the compound of interest to a complete and functional proteome. Thus, chances to discover its effective biological target are much higher than screening methodologies focussed on a particular sub-proteome. Even though such HTS of enzyme classes, e.g. kinases, is already offered as a commercial service by various companies, the costs of those efforts are still considerable and provide only limited information. Moreover, conclusions based on such results do not necessarily reflect biologically relevant effects due to cellular target localization and tissue specificity which may influence drug action *in vivo*.

After the pull-down, the proteins identified by chemical proteomics can be analysed either by immunoblotting or, after digestion, by liquid chromatography coupled with mass spectrometry (LC-MS). The benefits of this approach are on the one hand that this methodology actually measures a direct physical interaction. Moreover, the detection of potential interactors is not limited to a certain class of proteins, e.g. kinases. Furthermore, the bait encounters its potential targets in their natural state including all possible types of post-translational modifications (PTMs) and allosteric influences, such as those brought about by protein interaction partners or ligands and co-factors. On the other hand, the bait small molecule needs to be attached to the matrix which is inevitably linked to a chemical modification of the original compound. Hence some targets may not be detected because possible interactions sites are blocked by the required synthetic modification. This issue can be overcome by using several baits modified at different positions of the molecule but that also at least doubles the experimental effort. Additionally, the technical implementation of an immobilized small compound such as chemical functionalization can already provide a challenge in itself. Last but not least is the fact that as simple as the small molecule/bait strategy may seem, the proteins isolated during the experiment need to be validated as true interactors. It is important to consider that cells express a plethora of protein classes, some of which are highly abundant whereas others are only expressed in small numbers and/or for a limited amount of time. Moreover, various proteins can bind drugs unspecifically without exerting any relevant phenotypic effect and, notably, many small molecule compounds are targeted towards low abundance proteins to which they bind with high affinity. Thus, in order to eliminate all those interactors of dubious relevance, one can either do a reference pull-down with an inactive compound or preincubate the cells with the unmodified small molecule of interest assuming that the original compound will be at least a comparably competitive ligand for its targets so that the subsequent bead-associated experiment should only yield the unspecific binders.[18] Still the amount of data generated is tremendous, underscoring statistical filtering, large control sets and expert bioinformatics as an indispensable tool for any chemical proteomics approach. The reproducible, statistically significant characterization of a bona-fide binder is only the beginning, and not quite the end, as rigorous biochemical target validation is necessary. Usually, when available, a biochemical assay is used to demonstrate a dose-dependent effect. This is also important considering that the bait may pull out ensembles of several proteins so that direct and indirect interactors need to be distinguished, a crucial step required for functional understanding of the composition of the actual drug target. Impressive examples of the immobilized-compound methodology are the identification of a primary target of thalidomide teratogenicity (see example below)[19] and the extensive target profiling of the clinically used BCR-Abl inhibitors imatinib (Gleevec©), nilotinib (Tasigna©) and dasatinib (Sprycel©) revealing unanticipated off targets such as the oxidoreductase NQO2 and DDR1 tyrosine kinase.[20]

7.4.1.1 Example: Identification of a Primary Target of Thalidomide Teratogenicity by Chemical Proteomics

Thalidomide was originally developed as a mild sedative and, because of its antiemetic effects, was the drug of choice for pregnant women suffering from morning sickness in the 1950s. Probably one of the greatest tragedies in pharmacology, thalidomide was found to be teratogenic, causing severe limb defects and deformation to the new born. After the drug had been withdrawn from the market it incidentally resurfaced years later as thalidomide displayed beneficial effects in the treatment of leprosy and cancer. Some derivatives of thalidomide, such as lenalidomide, are in clinical trials for multiple myeloma. Still despite decades of extensive research the molecular target of thalidomide responsible for its teratogenicity had remained elusive. Using a chemical proteomics approach, Ito and colleagues were able to identify a primary target candidate of thalidomide teratogenicity.[19] By preparing an immobilized derivative of thalidomide (Figure 7.2) they discovered cereblon and the related Cul4A-DDB1-E3 ubiquitin ligase complex as essential mediators of thalidomide-induced teratogenicity.

7.4.2 Kinobeads

Poor specificity of some compounds in a drug pull-down setting can be exploited to obtain a broad-specificity reagent capable of simultaneously retrieving the repertoire of whole target classes in one go. Thus, the beads constituting the affinity matrix can also be used for a different chemical proteomics approach. For example, to evaluate the kinase specificity of a selected inhibitor the cells or lysates are incubated with the kinase inhibitor of interest, which consequently will bind its target kinases. In the next step, and conversely to the direct chemical proteomics approach, all remaining unbound kinases are extracted by kinobeads. The latter are prepared by loading the beads with several different kinase inhibitors that taken together cover the majority of the kinome, thus 'kinobeads'. Comparison of an untreated sample with the one in which the inhibitor has occupied all binding

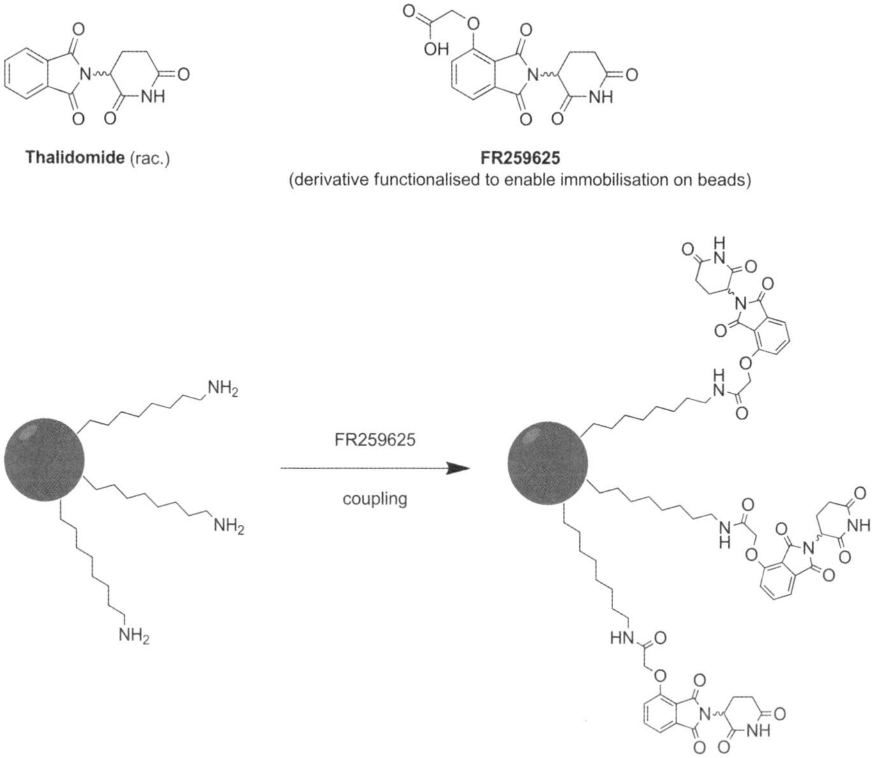

Figure 7.2 Chemical proteomics approach applied to the identification of a primary target of thalidomide teratogenicity.

sites of its target kinases yields a comprehensive selectivity profile (Figure 7.3). Bantscheff *et al.* have successfully applied this strategy to identify several unknown targets of clinical kinase inhibitors.[21] An important advantage of the kinobeads approach in comparison to the previously discussed single-drug strategy is that the compound of interest does not have to be chemically modified, which otherwise could lead to differences in the identified target spectra. Also, and in contrast to *in vitro* panel kinase screening, there is no need to express and purify the kinases of interest as the complete kinome of the cell line used is interrogated simultaneously in the same natural system. Quantitative MS detection methods such as SILAC (stable isotope labelling with amino acids in cell culture) or iTRAQ (isobaric tag for relative and absolute quantification) also allow for determination of K_d or IC_{50} values. General drawbacks of this method are that interaction profiles are limited to the particular sub-proteome which is covered by the beads, such as protein kinases, and the fact that it is very challenging to assure the whole sub-proteome is captured by the beads. An analogous approach for HDAC inhibitors has been developed recently.[22]

7.4.3 Activity-Based Protein Profiling (ABPP)

Activity-based protein profiling (ABPP) uses a probe which is directed against the active site of an enzyme (Figure 7.3). The probe consists of two essential parts – a reactive moiety, which forms a covalent bond with the desired protein, and a handle, e.g. a fluorophore or a biotin tag, allowing probe detection and visualization. After incubation with a small molecule or several compounds of interest, the probe is added to the cells or lysate and reacts specifically with the remaining protein targets whose active sites have not been occupied by the previously added compounds. Subsequently, all probe-bound proteins can be identified by direct in-gel fluorescence detection. Alternatively, the biotinylated probe can be used to enrich the protein class of interest with avidin-coated beads followed by tryptic digestion and MS analysis. Recently, high-throughput screening has been made possible by combining ABPP with fluorescence polarization in a homogenous assay.[23] As for the kinobead approach, ABPP in general enables identification of protein targets in an active state but is also limited to one distinct class of proteins. Furthermore, ABPP highly depends on the effectiveness of the probe and may thus not cover the complete sub-proteome of interest. A remarkable example of the power of ABPP is the identification of the biological relevance of serine hydrolases and accordant inhibitors in cancer.[24]

On top of all that, any proteomic approach requires a suitable biological system, i.e. a cell line, which expresses all the proteins of interest. Although it is possible to genetically engineer a corresponding cell line this again may not reflect the natural abundance, state of PTM and general cellular environment of a particular protein of interest.

7.4.4 Global Phosphoproteomics

Global quantitative phosphoproteomics is an excellent tool to uncover drug-induced cellular signalling changes in a systems-wide manner. Especially when used in conjunction with a direct proteomics approach or functional assays, this combination can provide the basis for exhaustive mapping of pathways as direct drug interactors and downstream effectors are captured equally.[16] Since phosphorylation is a key PTM in signal transduction, both in normal and malignant cells, a quantitative analysis of the phosphoproteome upon drug treatment may yield not only clues towards unknown side effects but also reveal alternative modes of action of small molecule compounds.[25] Phosphoproteomics greatly benefits from MS quantitation techniques like SILAC or iTRAQ which allow mapping of phosphorylation sites down to single amino acid residues of the previously digested peptide sequences. Comparison of untreated versus inhibitor-exposed cells uncovers, for example, all phosphorylation sites that are up- or down-regulated due to inhibitor action. For example, following a triple SILAC strategy, Pan and co-workers were able to describe MAPK and BCR-Abl signalling networks.[26]

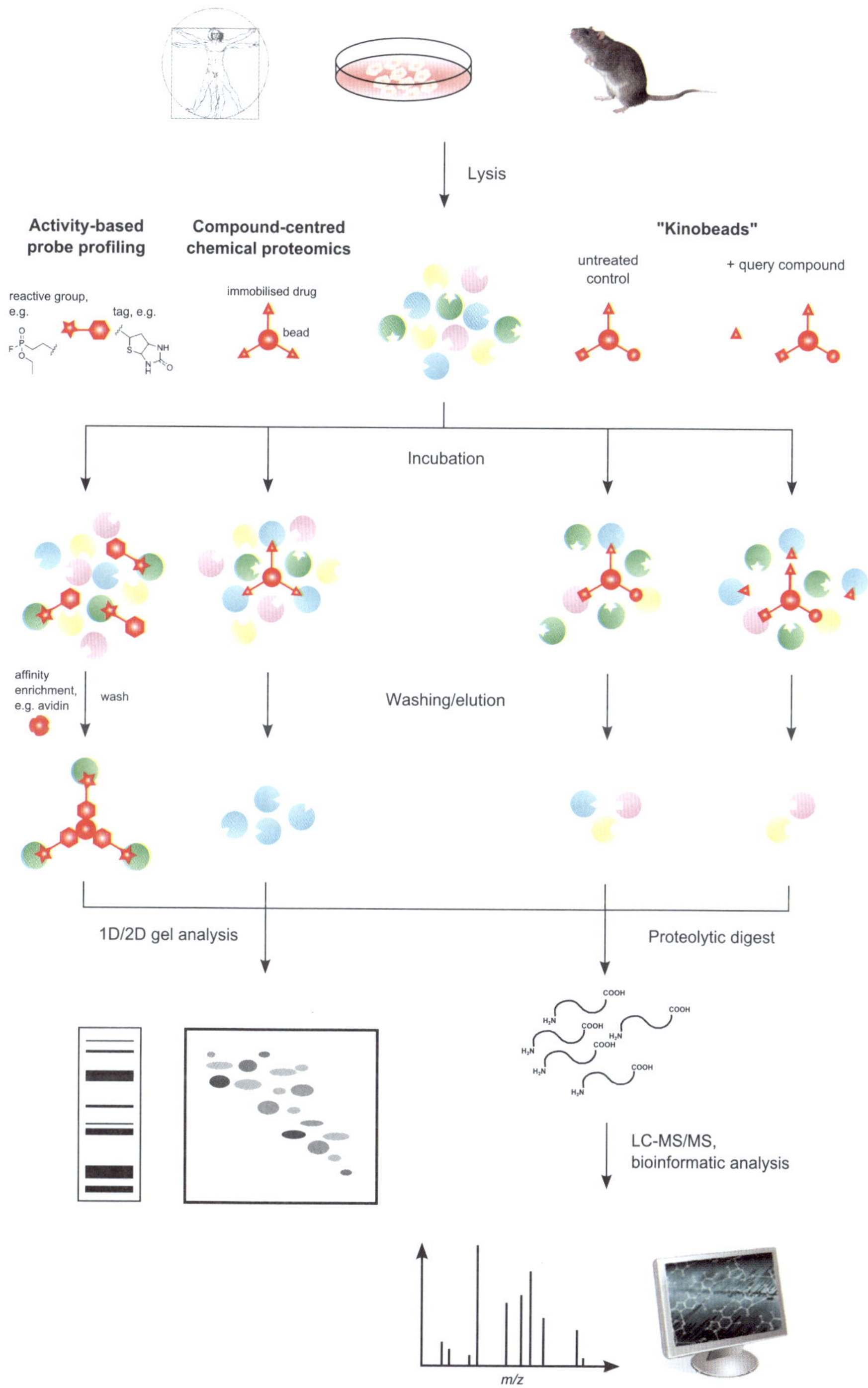

Figure 7.3 Comparison of chemical proteomics approaches.

7.4.5 Lysine Acetylation Profiling

Reversible acetylation of lysine residues is one of the major PTMs involved in cellular signalling and is highly conserved among species. Mediated by a complex interplay of lysine acetyltransferases (KATs) and lysine deacetylases (KDACs), it regulates important processes such as transcription by acetylating core histone tails. In view of the many diseases potentially treatable with KAT and KDAC modulators, and in particular the identification of the tumour suppressor p53 as an acetylation substrate, KDAC inhibitors have become prominent in cancer research. In a SILAC-based attempt to unveil the core lysine acetylome, Choudhary *et al.* discovered an unanticipated specificity of the two prominent and supposedly global KDAC inhibitors, suberoylanilide hydroxamic acid (SAHA) and MS-275.[27] Interestingly, SAHA was found to increase acetylation levels of chaperones HSP90α and HSPβ, whereas MS-275 was much more influential with regard to p53 acetylation.

7.4.6 Drug Affinity Responsive Target Stability (DARTS)

DARTS is a concept that has been introduced only recently and that is based on the notion that drug binding renders a drug target less susceptible to proteolysis.[28] Hence, parallel protease treatment in the presence and absence of the drug of interest yields differential degradation of the putative drug-binding protein target. In proof-of-concept experiments, this approach has been shown to successfully reveal well-studied small molecule–protein interactions like FK506 and rapamycin to FKBP12. Moreover, Lomenick *et al.* could identify the translation initiation machinery as a novel target for the natural product resveratrol. A clear advantage of this approach is that it does not require the modification of the drug of interest. However, the general applicability of this approach is so far unclear as the masking of protease sites by small molecule binding and the thereby decreased proteolysis might not be common for all putative drug targets. Furthermore, the approach is limited by the sensitivity of downstream mass spectrometry to an even higher extent than chemical proteomics as the extent of enrichment of the target protein over the background, achieved via differential proteolysis, is unclear.

7.5 Other Methods

7.5.1 Yeast Three-Hybrid (Y3H)

The yeast three-hybrid approach is based on the yeast two-hybrid (Y2H) approach that was introduced by Chien *et al.* and represents an extension suitable for interrogation of cDNA libraries for gene products interacting with a certain bioactive molecule of interest.[29] The initial Y2H system was developed to investigate protein–protein interactions. Protein A is fused to a DNA-binding domain whereas the candidate interactor (Protein B) features a transcriptional activator fusion. Upon an interaction of A and B, the proximity of the DNA binding domain and the transcriptional activator domain gets reconstituted, resulting in the expression of a target gene, e.g. driving a resistance marker that can counteract a certain selection pressure.

Licitra *et al.* introduced, as mentioned above, a third component in order to be able to investigate the target protein(s) of a small molecule of interest.[30] To do so, they made use of well-studied and tight small molecule–protein interactions such as dexamethasone–glucocorticoid receptor or methotrexate–DHFR to anchor the compound of interest via e.g. methotrexate to DHFR that itself is fused to a DNA-binding domain. The query cDNA library is again fused to a transcriptional activator domain, thus a fruitful ligand–protein interaction drives, just as in the Y2H system, the expression of a reporter gene. In a proof-of-principle situation, Licitra *et al.* have isolated overlapping clones of human FKBP12 when probing a Jurkat cDNA library with a dexamethasone ('anchor') FK506 ('bait') hybrid construct linked to glucocorticoid receptor-LexA DNA-binding domain fusion protein.[30] The Y3H approach was further exploited by Becker *et al.* in order to identify novel protein–interaction partners of various cyclin-dependent kinase inhibitors.[31]

In comparison to Y2H, Y3H (Figure 7.4) has not really been extensively used by the scientific community. In fact, the findings of Becker *et al.* remain the only reported ones that identified novel interactions, independent of a proof-of-principle situation. Traditionally, Y2H-derived data suffer from relatively high background noise. This drawback holds similarly true for Y3H. Moreover, Y3H requires chemical modification of the small molecule of interest in order to create the hybrid molecule that allows anchoring. This rather cumbersome effort prevents the Y3H approach from being applied in a high-throughput manner, thus maybe explaining its relatively rare appearance in the literature.

7.5.2 Protein Microarrays

With the evolution of high-throughput protein expression and purification techniques, the adaptation of DNA microarray technology for analysis of protein–protein or protein–small molecule interactions has provided a chip-based alternative staged at the high level of complexity required for target deconvolution approaches. In a typical experiment, the purified proteins are printed on glass slides which have been coated with, for example, an aldehyde-containing silane reagent or BSA-*N*-hydroxysuccinimide. As lysine residues (as well as aspartate and glutamate residues in case of BSA-*N*-hydroxysuccinimide) on the protein surface readily react with those reactive groups, up to 10 800 proteins can easily be immobilized on a single microscope glass slide.[32] The multitude of available reactive groups on the protein surface also allows binding of the proteins in various orientations thus exposing different sides of the protein to potential interactors. After incubation with a fluorescent or radioisotope-labelled derivative of the small molecule, interacting proteins can be identified by their corresponding spots on the chip. Advantages of the protein microarray strategy are the fact that all proteins are interrogated

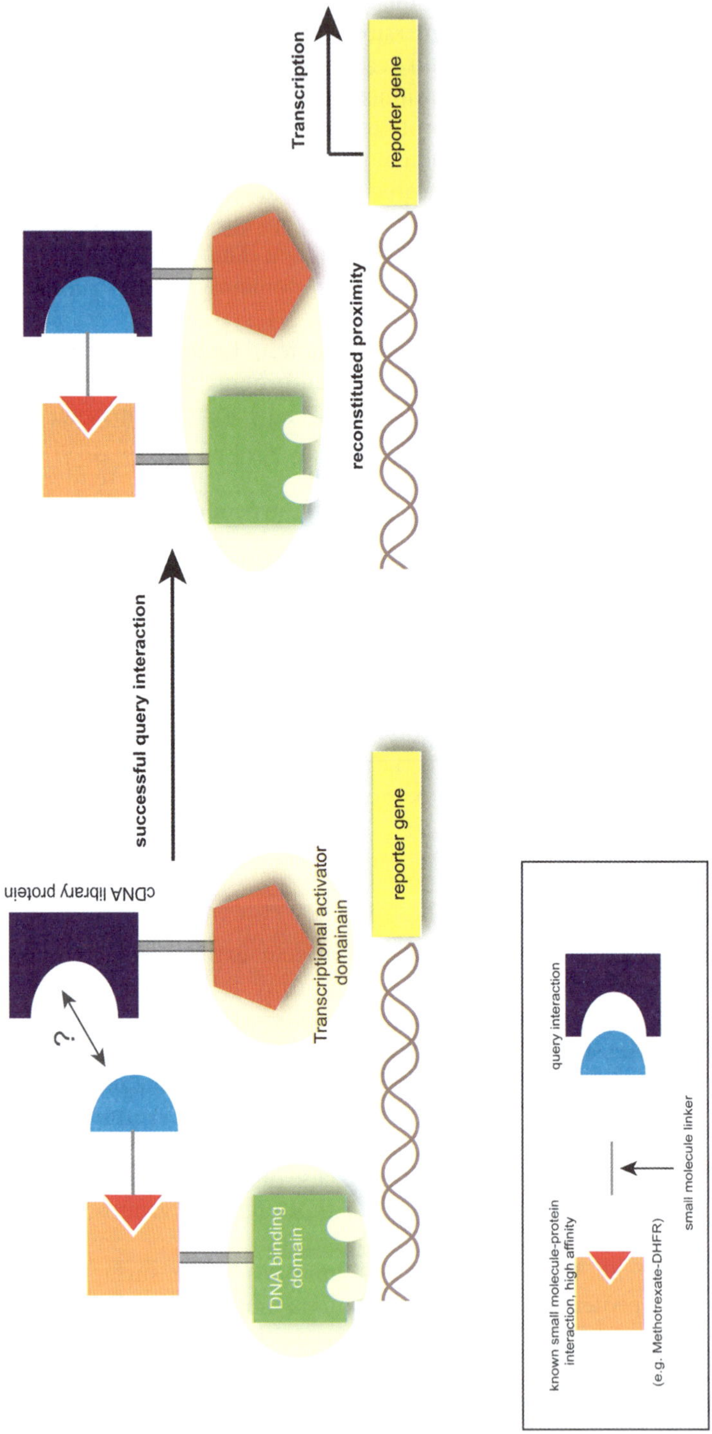

Figure 7.4 Yeast-three hybrid (Y3H).

simultaneously in a high density format and that naturally occurring differences in protein abundance are levelled out. This enables screening of an entire proteome in a few days and permits identification of lowly expressed protein targets. However, protein chips do not consider the state or changes of post-translational modifications as well as protein complex formation which may be essential for some interactions. An interesting study by Huang *et al.* exemplifying the use of protein microarrays for target deconvolution purposes, revealed yeast homologues of PTEN and Nir1p as players in the target-of-rapamycin (TOR) dependent nutrient response network.[33]

7.6 Conclusions

The research community has clearly abandoned its previous obsession with ultra-specific drugs. Too many times, these were selected to be highly specific through numerous counter-screens *in vitro* only to fail later in more physiological settings. On the other hand, multi-target, or 'pharmacological rich', drugs are increasingly being perceived as an opportunity rather than a threat. Much of this change has to do with a new understanding of systems-level action. The complete elimination of a node in the network, conceptually equal to strong and specific inhibition of a single target within a biological process, offers robust biological systems many ways to circumvent the roadblock.[34] In contrast, compounds or combinations thereof that simultaneously affect the activity of several components may affect the system in a more profound yet less toxic way. Ideally, these multi-target or 'systems drugs' can be designed from scratch at the drawing board. However, our still primitive overall understanding of what represents good chemical properties of a drug and our only rough appreciation of biological networks make this *de novo* approach highly challenging. The approaches described here, though passive analytical tools usually employed to already designed drugs, help decipher the real molecular impact of the action of individual drugs. The growth of this knowledge domain trains computer models, topological or statistical, with important new empirical data. It will be only a matter of years until our understanding of the systems-level action of drugs will start instructing a generation of new drugs that make systematic use of it. Then the cycle of planning and testing compounds in assays with high-content read-outs will be more efficient. There are reasons to hope that the new generation of multi-target drugs will lead the way to less toxicity and better specificity.

References

1. E. S. L. Chan and B. N. Cronstein, *Nat. Rev. Rheumatol.*, 2010, **6**, 175–178.
2. O. Hantschel, U. Rix, U. Schmidt, T. Burckstummer, M. Kneidinger, G. Schutze, J. Colinge, K. L. Bennett, W. Ellmeier, P. Valent and G. Superti-Furga, *Proc. Natl. Acad. Sci. USA*, 2007, **104**, 13283–13288.

3. E. A. Winzeler, D. D. Shoemaker, A. Astromoff, H. Liang, K. Anderson, B. Andre, R. Bangham, R. Benito, J. D. Boeke, H. Bussey, A. M. Chu, C. Connelly, K. Davis, F. Dietrich, S. W. Dow, M. El Bakkoury, F. Foury, S. H. Friend, E. Gentalen, G. Giaever, J. H. Hegemann, T. Jones, M. Laub, H. Liao, N. Liebundguth, D. J. Lockhart, A. Lucau-Danila, M. Lussier, N. M'Rabet, P. Menard, M. Mittmann, C. Pai, C. Rebischung, J. L. Revuelta, L. Riles, C. J. Roberts, P. Ross-MacDonald, B. Scherens, M. Snyder, S. Sookhai-Mahadeo, R. K. Storms, S. Veronneau, M. Voet, G. Volckaert, T. R. Ward, R. Wysocki, G. S. Yen, K. Yu, K. Zimmermann, P. Philippsen, M. Johnston and R. W. Davis, *Science*, 1999, **285**, 901–906.
4. S. J. Dixon and B. R. Stockwell, *Curr. Opin. Chem. Biol.*, 2009, **13**, 549–555.
5. T. R. Hughes, *Funct. Integr. Genomics*, 2002, **2**, 199–211.
6. (a) G. Giaever, D. D. Shoemaker, T. W. Jones, H. Liang, E. A. Winzeler, A. Astromoff and R. W. Davis, *Nat. Genet.*, 1999, **21**, 278–283; (b) G. Giaever, P. Flaherty, J. Kumm, M. Proctor, C. Nislow, D. F. Jaramillo, A. M. Chu, M. I. Jordan, A. P. Arkin and R. W. Davis, *Proc. Natl. Acad. Sci. USA*, 2004, **101**, 793–798.
7. Z. Yan, M. Costanzo, L. E. Heisler, J. Paw, F. Kaper, B. J. Andrews, C. Boone, G. Giaever and C. Nislow, *Nat. Methods*, 2008, **5**, 719–725.
8. A. M. Smith, R. Ammar, C. Nislow and G. Giaever, *Pharmacol. Ther.*, 2010, **127**, 156–164.
9. A. B. Parsons, R. L. Brost, H. Ding, Z. Li, C. Zhang, B. Sheikh, G. W. Brown, P. M. Kane, T. R. Hughes and C. Boone, *Nat. Biotechnol.*, 2004, **22**, 62–69.
10. (a) W. Lee, R. P. St Onge, M. Proctor, P. Flaherty, M. I. Jordan, A. P. Arkin, R. W. Davis, C. Nislow and G. Giaever, *PLoS Genet.*, 2005, **1**, e24; (b) L. Yu, A. Lopez, A. Anaflous, B. El Bali, A. Hamal, E. Ericson, L. E. Heisler, A. McQuibban, G. Giaever, C. Nislow, C. Boone, G. W. Brown and M. Bellaoui, *PLoS Genet.*, 2008, **4**, e1000284.
11. T. R. Hughes, M. J. Marton, A. R. Jones, C. J. Roberts, R. Stoughton, C. D. Armour, H. A. Bennett, E. Coffey, H. Dai, Y. D. He, M. J. Kidd, A. M. King, M. R. Meyer, D. Slade, P. Y. Lum, S. B. Stepaniants, D. D. Shoemaker, D. Gachotte, K. Chakraburtty, J. Simon, M. Bard and S. H. Friend, *Cell*, 2000, **102**, 109–126.
12. J. Lamb, E. D. Crawford, D. Peck, J. W. Modell, I. C. Blat, M. J. Wrobel, J. Lerner, J. P. Brunet, A. Subramanian, K. N. Ross, M. Reich, H. Hieronymus, G. Wei, S. A. Armstrong, S. J. Haggarty, P. A. Clemons, R. Wei, S. A. Carr, E. S. Lander and T. R. Golub, *Science*, 2006, **313**, 1929–1935.
13. H. Hieronymus, J. Lamb, K. N. Ross, X. P. Peng, C. Clement, A. Rodina, M. Nieto, J. Du, K. Stegmaier, S. M. Raj, K. N. Maloney, J. Clardy, W. C. Hahn, G. Chiosis and T. R. Golub, *Cancer Cell*, 2006, **10**, 321–330.
14. (a) T. R. Brummelkamp, A. W. Fabius, J. Mullenders, M. Madiredjo, A. Velds, R. M. Kerkhoven, R. Bernards and R. L. Beijersbergen, *Nat. Chem. Biol.*, 2006, **2**, 202–206; (b) K. Berns, H. M. Horlings, B. T. Hennessy,

M. Madiredjo, E. M. Hijmans, K. Beelen, S. C. Linn, A. M. Gonzalez-Angulo, K. Stemke-Hale, M. Hauptmann, R. L. Beijersbergen, G. B. Mills, M. J. van de Vijver and R. Bernards, *Cancer Cell*, 2007, **12**, 395–402.
15. H. Jiang, J. R. Pritchard, R. T. Williams, D. A. Lauffenburger and M. T. Hemann, *Nat. Chem. Biol.*, 2011, **7**, 92–100.
16. C. K. Hahn, J. E. Berchuck, K. N. Ross, R. M. Kakoza, K. Clauser, A. C. Schinzel, L. Ross, I. Galinsky, T. N. Davis, S. J. Silver, D. E. Root, R. M. Stone, D. J. DeAngelo, M. Carroll, W. C. Hahn, S. A. Carr, T. R. Golub, A. L. Kung and K. Stegmaier, *Cancer Cell*, 2009, **16**, 281–294.
17. U. Rix and G. Superti-Furga, *Nat. Chem. Biol.*, 2009, **5**, 616–624.
18. S.-E. Ong, M. Schenone, A. A. Margolin, X. Li, K. Do, M. K. Doud, D. R. Mani, L. Kuai, X. Wang, J. L. Wood, N. J. Tolliday, A. N. Koehler, L. A. Marcaurelle, T. R. Golub, R. J. Gould, S. L. Schreiber and S. A. Carr, *Proc. Natl. Acad. Sci. USA*, 2009, **106**, 4617–4622.
19. T. Ito, H. Ando, T. Suzuki, T. Ogura, K. Hotta, Y. Imamura, Y. Yamaguchi and H. Handa, *Science*, 2010, **327**, 1345–1350.
20. U. Rix, O. Hantschel, G. Durnberger, L. L. Remsing Rix, M. Planyavsky, N. V. Fernbach, I. Kaupe, K. L. Bennett, P. Valent, J. Colinge, T. Kocher and G. Superti-Furga, *Blood*, 2007, **110**, 4055–4063.
21. M. Bantscheff, D. Eberhard, Y. Abraham, S. Bastuck, M. Boesche, S. Hobson, T. Mathieson, J. Perrin, M. Raida, C. Rau, V. Reader, G. Sweetman, A. Bauer, T. Bouwmeester, C. Hopf, U. Kruse, G. Neubauer, N. Ramsden, J. Rick, B. Kuster and G. Drewes, *Nat. Biotechnol.*, 2007, **25**, 1035–1044.
22. M. Bantscheff, C. Hopf, M. M. Savitski, A. Dittmann, P. Grandi, A.-M. Michon, J. Schlegl, Y. Abraham, I. Becher, G. Bergamini, M. Boesche, M. Delling, B. Dumpelfeld, D. Eberhard, C. Huthmacher, T. Mathieson, D. Poeckel, V. Reader, K. Strunk, G. Sweetman, U. Kruse, G. Neubauer, N. G. Ramsden and G. Drewes, *Nat. Biotechnol.*, 2011, **29**, 255–265.
23. D. A. Bachovchin, S. J. Brown, H. Rosen and B. F. Cravatt, *Nat. Biotechnol.*, 2009, **27**, 387–394.
24. (a) K. P. Chiang, S. Niessen, A. Saghatelian and B. F. Cravatt, *Chem. Biol.*, 2006, **13**, 1041–1050; (b) J. Z. Long, W. Li, L. Booker, J. J. Burston, S. G. Kinsey, J. E. Schlosburg, F. J. Pavon, A. M. Serrano, D. E. Selley, L. H. Parsons, A. H. Lichtman and B. F. Cravatt, *Nat. Chem. Biol.*, 2009, **5**, 37–44; (c) D. K. Nomura, J. Z. Long, S. Niessen, H. S. Hoover, S.-W. Ng and B. F. Cravatt, *Cell*, 2010, **140**, 49–61.
25. J. Li, U. Rix, B. Fang, Y. Bai, A. Edwards, J. Colinge, K. L. Bennett, J. Gao, L. Song, S. Eschrich, G. Superti-Furga, J. Koomen and E. B. Haura, *Nat. Chem. Biol.*, 2010, **6**, 291–299.
26. C. Pan, J. V. Olsen, H. Daub and M. Mann, *Mol. Cell. Proteomics*, 2009, **8**, 2796–2808.
27. C. Choudhary, C. Kumar, F. Gnad, M. L. Nielsen, M. Rehman, T. C. Walther, J. V. Olsen and M. Mann, *Science*, 2009, **325**, 834–840.
28. B. Lomenick, R. Hao, N. Jonai, R. M. Chin, M. Aghajan, S. Warburton, J. Wang, R. P. Wu, F. Gomez, J. A. Loo, J. A. Wohlschlegel, T. M. Vondriska,

J. Pelletier, H. R. Herschman, J. Clardy, C. F. Clarke and J. Huang, *Proc. Natl. Acad. Sci. USA*, 2009, **106**, 21984–21989.
29. C. T. Chien, P. L. Bartel, R. Sternglanz and S. Fields, *Proc. Natl. Acad. Sci. USA*, 1991, **88**, 9578–9582.
30. E. J. Licitra and J. O. Liu, *Proc. Natl. Acad. Sci. USA*, 1996, **93**, 12817–12821.
31. F. Becker, K. Murthi, C. Smith, J. Come, N. Costa-Roldan, C. Kaufmann, U. Hanke, C. Degenhart, S. Baumann, W. Wallner, A. Huber, S. Dedier, S. Dill, D. Kinsman, M. Hediger, N. Bockovich, S. Meier-Ewert, A. F. Kluge and N. Kley, *Chem. Biol.*, 2004, **11**, 211–223.
32. G. MacBeath and S. L. Schreiber, *Science*, 2000, **289**, 1760–1763.
33. J. Huang, H. Zhu, S. J. Haggarty, D. R. Spring, H. Hwang, F. Jin, M. Snyder and S. L. Schreiber, *Proc. Natl. Acad. Sci. USA*, 2004, **101**, 16594–16599.
34. H. Kitano, *Science*, 2002, **295**, 1662–1664.

CHAPTER 8

Historical Strategies for Lead Generation

J. RICHARD MORPHY[†]

Stirling, UK
Email: richard.morphy@spcorp.com

8.1 Introduction

Multi-target drug discovery (MTDD) is based on the premise that compounds interacting with more than one target may have superior efficacy or safety. Designed multiple ligands (DMLs) are compounds that are prospectively and intentionally designed to interact specifically with multiple targets. The choices made at the start of the lead generation phase are crucial to the overall chance of success of any MTDD project, in terms of both the choice of lead generation approach and ultimately the choice of starting compound. Similarly to single target-focussed projects, there are two lead generation approaches to MTDD that medicinal chemists have historically followed – screening approaches and knowledge-based approaches. Screening can take the form of either large-scale high throughput screening of many thousands of compounds or focussed screening of smaller sets of more carefully selected compounds. There are also emerging approaches to MTDD lead generation that show much promise, including fragment-based and *in silico* screening. Both the historical and the emerging approaches are described in the sections below, starting with the well-established knowledge-based approach known as framework combination.

[†]Current address: Lilly Research Centre, Windlesham, Surrey, GU20 9PH, UK.

RSC Drug Discovery Series No. 21
Designing Multi-Target Drugs
Edited by J. Richard Morphy and C. John Harris

Published by the Royal Society of Chemistry, www.rsc.org

8.2 Historical Approaches

8.2.1 Framework Combination

The knowledge-based lead generation approach termed 'framework combination' integrates the frameworks and underlying pharmacophores of two molecules, each selective for a different target of interest, into a single molecule with dual activity. Framework combination relies upon SAR knowledge from previous single target drug discovery projects, which can be used to guide this integration process. DMLs arising from framework combination can be viewed as 'linked', 'fused' or 'merged' depending upon the extent to which the frameworks have been overlapped (Figure 8.1). In linked DMLs (also known as conjugates), the molecular frameworks are separated by a distinct linker group between the two components. Different regions of the molecule are typically responsible for the activity at the different targets. If the frameworks are essentially in contact without a discernable linker or much framework overlap, the DML can be viewed as fused. The frameworks in merged DMLs are overlapped by taking advantage of structural commonalities in the pharmacophores of the targets. In reality, the degree of overlap forms a continuum, with high molecular weight (MW) DMLs with lengthy linker groups at one extreme and small DMLs with highly merged frameworks at the other.

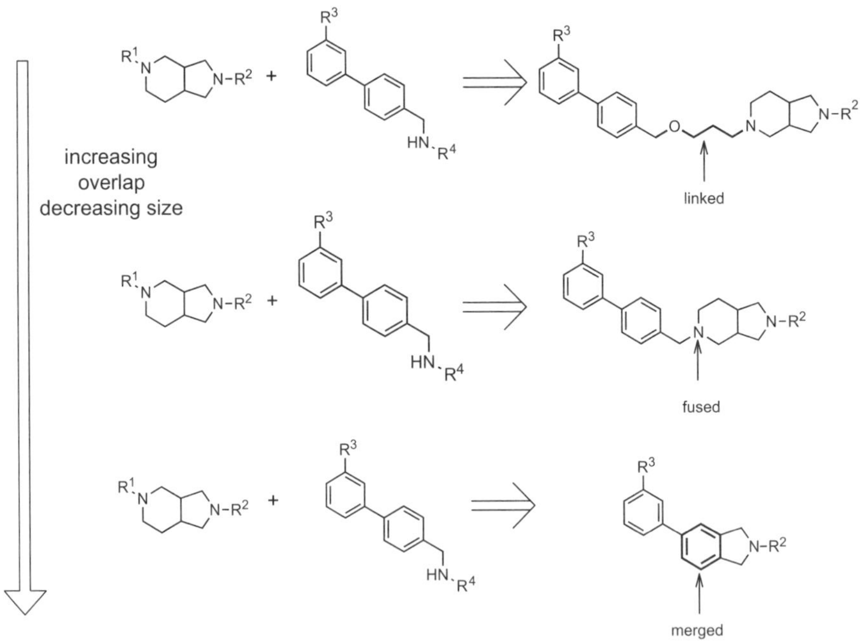

Figure 8.1 Framework combination is a knowledge-based approach to generate a dual ligand by integrating the frameworks of two molecules, each selective for a different target of interest. The dual ligands can be classified as linked, fused or merged depending upon the degree of integration of the pharmacophores.

Normally, medicinal chemists will aspire to maximize the degree of overlap in order to produce smaller and simpler molecules, which have a better chance of oral activity. However achieving this high degree of overlap will necessitate a high level of similarity of the two pharmacophores. There are a large number of literature examples of the framework combination approach being applied across a broad range of proteomic families like GPCRs, transporters, nuclear receptors, proteases and oxidases, for which extensive knowledge of selective ligand SARs allows the rational 'designing in' of multiple activities.

One example of a linked DML is the dual β2-agonist/muscarinic antagonist, GSK-961081 **1**, which has progressed to phase 2 trials as an inhaled treatment for chronic obstructive pulmonary disease (COPD) (Figure 8.2).[1] The large size of the molecule is in this case an advantage since the drug is designed to be administered not by an oral route but via an intranasal route. Likewise, linked DMLs have also been administered by other parenteral routes, such as an anti-thrombotic pentasaccharide drug given intravenously.[2]

Lowe *et al.* at Pfizer describe how a D_2 receptor ligand, dopamine itself **2**, was fused to the 5-HT ligand **3**, transforming the D_2-agonist activity of the endogenous ligand into an antagonist as well as providing dual $5HT_2$ activity (Figure 8.3).[3] The only overlapped atom is a basic nitrogen common to both the D_2 and $5HT_2$ pharmacophores. Further optimization of this lead compound provided the marketed drug Ziprasidone, which was launched in 2001 by Pfizer for the treatment of schizophrenia, as described in Chapter 10.

In Chapter 14, Youdim and colleagues describe a merged DML, the dual AChE/MAO-B inhibitor Ladostigil **7**, as a potential treatment for Alzheimer's disease and Parkinson's disease.[4] The structural frameworks of the Alzheimer's drug Rivastigmine **5** and the Parkinson's drug, Rasagiline **6**, were combined to produce the dual inhibitor (Figure 8.4). Reported SAR around this compound indicates that carbamate and propargylamine groups are key pharmacophoric elements responsible for the AChE and brain MAO inhibition, respectively. Ladostigil has shown efficacy in rhesus monkey cognition and neuroprotection models.[5] In Chapter 9, Murugesan and colleagues describe the discovery of

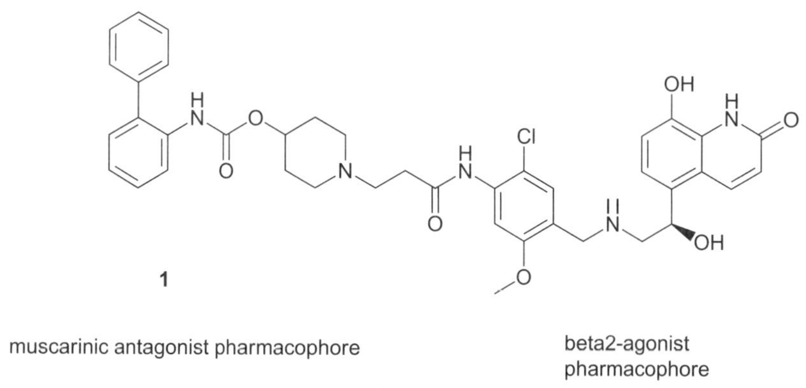

Figure 8.2 An example of a linked DML.

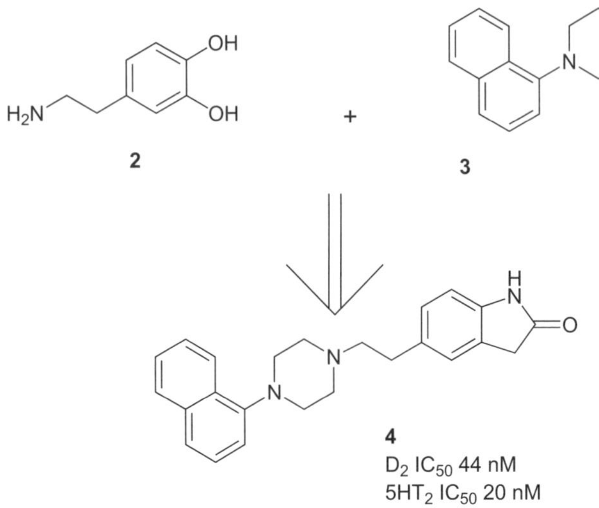

Figure 8.3 An example of a fused DML.

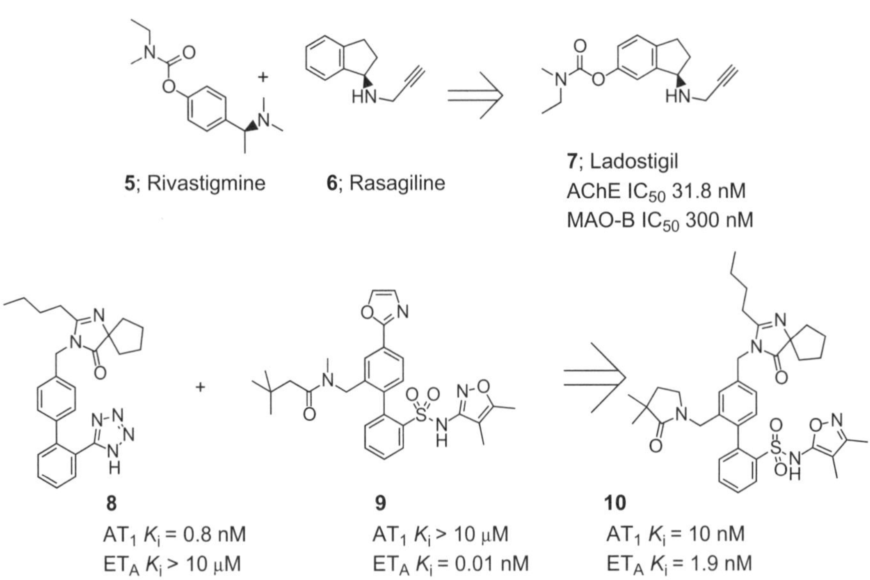

Figure 8.4 Examples of merged DMLs.

another merged DML, an AT_1/ET_A antagonist. The dual antagonist **10** was designed by overlapping the biaryl core found in both the selective AT_1 and ET_A antagonists, **8** and **9** respectively (Figure 8.4).[6]

Activity against certain targets can be highly dependent upon the presence of key groups and the introduction of such groups into known ligands for other

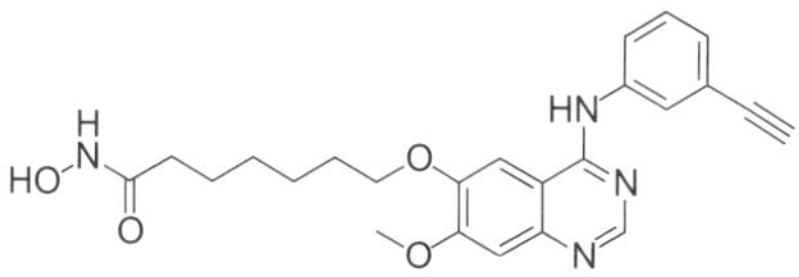

11

EGFR	IC_{50}	2.4 nM
erbB2	IC_{50}	15.7 nM
HDAC	IC_{50}	4.4 nM

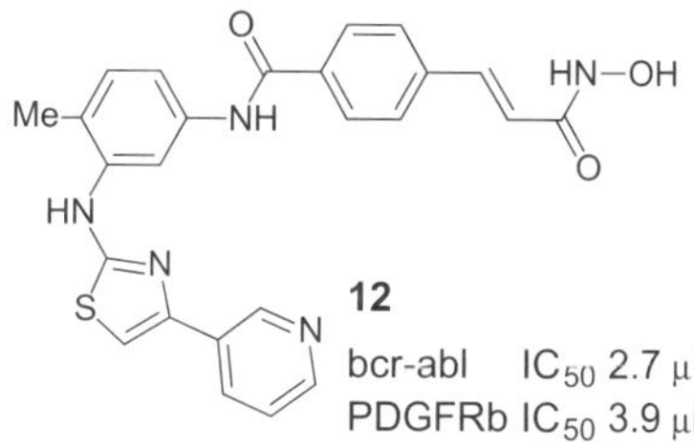

12

bcr-abl	IC_{50}	2.7 μM
PDGFRb	IC_{50}	3.9 μM
HDAC	IC_{50}	0.080 μM

Figure 8.5 Hydroxamic acid-containing HDAC/kinase inhibitors.

targets can often be enough to confer multi-target activity. For example, inhibition of zinc-containing enzymes can often be achieved by the introduction of a chelating group such as a hydroxamic acid. In Chapter 7, kinase inhibitors are described that also inhibit histone deacetylase (HDAC) enzymes (Figure 8.5). A triple inhibitor **11** of HDAC, EGFR and erbB2 has been reported with low nanomolar activity against all 3 enzymes.[7] In a similar fashion, compound **12** was obtained by adding a hydroxamic acid warhead to a marketed kinase inhibitor, imatinib, to provide an inhibitor of bcr-abl, PDGFRβ and HDAC.

8.2.2 Screening

Whilst framework combination has been historically popular, the screening of either diverse or focussed compound collections has now become the principal method of MTDD hit generation. Using modern high throughput screening (HTS) assays, it is perfectly possible to screen large, diverse compound sets against several targets of interest in parallel. However, it is common to screen first at one target of interest and then filter actives on the basis of activity at the other target(s).

One example of a DML derived from HTS is the dual SERT/NK_1 antagonist **13**, a profile of interest for treating depression (Figure 8.6).[8] Optimization of potency was achieved by subtle variations to the aromatic substituents. The ability to make such small changes during optimization highlights the value of having a high quality hit. Implementation of a HTS-based strategy for the discovery of a dual SERT/$5HT_{1A}$ blocker provided a high potency hit **15**.[9] HTS was performed using a 5-HT_{1A}-expressing cell line and actives were then triaged by testing for SERT blockade (Figure 8.6).

In focussed screening, smaller numbers of compounds that are already known to be active against one of the targets of interest are screened against another target. The enduring popularity of focussed screening as a lead generation method for MTDD could be due to a much higher probability of detecting a compound with the multiple profile of interest compared to screening compounds at random using HTS. Focussed screening has been

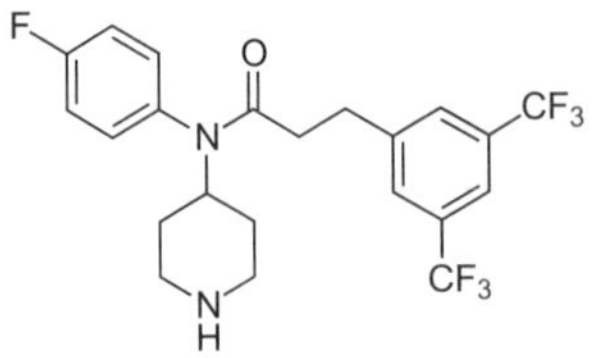

13
NK_1 pK_i 6.7
SERT pK_i 6.6

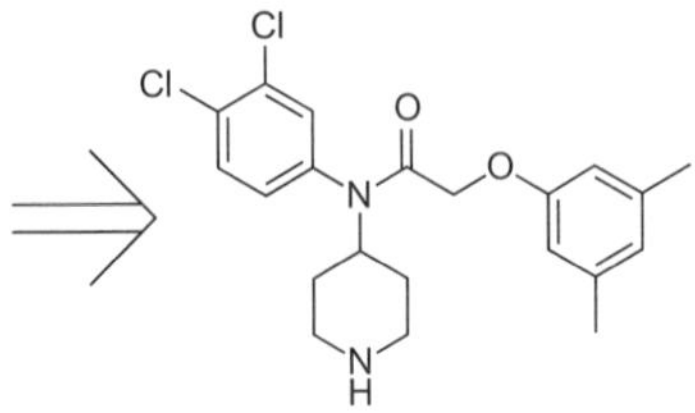

14
NK_1 pK_i 7.6
SERT pK_i 7.5

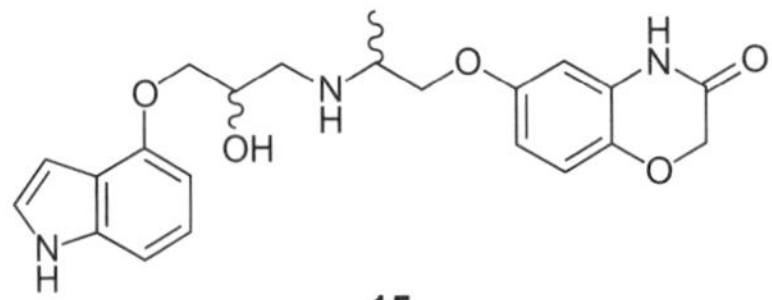

15
$5HT_{1A}$ pK_i 9.1
SERT pK_i 7.3
β2 pK_i 9.2

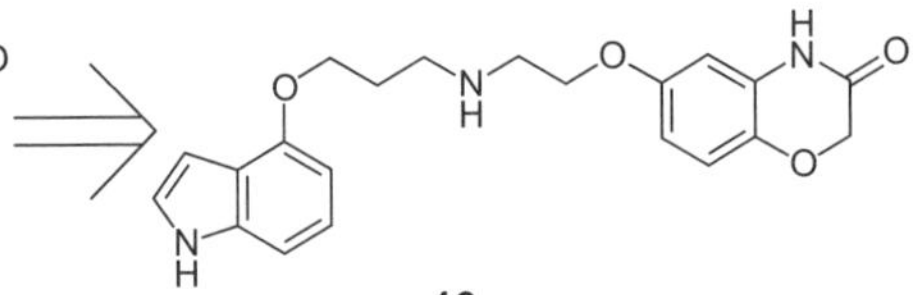

16
$5HT_{1A}$ pK_i 9.5
SERT pK_i 8.2
β2 pK_i <6.3

Figure 8.6 HTS-derived MTDD projects.

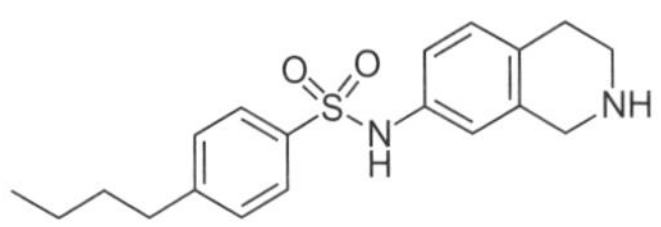

17
D_2 pK_i 6.0
D_3 pK_i 8.0
$5HT_{2A}$ pK_i 7.5
$5HT_{2C}$ pK_i 7.9
$5HT_6$ pK_i 7.6

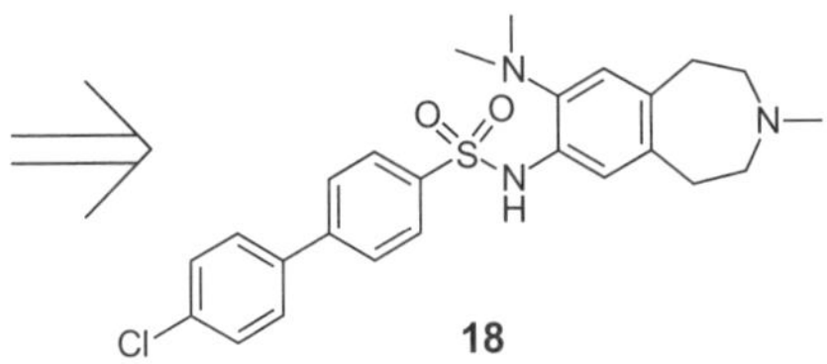

18
D_2 pK_i 7.3
D_3 pK_i 8.5
$5HT_{2A}$ pK_i 8.8
$5HT_{2C}$ pK_i 8.3
$5HT_6$ pK_i 8.1

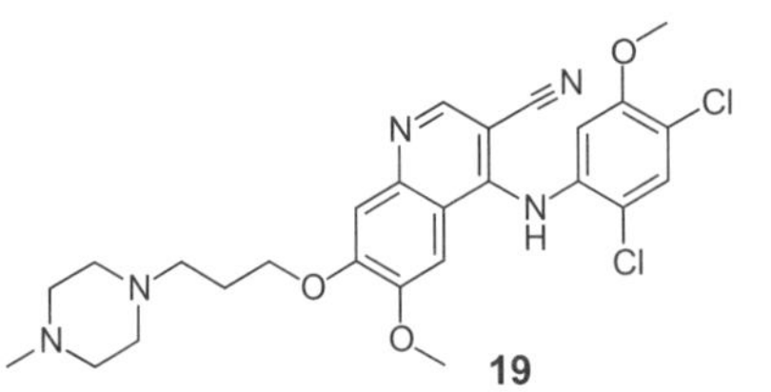

19
Src IC_{50} 3.8 nM
Abl IC_{50} 1.1 nM

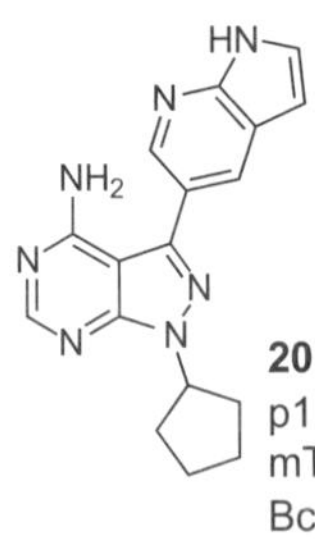

20
p110α IC_{50} 52 nM
mTOR IC_{50} 10 nM
Bcr-abl IC_{50} 18 nM
Src IC_{50} 14 nM
VEGFR2 IC_{50} 12 nM
PDGFR IC_{50} 2 nM

Figure 8.7 Focussed screening-derived MTDD projects.

particularly common for target classes with well-known cross-reactivities such as aminergic GPCRs and protein kinases.

A hit **17** (Figure 8.7) with activity at 5-aminergic GPCRs of interest for schizophrenia was discovered by focussed screening at GSK and optimized to give a more potent multi-target profile.[10]

DMLs for kinase targets are usually discovered serendipitously through cross-screening of ligands from selective kinase programmes against other kinases. The big advantage of cross-screening is the high hit rate due to the high similarity of the ATP binding site between kinases. The value of focussed screening for multi-kinase inhibitors is illustrated by the development of the dual ABL/SRC inhibitors starting from compounds designed as selective SRC inhibitors. For example, Boschelli *et al.* identified the potent dual inhibitor ABL/SRC Bosutinib **19** (Figure 8.7).[11] By screening a library of pyrazolopyrimidines, Apsel *et al.* identified a compound **20** that potently inhibited both protein and lipid kinases (Figure 8.7).[12]

8.2.3 Comparing Lead Generation Approaches

The screening and framework combination approaches to lead generation each have characteristic advantages and disadvantages that influence which one is best applied to a particular project (Table 8.1).

Table 8.1 Characteristic features of the screening and framework combination approaches.

Screening approaches	*Framework combination approach*
• Hit compound already has multiple activity, albeit probably sub-optimal.	• May be difficult to incorporate a 2nd activity whilst retaining the 1st activity and attaining acceptable physicochemical properties.
• Can provide complex multi-target profiles unavailable by framework combination.	• Typically only applicable to dual ligands.
• Can be employed if there is a lack of selective ligands and associated SAR information required for framework combination.	• Can be employed where selective ligands are known for each target, especially when extensive SAR information is available.
• Can deliver novel and unexpected chemotypes with good patentability, even for unusual target combinations.	• This knowledge-based approach tends to produce chemotypes related to those already known.
• The chance of success is generally low for unrelated targets where the pharmacophores are distinct.	• Linking together two independent ligands can provide DMLs for two targets with incompatible pharmacophores.
• Screening, particularly fragment screening, is more likely than framework combination to provide drug-like ligands with good oral pharmacokinetic properties.	• DMLs from framework combination tend to be larger and may have poor oral pharmacokinetics. Linked DMLs may still be useful as non-oral drugs or biochemical tools.

8.3 Emerging Approaches

8.3.1 Fragment Approach

Fragment-based drug discovery (FBDD) has emerged over recent years as a vehicle for providing compounds with high 'ligand efficiency' in terms of their binding energy per heavy atom.[13] Given the physicochemical issues in MTDD (see Section 5.3), the FBDD approach has been proposed as a potentially attractive route to more drug-like structures.[14] It would be expected from the molecular complexity model of Hann *et al.* that smaller ligands are more likely to bind to multiple targets than larger ones.[15] Less complex molecules are more likely to bind to multiple proteins due to the lower probability of a mismatch between ligand and one or more of the proteins of interest. Moreover, there is substantial experimental evidence for a higher hit rate for fragments in multi-target screening compared to the larger compounds typically screened in HTS.[14,16,17] The biophysical or biochemical methods that have been applied to single-target FBDD can also be considered for MTDD, for example NMR, X-ray crystallography, mass spectrometry, surface plasmon resonance (SPR) or high concentration screening. Following identification of a basic core scaffold capable of binding to multiple targets (Figure 8.8), subsequent 'growing' of the fragment could provide a ligand with higher affinity at each target whilst maintaining high ligand efficiency. Whilst it may be perfectly possible to identify a multi-target fragment hit, the challenge with the FBDD approach to MTDD will be the extent to which affinity can be improved and appropriately balanced for each target. As with single-target FBDD, biostructural information for one or preferably for all of the targets will be invaluable to guide the optimization process. If the targets are too dissimilar, it may not be possible to grow the fragment in a way that simultaneously improves affinity at each target. However, the extent to which affinity must be improved may be lower for a DML than for a target-selective agent if strong pharmacodynamic synergy exists between the targets.[14]

A recent report describes a fragment screening approach to pan-PPAR inhibitors.[18] Despite the superficial similarity between the PPAR nuclear receptors it has not been trivial to design pan-PPAR inhibitors even with the availability of crystallographic information. The indole fragment **23** was obtained by conducting a high concentration fragment screen and was found to bind weakly to all three PPARs (Figure 8.9). An X-ray structure of **23** bound to the ligand binding domain of PPARγ indicated an unoccupied pocket that was common to all three receptors. By extending into this pocket using an aryl

Figure 8.8 A fragment hit is grown to fit optimally to two target binding sites.

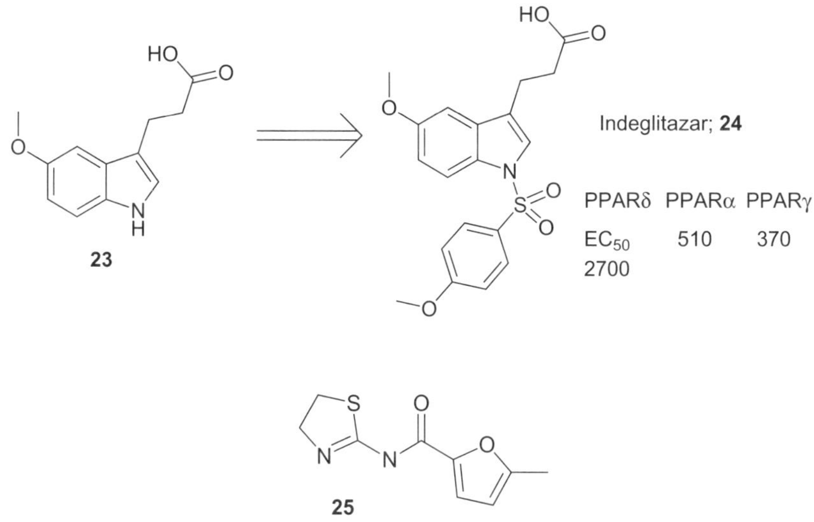

Figure 8.9 Fragment approaches to MTDD.

sulfonamide motif on the indole nitrogen, activity was substantially improved to give the clinical candidate, indeglitazar **24**.

An important question for the future is whether it is possible to apply the fragment approach to more distantly related targets from different target families. At the University of York, a STD NMR fragment screen was conducted against two very different enzyme targets, JNK3 kinase and β-glucosidase.[19] A fragment **25** (Figure 8.9) was found that bound to both these targets at a concentration of 100 μM and binding was confirmed crystallographically. This work is one of the first indications of the potentially wide scope of applicability of FBBD in MTDD.

8.3.2 Virtual Screening

Another increasingly popular approach to multi-target lead generation is virtual screening. Increasingly complex multi-step protocols mixing ligand-based and structure-based methods are being employed. For example, Knox *et al.* used a hybrid protocol to search for a dual Hsp90/tubulin inhibitor.[20] The first step employed a Bayesian model, trained with known Hsp90 ligands, and the hits were then docked into the X-ray structure of Hsp90. Finally the putative Hsp90 actives were matched to a tubulin pharmacophore model. Experimental testing then resulted in a compound with the desired dual activity. Virtual screening will become an ever more important means of identifying multi-target hits. Currently ligand-based virtual screening is generally regarded as a more robust multi-target hit generation method than the structure-based methods that suffer from lower throughput and poor consideration of solvation and conformational flexibility. Future developments should improve the speed and fidelity of docking-based methods, including conformational flexibility in both ligand and protein, since it is known that even closely related targets can adopt different conformations in binding the same ligand[21] and the same ligand can bind to different targets in very different conformations.[22] More details on such approaches can be found in Chapter 9.

8.3.3 Structure Guided Approaches

Increasingly biostructural information is being used to guide the generation of multi-target lead compounds. Bruncko *et al.* started from a selective Bcl-xL inhibitor **26** and used a structure-based approach to enhance Bcl-2 activity (Figure 8.10).[23] A hydrophobic biaryl group was added (**27**) which penetrated deeply into a groove on the surface of Bcl-2 thereby enhancing activity by more than 200-fold. Unexpectedly the same substituent also enhanced Bcl-xL activity, albeit to a lesser extent. This was rationalized on the basis of the crystallography that showed that there was a large conformational change in the Bcl-xL structure upon inhibitor binding. This induced fit involved the opening up of a similar groove to that observed for Bcl-2. As this example illustrates, access to three-dimensional protein structures enables a more rational approach to both the lead generation and optimization of DMLs. For more details of this programme, see Chapter 15.

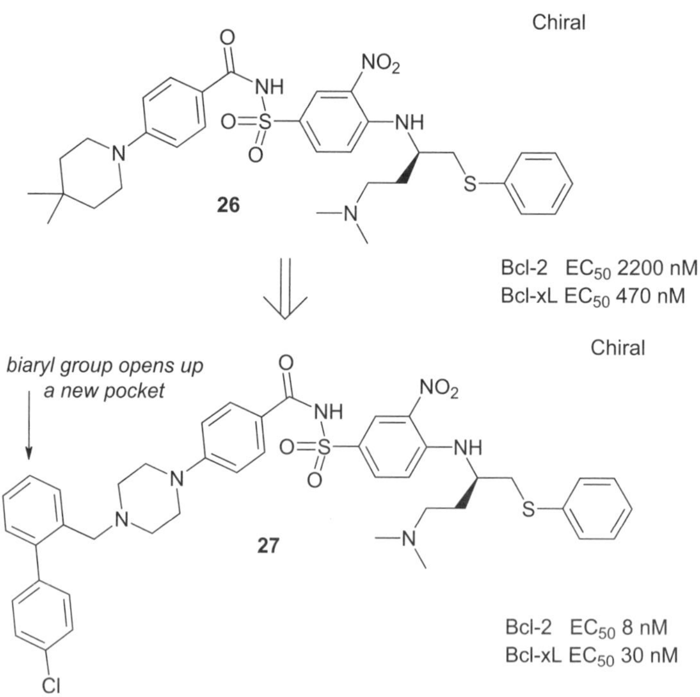

Figure 8.10 Structure-guided multi-target optimization.

8.3.4 Natural Products

Medicinal plants have been used as sources of traditional therapies for thousands of years. Their therapeutic efficacy may be due to a combination of individual ingredients and/or individual compounds with multi-target activity. Many examples of the multi-target compounds are known, for example resveratrol and curcumin are known to interact with large numbers of targets.[24,25] It has been suggested that natural products are more predisposed to exhibit multi-target activity than many synthetic compounds and therefore are potentially very useful starting points for MTDD projects.[26] For example, cinnamophilin was found to be a dual inhibitor of the thromboxane A_2 receptor and the thromboxane synthase enzyme.[27] For an alternative view see Chapter 6, Section 6.2.

8.4 Chemical Biology

The development of molecular probes that are selective for individual proteins is an area of intense current interest, especially in academic laboratories. By virtue of their ability to potently and selectively modulate proteins, suitable tool compounds can be used to explore their biological function *in vitro* and *in vivo*. Moreover, such compounds are often useful for validating targets for drug discovery programmes. Similarly, the development of high quality pharmacological tools to explore the potential therapeutic value of novel target combinations is an important future goal for academic research in this field.[28] In one such example, Portoghese *et al.* reported a range of linked DMLs, with variable length oligoglycyl-based linkers, designed to investigate the role of opioid heterodimers in pain.[29] Ligands, such as **28** (Figure 8.11), containing a delta-antagonist pharmacophore tethered to a kappa-agonist pharmacophore possessed significantly greater potency and selectivity compared to the individual opioids. With molecular weights well in excess of 1000, clearly such compounds do not possess the physicochemical properties commonly associated with oral bioavailability. However, in the case of probe development, the wider selectivity profile required for the validation of target combinations will be more

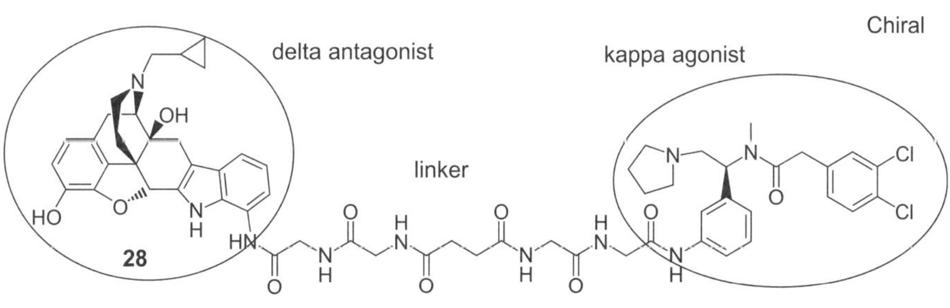

Figure 8.11 Linked DML as a tool compound for exploring opioid receptor heterodimerization.

important than oral developability criteria. Peptide DML tools have also been reported for a number of combinations, including bifunctional compounds, acting as opioid agonists and NK1 antagonists,[30] and trifunctional ligands targeting opioid, cholecystokinin and melanocortin receptors.[31]

8.5 Factors Influencing the Feasibility of MTDD

A study of the historical DML literature indicates that some targets feature prominently, which is likely due to a combination of perceived disease relevance and medicinal chemistry tractability.[32] Combinations involving the serotonin transporter (SERT) have been particularly popular (Figure 8.12). SERT inhibition has been combined with inhibition of one or both of the noradrenaline and dopamine transporters **29**[33] or **30**,[34] with antagonism of the histamine H_3 receptor **31**,[35] inhibition of enzymes such as acetylcholinesterase **32**[36] or phosphodiesterase4 **33**[37] or peptide GPCRs **14**/**34** (Figure 8.13).[8,38] The popularity of SERT combinations stems partly from the role of SERT in a range of CNS diseases, especially psychiatry, but it also appears from a medicinal chemistry perspective to be relatively feasible to combine SERT inhibition with activities against other targets. The binding site typically occupied by SERT ligands tolerates a wide diversity of chemotypes and this flexibility makes the design task faced by medicinal chemists a little easier than for many other combinations. SAR flexibility is thus an important concept for medicinal chemists engaged in MTDD and seems to apply to some extent to monoamine-binding targets in general.

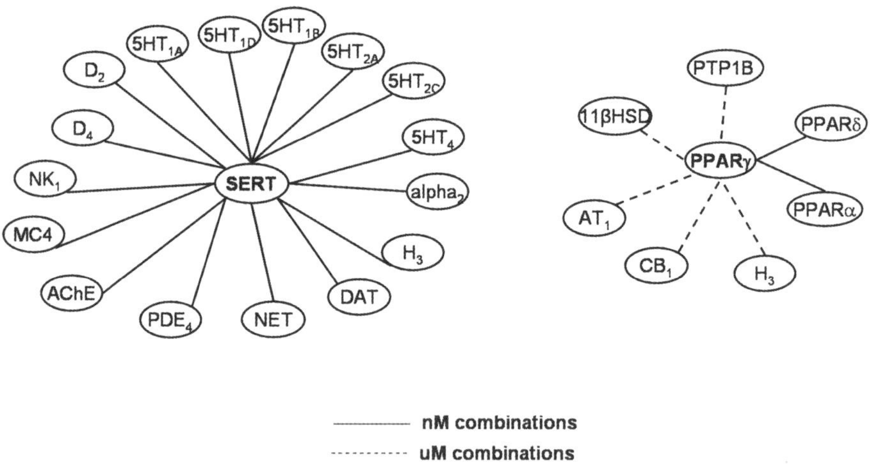

Figure 8.12 (A) Many compounds have been designed that inhibit the serotonin transporter (SERT) and another target with nanomolar affinity. (B) In contrast it has proven much more challenging to combine potent PPARγ agonism with other activities and, except in the case of the closely related PPARα and PPARδ receptors, the 'micromolar barrier' was not overcome.

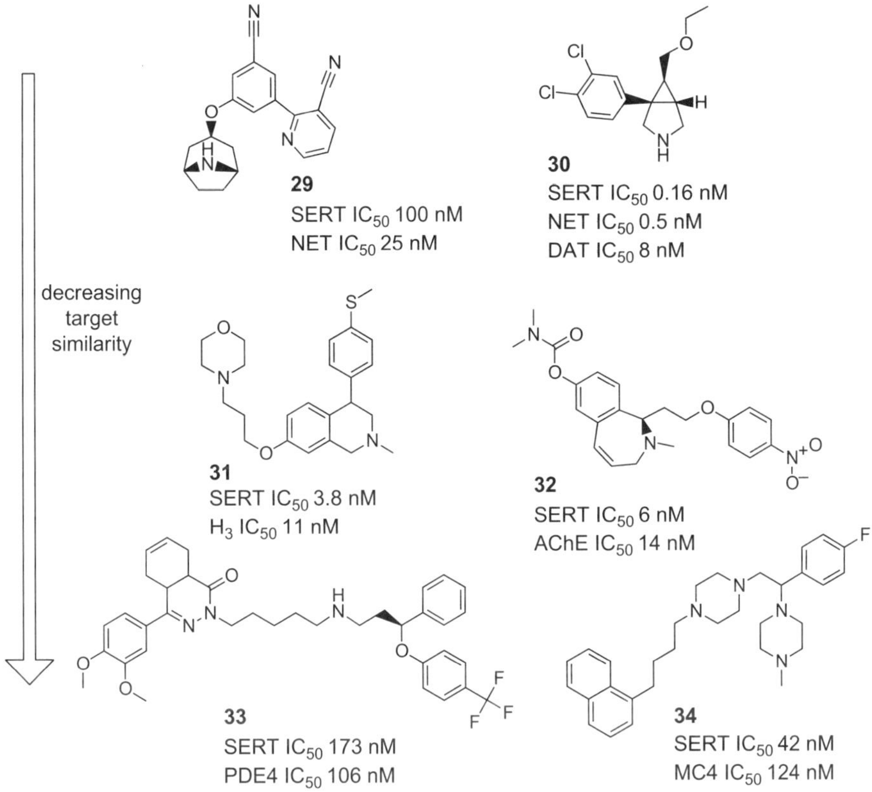

Figure 8.13 Serotonin transporter (SERT) combinations.

Extensive efforts have been made to combine agonism of the PPARγ receptor with other activities. Again this is due to the clinical validation of this target for treating a disease (diabetes) and a desire to enhance efficacy and mitigate side effects. Many attempts have been made to add other activities, such as inhibition of enzymes such as 11βHSD,[39] γ-secretase[40] and PTP1B[41] and antagonism of GPCRs, the H_3 receptor,[42] the AT1 receptor[43] and the CB1 receptor[44] (Figure 8.14). These efforts have generally met with much less success than for the SERT combinations above. In some cases, it did not prove possible to introduce the second activity at all and in most others, only very weak activity at the second target was obtained. In fact the only PPAR combinations where submicromolar activity was obtained for all targets was the combination of PPARγ with other PPAR receptors, as in the case of aleglitazar **36** (Figure 8.14).[45]

One conclusion from these observations is that whereas SAR flexibility for SERT inhibition is high, it is rather low for PPARγ agonism and as a consequence medicinal chemists have struggled. Why is this? In several cases, targets with incompatible pharmacophores appear to have been chosen. Trying to combine a target that binds a polar amine with another that binds a

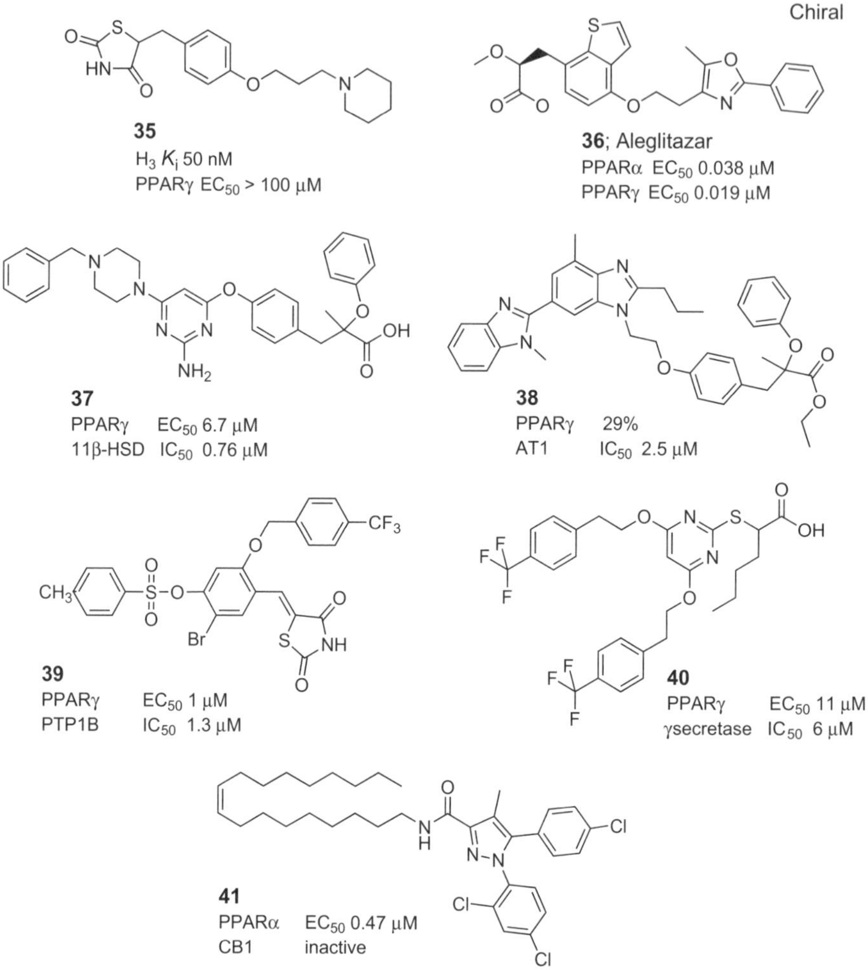

Figure 8.14 PPARγ combinations.

lipophilic acid will always be a major challenge. It may, however, be possible to combine PPAR agonism with more potent activity at other targets that bind similarly lipophilic and/or acidic compounds as their natural ligands.

Whereas the structural diversity of SERT inhibitors is high, the diversity of PPAR agonists is relatively narrow, typically containing a linear arrangement of polar head group, lipophilic linker and acidic tail. In the case of monoamine targets, the binding sites appear to tolerate a wide breadth of scaffolds and physicochemical properties. The SAR flexibility is probably related to the presence of a basic nitrogen that binds to a key aspartate residue in the target. By contributing a large proportion of the binding energy, this anchor group may allow greater flexibility elsewhere in the molecule. This can explain why there are so many SERT combinations where the ligands have potent nanomolar activity at all targets, even from very different target families. A similar phenomenon occurs with some enzyme inhibitors where a critical warhead group dominates binding, such as hydroxamic acids for zinc-containing enzymes or a nitrile groups for cysteine proteases.

Other targets appear to be analogous to SERT in terms of the feasibility for combinations. There are many combinations involving acetylcholinesterase (AChE) with nanomolar potency at all targets. Some particularly challenging targets have been combined with AChE. For example, dual inhibitors of AChE and beta-site amyloid precursor protein cleavage enzyme (BACE) have also been reported as another possible approach to Alzheimer's disease.[46]

The published literature supports the intuitive perception that it should be significantly easier to combine closely related targets than more distant ones. The concept of a 'micromolar barrier' comes into play for more distant targets, namely it will often be possible to obtain micromolar multi-target activity but breaking through to the nanomolar level for all targets can be a real hurdle. This concept, which is illustrated by some of the PPAR combinations in Figure 8.14, may or may not be an issue for efficacy, depending upon the extent to which there is synergy between those targets.

Given the embryonic nature of the MTDD field, it is important to note that surprising observations often occur. Targets that might be expected to be easy to combine can prove to be surprisingly difficult and yet the literature also contains many unusual combinations. It has been unexpectedly challenging to combine potent activity at the histamine receptors, H_1 and H_3, due to differences in the pharmacophores.[47] On the other hand, it was possible to combine activity at the H_1 and TxA_2 receptors, even though the former has a polar amine as the natural ligand and the latter a lipophilic acid.[48]

Although the DML field has expanded enormously over recent years, there are surely many new combinations awaiting discovery. Proteins that are lenient in terms of their binding requirements and ligands that can adopt multiple binding modes, for instance via different conformations, would likely increase the feasibility of a MTDD project. Given the number of possible permutations, it might be expected that, even taking account of the medicinal chemistry challenges in this field, there will be a larger number of druggable combinations than there are druggable single targets. The *in silico* prediction of novel druggable combinations for which it may be possible to find a multi-target agent is an area of intensive research at present.[49–52] These methods are further described in Chapter 4. Since DML projects are resource hungry, there is a real need for new ways of assessing the chance of ultimate success at an earlier stage of a project. Although similarity based methods have great potential in this regard, it is still early days for such predictive approaches and it is difficult to know how generally useful they will be for assessing feasibility of those unexpected combinations where the targets are not obviously related by structure or function and yet it is still possible to obtain a drug-like DML, such as the dual NK1/SERT inhibitors and AChE/BACE described earlier.

8.6 Summary

A wide range of different lead generation approaches are available for MTDD, with some, such as framework combination and focussed screening, being well established. New levels of sophistication need to be applied to such methods to deliver the higher quality starting compounds that are essential for success in lead optimization. Instead of merely overlapping the frameworks of selective ligands, a more flexible approach would involve hybridization at the level of the underlying pharmacophores which could result in compounds that differ structurally from either starting compound. As the two main emerging approaches in MTDD, fragment and virtual screening show great promise for generating tractable hit matter. Given the added challenges of multiple ligand projects in general, it makes good sense for medicinal chemists to deploy multiple lead generation strategies to increase the overall chances of success. Some target combinations will be much more feasible than others and given the additional resource demands of MTDD projects, new methods for assessing feasibility are urgently needed to ruthlessly assess the chance of success at the outset. To complement a medicinal chemist's personal experience, computational tools are already emerging, but it is early days for such methods and current feasibility predictions lack precision. Complex diseases like cancer and CNS diseases will continue to be prominent areas of emphasis in MTDD and new target classes, such as the epigenetic targets, can be expected to emerge.[53]

References

1. P. Norman, *Expert Opin. Ther. Pat.*, 2008, **18**, 1091–1096.
2. R. C. Buijsman, J. E. Basten, T. G. van Dinther, G. A. van der Marel, C. A. van Boeckel and J. H. van Boom, *Bioorg. Med. Chem. Lett.*, 1999, **9**, 2013–2018.
3. J. Lowe, T. Seeger, A. Nagel, H. Howard, P. Seymour, J. Heym, F. Ewing, M. Newman and A. Schmidt, *J. Med. Chem.*, 1991, **34**, 1860–1866.
4. J. Sterling, Y. Herzig, T. Goren, N. Finkelstein, D. Lerner, W. Goldenberg, I. Miskolczi, S. Molnar, F. Rantal, T. Tamas, G. Toth, A. Zagyva, A. Zekany, G. Avian, A. Gross, R. Friedman, M. Razin, W. Huang, B. Krais, M. Chorev, M. B. Youdim and M. Weinstock, *J. Med. Chem.*, 2002, **45**, 5260–5270.
5. Y. Sagi, N. Drigues and M. B. Youdim, *Br. J. Pharmacol.*, 2005, **146**, 543–552.
6. N. Murugesan, J. Tellew, Z. Gu, B. Kunst, L. Fadnis, L. Cornelius, R. Baska, Y. Yang, S. Beyer, H. K. Monshizadegan, K. E. Dickinson, B. Panchal, M. Valentine, S. Chong, R. Morrison, K. Carlson, J. Powell, S. Moreland, J. Barrish, M. Kowala and J. Macor, *J. Med. Chem.*, 2002, **45**, 3829–3835.
7. Poster from Curis Inc. at 235th American Chemical Society National Meeting, New Orleans, Louisiana, April 6–10, 2008.

8. T. Ryckmans, L. Balançon, O. Berton, C. Genicot, Y. Lamberty, B. Lallemand, P. Pasau, N. Pirlot, L. Quéré and P. Talaga, *Bioorg. Med. Chem. Lett.*, 2002, **12**, 261–264.
9. P. Atkinson, S. Bromidge, M. Duxon, M. Laramie, L. Gaster, M. Hadley, B. Hammond, C. Johnson, D. Middlemiss, S. North, G. Price, H. Rami, J. Riley, C. Scott, T. Shaw, K. Starr, G. Stemp, K. Thewlis, D. Thomas, M. Thompson, A. Vong and J. Watson, *Bioorg. Med. Chem. Lett.*, 2005, **15**, 737–741.
10. V. Garzya, I. T. Forbes, A. D. Gribble, M. S. Hadley, A. P. Lightfoot, A. H. Payne, A. B. Smith, S. E. Douglas, D. G. Cooper, I. G. Stansfield, M. Meeson, E. E. Dodds, D. N. Jones, M. Wood, C. Reavill, C. A. Scorer, A. Worby, G. Riley, P. Eddershaw, C. Ioannou, D. Donati, J. J. Hagan and E. A. Ratti, *Bioorg. Med. Chem. Lett.*, 2007, **17**, 400–405.
11. D. Boschelli, Y. Wang, S. Johnson, B. Wu, F. Ye, A. Sosa, J. Golas and F. Boschelli, *J. Med. Chem.*, 2004, **47**, 1599–1601.
12. B. Apsel, J. A. Blair, B. Gonzalez, T. M. Nazif, M. E. Feldman, B. Aizenstein, R. Hoffman, R. L. Williams, K. M. Shokat and Z. A. Knight, *Nat. Chem. Biol.*, 2008, **4**, 691–699.
13. M. Congreve, G. Chessari, D. Tisi and A. J. Woodhead, *J. Med. Chem.*, 2008, **51**, 3661–3680.
14. R. Morphy and Z. Rankovic, *Drug Discovery Today*, 2007, **12**, 156–160.
15. M. M. Hann, A. R. Leach and G. Harper, *J. Chem. Inf. Comput. Sci.*, 2001, **41**, 856–864.
16. G. V. Paolini, R. H. B. Shapland, W. P. van Hoorn, J. S. Mason and A. L. Hopkins, *Nat. Biotechnol.*, 2006, **24**, 805–815.
17. M. L. Radhakrishnan and B. Tidor, *J. Phys. Chem. B*, 2007, **111**, 13419–13435.
18. D. R. Artis, J. J. Lin, C. Zhang, W. Wang, U. Mehra, M. Perreault, D. Erbe, H. I. Krupka, B. P. England, J. Arnold, A. N. Plotnikov, A. Marimuthu, H. Nguyen, S. Will, M. Signaevsky, J. Kral, J. Cantwell, C. Settachatgull, D. S. Yan, D. Fong, A. Oh, S. Shi, P. Womack, B. Powell, G. Habets, B. L. West, K. Y. Zhang, M. V. Milburn, G. P. Vlasuk, K. P. Hirth, K. Nolop, G. Bollag, P. N. Ibrahim and J. F. Tobin, *Proc. Natl. Acad. Sci. USA*, 2009, **106**, 262–267.
19. K. Bright, *PhD Thesis*, University of York, 2010.
20. A. J. Knox, T. Price, M. Pawlak, G. Golfis, C. T. Flood, D. Fayne, D. C. Williams, M. J. Meegan and D. G. Lloyd, *J. Med. Chem.*, 2009, **52**, 2177–2180.
21. T. W. Wong, F. Y. Lee, C. Yu, F. R. Luo, S. Oppenheimer, H. Zhang, R. A. Smykla, H. Mastalerz, B. E. Fink and J. T. Hunt, *Clin. Cancer Res.*, 2006, **12**, 6186–6193.
22. P. Smid, H. K. Coolen, H. G. Keizer, R. van Hes, J. P. de Moes, A. P. den Hartog, B. Stork, R. H. Plekkenpol, L. C. Niemann, C. N. Stroomer, M. T. Tulp, H. H. van Stuivenberg, A. C. McCreary, M. B. Hesselink, A. H. Herremans and C. G. Kruse, *J. Med. Chem.*, 2005, **48**, 6855–6869.
23. M. Bruncko, T. K. Oost, B. A. Belli, H. Ding, M. K. Joseph, A. Kunzer, D. Martineau, W. J. McClellan, M. Mitten, S. C. Ng, P. M. Nimmer,

T. Oltersdorf, C. M. Park, A. M. Petros, A. R. Shoemaker, X. Song, X. Wang, M. D. Wendt, H. Zhang, S. W. Fesik, S. H. Rosenberg and S. W. Elmore, *J. Med. Chem.*, 2007, **50**, 641–662.
24. A. Shehzad and Y. S. Lee, *Drugs Future*, 2010, **35**, 113.
25. M. Athar, J. Ho Back, L. Kopelovich, D. R. Bickers and A. L. Kim, *Arch. Biochem. Biophys.*, 2009, **486**, 95–102.
26. K. Park, S. Lee, H. Ahn and D. Kim, *Mol. BioSyst.*, 2009, **5**, 844–853.
27. S. M. Yu, T. S. Wu and C. M. Teng, *Br. J. Pharmacol.*, 1994, **111**, 906–912.
28. R. Garcia-Sernaand and J. Mestres, *Drug Discovery Today*, 2011, **16**, 99–106.
29. D. J. Daniels, A. Kulkarni, Z. Xie, R. G. Bhushan and P. S. Portoghese, *J. Med. Chem.*, 2005, **48**, 1713–1716.
30. T. Yamamoto, P. Nair, N. E. Jacobsen, V. Kulkarni, P. Davis, S. Ma, E. Navratilova, H. I. Yamamura, T. W. Vanderah, F. Porreca, J. Lai and V. J. Hruby, *J. Med. Chem.*, 2010, **53**, 5491–5501.
31. Y. Lee, S. Fernandes, V. Kulkarani, A. Mayorov, P. Davis, S. Mab, K. Brown, R. J. Gillies, J. Lai, F. Porreca and V. J. Hruby, *Bioorg. Med. Chem. Lett.*, 2010, **20**, 4080–4084.
32. R. Morphy and Z. Rankovic, *J. Med. Chem.*, 2005, **48**, 6523–6543.
33. D. Angus, M. Bingham, D. Buchanan, N. Dunbar, L. Gibson, R. Goodwin, A. Haunsø, A. Houghton, M. Huggett, R. Morphy, S. Napier, O. Nimz, J. Passmore and G. Walker, *Bioorg. Med. Chem. Lett.*, 2011, **21**, 271–275.
34. F. Micheli, P. Cavanni, D. Andreotti, R. Benedetti, B. Bertani, M. Bettati, L. Bettelini, G. Bonanomi, S. Braggio, R. Carletti, A. Checchia, L. Zonzini and R. Di Fabio, *J. Med. Chem.*, 2010, **53**, 4989–5001.
35. M. A. Letavic, J. M. Keith, J. A. Jablonowski, E. M. Stocking, L. A. Gomez, K. S. Ly, J. M. Miller, A. J. Barbiera, P. Bonaventure, T. W. Lovenberg and N. I. Carruthers, *Bioorg. Med. Chem. Lett.*, 2007, **17**, 1047–1051.
36. N. Toda, K. Tago, S. Marumoto, K. Takami, M. Ori, N. Yamada, K. Koyama, S. Naruto, K. Abe, R. Yamazaki, T. Hara, A. Aoyagi, Y. Abe, T. Kaneko and H. Kogen, *Bioorg. Med. Chem.*, 2003, **11**, 4389–4415.
37. J. R. Cashman, T. Voelker, R. Johnson and A. Janowsky, *Bioorg. Med. Chem.*, 2009, **17**, 337–343.
38. S. Chaki, Y. Oshida, S. Ogawa, T. Funakoshi, T. Shimazaki, T. Okubo, A. Nakazato and S. Okuyama, *Pharmacol. Biochem. Behav.*, 2005, **82**, 621–626.
39. Y. Ye, Z. Zhoua, H. Zoua, Y. Shena, T. Xua, J. Tanga, H. Yinb, M. Chenb, Y. Lenga and J. Shen, *Bioorg. Med. Chem.*, 2009, **17**, 5722–5732.
40. M. Hieke, J. Ness, R. Steri, M. Dittrich, C. Greiner, O. Werz, K. Baumann, M. Schubert-Zsilavecz, S. Weggen and H. Zettl, *J. Med. Chem.*, 2010, **53**, 4691–4700.
41. B. R. Bhattarai, B. Kafle, J. S. Hwang, S. W. Ham, K. H. Lee, H. Park, I. O. Han and H. Cho, *Bioorg. Med. Chem. Lett.*, 2010, **20**, 6758–6763.
42. K. Sander, Y. von Coburg, J. C. Camelin, X. Ligneau, O. Rau, M. Schubert-Zsilavecz, J. C. Schwartz and H. Stark, *Bioorg. Med. Chem. Lett.*, 2010, **20**, 1581–1584.

43. C. S. Mizuno, A. G. Chittiboyina, F. H. Shah, A. Patny, T. W. Kurtz, H. A. Pershadsingh, R. C. Speth, V. T. Karamyan, P. B. Carvalho and M. A. Avery, *J. Med. Chem.*, 2010, **53**, 1076–1085.
44. M. Alvarado, P. Goya, M. Macías-González, F. J. Pavón, A. Serrano, N. Jagerovic, J. Elguero, A. Gutiérrez-Rodríguez, S. García-Granda, M. Suardíaz and F. Rodríguez de Fonseca, *Bioorg. Med. Chem.*, 2008, **16**, 10098–10105.
45. A. Bénardeau, J. Benz, A. Binggeli, D. Blum, M. Boehringer, U. Grether, H. Hilpert, B. Kuhn, H. P. Märki, M. Meyer, K. Püntener, S. Raab, A. Ruf, D. Schlatter and P. Mohr, *Bioorg. Med. Chem. Lett.*, 2009, **19**, 2468–2473.
46. L. Piazzi, A. Cavalli, F. Colizzi, F. Belluti, M. Bartolini, F. Mancini, M. Recanatini, V. Andrisano and A. Rampa, *Bioorg. Med. Chem. Lett.*, 2008, **18**, 423–426.
47. R. Aslanian, M. wa Mutahi, N.-Y. Shih, J. J. Piwinski, R. West, S. M. Williams, S. She, R.-L. Wu and A. Hey, *Bioorg. Med. Chem. Lett.*, 2003, **13**, 1959–1961.
48. E. Ohshima, H. Takami, H. Harakawa, H. Sato, H. Obase, I. Miki, A. Ishii, H. Ishii and Y. Sasaki, *J. Med. Chem.*, 1993, **36**, 417–420.
49. A. Bender, *J. Chem. Inf. Model.*, 2006, **46**, 2445.
50. M. J. Keiser, B. L. Roth, B. N. Armbruster, P. Ernsberger, J. J. Irwin and B. K. Shoichet, *Nat. Biotechnol.*, 2007, **25**, 197–206.
51. G. V. Paolini, R. H. B. Shapland, W. P. van Hoorn, J. S. Mason and A. L. Hopkins, *Nat. Biotechnol.*, 2006, **24**, 805–815.
52. D. Vidal and J. Mestres, *Mol. Inf.*, 2010, **29**, 543–551.
53. A. Main, D. Cheng, M. T. Bedford, S. Valente, A. Nebbioso, A. Perrone, G. Brosch, G. Sbardella, F. De Bellis, M. Miceli and L. Altucci, *J. Med. Chem.*, 2008, **51**, 2279–2290.

CHAPTER 9

In Silico Lead Generation Approaches in Multi-Target Drug Discovery

XIAOHOU MA AND YUZONG CHEN*

Bioinformatics and Drug Design Group, Department of Pharmacy, Centre for Computational Science and Engineering, National University of Singapore, Blk S16, Level 8, 3 Science Drive 2, Singapore 117543
*Email: phacyz@nus.edu.sg

9.1 Introduction to *In Silico* Screening Methods

A number of *in silico* methods have been developed and used extensively as lead discovery tools against individual targets. The most popular *in silico* methods are molecular docking,[1] pharmacophore matching,[2] the use of structure–activity relationships (SAR)[3] and quantitative structure–activity relationship (QSAR),[4–6] similarity searching,[7] machine learning,[8] and combinations of these methods.[9,10] Some of these methods have recently been explored for multi-target drug discovery (MTDD) to find drug leads that simultaneously interact with multiple targets. To facilitate their optimal use in MTDD, the methodology, advantages, limitations and strategies for improving the *in silico* methods are first described before describing their applications in multi-target lead discovery.

9.1.1 Molecular Docking

Molecular docking is a widely used *in silico* method for identifying drug leads by geometrically docking small molecules to the 3D structure of a pre-selected

RSC Drug Discovery Series No. 21
Designing Multi-Target Drugs
Edited by J. Richard Morphy and C. John Harris

Published by the Royal Society of Chemistry, www.rsc.org

target site followed by the optimization of binding configurations and evaluation of binding feasibility based on complementary molecular interactions.[1,11] This method does not require knowledge of the structural features and molecular frameworks of active compounds, but may have limited capability in accounting for target structural flexibility[12] and certain physicochemical features important for drug binding.[1] These limitations arise from inadequate sampling of target conformational space, the use of gas-phase algorithms and parameters in scoring, inadequate modelling of solvation and entropic effects, and the use of simplified cut-off distances for non-bonded interactions.

To improve performance, the molecular dynamics-enhanced docking approach has been used to attempt to account for target flexibility.[13] In this approach, all-atom molecular dynamics simulations are used to generate an ensemble of multiple conformers that can be regarded conceptually as defining the thermodynamic equilibrium state in solution, and this ensemble is subsequently used in flexible docking and binding evaluation.[14] Several other target flexibility approximation methods have been introduced for improving docking performance.[12] These include soft docking (where ligands are allowed to partially penetrate binding site surface),[15] rotamer-based flexible docking (conformation of binding-site residues are sampled by backbone-dependent or backbone-independent rotamer libraries[16]), combined protein grid docking (multiple structures and conformations are combined into an energy-weighted or geometry-weighted average representation),[17,18] united description of receptor docking (multiple structures and conformations are superimposed and an average structure is generated from conserved features),[19] combined ligand conformational sampling,[20] receptor side-chain conformational sampling,[21] and receptor backbone energy minimization.[22]

Fragment-based drug discovery has in recent years been increasingly used for lead discovery.[23,24] Virtual fragment libraries can be constructed by computational methods, for example to deconstruct drugs and active compounds into their constituent fragments[25] which can then be filtered by fragment-like or lead-like guidelines such as 'Rule of 3'.[26] Such methods have been used for identifying potential hits from virtual fragment libraries. For instance, molecular docking has been used to screen large libraries of fragments against CTX-M–lactamase leading to the identification of several millimolar-range inhibitors with a high hit rate.[27] Molecular docking has also been used for identifying virtual fragment hits against other targets such as thrombin, factor Xa, cathepsin D, TGT and CDK4.[28]

9.1.2 Pharmacophores

The pharmacophore method has also been extensively used for identifying drug leads by matching molecules to an assembly of steric and electronic features necessary to ensure optimal interactions with a specific target and to trigger (or to block) its biological response.[29] Pharmacophoric features can be derived by either ligand-based or structure-based methods.[2] Ligand-based methods generate pharmacophoric features by superposing a set of active molecules and subsequently extracting their common chemical features essential for binding activity. Structure-based methods construct pharmacophoric features by probing possible interaction points between target and ligands. Pharmacophore methods complement molecular docking by enabling the selection of drug leads with higher structural flexibility, while molecular docking provides more accurate evaluation of specific binding interactions.

The quality of pharmacophore models is affected by several factors, including sensitivity to training datasets,[2] selection of a few relevant chemical features among many possible choices,[2,30] quality of conformational sampling and molecular overlay, selection of anchoring points, consideration of tautomeric and protonation status of compounds, and the general problem of estimating ligand binding affinities.[31] Moreover, pharmacophore modelling cannot easily be automated, making it harder for a non-expert to develop high-performance models.[32] Various strategies have been introduced for facilitating the construction of good pharmacophore models. For instance, the insufficient conformational sampling problem can be reduced by using pre-enumerating methods (multiple conformations are pre-computed) or on-the-fly methods (conformational analysis is carried out during the modelling process).[2] Relevant chemistry can be selected based on the frequency of occurrence of features known to be important for binding to a specific target.[30] Anchoring points can be selected by a knowledge-based approach that identifies prioritized target atoms using a machine learning model trained from known ligand binding pockets.[33] Further prioritization can be carried out by evaluating such important properties as drug-likeness, bioactivity and synthetic accessibility.[34]

9.1.3 QSAR

Another popular *in silico* method, the QSAR method, identifies drug leads by estimating binding activity based on a statistically significant correlation between molecular structures and activities.[4–6] Molecular structures are typically represented by specific sets of structural and physicochemical properties or molecular descriptors of the entire molecules, their substituents, or their electrostatic fields considered most relevant to their binding activities. The quality of QSAR models can be affected by a number of factors, including compatibility of concepts, representativeness of assay data, influence of data outliers, selection of molecular descriptors, fitness of developed quantitative relationships, occurrence of chance correlations, the starting geometry in 3D-QSAR and problems arising from multiple solutions; good QSAR models can be developed by careful attention to these factors.[35]

9.1.4 Machine Learning Methods

Machine learning methods have recently been explored as potential drug lead discovery tools.[8] These methods identify drug leads by the statistical analysis of intrinsic correlations between molecular structures and activity status based on the structural and physicochemical profiles of known active and inactive compounds;[36] regression methods can be incorporated for estimating activity levels.[37] Machine learning methods utilize nonlinear supervised learning algorithms to develop statistical models capable of predicting a more diverse spectrum of structures and physicochemical properties than conventional QSAR models, which is particularly useful for screening large compound libraries and for identifying novel scaffolds, nicely complementing conventional lead discovery tools.[36,37] Machine learning methods include logistic regression, linear discriminant analysis, k-nearest neighbour, binary kernel discrimination, naive Baysian methods, decision trees, random forests, artificial neural networks (including multilayer perceptron, radial basis functions and probabilistic neural networks) and support vector machines.[37,38]

The performance of machine learning methods depends on several factors, including training set diversity, their ability to deal with imbalanced datasets (inactive compounds typically outnumber active compounds), and applicability domain and parameter ranges in covering active and inactive chemical space.[37] Certain strategies have been introduced to aid the development of good machine learning models. For example, computational methods for re-adjusting biased shift of hyperplane have been used to gain improvements in the case of imbalanced datasets.[39] Several feature selection methods have been used for selecting appropriate sets of molecular descriptors that optimally distinguish active and inactive compounds.[37,38,40–42] Machine learning models capable of screening large compound libraries with good yields of actives and low false-hit rates can be developed by using highly diverse training datasets composed of predicted inactive compounds.[43–45]

9.2 Application of *In Silico* Screening to MTDD

The historical MTDD strategies described in Chapter 8 can result in larger and more complex non-drug like molecules.[46,47] *In silico* screening, like fragment-based drug discovery, has the potential to deliver compounds with a better balance of physicochemical properties and binding affinities. As binding sites become more dissimilar, it is increasingly difficult to improve and balance adequately the binding affinities needed to achieve acceptable *in vivo* efficacy and safety. One way to reduce this difficulty is to explore synergistic targets, such that multi-target agents with modest activity at one or more of the relevant targets may still produce similar or better *in vivo* effects compared with higher-affinity target-selective compounds.[48]

An *in silico* screening strategy is practically useful if the retrieval rates against individual targets are sufficiently high and the false-hit rates are sufficiently low. High individual retrieval rates can compensate for the reduced collective retrieval rates against multiple targets. For instance, if the retrieval rate against each individual target is at a level of 50% ~ 70%, the collective retrieval rate for identifying multi-target agents against two targets may be reduced to 25%–49% assuming the searches are independent of each other. Low false-hit rates are critical for high enrichment in searching for multi-target agents as these are expected to be significantly fewer in number and more sparsely distributed in chemical space than single-target agents. Application of *in silico* methods to multi-target lead discovery will offer a stringent test of the potential and limitations of such methods in MTDD.

Although molecular docking and pharmacophore methods can be used independently for multi-target lead discovery (Figure 9.1), the two methods have more typically been combined for identifying multi-target agents[49,50] (Figure 9.2). For instance, in identifying dual inhibitors of two anti-inflammatory targets, PLA2 and LTA4H-h, pharmacophore models for each target were first generated followed by the identification of a common pharmacophore for both targets. Compounds accommodated by the common pharmacophore were subsequently docked to both targets and the binding conformations were optimized and analysed to select multi-target virtual hits.[49] A similar procedure was used to identify multi-target inhibitors against LTA-4H and LTC4S.[50] The molecular dynamics-enhanced molecular docking method has also been used in virtual screening against individual targets in HIV and its associated opportunistic pathogens to find multi-target agents such as KNI-764 that inhibit both HIV-1 protease and malarial plasmepsin II enzyme.[51]

Multi-target QSAR models for identifying multi-target agents[52] and active agents against multiple bacterial,[53] fungal[54,55] and viral[53] species have been developed by incorporating multi-target or species variations of binding site features into the multi-target dependent molecular descriptors or species-dependent

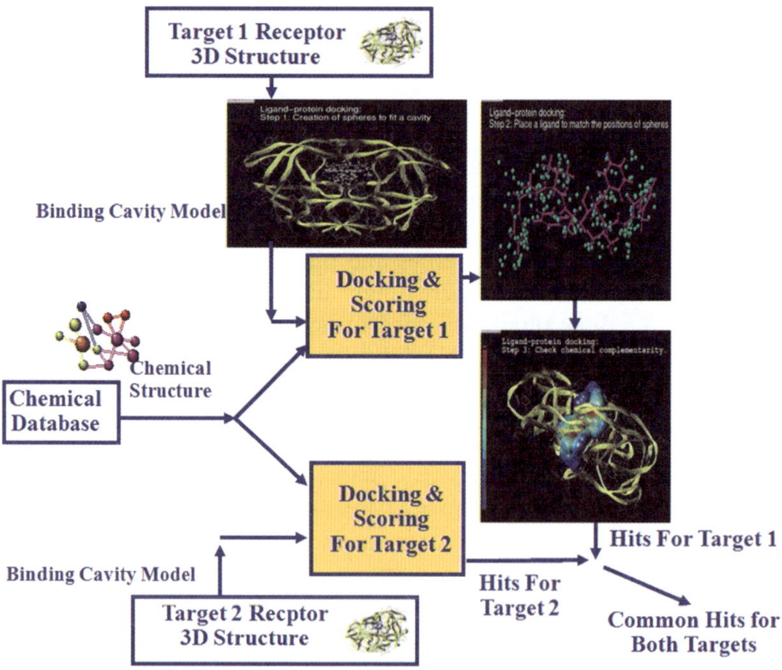

Figure 9.1 Molecular docking strategy for MTDD.

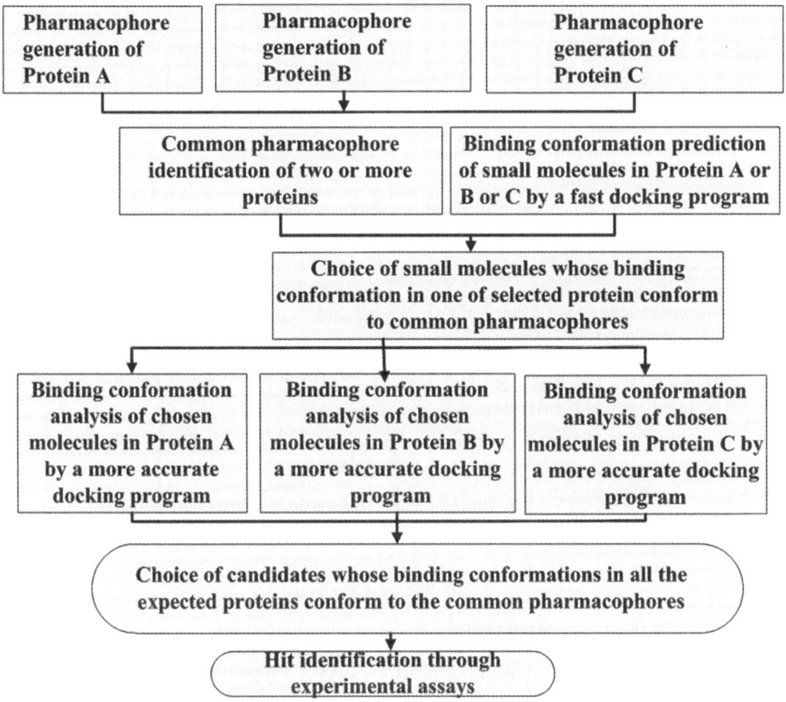

Figure 9.2 Combined pharmacophore and molecular docking strategy for MTDD.

molecular descriptors, and stochastic Markov drug-binding process models. These multi-target QSAR models achieved high retrieval rates of 72% ~ 85% and moderately low false-hit rates of 15% ~ 28%.[53–55] The development of multi-target QSAR models may be limited by inadequate ligand binding data for some of the targets or species. Moreover, the molecular weights of the screening compounds need to be in a certain range for accurate computation of multi-target dependent or species-dependent molecular descriptors.[55]

A series of machine learning methods based on artificial neural networks (ANN) have been explored for developing multi-target QSAR models against different parasite species.[56] These ANN methods relate molecular structures (represented by molecular descriptors) with activities by nonlinear mapping. This nonlinear mapping is distinguished from the linear mapping of conventional multi-target QSARs[53–55] by its ability to derive non-linear relationships that linear models may fit poorly due to high data complexity. The best performing ANN model, probabilistic neural networks, achieves a retrieval rate of 87%,[56] which is not much better than the rates of 72% ~ 85% of conventional multi-target QSAR models.[53–55] The retrieval rates of machine learning methods have been found to be sensitively dependent on training dataset diversity.[57] Hence, limited training dataset diversity may be one possible reason for the modest improvement of these machine learning models over conventional multi-target QSAR models.

Another machine learning method, support vector machine (SVM), has been explored as a potential tool for the discovery of multi-kinase inhibitors (Figures 9.3 and 9.4) against 11 combinations of 9 anticancer kinase targets

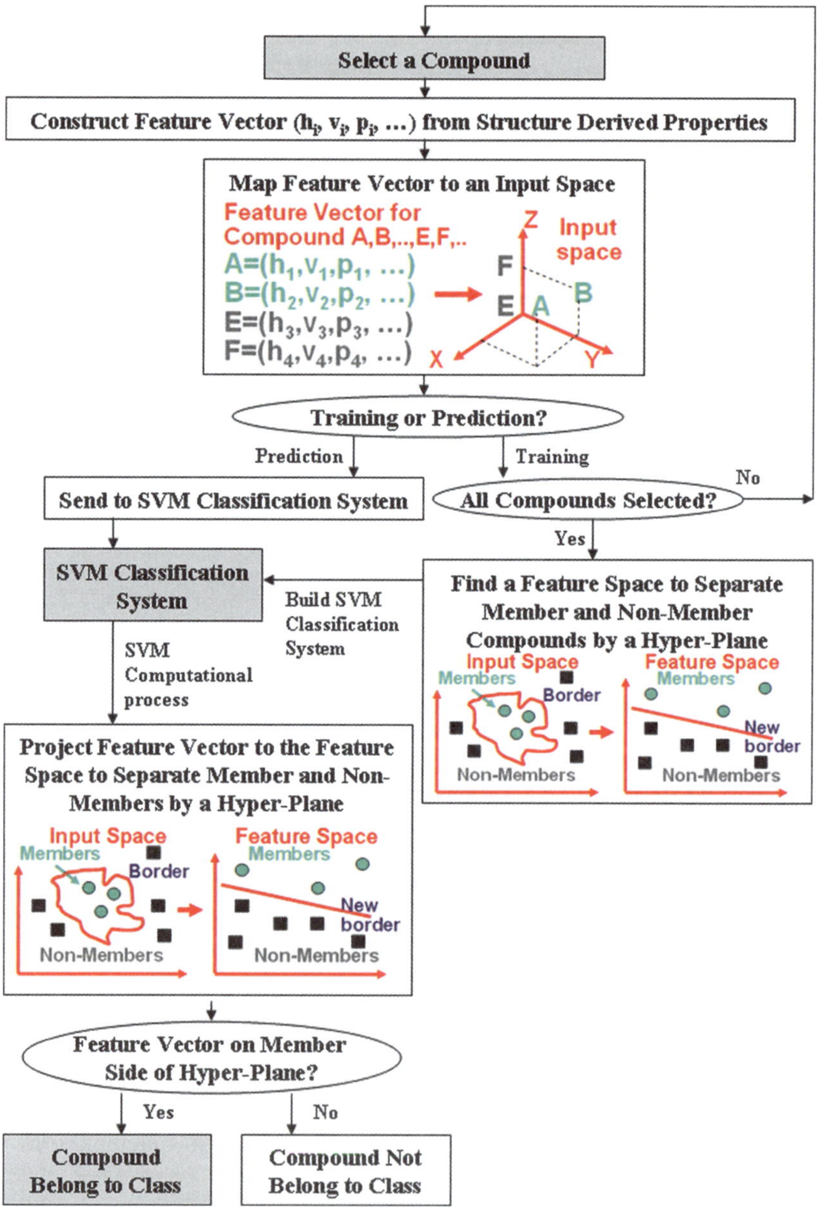

Figure 9.3 Illustration of training a SVM virtual screening model to search for single-target inhibitors.

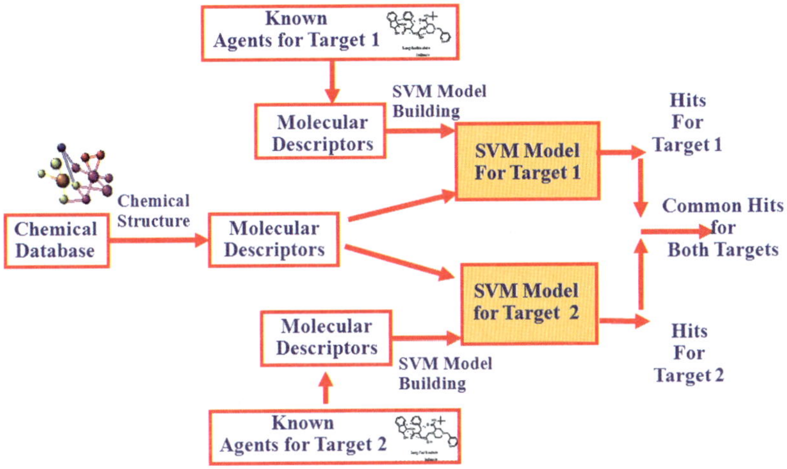

Figure 9.4 Illustration of using SVM methods to search for multi-target inhibitors.

(EGFR, VEGFR, PDGFR, Src, FGFR, Lck, CDK1, CDK2, GSK3).[45] The 11 kinase pairs are EGFR-PDGFR, EGFR-FGFR, EGFR-Src, VEGFR-Lck, PDGFR-FGFR, PDGFR-Src, Src-Lck, CDK1-CDK2, CDK1-GSK3, CDK2-GSK3 and CDK1-VEGFR. The first 7 pairs are intra-PTK group, the 8th to 10th are intra-CMGC group and the 11th is inter-PTK-CMGC group, representative of different types of kinase pairs frequently co-expressed or co-activated in various cancers[58,59] and co-targeted by drugs with good anticancer efficacies.[60,61] The SVM of each of the 9 kinases has been trained with published compounds reported to selectively inhibit only one of the kinases, by using algorithms described in the literature.[43,44] These SVM tools were used to search for published dual inhibitors of the 11 kinase pairs to test their collective retrieval rates. These SVM tools correctly identified 26.8%–57.3% (majority >36%) of the 56–230 intra-kinase-group dual inhibitors and 12.2% of the 41 inter-kinase-group dual inhibitors.[45] This suggests that machine learning methods may be successfully combined with other *in silico* methods for screening large compound libraries in order to identify novel multi-target scaffolds.

9.3 Perspectives

In silico methods have been increasingly explored and have shown promising potential as multi-target lead discovery tools. The capability of these methods may be further enhanced by incorporating knowledge of newly discovered selective multi-target agents from the current and future drug discovery efforts,[60,61] and by the further development and improvement of *in silico* methods into more suitable methods for MTDD.[43,44,62–66]

References

1. D. B. Kitchen, H. Decornez, J. R. Furr and J. Bajorath, *Nat. Rev. Drug Discovery*, 2004, **3**, 935–949.
2. S. Y. Yang, *Drug Discovery Today*, 2010, **15**, 444–450.
3. A. D. Andricopulo and C. A. Montanari, *Mini-Rev. Med. Chem.*, 2005, **5**, 585–593.
4. A. Z. Dudek, T. Arodz and J. Galvez, *Comb. Chem. High Throughput Screening*, 2006, **9**, 213–228.
5. R. D. Clark, *Curr. Top. Med. Chem.*, 2009, **9**, 791–810.
6. J. Verma, V. M. Khedkar and E. C. Coutinho, *Curr. Top. Med. Chem.*, 2010, **10**, 95–115.
7. P. Willett, *Drug Discovery Today*, 2006, **11**, 1046–1053.
8. H. Li, C. W. Yap, C. Y. Ung, Y. Xue, Z. R. Li, L. Y. Han, H. H. Lin and Y. Z. Chen, *J. Pharm. Sci.*, 2007, **96**, 2838–2860.
9. I. Muegge, *Mini-Rev. Med. Chem.*, 2008, **8**, 927–933.
10. A. Jennings and M. Tennant, *Curr. Pharm. Des.*, 2005, **11**, 335–344.
11. R. Gozalbes, L. Simon, N. Froloff, E. Sartori, C. Monteils and R. Baudelle, *J. Med. Chem.*, 2008, **51**, 3124–3132.
12. H. Alonso, A. A. Bliznyuk and J. E. Gready, *Med. Res. Rev.*, 2006, **26**, 531–568.
13. J. H. Lin, A. L. Perryman, J. R. Schames and J. A. McCammon, *J. Am. Chem. Soc.*, 2002, **124**, 5632–5633.
14. R. E. Amaro, R. Baron and J. A. McCammon, *J. Comput.-Aided Mol. Des.*, 2008, **22**, 693–705.
15. A. M. Ferrari, B. Q. Wei, L. Costantino and B. K. Shoichet, *J. Med. Chem.*, 2004, **47**, 5076–5084.
16. S. B. Nabuurs, M. Wagener and J. de Vlieg, *J. Med. Chem.*, 2007, **50**, 6507–6518.
17. H. B. Broughton, *J. Mol. Graphics Modell.*, 2000, **18**, 247–257, 302–244.
18. C. M. Venkatachalam, X. Jiang, T. Oldfield and M. Waldman, *J. Mol. Graphics Modell.*, 2003, **21**, 289–307.
19. H. Claussen, C. Buning, M. Rarey and T. Lengauer, *J. Mol. Biol.*, 2001, **308**, 377–395.
20. D. M. Lorber and B. K. Shoichet, *Protein Sci.*, 1998, **7**, 938–950.
21. C. N. Cavasotto and R. A. Abagyan, *J. Mol. Biol.*, 2004, **337**, 209–225.
22. W. Sherman, T. Day, M. P. Jacobson, R. A. Friesner and R. Farid, *J. Med. Chem.*, 2006, **49**, 534–553.
23. P. J. Hajduk and J. Greer, *Nat. Rev. Drug Discovery*, 2007, **6**, 211–219.
24. M. M. Hann, A. R. Leach and G. Harper, *J. Chem. Inf. Comput. Sci.*, 2001, **41**, 856–864.
25. M. Vieth, M. G. Siegel, R. E. Higgs, I. A. Watson, D. H. Robertson, K. A. Savin, G. L. Durst and P. A. Hipskind, *J. Med. Chem.*, 2004, **47**, 224–232.
26. M. Congreve, R. Carr, C. Murray and H. Jhoti, *Drug Discovery Today*, 2003, **8**, 876–877.
27. Y. Chen and B. K. Shoichet, *Nat. Chem. Biol.*, 2009, **5**, 358–364.

28. M. Congreve, G. Chessari, D. Tisi and A. J. Woodhead, *J. Med. Chem.*, 2008, **51**, 3661–3680.
29. W. Camille-Georges, C. R. Ganellin, P. Lindberg, L. A. Mitscher, Glossary of Terms Used in Medicinal Chemistry (IUPAC Recommendations 1997), *Annu. Rep. Med. Chem.*, 1998, **33**, 385–395.
30. J. Zou, H. Z. Xie, S. Y. Yang, J. J. Chen, J. X. Ren and Y. Q. Wei, *J. Mol. Graphics Modell.*, 2008, **27**, 430–438.
31. D. Horvath, *Methods Mol. Biol.*, 2011, **672**, 261–298.
32. P. Tiikkainen, P. Markt, G. Wolber, J. Kirchmair, S. Distinto, A. Poso and O. Kallioniemi, *J. Chem. Inf. Model.*, 2009, **49**, 2168–2178.
33. C. Barillari, G. Marcou and D. Rognan, *J. Chem. Inf. Model.*, 2008, **48**, 1396–1410.
34. Q. Huang, L. L. Li and S. Y. Yang, *J. Mol. Graphics Modell.*, 2010, **28**, 775–787.
35. T. Scior, J. L. Medina-Franco, Q. T. Do, K. Martinez-Mayorga, J. A. Yunes Rojas and P. Bernard, *Curr. Med. Chem.*, 2009, **16**, 4297–4313.
36. X. H. Ma, J. Jia, F. Zhu, Y. Xue, Z. R. Li and Y. Z. Chen, *Comb. Chem. High Throughput Screening*, 2009, **12**, 344–357.
37. C. W. Yap, H. Li, Z. L. Ji and Y. Z. Chen, *Mini-Rev. Med. Chem.*, 2007, **7**, 1097–1107.
38. A. Koike, *SAR QSAR Environ. Res.*, 2006, **17**, 497–514.
39. M. P. Brown, W. N. Grundy, D. Lin, N. Cristianini, C. W. Sugnet, T. S. Furey, M. Ares, Jr. and D. Haussler, *Proc. Natl. Acad. Sci., USA*, 2000, **97**, 262–267.
40. H. H. Lin, L. Y. Han, C. W. Yap, Y. Xue, X. H. Liu, F. Zhu and Y. Z. Chen, *J. Mol. Graphics Modell.*, 2007, **26**, 505–518.
41. Y. Cong, X. G. Yang, W. Lv and Y. Xue, *J. Mol. Graphics Modell.*, 2009, **28**, 236–244.
42. W. Lv and Y. Xue, *Eur. J. Med. Chem.*, 2010, **45**, 1167–1172.
43. L. Y. Han, X. H. Ma, H. H. Lin, J. Jia, F. Zhu, Y. Xue, Z. R. Li, Z. W. Cao, Z. L. Ji and Y. Z. Chen, *J. Mol. Graphics Modell.*, 2008, **26**, 1276–1286.
44. X. H. Liu, X. H. Ma, C. Y. Tan, Y. Y. Jiang, M. L. Go, B. C. Low and Y. Z. Chen, *J. Chem. Inf. Model.*, 2009, **49**, 2101–2110.
45. X. H. Ma, R. Wang, C. Y. Tan, Y. Y. Jiang, T. Lu, H. B. Rao, X. Y. Li, M. L. Go, B. C. Low and Y. Z. Chen, *Mol. Pharm.*, 2010, **7**, 1545–1560.
46. R. Morphy, *J. Med. Chem.*, 2006, **49**, 2969–2978.
47. R. Morphy and Z. Rankovic, *J. Med. Chem.*, 2006, **49**, 4961–4970.
48. J. Jia, F. Zhu, X. Ma, Z. Cao, Y. Li and Y. Z. Chen, *Nat. Rev. Drug Discovery*, 2009, **8**, 111–128.
49. D. Wei, X. Jiang, L. Zhou, J. Chen, Z. Chen, C. He, K. Yang, Y. Liu, J. Pei and L. Lai, *J. Med. Chem.*, 2008, **51**, 7882–7888.
50. S. Thangapandian, S. John, S. Sakkiah and K. W. Lee, *J. Chem. Inf. Model.*, 2011, **51**, 33–44.
51. J. C. Clemente, L. Govindasamy, A. Madabushi, S. Z. Fisher, R. E. Moose, C. A. Yowell, K. Hidaka, T. Kimura, Y. Hayashi, Y. Kiso,

M. Agbandje-McKenna, J. B. Dame, B. M. Dunn and R. McKenna, *Acta Crystallogr., Sect. D: Biol. Crystallogr.*, 2006, **62**, 246–252.
52. D. Vina, E. Uriarte, F. Orallo and H. Gonzalez-Diaz, *Mol. Pharm.*, 2009, **6**, 825–835.
53. F. J. Prado-Prado, E. Uriarte, F. Borges and H. Gonzalez-Diaz, *Eur. J. Med. Chem.*, 2009, **44**, 4516–4521.
54. H. Gonzalez-Diaz and F. J. Prado-Prado, *J. Comput. Chem.*, 2008, **29**, 656–667.
55. H. Gonzalez-Diaz, F. J. Prado-Prado, L. Santana and E. Uriarte, *Bioorg. Med. Chem.*, 2006, **14**, 5973–5980.
56. F. J. Prado-Prado, X. Garcia-Mera and H. Gonzalez-Diaz, *Bioorg. Med. Chem.*, 2010, **18**, 2225–2231.
57. X. H. Ma, R. Wang, S. Y. Yang, Z. R. Li, Y. Xue, Y. C. Wei, B. C. Low and Y. Z. Chen, *J. Chem. Inf. Model.*, 2008, **48**, 1227–1237.
58. I. Gockel, M. Moehler, K. Frerichs, D. Drescher, T. T. Trinh, F. Duenschede, T. Borschitz, K. Schimanski, S. Biesterfeld, K. Herzer, P. R. Galle, H. Lang, T. Junginger and C. C. Schimanski, *Oncol. Rep.*, 2008, **20**, 845–850.
59. J. M. Stommel, A. C. Kimmelman, H. Ying, R. Nabioullin, A. H. Ponugoti, R. Wiedemeyer, A. H. Stegh, J. E. Bradner, K. L. Ligon, C. Brennan, L. Chin and R. A. DePinho, *Science*, 2007, **318**, 287–290.
60. M. Krug and A. Hilgeroth, *Mini-Rev. Med. Chem.*, 2008, **8**, 1312–1327.
61. A. L. Gill, M. Verdonk, R. G. Boyle and R. Taylor, *Curr. Top. Med. Chem.*, 2007, **7**, 1408–1422.
62. A. Bender, J. L. Jenkins, M. Glick, Z. Deng, J. H. Nettles and J. W. Davies, *J. Chem. Inf. Model.*, 2006, **46**, 2445–2456.
63. A. Givehchi, A. Bender and R. C. Glen, *J. Chem. Inf. Model.*, 2006, **46**, 1078–1083.
64. S. Renner, S. Derksen, S. Radestock and F. Morchen, *J. Chem. Inf. Model.*, 2008, **48**, 319–332.
65. D. Erhan, J. L'Heureux, S. Y. Yue and Y. Bengio, *J. Chem. Inf. Model.*, 2006, **46**, 626–635.
66. H. Dragos, M. Gilles and V. Alexandre, *J. Chem. Inf. Model.*, 2009, **49**, 1762–1776.

CHAPTER 10

The Challenges of Multi-Target Lead Optimization

J. RICHARD MORPHY†

Stirling, UK
Email: richard.morphy@spcorp.com

10.1 Introduction

Multi-target drug discovery (MTDD) is based on the premise that compounds interacting with more than one target may have superior efficacy or safety. Designed multiple ligands (DMLs) are compounds that are prospectively and intentionally designed to interact specifically with multiple targets. Given the multi-parameter nature of multi-target lead optimization, the need for a high quality lead compound is paramount. The decisions taken during the lead generation phase of a MTDD project are a critical determinant of the success, or failure, of lead optimization. However, it is realistic to expect that in most lead optimization projects, a substantial amount of work remains to be done, improving activity at one or more of the targets being one possible goal, thereby adjusting the balance of activities to maximize efficacy and safety. In addition, removing undesired activities associated with side effects is often required. Given that in most cases, an orally administered drug will be the goal, the third main challenge will be attaining the required physicochemical property and pharmacokinetic profile.

†Current address: Lilly Research Centre, Windlesham, Surrey, GU20 9PH, UK.

RSC Drug Discovery Series No. 21
Designing Multi-Target Drugs
Edited by J. Richard Morphy and C. John Harris

Published by the Royal Society of Chemistry, www.rsc.org

10.2 Optimization of the Activity Profile

Since most currently marketed multi-target drugs were serendipitously discovered and not prospectively designed, the extent to which inhibition of each target contributes to the overall therapeutic effect is unclear. It is quite possible that the efficacy of some of these drugs is driven primarily by targets other than the designated targets. Establishing an optimal ratio of the desired activities for a DML is highly desirable since inadequate activation or inhibition of one or more targets will reduce efficacy. Moreover, a suboptimal activity ratio might introduce a safety liability or, in the case of an anti-cancer or anti-infective drug, the emergence of therapeutic resistance. Despite its importance to efficacy and safety, the activity ratio is perhaps the most difficult and poorly understood aspect of lead optimization. Few publications even discuss this issue, suggesting it has been quietly ignored altogether. The aim of most DML projects has been to obtain roughly equivalent *in vitro* activity for each target, with the assumptions that firstly this will lead to a similar level of modulation of each target *in vivo* and secondly that an equal level of modulation is desirable, which may or may not be the case.

In the antidepressant area, equal modulation of the serotonin and noradrenaline transporters has been associated with improved efficacy and safety. Agents evolved from the mostly selective serotonin transporter (SERT) inhibitor fluoxetine **1** towards the dual SERT/norepinephrine transporter (NET) inhibitor venlafaxine **2** and most recently duloxetine **3** (Figure 10.1). Venlafaxine, despite being classified as a dual SERT/NET blocker (SNRI), is 30-fold selective for SERT, suggesting that it behaves as a multiple ligand *in vivo* only at high doses,[1] whereas duloxetine has a more potent and balanced *in vitro* profile which is regarded as advantageous for treating depression.[2] Duloxetine has also been approved for the treatment of neuropathic pain. Agents that have even lower SERT:NET ratios may also be of interest for treating pain, given that efficacy may be driven predominantly via NET and if SERT inhibition is too high, side effects such as nausea and sexual dysfunction may arise.[3]

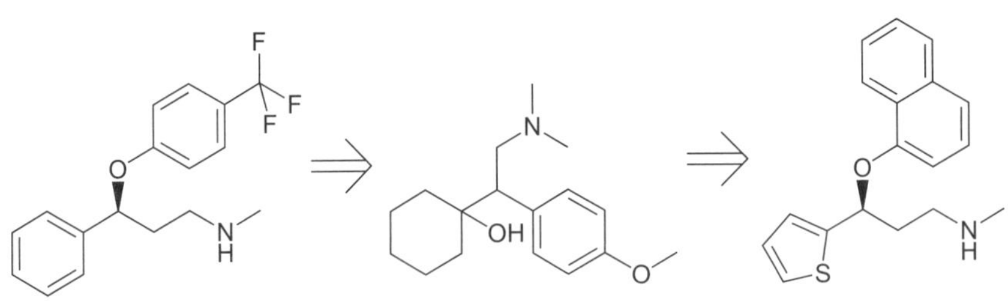

1; Fluoxetine
SERT K_i 0.8 nM
NET K_i 240 nM
NET/SERT ratio 300

2; Venlafaxine
SERT K_i 82 nM
NET K_i 2480 nM
NET/SERT ratio 30

3; Duloxetine
SERT K_i 0.8 nM
NET K_i 7.5 nM
NET/SERT ratio 9.4

Figure 10.1 The historical trend from SERT-selective to dual SERT/NET inhibition.

The clinical development of atypical antipsychotic drugs represents another example of knowledge generated during clinical studies helping researchers to identify an optimal activity profile. In this case, clinical studies showed that an improved efficacy to safety ratio could be provided by an asymmetric ratio of activities. The unsurpassed efficacy of the schizophrenia drug, clozapine, has been associated with a specific 10:1 ratio of serotonin $5HT_2$ and dopamine D_2 antagonism. Lower activity at D_2 is associated with a lower risk of extrapyramidal side effects (EPS). Thus, lower D_2 than $5HT_2$ activity was targeted by John Lowe and colleagues at Pfizer during the development of the marketed schizophrenia drug, ziprasidone (see Chapter 16).[4]

Clearly, clinical feedback is invaluable in guiding researchers towards the optimal activity profile, but this insight is usually unavailable even for well-advanced target combinations let alone more novel target combinations. The 1:1 affinity ratio normally targeted by medicinal chemists provides a clear but often arbitrary goal, since without extensive testing in predictive animal models, and preferably clinical feedback, determining the optimal balance of activities will be guesswork. In the absence of this knowledge in the early stages of a MTDD programme, how can we approach this difficult area of the activity balance in a more rational and systematic way?

If selective ligands for the targets of interest are available, useful information regarding the optimal activity ratio might be obtained by combining such ligands or even combining a partially optimized DML and a selective ligand. Isobolographic analysis of a functional readout in a disease relevant *in vitro* or *in vivo* assay can provide clues as to the relative balance required for the maximum achievable efficacy in a preclinical model.[5] The increasing information from combining selective ligands in clinical studies can inform the discovery of DMLs. In such cases, combinations can be regarded as a trailblazer for DMLs.

However, the knowledge gap between a simple target affinity ratio in an *in vitro* assay and efficacy in the clinic is a huge one to bridge so, as in a typical single target project, the key will be to devise a flowchart of progressively more sophisticated and 'physiological' assays and models to systematically build confidence in the approach. Medicinal chemists need to consider the activity balance in both binding and functional assays to avoid reaching misleading conclusions. For instance, in the area of multi-kinase inhibitors, the ratio in a whole cell assay that employs a physiologically relevant form of the kinase will need to be considered alongside a biochemical kinase assay where often only a truncated kinase domain is used. This necessity is illustrated by examples where activity ratios in biochemical kinase assays differ markedly from those in cellular assays.[6] For monoamine GPCRs, functional assays are a key part of the screening cascade since small structural changes can convert for instance a pan-agonist into a mixed agonist/antagonist DMLs or vice versa.[7] Also, since functional efficacy can be highly assay dependent, it can be crucial to measure the activity balance in both recombinant and native functional assays.

Key pieces of information are the target occupancies at the desired targets needed for efficacy, relative to the target occupancies at desired or undesired targets that causes adverse effects. In the case of antipsychotic drugs, such as ziprasidone, D_2 receptor occupancy above 65% is associated with acceptable antipsychotic efficacy whereas D_2 occupancy above 80% risks the undesirable motor effects of EPS.[8] The *in vivo* target occupancy ratio is an extremely valuable piece of information to go alongside the *in vitro* affinity ratio. How these two ratios relate to each other will depend upon numerous factors such as the distribution of the targets, the free compound concentration in the vicinity of those targets, the target densities and if the DML is competitive, the relative concentrations and affinities of the endogenous ligands that must be displaced. More often than not, the affinity ratio will be of little value when viewed in isolation.

Biomarkers can help unravel the complex relationships between the affinity ratio, the occupancy ratio and the maximum efficacy achievable for a given target combination. Biomarkers relating to target engagement, efficacy and safety can indicate whether an agent is modulating each intended target to an appropriate extent in patients or indeed whether unintended targets are also being hit. Our lack of knowledge of how the current generation of multi-target drugs are actually working greatly complicates clinical development. Although biomarker development is more complex where multiple targets are involved, it will be crucial for the future development of MTDD, facilitating dose calculations and the identification of groups of patients who are more likely to respond.

In some cases, the most tractable way forward for assessing the optimal ratio may simply be to be test compounds with a range of activity ratios, if available, in disease-relevant models. In Chapter 17, Robert Weikert describes a triple blocker of the serotonin, noradrenaline and dopamine transporters. The addition of moderate DAT inhibition may improve antidepressant efficacy and side effect liability, but too high a level of inhibition could introduce abuse potential, so it was felt that higher SERT than DAT occupancy may be desirable. Since the optimal occupancy ratio was uncertain, the group set out to discover compounds with a range of SERT:DAT activity ratios and evaluate these in preclinical models. It was known from historical SSRI data that >80% SERT occupancy was associated with clinical efficacy, so DAT occupancy was determined for a range of compounds at doses that gave 80% SERT occupancy. Perhaps unsurprisingly the relationship between affinity and occupancy ratios was complex with compounds with similar affinity ratios having very different occupancy ratios and vice versa. In this case, PET ligands were used as target engagement biomarkers to determine the occupancies. By providing the occupancy ratios associated with efficacy and adverse effects, mediated through on-target and off-target activities, PET ligands can help bridge the gap between animal models and clinical studies.

Target modulation biomarkers can then be used to demonstrate that occupancy has the desired downstream effect. For example, monitoring the relative levels of neurotransmitters like serotonin and noradrenaline can be used to monitor the *in vivo* effects of SNRIs like duloxetine. Factors such as the receptor reserves for each target will then influence how the occupancy ratio translates into the pharmacodynamic effect due to modulation of each target. For instance, where the receptor reserve is high, low potency agonists may be surprisingly efficacious. If a mixed agonist:antagonist profile is desired, the receptor occupancy for the agonist target could be significantly lower than for the antagonist target. DML feasibility may well be higher if lower receptor occupancy is required. If the interaction between the targets is synergistic rather than purely additive, a lowering of the receptor occupancy associated with the required efficacy may occur and the extent of this effect may differ between targets.

For complex DML profiles, the relative contribution of each target activity to the overall efficacy or safety profile is almost always unknown. Studies in knockout animals may provide useful mechanistic information, as might the use of RNAi or, in the case of agonist DMLs, the use of target-selective antagonists.

Even if information concerning the optimal balance of activities is available, achieving that ratio may be far from straightforward. An important consideration is how flexible and how similar the SARs are for the targets of interest. We have already seen (Chapter 8) how some targets such as SERT have high SAR flexibility and have been combined with many other targets in high potency combinations. For example, the flexibility of the SERT SAR helped in the lead optimization of a dual serotonin and noradrenaline reuptake inhibitor.[9] In some cases multi-target SARs run roughly in parallel, such as in the case of D_1 and $5HT_{2A}$,[10] whereas in the case of optimization of a dual H_2:gastrin and H_1/NK_1 leads, the SARs run antiparallel with changes that favour one target being detrimental for the other.[11,12] In these latter two cases, it was possible to construct a molecule with essentially two independent pharmacophores present but such molecules tend to have non-drug-like physicochemical properties. Another example of divergent SARs is provided by the dual A_{2A} antagonist/MAO-B inhibitor described in Chapter 18.

As the number of targets to be balanced increases, the complexity of the task for a medicinal chemist increases supra-proportionally. It is therefore not surprising that the vast majority of reported DMLs are dual ligands. More complex multiple activity profiles have been achieved, particularly for targets from families with conserved binding sites, such as monoamine transporters, monoamine GPCRs, proteases or kinases. In another example of an atypical antipsychotic, Garzya *et al.* desired a molecule that had five activities regarded as being critical for efficacy, D_2, D_3, $5HT_{2A}$, $5HT_{2C}$ and $5HT_6$ receptor antagonism, with D_2 functional potency again being lower than the other activities to avoid EPS.[13] Careful optimization of the focussed screening hit **4** produced a DML **5** with the optimal balance of affinities (Figure 10.2).

One important consideration is the possibility of forming active metabolites with a different activity profile from the parent. Perhaps the most simple outcome would be to prioritize DMLs where the metabolites are inactive at each target since the chance that they will possess an identical occupancy ratio to the parent is usually low. However, serendipitous differences in profile can sometimes be advantageous. For example, the active metabolite of ladostigil (Chapter 18) may contribute to the overall neuroprotective activity of its parent.

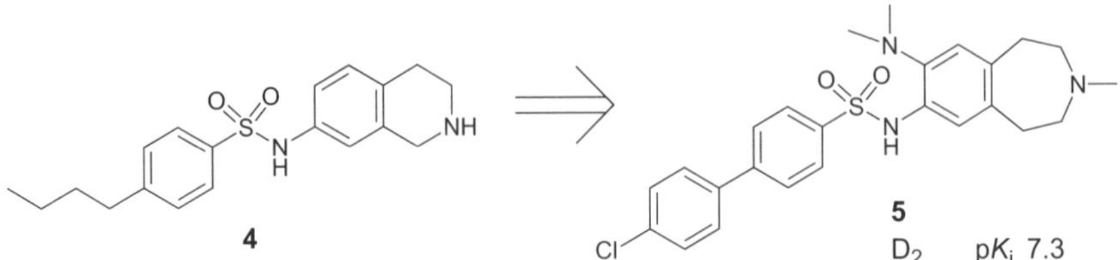

4
D_2 pK_i 6.0
D_3 pK_i 8.0
$5HT_{2A}$ pK_i 7.5
$5HT_{2C}$ pK_i 7.9
$5HT_6$ pK_i 7.6

5
D_2 pK_i 7.3
D_3 pK_i 8.5
$5HT_{2A}$ pK_i 8.8
$5HT_{2C}$ pK_i 8.3
$5HT_6$ pK_i 8.1

Figure 10.2 Optimization of a DML profile to enhance efficacy and safety.

10.3 Wider Selectivity

In addition to adjusting the ratio of desired activities, optimizing wider selectivity against a broad panel of targets is often required. Indeed a popular historical approach to MTDD has been to start with a non-selective compound that possesses undesired activities in addition to those associated with the disease and to attempt to rationally 'design out' the side activity.

The potential to remove these unintended side activities needs to be explored at the hit-to-lead stage of a MTDD project. Where many closely related targets and anti-targets exist, the task of achieving wider selectivity will be particularly intricate. For multi-target projects involving aminergic GPCR ligands and multi-kinase inhibitors (MKIs), gaining selectivity is perhaps the highest hurdle facing medicinal chemists. Given the conserved nature of the binding sites for these protein families, it can be very difficult, if not impossible, to rationally design a DML with absolute selectivity for the desired targets. This reality has lead to a pragmatic approach whereby DMLs are deemed to be sufficiently selective to be progressed into toxicity studies. Even if high selectivity cannot be achieved, it is still worthwhile to determine if particular off-target activities are detrimental before terminating the development of an otherwise promising lead compound.

Studying clinical drugs can reveal which anti-targets may be modulated without overt safety concerns and which pose a higher safety risk and should be avoided. For instance, certain kinase anti-targets have been associated with cardiotoxicity and should be avoided.[14] Even if a compound is found to inhibit a target *in vitro* it may not be a problem *in vivo* if there is a sufficient difference in the exposure required to achieve efficacy compared to that producing unacceptable side effects. When screened against large kinase panels, some multi-kinase drugs have been found to highly promiscuous and yet have an acceptable safety profile for treating cancer. For instance sunitinib bound $>15\%$ of kinases tested with $K_d < 100$ nM.

To avoid arriving at erroneous conclusions, the selectivity of DMLs should be evaluated not just at the protein level, but also at the cell/tissue level and preferably also at the level of the whole organism. For example, kinase inhibitors are typically evaluated initially at the protein level using enzymatic or binding assays. It is however necessary to confirm on- versus off-target kinase activity in a more physiological context through further analysis in cell-based and, ultimately, *in vivo* settings.

In a majority of the published studies describing DML specificity, only a small number of anti-targets were selected, typically closely related family members. This may lead to erroneous conclusions about an agent's perceived selectivity. Small structural differences can make significant differences to the selectivity profile so, as well as looking at a wide anti-target panel at the start of a project, selectivity should be closely monitored during the optimization process. Ideally large and diverse anti-target panels are needed to ensure the safety of DMLs. However, it is an expensive, if not impractical, task to determine full selectivity profiles for large numbers of compounds in a lead optimization series. Even with the largest and most diverse commercially available panels, there are likely to be targets that are missed. Unanticipated activities, even for well-studied drugs like imatinib, are still being found via panel screening.[15] One area of research that is attracting increasing interest is the *in silico* prediction of polypharmacology profiles, described in detail in Chapter 4. These methods are currently not sufficiently precise to represent an alternative to panel-based selectivity screening but they do have the potential to direct experimental selectivity screening towards possible areas of concern or opportunity.

There are as yet few literature examples of a prospective approach to rationally remove side activities, although such an approach is undoubtedly occurring in many laboratories. A favoured strategy for discovering complex profile DMLs could be to first identify a non-selective inhibitor and then attempt to 'design out' undesired activities. Undesired activities can fall into two general categories, those that may be removed via changes to the global physicochemical properties of the molecule and those that require more subtle changes to specific regions of the molecule. In the former category, binding to promiscuous proteins such as the cytochrome P450s and the hERG channel frequently correlates with lipophilicity so reducing the global $c\log P$ of a molecule may be a profitable approach. In Chapter 17, a successful example of the designing out of hERG and CYP activity is described for a dual SERT/NET inhibitor.

The removal of binding to a closely related protein is less likely to be solved by such an approach and instead benefits from a more precise understanding of the differences in the pharmacophores between the desired and undesired targets. In the case of soluble proteins like kinases, biostructural information can be extremely useful for rationally removing side activities, as illustrated in the following example.

Heerding *et al.* described the discovery of a pan-AKT inhibitor. The starting compound **6** was a modest inhibitor of the AKTs, with only micromolar potency for AKT2 and poor selectivity over other AGC family kinases (Figure 10.3). A significant feature of this work is that a docking model of **6** at

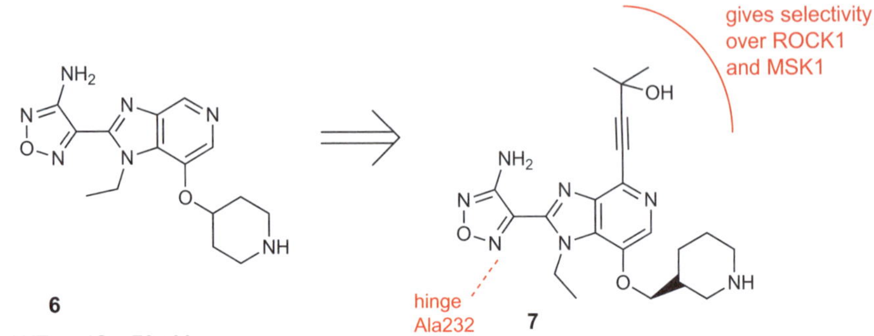

6

AKT_1 IC_{50} 79 nM
AKT_2 IC_{50} 1000 nM
AKT_3 IC_{50} 398 nM
MSK_1 IC_{50} 21 nM
$ROCK_1$ IC_{50} 8 nM

7

AKT_1 IC_{50} 2 nM
AKT_2 IC_{50} 13 nM
AKT_3 IC_{50} 9 nM
MSK_1 IC_{50} 8000 nM
$ROCK_1$ IC_{50} 890 nM

Figure 10.3 Optimization of wider selectivity using structure-based drug design.

the AKT2 active site was used to guide potency improvements at the AKTs and, at the same time, reduce activity at ROCK and RSK1 by incorporation of an additional methyl-3-butyn-2-ol group into **6**. Compound **7** was co-crystallized with AKT2, which confirmed the binding mode predicted by the docking study. Although compound **7** had poor oral exposure, it was progressed into clinical trials as an intravenous agent (GSK690693) to treat patients with solid tumours or hematological malignancies. This example shows how a detailed understanding of structural differences in the desired and undesired kinase binding sites can assist medicinal chemists to remove undesired off-target activities. However compound **7** still possessed undesired activity at the PKA and PKC isozymes, clearly illustrating the extent of the selectivity challenge in the kinase area. Moreover, Bamborough *et al.* screened 577 diverse compounds versus 203 protein kinases and found that two-thirds of the compounds bound to no less than ten kinases.[16]

This strategy of identifying anti-targets associated with toxicity and then rationally designing out those activities is also relevant to aminergic GPCR ligands for treating CNS disorders. Older schizophrenia drugs such as clozapine might be cleaned of undesired activities such as adrenergic alpha$_1$ receptor antagonism associated with CV side effects and the histamine H_1 receptor antagonism associated with weight gain.[17] Bonnert *et al.* reported a successful example of designing out adrenergic alpha$_1$ receptor activities from a dual dopamine D_2/adrenergic beta$_2$ agonist.[18] Similarly, Atkinson *et al.* discussed how undesired adrenergic beta$_2$ receptor activity was effectively removed from a 5-HT_{1A}/SERT ligand **8** by using knowledge of the anti-target pharmacophore (Figure 10.4).[19]

One concern often associated with multi-target agents is whether by virtue of their promiscuous character they are more likely to hit a wider range of

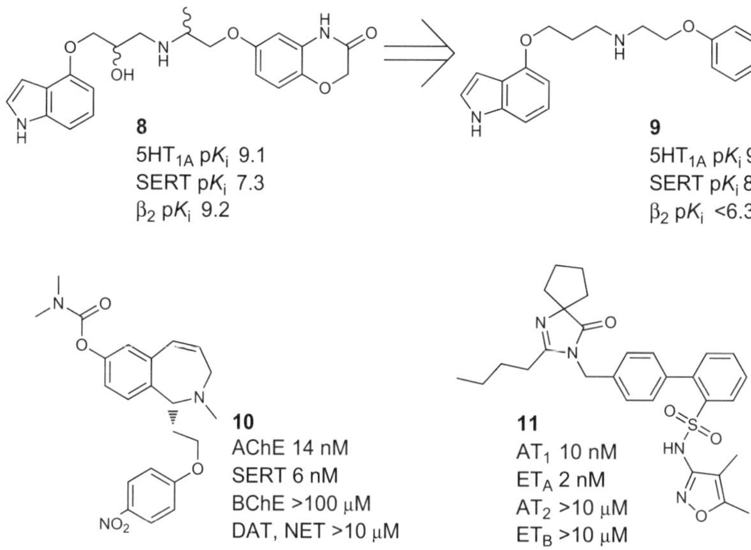

Figure 10.4 Optimization of DML selectivity.

anti-targets. The medicinal chemistry challenge is made somewhat easier if the desired targets are pharmacophorically similar to each other but pharmacophorically dissimilar to the undesired targets. If the desired targets are more dissimilar, it may only be possible to bridge such distant targets with highly promiscuous ligands that also hit more closely related undesired targets. Nonetheless, there are several examples that provide optimism to the medicinal chemist that surprising activity and selectivity profiles can be achieved. Kogen *et al.* reported a dual acetylcholinesterase (AChE)/SERT blocker such as **10** which possessed high selectivity over several homologous targets, including butyrylcholinesterase and the norepinephrine/dopamine transporters (NET/DAT) (Figure 10.4).[20] Similarly the AT_1/ET_A antagonist **11** described by Natesan Murugesan in Chapter 19 was selective over the closely related AT_2 and ET_B receptors.

10.4 Physicochemical Properties

It has been reported that DMLs reported in the literature tend towards higher molecular weight and lipophilicity than either marketed drugs or preclinical compounds in general (Figure 10.5).[21,22] This is a critical issue during lead optimization since such properties are often associated with poor oral absorption, high rates of metabolism and undesired polypharmacology.[23,24]

The generally less favourable physicochemical properties of historical DMLs can be rationalized by the historical popularity of the framework combination strategy whereby the molecular frameworks from two selective ligands are

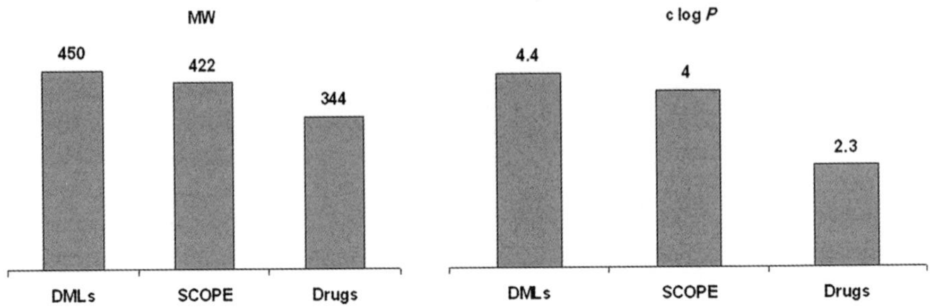

Figure 10.5 Median molecular weight and c log *P* values for designed multiple ligands (DMLs) are higher than those for oral drugs or a general set of pre-clinical compounds from the SCOPE database.[21]

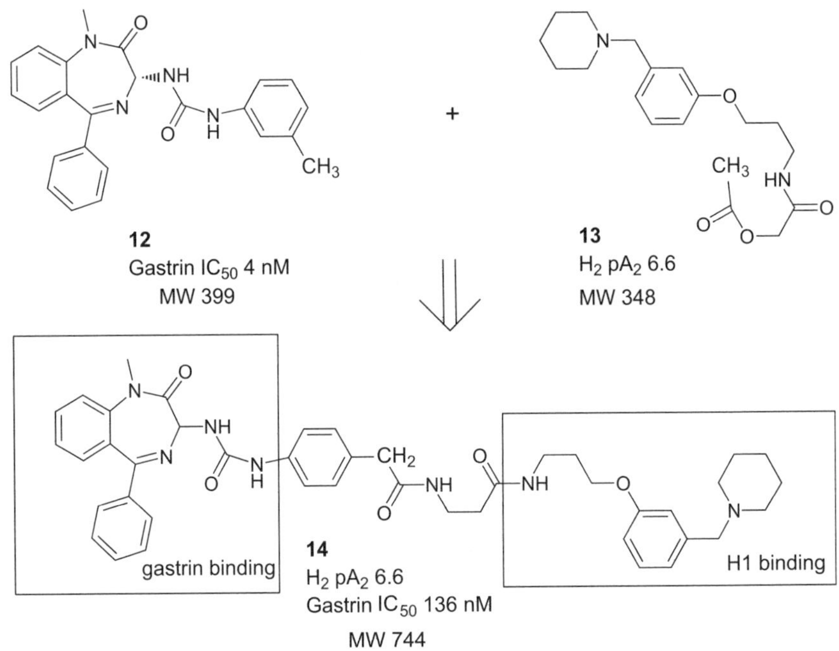

Figure 10.6 An example of a 'fused' DML with non-overlapping pharmacophores, a high MW and low oral absorption.

combined. Since the selective starting compounds are often already drug-like in size, and the extent to which the frameworks are overlapped is generally low, it is not surprising that this approach frequently leads to large property increases which compromise oral bioavailability. This is well illustrated by an example shown in Figure 10.6. Here the framework of a selective gastrin receptor antagonist **12** was combined with that of a histamine H_2 ligand **13**,[25] resulting in a DML **14** with a single carbon atom overlap. The incompatibility of the

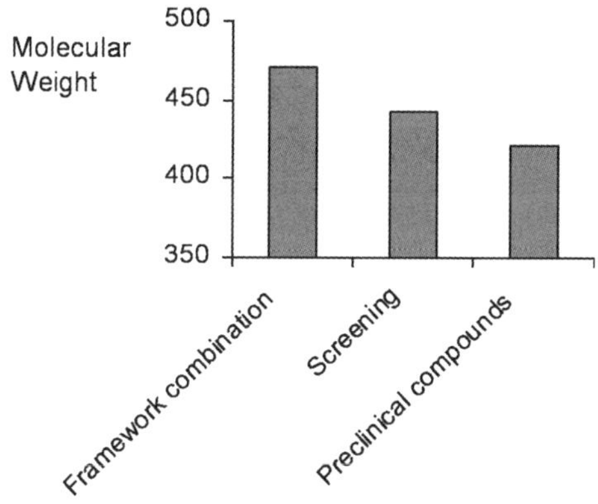

Figure 10.7 Median MW of DMLs derived via framework combination and screening compared to a general set of preclinical compounds.

hydrophobic gastrin pharmacophore with the hydrophilic H_2 pharmacophore produces regions that are only relevant for binding at one of the targets, resulting in a molecule with high molecular weight (MW 744) and compromised oral absorption.

For the framework combination approach to be successful, it is important that the size and complexity of the selective ligands is minimized whilst the degree of the framework overlap is maximized. By following these guidelines, oral drugs designed using this strategy can reach the market, as illustrated by the discovery of ziprasidone, which was launched in 2001 for the treatment of schizophrenia (Chapter 16).

In comparison to framework combination, screening-derived DMLs are frequently smaller molecules with more attractive physicochemical properties (Figure 10.7). The starting compounds obtained via screening already possess multi-target activity, so that achieving the desired DML profile during lead optimization involves addition of only modestly sized groups or modifying the existing functionality. This typically has a less detrimental effect on the physicochemical properties of the molecule than the combination of two drug-like frameworks. Fragment screening has been proposed as a way of obtaining DMLs with better properties than those available using framework combination (Chapter 8).[26]

It has become apparent over recent years that the physicochemical properties of ligands are greatly influenced by the target gene family for which they were designed.[21] A similar target family related effect has also been reported for DMLs.[22] Since the optimal molecular properties that determine absorption across the gut wall are independent of target family, this renders some target combinations more amenable to drug discovery than others. The ligands for transporters, monoamine GPCRs and oxidases generally possess favourable physicochemical properties, hence the feasibility of such targets for DML projects will be relatively high (Figure 10.8). In contrast,

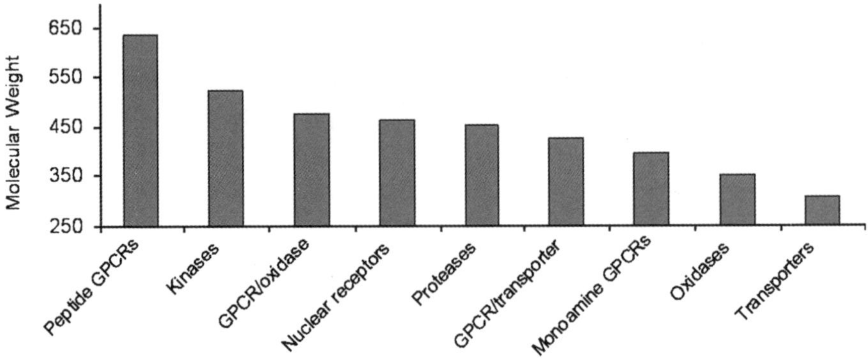

Figure 10.8 Median MW of DMLs classified according to proteomic target family.

peptide and protein binding targets typically exhibit high property values and tractability will be lower.

Nevertheless, there are still encouraging examples in the literature suggesting that even 'difficult' target family combinations such as peptide GPCRs can sometimes be successfully addressed with a creative framework combination approach. Such examples are the dual AT_1/ET_A antagonist described in Chapter 19 and the Bcl-Xl/Bcl2 antagonist described in Chapter 15, which have a good oral bioavailability. In the former case, a strong focus during optimization on structural simplification of the DML lead was the key to success. In the case of the dual Bcl-Xl/Bcl-2 antagonist, a reduction in molecular weight was not compatible with inhibiting this protein–protein interaction but subtle variations in various regions of the molecule improved oral bioavailability. The large size and lipophilicity of DMLs for peptide targets increases the chance that such agents will be substrates for P-glycoprotein (Pgp) and other efflux transporters at the blood–brain barrier. This raises the bar still further for medicinal chemists when the goal is a brain-penetrant DML.

Although aminergic target combinations typically have lower MW ligands that peptidergic ones, issues with high lipophilicity are surprisingly common. This can translate into undesired polypharmacology; for instance, interactions with cytochrome P450s and ion channels. Reducing $c\log P$ for multi-target ligands has been shown to have a beneficial effect on off-target activities such as hERG blockade.[27] Similarly, many multi-kinase inhibitors display high lipophilicity as a consequence of medicinal chemists exploiting the primarily hydrophobic binding pockets adjacent to the small polar hinge recognition site to improve selectivity over other kinases.[28] Consequently there are many examples in the kinase inhibitor literature of trying to reduce lipophilicity during lead optimization. For example, the lipophilicity of lapatinib analogues needed to be reduced to reduce plasma protein binding.[29]

The inherently challenging physicochemical property profiles of DMLs are less problematic if the goal of a project is a parenterally administered drug or a biochemical probe rather than an oral drug. The development of high quality pharmacological tools to explore and validate the potential therapeutic value of novel target combinations is an important area of future research in this field (Chapter 8). Establishing the ground rules for designing such chemical probes is the subject of much current interest within the chemical biology community.[30]

10.5 Summary

Although the field of MTDD is still relatively new, the prospective design of multiple ligands as a means of discovering drugs with superior efficacy and safety profiles is becoming an increasingly tractable task for medicinal chemists. Lead optimization projects where the activity ratio, wider selectivity and physicochemical properties can be most readily fine tuned are most likely to deliver DMLs that combine the optimal PD and PK properties.

References

1. S. M. Stahl, *Biol. Psychol.*, 2002, **52**, 1166.
2. F. P. Bymaster, E. E. Beedle, J. Findlay, P. T. Gallagher, J. H. Krushinski, S. Mitchell, D. W. Robertson, D. C. Thompson, L. Wallace and D. T. Wong, *Bioorg. Med. Chem. Lett.*, 2003, **13**, 4477.
3. D. Angus, M. Bingham, D. Buchanan, N. Dunbar, L. Gibson, R. Goodwin, A. Haunsø, A. Houghton, M. Huggett, R. Morphy, S. Napier, O. Nimz, J. Passmore and G. Walker, *Bioorg. Med. Chem. Lett.*, 2011, **21**, 271.
4. J. A. Lowe, *Curr. Med. Chem.*, 1994, **1**, 50.
5. A. A. Borisy, P. J. Elliott, N. W. Hurst, M. S. Lee, J. Lehár, E. R. Price, G. Serbedzija, G. R. Zimmermann, M. A. Foley, B. R. Stockwell and C. T. Keith, *Proc. Natl. Acad. Sci. USA*, 2003, **100**, 7977.
6. K. Scotlandi, M. C. Manara, G. Nicoletti, P. L. Lollini, S. Lukas, S. Benini, S. Croci, S. Perdichizzi, D. Zambelli, M. Serra, C. García-Echeverría, F. Hofmann and P. Picci, *Cancer Res.*, 2005, **65**, 3868.
7. R. W. Feenstra, A. van den Hoogenband, C. N. Stroomer, H. H. van Stuivenberg, M. T. Tulp, S. K. Long, J. A. van der Heyden and C. G. Kruse, *Chem. Pharm. Bull.*, 2006, **54**, 1326.
8. M. L. Wadenberg, A. Soliman, S. C. VanderSpek and S. Kapur, *Neuropsychopharmacology*, 2001, **25**, 633.
9. P. V. Fish, M. D. Andrews, M. J. Fray, A. Stobie, F. Wakenhut and G. A. Whitlock, *Bioorg. Med. Chem. Lett.*, 2009, **19**, 2829
10. C. Enzensperger, T. Görnemann, H. H. Pertz and J. Lehmann, *Bioorg. Med. Chem. Lett.*, 2008, **18**, 3809.
11. R. J. Vaz, G. D. Maynard, E. M. Kudlacz, L. D. Bratton, J. M. Kane, S. A. Shatzer and R. W. Knippenberg, *Bioorg. Med. Chem. Lett.*, 1997, **7**, 2825.
12. Y. Kawanishi, S. Ishihara, T. Tsushim, K. Seno, M. Miyagoshi, S. Hagishita, M. Ishikawa, N. Shima, M. Shimamura and Y. Ishihar, *Bioorg. Med. Chem. Lett.*, 1996, **6**, 1427.

13. V. Garzya, I. T. Forbes, A. D. Gribble, M. S. Hadley, A. P. Lightfoot, A. H. Payne, A. B. Smith, S. E. Douglas, D. G. Cooper, I. G. Stansfield, M. Meeson, E. E. Dodds, D. N. C. Jones, M. Wood, C. Reavill, C. A. Scorer, A. Worby, G. Riley, P. Eddershaw, C. Ioannou, D. Donati, J. J. Hagan and E. A. Ratti, *Bioorg. Med. Chem. Lett.*, 2007, **17**, 400.
14. T. Force, D. S. Krause and R. A. Van Etten, *Nat. Rev. Cancer*, 2007, **7**, 332.
15. M. Bantscheff, D. Eberhard, Y. Abraham, S. Bastuck, M. Boesche, S. Hobson, T. Mathieson, J. Perrin, M. Raida, C. Rau, V. Reader, G. Sweetman, A. Bauer, T. Bouwmeester, C. Hopf, U. Kruse, G. Neubauer, N. Ramsden, J. Rick, B. Kuster and G. Drewes, *Nat. Biotechnol.*, 2007, **25**, 1035.
16. P. Bamborough, D. Drewry, G. Harper, G. K. Smith and K. Schneider, *J. Med. Chem.*, 2008, **51**, 7898.
17. Y. Von Coburg, T. Kottke, L. Weizel, X. Ligneau and H. Stark, *Bioorg. Med. Chem. Lett.*, 2009, **19**, 538.
18. R. V. Bonnert, R. C. Brown, D. Chapman, D. R. Cheshire, J. Dixon, F. Ince, E. C. Kinchin, A. J. Lyons, A. M. Davis, C. Hallam, S. T. Harper, J. F. Unitt, I. G. Dougall, D. M. Jackson, K. McKechnie, A. Young and W. T. Simpson, *J. Med. Chem.*, 1998, **41**, 4915.
19. P. Atkinson, S. Bromidge, M. Duxon, M. Laramie, L. Gaster, M. Hadley, B. Hammond, C. Johnson, D. Middlemiss, S. North, G. Price, H. Rami, J. Riley, C. Scott, T. Shaw, K. Starr, G. Stemp, K. Thewlis, D. Thomas, M. Thompson, A. Vong and J. Watson, *Bioorg. Med. Chem. Lett.*, 2005, **15**, 737.
20. N. Toda, K. Tago, S. Marumoto, K. Takami, M. Ori, N. Yamada, K. Koyama, S. Naruto, K. Abe, R. Yamazaki, T. Hara, A. Aoyagi, Y. Abe, T. Kaneko and H. Kogen, *Bioorg. Med. Chem.*, 2003, **11**, 4389.
21. J. R. Morphy, *J. Med. Chem.*, 2006, **49**, 2969.
22. J. R. Morphy and Z. Rankovic, *J. Med. Chem.*, 2006, **49**, 4961.
23. C. A. Lipinski, F. Lombardo, B. W. Dominy and P. J. Feeney, *Adv. Drug Delivery Rev.*, 1997, **23**, 3.
24. P. D. Leeson and B. Springthorpe, *Nat. Rev. Drug Discovery*, 2007, **6**, 881.
25. Y. Kawanishi, S. Ishihara, T. Tsushim, K. Seno, M. Miyagoshi, S. Hagishita, M. Ishikawa, N. Shima, M. Shimamura and Y. Ishihar, *Bioorg. Med. Chem. Lett.*, 1996, **6**, 1427.
26. R. Morphy and Z. Rankovic, *Drug Discovery Today*, 2007, **12**, 156.
27. G. A. Whitlock, P. V. Fish, M. J. Fray, A. Stobie and F. Wakenhut, *Bioorg. Med. Chem. Lett.*, 2008, **18**, 2896.
28. R. Morphy, *J. Med. Chem.*, 2010, **53**, 1413.
29. A. G. Waterson, K. G. Petrov, K. R. Hornberger, R. D. Hubbard, D. M. Sammond, S. C. Smith, H. D. Dickson, T. R. Caferro, K. W. Hinkle, K. L. Stevens, S. H. Dickerson, D. W. Rusnak, G. M. Spehar, E. R. Wood, R. J. Griffin and D. E. Uehling, *Bioorg. Med. Chem. Lett.*, 2009, **19**, 1332.
30. T. I. Oprea, C. G. Bologa, S. Boyer, R. F. Curpan, R. C. Glen, A. L. Hopkins, C. A. Lipinski, G. R. Marshall, Y. C. Martin, L. Ostopovici-Halip, G. Rishton, O. Ursu, R. J. Vaz, C. Waller, H. Waldmann and L. A. Sklar, *Nat. Chem. Biol.*, 2009, **5**, 441.

CHAPTER 11

Combination Agents Versus Multi-Targeted Agents – Pros and Cons

JOSE G. MONZON AND JANET DANCEY*

National Cancer Institute of Canada, Clinical Trials Group, 10 Stuart Street, Kingston, ON K7L 3N6, Canada
*Email: jdancey@ctg.queensu.ca

11.1 Introduction

Although the vast majority of diseases are multi-factorial in nature, most modern drug discovery is based on identifying a drug that acts on a single derangement felt to be involved in disease development or progression. Due to the multi-factorial nature of most diseases, a selective compound for a single target rarely achieves the desired effect and is often combined with standard treatments or other novel targeted agents to improve effectiveness. This could not be truer for novel anti-cancer molecularly targeted therapeutics (MTTs).

Most curative cancer treatment is based on identification of effective drug combinations. The success of combinations is likely due to the fact that cancer is a heterogeneous disease among patients and within the same patient. Cancer cells are genotypically and phenotypically complex and adaptive. There may be *de novo* protective mechanisms that render individual drugs ineffective. In addition, acquired resistance occurs with almost all agents over time unless the therapy is curative. Historically, the goal of cytotoxic agents was to maximize

RSC Drug Discovery Series No. 21
Designing Multi-Target Drugs
Edited by J. Richard Morphy and C. John Harris

Published by the Royal Society of Chemistry, www.rsc.org

tumor cell kill. The limited selectivity of conventional cytotoxic cancer drugs was based on their disruption of the frequent cell division and DNA replication of cancer cells relative to most normal cells. Most cytotoxic cancer drugs act by inhibiting synthesis of DNA precursors, damaging the DNA template, or disrupting chromosomal segregation. However, rapidly dividing normal tissues, such as those of the bone marrow, gastrointestinal tract, and hair follicles were also affected. Ultimately, these side effects would result in suboptimal dosing because of normal tissue toxicity, resulting in reduced efficacy, drug resistance, and decreased quality of life for patients. In contrast, the goals of rational combinations of MTTs are to achieve durable tumor control, which may lead to better therapeutic outcome through simultaneous blockade of cancer-relevant targets in properly selected patients.

Following decades of research, a plethora of genes have been identified that are differentially expressed in cancer cells with the potential to act as molecular targets for anti-cancer drugs. Numerous molecularly targeted agents are now approved (see Table 11.1) and are being developed, with the hopes that they have improved anti-cancer activity and fewer side effects. One of the main differences between the development of conventional cytotoxic agents and newer targeted agents is in the way they are designed. Cytotoxic agents were discovered empirically by screening several different natural or synthetic compounds for their anti-cancer properties. Screening was usually done in rapidly proliferating human or murine cancer cell lines. Now, a more rational approach to drug

Table 11.1 Approved molecularly targeted agents.

Agent	*Target*	*Tumour type*	*Agent class*
Imatinib	BCR-ABL chromosomal translocation, PDGFR, C-KIT	CML, CMML, ALL, DFSP, GIST	TKI
Dasatinib	BCR-ABL chromosomal translocation	CML, ALL	TKI
Erlotinib	EGFR	NSCLC	TKI
Cetuximab	EGFR	CRC, Head and Neck	Monoclonal antibody
Panitmumab	EGFR	CRC	Monoclonal antibody
Trastuzumab	HER2	Breast	Monoclonal antibody
Lapatinib	HER2	Breast	TKI
Bevacizumab	VEGF	CRC, NSCLC	Monoclonal antibody
Sunitinib	VEGFR	Kidney	TKI
Sorafenib	VEGFR	Kidney	TKI
Temsirolimus	mTOR	Kidney	Rapamycin analogue
Azacitidine	DNA methyltransferase	MDS	Pyrimidine analogue
Decitibine	DNA methyltransferase	MDS	Pyrimidine analogue
Vorinostat	Histone deacetylase	CTCL	Hydroxamic acid
Bortezemib	Proteosome	Multiple Myeloma	Proteosome inhibitor

Abbreviations: ALL, acute lymphocytic leukemia; BCR-ABL, break-point cluster region-Abelson; CML, chronic myelogenous leukemia; CMML, chronic myelomonocytic leukemia; CRC, colorectal cancer; CTCL, cutaneous, T-cell leukemia; DFSP, dermatofibrosarcoma protuberans; EGFR, epidermal growth factor receptor; HER2, human epidermal growth factor receptor 2; MDS, myelodysplastic syndrome; mTOR, mammalian target of rapamycin; NSCLC, non-small cell lung cancer; TKI, tyrosine kinase inhibitor; VEGF, vascular endothelial growth factor; VEGFR, VEGF receptor.

design is being pursued. In contrast to conventional chemotherapy agents, most targeted agents do not directly damage DNA or interfere with its replication, but rather disrupt the function of abnormal cellular signaling cascades of tumor cells that promote cancer or stromal cell division and survival. These agents more frequently inhibit cell proliferation rather than induce apoptosis and result in inhibition of tumor growth rather than induce tumor regression.

Previously, the norm has been to evaluate these targeted drugs individually and in combination with standard cytotoxic agents. Currently, the emergence of numerous targeted agents in a relatively short period of time has resulted in attempts to combine multiple targeted agents, even in the absence of clinically relevant single agent activity. As the number of drug combinations is limitless, a strategy for determining the most promising combinations and prioritizing their evaluation is crucial. To do so requires greater knowledge of these targeted agents and their combinations in regards to tumor biology, mechanisms of interaction between the agents and their reported targets, mechanisms of resistance, and improved assessment of their actions in preclinical and clinical settings. In addition, individual agents may be designed to relatively and selectively inhibit a specific target or more broadly inhibit multiple targets at clinically achievable drug exposures. Thus, there is the potential for one agent to inhibit multiple potentially relevant targets. Such multi-targeted agents may be easier to develop than novel drug combinations. There are a number of advantages and disadvantages to trying to create a molecule that can inhibit more than one cancer relevant target versus combining individual agents that are relatively selective for specific targets. In this chapter, we will review the rationale for combination therapy in cancer, the relevance of combination strategies, and the strengths and weaknesses of selective and multi-targeted agents as combinations for cancer therapy.

11.2 Principles of Combination Chemotherapy for the Treatment of Cancer

Combination therapy is an important treatment modality in many disease settings, including hypertension, dyslipidemia, tuberculosis, human immunodeficiency virus (HIV) infections, and cancer. Multi-agent cancer therapies are based on the assumption that combining agents may result in increased therapeutic benefit by overcoming mechanisms of resistance or enhancing the vulnerability of the cancer to individual agents. Prior to the advent of molecular biology, when the genetic underpinnings of cancer could not be studied, classical cytotoxic cancer agent combinations were designed based on empirical evidence of activity, non-overlapping toxicity of the individual agents, and on theoretical/mathematical models of tumor cell kinetics and drug resistance. Based on the clinical results of cytotoxic regimens on cancer patients, several principles emerged in regards to combining traditional cytotoxic agents. Generally, these principles can also be applied to the combination of MTTs, but with certain caveats (discussed below).

The development of cytotoxic combination therapies for cancer was based on three postulates: (1) the cumulative logarithmic cell kill as individual agents are combined; (2) the inverse relationship of drug effectiveness to tumor burden; and (3) the intrinsic mutation rate of cancer cells increases the probability that even relatively small tumors will have clones with mutations that could render them resistant to individual drugs. From these postulates much of modern cytotoxic therapy has been developed, based on the following principles: that drugs in a combination (1) should be individually active; (2) should have different mechanisms of action; (3) should have non-overlapping mechanisms of resistance; (4) should have non-overlapping toxicities; and (5) should be administered at maximum doses and schedules. The potential relevance of these principles to the development of targeted agent combinations will be demonstrated and discussed in the following sections.[1–3]

11.2.1 Principle #1: All Drugs Must be Active as Single Agents

The principle that drugs should be individually active is based on the desire to maximize tumor cell kill. It was initially postulated that cancer cell growth was logarithmic and that combination chemotherapy regimens should induce multiplicative log kills.[4] This theory states that a specific dose of chemotherapy would produce an associated log cell kill that was independent of the number of cells in the starting tumor. For instance, a specific dose of chemotherapy that could kill one log of cells would result in 90% reduction of the original tumor cell number. Each additional agent to a regimen would result in the addition of a log kill: two agents would result in a two-log kill and a 99% decrease in cell number and three agents would result in a three-log kill and a 99.9% decrease in cell number. This multiplicative log kills model rationalized the implementation of multiple agents in the treatment of cancer. This theory was favored for its simplicity and ability to model cancer cell growth rate, tumor bulk, and the multiplicative log cell kill of combination cytotoxic regimens in a murine leukemia model.[4] However, it was soon apparent that logarithmic growth was the exception rather than the rule, and for most other cancers (in particular solid tumors) a sigmoidal Gompertzian growth curve was the norm.[5–9]

The Gompertzian growth model predicts that as a tumor gets larger, the doubling time gets longer and the growth fraction gets smaller. Based on this decrease in cell production and lower growth fraction, a larger tumor theoretically responds more poorly to a given dose of cytotoxic chemotherapy than a smaller tumor. The Norton-Simon model embraced the concept of Gompertzian growth to explain clinically observed phenomena and rationalize multiple agent treatment strategies.[10–12] The Norton-Simon model proposes that a tumor is composed of populations of faster-growing cells, which are sensitive to therapy, and slower-growing, more resistant cells. The proportion of slower proliferating and thus resistant cells increases as a tumor gets larger. The model predicts that the log cell kill will be greater for smaller cancers, that only therapy that completely eradicates all tumor cells will be curative, and that this is most likely to occur with sequential, non-cross-resistant regimens at high doses and alternating regimens over more than one cycle. The initial regimen must be effective enough to result in a low residual tumor burden and is followed by one or more non-cross-resistant treatments to eradicate the remainder of the cancer.

Consistent with the Norton-Simon hypothesis, clinical testing has demonstrated that combining ineffective drugs has rarely produced effective regimens for conventional cytotoxic agents. Notable exceptions have been the combinations of 5-fluoruracil with leucovorin or oxaliplatin in colon carcinoma, where the activity of the combination is greater than the additive effects of the individual agents.[13,14] MTTs may also be exceptions to this principle. MTTs typically are cytostatic as opposed to cytotoxic, making their evaluation more complicated. The evidence required from preclinical studies evaluating MTTs must be weighed differently than traditional cytotoxic agents, as cancer cell death *in vitro* or in animal studies may not be the most accurate measure of a targeted agent's efficacy. Rather, MTTs may be evaluated based on their ability to act how they were designed. For instance, if the targeted agent was extremely effective at inhibiting a particular growth-promoting pathway felt to be crucial in cancer development, but did not have a cytotoxic effect in preclinical models, the agent should be combined with other agents before being deemed ineffective. This additional evaluation is warranted as the effectiveness of inhibition of multiple relevant cancer pathways may lead to a greater than additive therapeutic effect. As a result, we can argue that if a MTT is not active as a single agent in preclinical studies, it should not be discarded, but rationally combined with other agents and the combination tested in preclinical models. Whether this is feasible is another issue, as additional costs and lengthened developmental timelines may be prohibitive. Indeed, all MTTs that have been approved for cancer treatments to date also demonstrate single agent activity in clinical trials.

11.2.2 Principle #2: Drugs Should be Chosen for Non-Overlapping Toxicity

This principle is particularly true for traditional cytotoxic agents, as patients may be able to tolerate the maximum tolerated doses of each drug, without requiring dose reductions, and benefit from the additive advantages of the drug combination. This principle also applies to the combination of MTTs; however, toxicity may be more difficult to avoid, as combining agents directed towards the same target, pathway, or collateral pathways may produce greater mechanism-based or off-targeted toxicities and chronic schedules may lead to intolerable lesser grades of toxicities that require dose/schedule adjustments. For example, combinations of agents affect certain targets: vascular endothelial growth factor (VEGF) and its receptor (VEGFR) produce greater target-specific hypertension and proteinuria that requires dose modifications of the individual agents.[15] Significantly increased skin and gastrointestinal toxicity has required dose and/or schedule modifications when growth factor receptor inhibitors have been combined with downstream cytoplasmic kinase inhibitors.[16] Dose and schedule modifications required to limit toxicity may affect the activity of the drugs in combination as many MTTs reversibly inhibit the activation of protein kinases and the degree of target inhibition is usually proportional to drug exposure.

11.2.3 Principle #3: Drugs Should be Chosen for Different Synergistic Mechanisms of Action

This principle holds true for both cytotoxic agents and MTTs. Combinations of drugs with synergistic mechanisms of action should minimize drug resistance and maximize cellular effects. With MTTs, this principle takes a different spin where synergistic mechanisms of action may translate into combining agents acting on the same target, or acting on targets in the same pathway, or two different pathways or processes involved in the neoplastic process. Synergy is difficult to demonstrate clinically with cytotoxic agents, but this may be observed with MTTs particularly in combinations that exploit cancer specific vulnerabilities. The striking activity of PARP inhibitors which inhibit DNA repair when combined with a platin agent in patients with BRCA mutations is an example of such synergy.[17]

11.2.4 Principle #4: Drugs Should be Chosen That Have Different Mechanisms or Patterns of Resistance

Cancer cells may be resistant to agents through intrinsic mechanisms or through adaption from exposure to sub-lethal concentrations of agents. Intrinsic mechanisms include genetic mutations and phenotypic alterations that render cancer cells resistant. The Goldie-Coldman hypothesis is a mathematic model that predicts that tumor cells mutate to a resistant phenotype at a rate dependent on their intrinsic genetic instability.[18] The probability that a cancer would contain drug-resistant clones depends on the mutation rate and the size of the tumor. According to this hypothesis, even the smallest detectable cancers would contain at least one drug-resistant clone; therefore, the best chance of cure would be to use all effective chemotherapy drugs; in practice, this has meant using two or more different non-cross-resistant chemotherapy regimens in alternating cycles at maximum tolerable doses and schedules. In additional, cells may acquire a multi-drug resistant phenotype through over expression of drug efflux proteins, drug metabolism enzymes, or other means, which are potentially relevant to the efficacy of any drug.

Molecularly targeted agents are not immune to drug resistance. In fact, mechanisms of resistance may be more complex for MTTs. Within a specific patient, the target of an agent may be irrelevant to cancer cell proliferation and survival and thus inhibition within an individual patient will not induce an anti-cancer effect; the target may have a mutation that impairs drug binding; compensatory pathways may circumvent the effect of target inhibition; and multi-drug resistant phenotype may impede the ability of the drug to enter and be retained within the cell to get to the target (see Table 11.2). An excellent example of MTT resistance with a biological basis was demonstrated in

Table 11.2 Clinical Trials for Combinations of MTAs.

	Targets	*Clinical trials*	*Tumour types*
Increased inhibition of one target	VEGF + VEGFR	Bevacizumab + Sorafenib*	Kidney
	VEGF + VEGFR	Bevacizumab + Cedarinib	Phase 1
	EGFR + EGFR TKI	C225 + Erlotinib*	Colon
	HER2	Trastuzumab + Lapatinib	Breast
Linear target inhibition	VEGF + mTOR	Bevacizumab + Temsirolimus*	Kidney, neuroendocrine, hepatocellular, ovarian, endometrial
	VEGF + mTOR	Bevacizumab + Everolimus	Kidney, neuroendocrine
	VEGF + mTOR	Sorafenib + Temsirolimus	Melanoma, glioblastoma
	EGFR + mTOR	Erlotinib + Temsirolimus	Lung, glioblastoma
	Her-2 + mTOR	Trastuzumab + Everolimus	Breast
	EGFR + mTOR	EGFR TKI + Temsirolimus	NSCLC, glioblastoma
	Her-2 + CDK	Trastuzumab + flavopiridol*	Breast
	IGF-1R + MEK	IMC-A12 + AZD6244	Phase 1
	EGFR + MEK	Erlotinib + AZD6244	Lung
	IGF-1R + mTOR	IMC-A12 + Temsirolimus	Phase I, breast, sarcoma
Parallel pathways	VEGR + EGFR	Bevacizumab + C225*, Erlotinib	Colon, pancreas, kidney
	VEGR + EGFR	Bevscizumab + Cetuximab	Colon, pancreas
	VEGF + PDGF/CD117	Bevacizumab + Imatinib*	Melanoma, gastrointestinal stromal tumour
	EGFR + IGF-1R	IMC-A12 + Erlotinib *	NSCLC
	Her-2 + Her-1	Trastuzumab + gefitinib*	Breast
	mTOR + MEK	AZD6244 + Deforolimus	Phase I
Other	HDAC + VEGF	SAHA + Bevacizumab*	Kidney
	iMID + Proteosome I	Revlimid + Bortezomib*	Multiple myeloma, non-Hodgkins lymphoma, chronic lymphocytic lymphoma
	HDAC + proteasome	SAHA + Bortezomib*	Pancreatic, sarcoma
	HDAC + methylation	SAHA + Azacitadine*	Myelodysplastic syndrome, multiple myeloma
	Vaccine + modulator	Vaccine + anti-CTLA4 Ab*	Melanoma, prostate

Abbreviations: Ab, antibody; CDK, cyclin dependent kinase; CD117, cluster of differentiation 117; EGFR, epidermal growth factor receptor; mTOR, mammalian target of rapamycin; PDGFR, platelet derived growth factor receptor; HDAC, histone deacetylase; Her, human epidermal growth factor receptor; IGF-1R, insulin growth factor- 1 receptor; iMID, immunomodulatory drug; raf, rapidly accelerated fibrosarcoma; TKI, tyrosine kinase inhibitor; VEGF, vascular endothelial growth factor; VEGFR, VEGF receptor.

patients with metastatic colorectal cancer (CRC) treated with cetuximab or panitumumab, anti-epidermal growth factor receptor monoclonal antibodies (anti-EGFR moAb).[19–21] Several derangements activating the EGFR pathway have been identified in CRC, making the anti-EGFR moAbs attractive agents to test in patients with this disease. It soon became clear that not all patients with mCRC responded to cetuximab.[20,21] In fact, the presence of a mutation in the *KRAS* gene correlated with lack of objective response rate and improvement in survival (both progression free and overall survival) in patients with mCRC who received the anti-EGFR moAb alone, in combination with standard cytotoxic agents, or with the VEGF inhibitor bevacizumab[20,22,23] and panitumumab.[19,24–26] Activating mutations in the *KRAS* gene encode a constitutively active protein that acts downstream to the EGFR. As a consequence, the activation of the signaling pathway is no longer reliant on the activation of the EGFR, ultimately bypassing the effects of EGFR inhibitors like cetuximab. As a result of this mechanism of resistance, panitumumab, and cetuximab are approved in Europe and North America for patients with mCRC and *KRAS* wild-type tumors, in the first or second-line setting as monotherapy or in combination with standard chemotherapy. As the understanding of the molecular basis of cancer evolves, so do the molecular discoveries of drug resistance. Efforts to understand the molecular basis of resistance of MTTs will aide in the rational design of MTT combinations.

11.2.5 Principle #5: Drugs Should be Administered at the Optimum Dose and Schedule

Typically, traditional cytotoxic agents are titrated to maximum tolerable organ toxicity and these doses are maintained when they are combined as long as there is no overlapping toxicity. The rationale for dosing cytotoxics to maximum tolerability is that tumor cell tolerance and normal cell tolerance are often closely related. Conversely, MTTs may induce limited target-specific toxicity and should theoretically be dosed to maximize biological effect on the target as opposed to maximum tolerance. Whether this translates into decreased efficacy is unknown.

By design, molecularly targeted agents tend to be less toxic but may also be less effective when used as single agents compared to traditional cytotoxic agents. They act on targets within cellular pathways that are relevant to cancer growth and, on a molecular level, are involved in gene expression, growth regulation, cell cycle control, apoptosis, and angiogenesis. Cancer cells may be more reliant on these pathways and thus more vulnerable to specific inhibitors; however, these pathways are also relevant to normal tissue homeostasis and function. Thus individual MTT generally do cause toxicity; however, the specificity of action of the agents generally results in less collateral damage to crucial cellular machinery as compared to traditional cytotoxic agents.

MTTs can be administered intermittently or continuously; however, most schedules are designed to maximize target inhibition using continuous schedules. Continuous dosing is required as the effect of the drug on the target is exposure dependent and, for many agents, reversible. For instance many tyrosine kinase inhibitors reversibly inhibit the target kinase and their actions can be negated with the competitive inhibition of ATP (adenosine triphosphate). Continuous dosing schedules that may be optimal for individual agents may not be tolerable for combinations as prolonged duration of even relatively mild side effects may lead to patient intolerability.

In summary, it remains to be determined whether these principles that have been derived from experience with combinations of standard cancer cytotoxic agents apply to newer molecularly targeted agents. Certainly, failing to fulfill one or more of the above criteria does not preclude the development of a particular combination of agents. Key features of targeted agents that distinguish them from traditional cancer therapies are that susceptibility to individual agents may be dependent on cancer-specific vulnerabilities found in only subsets of patients; these agents are more likely to induce a cytostatic response that may be prolonged and thus render cancer a chronic disease and that specific combinations of targeted agents may result in synergistic effect analogous to synthetic lethality.

11.3 Comparison of Combinations of Single Target Drugs Versus Multi-Targeted Agents – The Pros and Cons of Each Approach

Until now, we have suggested that MTTs typically act on one target and that combining MTTs has the intention of improving efficacy and reducing the risk of drug resistance. However, a reasonable alternative to developing combinations of targeted agents is to develop a single agent that has multiple targets, which might address concerns that an agent with a limited spectrum of target inhibition is less likely to be effective. In fact, there are numerous agents now available that have been designed to act on multiple molecular targets involved in cancer. There are specific scientific, clinical, and regulatory considerations for multi-targeted versus relatively selective targeted combinations that in certain circumstances may favor one approach over the other; these will be discussed further in this section.

Two general subclasses of multi-targeted agents exist and they differ in regards to how they act on the targets and the induced cellular effects. One class has been designed to have potent activity on several different targets. The second class has potent activity for a single target but has effects on a broad number of additional cellular components. A good example of the former is the multi-targeted kinase inhibitor, sunitinib. Sunitinib is an ATP-mimetic, which binds to the ATP binding pocket of several protein kinases, inhibiting enzyme autophosphorylation and activation. Examples of the second subclass are agents that target protein metabolism such as the proteosome, DNA methylation, or histone deacetylation. By inhibiting cellular processes that regulate multiple targets, these agents can affect multiple cellular processes that can have broad anti-tumor effects. However, their ability to inhibit specific cellular targets relevant to cancer progression or survival may not be predictable within individual cancer patients and they may induce greater normal tissue toxicity as single agents or within drug combinations.

Multi-targeted kinases, such as sunitinib, were initially identified for their ability to inhibit a panel of known and relevant protein kinases using high-throughput analyses.[27] Each compound screened is selected for the unique inhibitory profile against a number of kinases. These agents inhibit multiple kinases and are thus called 'promiscuous' agents. Such agents may be active across a number of different cancer types; however, they often have greater toxicity due to multiple target and off-target effects in normal tissue. The potential advantage of an agent such as sunitinib, which inhibits c-kit, VEGFR (vascular endothelial growth factor receptor), and PDGFR (platelet-derived growth factor receptor) as well as other kinases, is that it may be developed successfully in more than one clinical indication. For example, sunitinib is approved for the treatment of patients with renal cell carcinoma,[28] which may be driven by aberrant VEGF production, as well as gastro-intestinal stromal tumors (GIST),[29] which may be driven by mutations in c-KIT or PDGFR. In addition, the spectrum of kinase inhibition within a tumor may result in greater therapeutic effect. Sunitinib, with its broader spectrum of kinase inhibition, has greater activity in renal cell carcinoma than the monoclonal antibody bevacizumab, which relatively specifically inhibits the VEGF pathway. However, as these agents may be 'jack of all trades, but master of none' in regards to the targets they inhibit, they may not optimally inhibit all the specific individual targets that are particularly relevant to a cancer type or within an individual cancer patient at clinically achievable concentrations and exposures. It is not yet possible to develop kinase inhibitors with a specific kinase-inhibitory profile with optimal potency and therapeutic index for each of the multiple cancer relevant targets. Therefore, even multi-targeted kinase inhibitors might need to be combined with other targeted agents for maximal therapeutic effect. Such multi-targeted agents could also preclude the 'validation' of individual targets, as the effectiveness of the agent could be due to its interaction with any or all of its proposed targets or even result from off-target effects.

The second class of multi-targeted agents act on a crucial mechanism that may result in changes in multiple potential targets. For example, demethylating agents remove methyl groups from cytosine- and guanine-rich areas of DNA, reversing the transcriptional silencing of genes, including tumor suppressor genes (reviewed in ref. 30). The advantage of such agents is that they may alter expression of multiple gene products; however, their ability to influence the expression of specific proteins is not necessarily predictable.

In addition to scientific issues of predictable target modulation and clinical issues of therapeutic index due to target and off-target toxicity, there are other clinical and regulatory advantages and disadvantages to the development and evaluation of multi-targeted agents. One potential advantage of multi-targeted agents over using multiple single targeted agents is that a multi-targeted agent reduces the number of drugs a patient has to take and therefore decreases the risk of drug–drug interactions and patient non-compliance. An agent with multiple targets might have activity in a broader spectrum of cases due to the molecular heterogeneity amongst the same types of cancers amongst different patients. A single agent with multiple targets might be easier to develop given that the regulatory requirements for demonstrating activity and safety of a combination are more arduous than for a single agent. A disadvantage of multi-targeted agents is the potential that combinations of multi-targeted kinase inhibitors might lead to increased toxicity because the cumulative target and off-target inhibition may be broader, and perhaps cause less predictable effects on cellular functions than combinations of target-specific agents.

By contrast, combinations of more specific targeted agents might be more suitable for regimens tailored to individual patients based on the molecular profile of their tumor. As opposed to multi-targeted agents, doses/schedules of the agents within a combination may be tailored for desired concentrations to enhance target inhibition and/or to optimize interactions between agents. The toxicity of the combination might also be more predictable because of the limited off-target effects. More specific targeted agents might allow greater flexibility for tailoring regimens to specific patients and molecular profiles of their cancers. Thus multi-targeted agents may be more likely to have single agent activity in one or more cancer indications; however, more specific MTTs may be better tolerated in combination with other MTTs or standard cancer therapies.

11.4 Defining which Targeted Agents to Combine

Determining which agents to combine is based on the principles described in Section 11.3 on and the molecular knowledge of the presence and relevance of the target to the pathogenesis and growth of a cancer in a given cancer patient or cancer indication. Other factors for consideration include the pharmacokinetics and pharmacology of the agents alone and in combinations. Better understanding of the pharmacokinetics of the single agents and the combinations will help to determine whether there exists the potential for favorable or unfavorable interactions between the two agents. Tolerability and associated inhibition of pharmacodynamic endpoints for the agents will also help determine which combinations of MTTs to advance to phase 3 clinical trials.

A better understanding of the molecular context for activity or resistance to individual drugs and their combinations would aid prioritization for development. The activity of agents in combinations should result in at least additive and ideally synergistic anti-cancer activity. One promising approach to selecting agents is based on synthetic lethal screening. This screen utilizes small interfering RNA to identify the genetic transcripts in cancer cells which, when expression is lessened, results in sensitization to anti-cancer agents. Combinations of agents can be rationalized in this way, as drugs inhibiting the expression of these genes may improve the effectiveness of the drug of interest.[17]

Several potential combinations have been identified, but whether these will translate into clinically relevant combination treatments remains to be seen.

Essentially, there are at least three broad approaches to selecting targeted agents for testing in combination to improve efficacy (see Table 11.2). Assuming that the agents have acceptable pharmacology and effect on target, the target of the second agent in the combination can enhance the activity of the first by: (1) more effectively inhibiting the same target; (2) inhibiting additional targets within the same pathway; or (3) inhibiting a different pathway or cellular process that is involved in the pathogenesis or drug resistance of the specific cancer type. These strategies have been proposed on the basis of our limited understanding of cancer targets/pathways, the availability of the agents, and limited preclinical experiments that suggest at least additive anti-tumor effects of these combination approaches.

11.4.1 Examples and Rationale Behind Combination MTTs

Mechanistically, there are several examples of how two MTTs were rationally combined (see Table 11.2). Although there is no approved combination of MTTs for any cancer, approval for an MTT combination will certainly occur in the near future based on some of the promising data presented below.

11.4.1.1 Combination MTTs Inhibiting the Same Target

Approximately 18–30% of invasive breast cancers exhibit gene amplification or overexpression of HER2 (human epidermal growth factor receptor-2, also known as ErbB2).[31–33] Overexpression and gene amplification of HER2/neu is an adverse prognostic indicator associated with decreased disease-free survival and overall survival in patients with breast cancer. Trastuzumab is a humanized monoclonal antibody targeting the extracellular domain of HER2 and has demonstrated clinical benefit in both the metastatic and adjuvant breast cancer treatment settings.[31,34,35,85]

Preclinical studies of HER2-positive cell lines demonstrated a synergistic interaction of a novel HER2 inhibitor, lapatinib, when combined with trastuzumab, suggesting that dual blockade may be of clinical benefit.[36,37] Moreover, these studies demonstrated that these agents did not have overlapping resistance mechanisms. Lapatinib (see also Chapter 12) is an oral small molecule that inhibits the intracellular kinase domain of both HER2 and the EGFR. Enhanced activity of the combination was suggested by a recent study where the combination of lapatinib and trastuzumab was superior to lapatinib alone in regards to improving progression-free survival, clinical response, and overall survival in patients with HER2-positive metastatic breast cancer, who had previously progressed on trastuzumab treatment.[38] The favorable effect of the combination may be due to enhanced inhibition of the primary target, HER2, either by overcoming mechanisms of resistance (resistance to trastuzumab with or without lapatinib) or enhanced effect on the target (greater HER2 inhibition) or the effect of combining both HER2 and EGFR inhibition. Trastuzumab and lapatinib is an example of a successful combination that demonstrates good therapeutic index, minimal additional toxicities, and increased clinical effect.

This study exemplifies that two MTTs targeting the same receptor, but different functional domains, and with different mechanisms of action and non-overlapping resistant mechanism can enhance efficacy over the single agent activity in heavily pre-treated cancer patients. Moreover, this example exemplifies a disadvantage associated with multi-targeted agents such as lapatinib. Although the majority of the benefit with the addition of lapatinib in this combination is believed to be due to its inhibition of the HER2 receptor based on the preclinical data, the contribution of lapatinib-induced EGFR inhibition to anti-tumor effect in this setting is unknown but unlikely. A phase I/II clinical trial assessing the combination of the EGFR inhibitor gefitinib with trastuzumab in patients with metastatic breast cancer did not identify a favorable interaction between the agents.[39] The EGFR inhibitory effect lapatinib does cause additional gastrointestinal and skin toxicity that may be intolerable in some patients.

11.4.1.2 Combination MTTs Inhibiting Additional Targets Within the Same Pathway

The mammalian target of rapamycin (mTOR) functions at a convergence point downstream to many growth factor receptor pathways, including the EGFR and the VEGFR pathways.[40] Therefore, mTOR inhibitors such as temsirolimus, everolimus, and sirolimus have been combined in clinical trials with other agents that target EGFR and VEGFR signaling. For instance, a phase 2 study evaluated the combinations of bevacizumab, a VEGF inhibitor, and everolimus in patients with advanced clear cell renal carcinoma.[41] The results demonstrated that the combination is active with an overall response rate of 30% and 23% in untreated and previously treated patients respectively. Unfortunately, increased grade 3 and 4 proteinuria was observed with the combination. Everolimus is not known to cause proteinuria and this grade of proteinuria is more than double what is observed with bevacizumab alone.[42] Also, preliminary results from a Phase 1 trial evaluating the combination of bevacizumab and temsirolimus demonstrate partial responses in more than 50% of patients.[43] However, when this combination was compared to bevacizumab alone or interferon alone in a randomized phase 2 study, there was no significant synergy or additivity with the combination regimen.[44] In fact, the toxicity of the temsirolimus and bevacizumab combination was more than expected and led to a high patient drop-out rate in this study. This significant increased toxicity seen with the combination of everolimus or temsirolimus and bevacizumab also was observed with the combinations of temsirolimus with small molecule inhibitors of the VEGF receptor, sunitinib or sorafenib.[45] Toxicity was observed, despite dose reductions and schedule modifications, resulting in waning interest in the dual inhibition of the mTOR and VEGF pathways in these cancers. These examples demonstrate that combining MTTs that act on different targets in the same pathway may not translate into improved efficacy and may come at the expense of worsening toxicity. The side effect profiles observed as single agents worsened with the addition of a second MTT and this increased toxicity may be due, in part, to the more complete inhibition of cellular pathways in both tumor and normal tissues.

11.4.1.3 Combination MTTs Inhibiting a Different Pathway or Cellular Process that are Involved in the Pathogenesis or Drug Resistance of the Specific Cancer Type

Both bevacizumab with chemotherapy[46–48] and erlotinib[49] as a single agent are approved for treatment of patients with advanced non-small cell lung cancer (NSCLC). Erlotinib is an oral agent inhibitor that binds the intracellular tyrosine kinase domain of the EGFR, competitively inhibiting ATP-binding and inhibits receptor autophosphorylation. Preclinical data supported the clinical evaluation of the combination of erlotinib and bevacizumab. Phase 1 and 2 studies of the combination demonstrated clinical activity, no unexpected safety concerns, and reasonable tolerability at full doses for the individual agents.

Based on the preclinical and early clinical trials,[50,51] a randomized phase 3 trial was performed comparing maintenance bevacizumab with or without erlotinib in patients with locally advanced, recurrent, or metastatic NSCLC that had just completed treatment with chemotherapy and bevacizumab.[52] The trial was discontinued early because it had met the primary endpoint by the second planned interim efficacy analysis. The median progression-free survival was improved by just over one month for the combination of bevacizumab and erlotinib (4.8 months) over bevacizumab alone. Moreover, the safety profile for the combination was consistent with the known profiles for the single agents. In contrast to this favorable result seen in a clinical disease setting where both MTTs are individually active, the combination of erlotinib and bevacizumab has also been tested in the phase 2 trials in patients with breast,[53] renal,[54] colorectal,[55] gynecological,[56] mesothelioma,[57] and pancreatic cancer,[58,59] Unfortunately, all trials have yielded disappointing results. Of note in most of these disease settings, one or both agents have been shown to lack single-agent clinical activity, suggesting that, at best, the combination yields additive antitumor activity and does not reverse intrinsic mechanisms of resistance to the individual agents.

In another example, researchers have attempted dual inhibition of both the VEGF and EGFR pathways by combining bevacizumab and chemotherapy with either cetuximab[60] or panitumumab,[25] in hopes of improving outcomes in patients with mCRC. Cetuximab and panitumumab are anti-EGFR monoclonal antibodies (moAbs) that bind to the extracellular portion of the receptor, preventing ligand binding and receptor dimerization. Unfortunately, these trials did not demonstrate any clinical benefit with the addition of an anti-EGFR moAb and may have been a

Table 11.3 Combinations that were not tolerable or active in phase 3 trials.

Targets	*Combinations*	*Indications*
VEGR + EGFR	Bevacizumab + chemotherapy + Cetuximab	Colon* *worse PFS and OS*
	Bevacizumab + Erlotinib	Pancreatic, kidney, breast
	Erlotinib + sorafenib	GBM
VEGF + PDGFR	Bevacizumab + Imatinib	Kidney
mTOR + estrogen	CCI-779 + aromatase inhibitor	Breast*
mTOR + immunomodulator	CCI-779 + INFα	Kidney*
mTOR + EGFR	CCI-779 + Erlotinib	GBM
mTOR + VEGF	CCI-779 + sorafenib	

*Individual agents validated in this setting, but combinations no better.
Abbreviations: EGFR, epidermal growth factor receptor; mTOR, mammalian target of rapamycin; PDGFR, platelet derived growth factor receptor; VEGF, vascular endothelial growth factor; VEGFR, VEGF receptor.

detriment to survival. This was surprising, because efficacy data in early phase clinical trials were promising for the combination of cetuximab and bevacizumab with chemotherapy.[61] Moreover, individually these agents have demonstrated benefit in colorectal carcinoma as single agents or in combination with standard chemotherapy.[25,58,62–68] A possible explanation was that the combination resulted in too much toxicity and/or antagonism between the two MTTs.

Collectively, these clinical results demonstrate that tolerability with non-overlapping side effects and improved efficacy with complementary mechanisms of action are achievable with some combinations of MTTs in certain disease settings. However, the results also demonstrate that greater understanding of molecular and cellular determinants of cancer responsiveness to MTTs is required to be able to prioritize MTTs for testing in combinations. Promising preclinical and early phase clinical trial data does not always translate into an observed clinical benefit in phase 3 clinical trials (see Table 11.3). Despite researchers' efforts to use the best available data to rationalize the combination of MTTs, some trials are still designed empirically, without adequate knowledge of optimal dose, schedule, and target tumor/population.

11.5 Preclinical Evaluation of Combinations

Preclinical combination studies involving *in vitro* cell lines and *in vivo* human tumor models have an essential role in the development of cancer therapies because they provide important insights into mechanism of action and interaction between agents. In general, *in vitro* studies allow the rapid assessment of synergy and the probing of mechanistic interactions. In addition to assessing anti-tumor activity and mechanism of action, *in vivo* studies allow the assessment of additional factors such as the effects of the agent on host, interactions with stromal elements, drug–drug interactions, and pharmacokinetics. However, the negative outcomes for most combinations thus far studied could not have been easily predicted by the preclinical studies. Preclinical studies have been designed with limited consideration of the predictive value of the model or of how the agents are likely to be used in cancer patients. Hence, improvements to the experimental design of preclinical studies to identify synergy or mechanisms of action are needed to enhance clinical development decisions.

Currently, laboratory evaluations can provide insight into the mechanism of action and interaction between the agents within the constraints of the experiment. However, to improve the predictability of preclinical studies and translation of the results into rationally designed clinical studies, several variables must be considered, including the origin of the tumor (that is, cell line versus patient biopsy); target/receptor status of the tumor; the site of tumor implantation (for example, subcutaneous, intraperitoneal, or orthotopic); the size of tumor at the onset of agent treatment; growth rate and growth characteristics; agent dose, formulation, scheduling, and route of administration; and the experimental endpoints.[69]

11.5.1 Factors that Limit the Applicability of *In Vitro* Studies

Even when preclinical evaluations of combinations for synergy have been meticulously performed, there remains a lack of strong correlation between laboratory and clinical outcomes because of the inherent limitations of *in vitro* and *in vivo* models. Cancer cell lines show considerable alterations in biological properties and chemosensitivity patterns when compared with the original tumors from which the cell lines are derived.[70–72] Moreover, cancer cell lines are grown in artificial environments that require additional growth factors and medium, ultimately altering the cell line's molecular characteristics. *In vitro* and *in vivo* cell lines have often been used without molecular characterization for the particular target/or pathway of interest or its manifestation in human cancer specimens. The lack of characterization of tumor cell lines and their correlation to human tumors limits the value of such models for predicting the clinical activity of combinations of anticancer agents. Furthermore, limiting experimentation to a few representative cell lines cannot reflect the diversity of human cancers and therefore cannot be expected to be predictive of clinical outcome.

To address some of these issues, an *in vitro* panel of 60 cell lines (National Cancer Institute's 60, NCI60), of different tumor types and known molecular characteristics, has been assembled to enable high-throughput testing of all possible combinations of MTTs at drug exposures/concentrations that are achievable in humans.[73,74] In this way, the molecular basis of response and resistance can be studied and only the most effective combinations are chosen for development. The strengths of this approach are that it is unbiased in regards to the rationale of the combination and has the ability to screen several combinations quickly and across a number of cancer types and help elucidate the molecular underpinnings of response and resistance. Due to its inclusivity, this limits the chance of missing an effective combination that would not have normally been combined. Preliminary data generated using the NCI60 panel has identified some potential synergistic combinations of MTTs and clinical studies are currently underway to determine the relevance of these combinations.[75]

For *in vitro* assessment of synergy a number of principles have been proposed to improve quality and interpretability. Using these principles a microcomputer fluorescence-based cytotoxicity assay for preclinical testing of combination agents (DIMSCAN)[76] has been developed. The DIMSCAN assay is a semi-automatic microscopy system that quantifies relative total or viable cell numbers in tissue culture multiwall plates. As such it is a high-throughput cytotoxicity screen that can be used to test MTT combinations. The DIMSCAN assay has the following ideal attributes: a wide dynamic range, ideally 3–4 logs of cell kill; several cell lines are employed in the panel, including drug-resistant lines; major mechanisms of resistance can be identified and used to structure the cell line panel; drugs are dosed at clinically achievable levels and schedules; and the combination MTTs can be tested under hypoxic conditions, because hypoxia might antagonize drug action;[77] and possible caveats related to culture conditions should be considered, such as effects of growth-factor supplementation or protein-binding differences between culture conditions and the clinical population.[78]

11.5.2 Factors that Limit the Applicability of *In Vivo* Studies

In some respects, *in vivo* models may better approximate human cancers; however, there remain important differences between *in vivo* xenograft models and cancer in patients. A xenograft is the transplantation of living cells from one species to another. In the cancer research realm, this often implies the transplantation of human tumors in immunologically compromised mice. Importantly, these models can evaluate response, pharmacokinetics, and pharmacodynamics of a drug and combination of drugs. There are limitations to xenograft models. For example, when human tumors are injected into subcutaneous sites, they do not generally metastasize, and therefore are not a good model for evaluating anti-metastatic strategies. For anti-angiogenic, anti-vascular, or immunological strategies, the targets are the host components and are therefore of murine, not human, origin. These factors are not unique to the evaluation of drug combinations, as they also need to be considered for single-agent experiments. Drug interactions can also be studied *in vivo*, and demonstrations of synergy in mouse xenografts have been reported.[79,80] However, the variables that can influence the outcomes of synergy studies in xenograft studies are numerous, requiring large numbers of animals to achieve statistically valid results, and relatively few laboratories have the resources to perform such evaluations.[81] Resource constraints have therefore led to a strategy of defining synergy in robust tissue culture models, and then confirming a beneficial interaction between the two drugs in more limited xenograft studies.

Correlations between results achieved in the laboratory and in the clinic may be improved if *in vivo* and *in vitro* experiments were designed to better reflect the likely clinical trial experiment. *In vivo* and *in vitro* studies should use the most appropriate tumor models with known target/receptor status of the tumor, using doses/schedules of the agents (both alone and in combination) that result in achievable concentrations/exposures in patients.[69] As most agents and combinations are tested in patients with advanced or metastatic disease that is resistant to standard agents, the use of such models to demonstrate activity of an investigational combination would be ideal. Finally, if a combination is to be compared with a standard treatment regimen in a clinical trial, non-clinical experiments might be improved if an active control representative of the standard drug treatment were included.

11.6 Challenges in the Clinical Development of Drug Combinations

Optimal clinical testing of MTTs requires addressing regulatory issues, trial design, and pharmacodynamic assessment challenges. For instance, regulatory agencies generally require the demonstration of safety of the individual agents prior to first in-human testing of the combination and the demonstration of the contributions of the individual agents to efficacy of the combination for regulatory submission. These requirements have delayed the assessment of novel combinations of MTTs until after the agents have been assessed as single agents and in combination with standard therapies. In addition, regulatory, and intellectual property issues may delay the testing of investigational agents from different companies developing several unique MTTs. Combination trials may not only generate new intellectual property, but may also generate novel toxicities and adverse outcomes. Which company takes responsibility for such matters becomes more complicated once more than one pharmaceutical company is involved.

To facilitate such trials of novel MTT combinations, the Cancer Therapy Evaluation Program (CTEP) of the National Cancer Institute (NCI) has developed data-sharing and patent rights language that has been accepted by many industry collaborators. Essentially, the language fosters that each collaborator shall receive non-excusive royalty-free licensing to the intellectual property on the combination for all purposes including that for commercial use. Pharmaceutical companies are also entering agreements to develop novel combinations, as exemplified by the AstraZeneca and Merck agreement to test the combination of two MTTs, an AKT and MEK inhibitor, each owned by one of the companies.[82] The Food and Drug Administration (FDA) also recently announced draft guidance on the development of new molecular entities, providing a regulatory pathway for development of combinations.[71b] Continued efforts to encourage collaboration among pharmaceutical companies are needed to facilitate trials combining MTTs.[82]

Another challenge to combination therapy with MTTs is the heightened potential for acute or delayed toxicities, target and off-target toxicities, and pharmacokinetic interactions. As a result of these issues, new trial designs and endpoints may be required to assess novel MTT combinations compared to traditional designs used to assess effects of MTTs alone or in combinations with cytotoxic agents. Experience to date with certain drug combinations suggests that cumulative acute and delayed toxicity may be greater as multi-targeted agents are combined. For example, phase 1 trials of the mTOR inhibitor, temsirolimus, with VEGFR inhibitors sunitinib,[16] sorafenib,[82] and bevacizumab[43] suggest that the broader the kinase inhibitor profile, the less likely it may be tolerated in combination. Sunitinib and temsirolimus were intolerable at even half the recommended doses of the individual agents, the sorafenib combination required modifications either to one or both agents, while the bevacizumab combination appeared to be tolerable at full doses of both agents. However, even this last combination was not well tolerated and did not lead to improved outcomes in renal cell carcinoma patients. These results suggest that concurrent administration of drugs in combination may require modifications in the doses/schedules that are used when the agents are used individually, and that traditional phase 1 and 2 trial designs and endpoints may underestimate the severity toxicity of a combination. Unfortunately, the adverse effects of the combination may only become apparent with negative results from late phase trials.

Clinical trials of MTTs combinations have clearly shown that tolerable schedules and doses for combinations may be different than for the single agents. When such modifications are required, there are no clear 'rules' to guide how doses and schedules should be modified. Additional laboratory experimentation, and pharmacokinetic and pharmacodynamic assessments may provide insights into how drug dose and schedule modifications may lead to improved efficacy. For example, Solit and colleagues found that high dose/short duration of MTTs in combination may result in improved activity in certain models.[83] Moreover, like several other drugs, small molecule kinase inhibitors are often substrates for the cytochrome p450 system of hepatic metabolism, increasing the potential for pharmacokinetic interaction among other MTTs and drugs.[84] Pharmacokinetic and pharmacodynamic assessment of effectiveness of target and pathway modulation in relevant tumor or normal tissues may provide assurance that effective target inhibition is retained. Careful clinical evaluation of these toxicities and pharmacokinetic interactions, and how these affect the administration of combination therapies, are required to determine whether modifications in dose and/or schedule are warranted to achieve improved outcomes. In addition, alternative approaches must be developed to evaluate late or cumulative toxicity, as it is likely that these will affect patient tolerability and compliance.

Pharmacodynamic markers must be developed early in the drug design process. Although MTTs have a molecular target by design, complete understanding of the involved pathways is often uncommon. Hence, agent-specific and pathway-specific biomarkers that predict for MTT effect are not typically incorporated into the first clinical trials in humans. This is particularly important when a lack of efficacy is demonstrated for a particular agent. Moreover, it is important to confirm the effect on targets, especially if the dose and/or schedule have been modified and the mechanism of action/interaction of the combination is unknown. Only with biomarker knowledge can we predict whether the target and pathway responded to the drug appropriately and whether other cellular alterations resulted in drug resistance or toxicity. Moreover, this information is useful when the MTTs are well tolerated and there may not be a maximum tolerated dose, instead the agent would be titrated to biological effect and target modulation. Unfortunately, the ascertainment of pharmacodynamics data leads to increasing complexity, costs, and patient interventions/biopsies.

11.7 Future Directions

We have yet to fully appreciate the benefit promised by combination MTTs. To ensure the rational approach to the implementation of such combinations in the clinic requires the evaluation of the MTTs as single agents and in combination, in regards to mechanisms of action, clinical pharmacology, toxicity profile, and anti-tumor effect. This information is obtained with preclinical evaluation of combinations performed systematically on representative and exhaustive *in vitro* studies, and screened and confirmed using *in vivo* studies. Ideally biomarkers that measure target and pathway specific outcomes should also be identified to help identify those patients that would and those who would not benefit from these combinations. Whether combinations will be progressed to more advanced clinical trials will be based on the achievement of certain objectives: tolerability at concentrations/exposures associated with significant anti-tumor activity, demonstration of mechanism of interaction for synergy, and demonstration of anti-tumor activity for the combination that exceeds that of single agents and/or other standard treatments.

Our current ability to move beyond the empirical selection and evaluation of molecularly targeted agents and their combinations is often limited by inadequate knowledge of the biology and molecular features of specific tumors, inadequate understanding of the mechanisms of action/resistance for individual agents and their combinations, as well as inadequate tools to measure and compare treatment effects in laboratory models that can be predictively used in clinical development. To overcome these barriers, future studies should continue to focus on several essential tasks: a systematic approach to non-clinical studies of targeted agents and combinations; improved efforts to molecularly characterize the targets, pathway, and mechanism of drug resistance and toxicity in both preclinical and clinical settings; and the development of reliable assays for clinical use for either patient selection and/or measuring pharmacodynamic effects. Such an approach could optimize the application of the many molecularly targeted agents now emerging, and fulfill the expectation of further therapeutic success.

References

1. V. T. DeVita, Jr., R. C. Young and G. P. Canellos, *Cancer*, 1975, **35**, 98–110.
2. V. T. DeVita and P. S. Schein, *N. Engl. J. Med.*, 1973, **288**, 998–1006; doi:10.1056/NEJM197305102881905.
3. E. Frei, 3rd, *Cancer Res.*, 1972, **32**, 2593–2607.
4. H. E. Skipper, F. M. Schabel, Jr. and W. S. Wilcox, *Cancer Chemother. Rep.*, 1964, **35**, 1–111.
5. V. P. Collins, R. K. Loeffler and H. Tivey, *Am. J. Roentgenol. Radium Ther. Nucl. Med.*, 1956, **76**, 988–1000.
6. R. Demicheli, *Eur. J. Cancer*, 1980, **16**, 1603–1608.
7. J. S. Spratt, R. A. Greenberg and L. S. Heuser, *Cancer Res.*, 1986, **46**, 970–974.
8. P. W. Sullivan and S. E. Salmon, *J. Clin. Invest.*, 1972, **51**, 1697–1708; doi:10.1172/JCI106971.
9. M. Tubiana, *Acta Oncol.*, 1989, **28**, 113–121.
10. L. Norton and R. Simon, *Cancer Treat. Rep.*, 1977, **61**, 1307–1317.
11. L. Norton and R. Simon, *Cancer Treat. Rep.*, 1986, **70**, 163–169.
12. L. Norton, *Cancer Res.*, 1988, **48**, 7067–7071.
13. A. de Gramont, A. Figer, M. Seymour, M. Homerin, A. Hmissi, J. Cassidy, C. Boni, H. Cortes-Funes, A. Cervantes, G. Freyer, D. Papamichael, N. Le Bail, C. Louvet, D. Hendler, F. de Braud, C. Wilson, F. Morvan and A. Bonetti, *J. Clin. Oncol.*, 2000, **18**, 2938–2947.
14. S. Giacchetti, B. Perpoint, R. Zidani, N. Le Bail, R. Faggiuolo, C. Focan, P. Chollet, J. F. Llory, Y. Letourneau, B. Coudert, F. Bertheaut-Cvitkovic, D. Larregain-Fournier, A. Le Rol, S. Walter, R. Adam, J. L. Misset and F. Levi, *J. Clin. Oncol.*, 2000, **18**, 136–147.
15. N. S. Azad, E. M. Posadas, V. E. Kwitkowski, S. M. Steinberg, L. Jain, C. M. Annunziata, L. Minasian, G. Sarosy, H. L. Kotz, A. Premkumar, L. Cao, D. McNally, C. Chow, H. X. Chen, J. J. Wright, W. D. Figg and E. C. Kohn, *J. Clin. Oncol.*, 2008, **26**, 3709–3714; doi:10.1200/JCO.2007.10.8332.
16. P. H. Patel, P. L. Senico, R. E. Curiel and R. J. Motzer, *Clin. Genitourin. Cancer*, 2009, 24–27; doi:10.3816/CGC.2009.n.004.
17. A. Ashworth, *J. Clin. Oncol.*, 2008, **26**, 3785–3790; doi:10.1200/JCO.2008.16.0812.
18. J. H. Goldie and A. J. Coldman, *Cancer Treat. Rep.*, 1979, **63**, 1727–1733.
19. R. G. Amado, M. Wolf, M. Peeters, E. Van Cutsem, S. Siena, D. J. Freeman, T. Juan, R. Sikorski, S. Suggs, R. Radinsky, S. D. Patterson and D. D. Chang, *J. Clin. Oncol.*, 2008, **26**, 1626–1634; doi:10.1200/JCO.2007.14.7116.
20. C. S. Karapetis, S. Khambata-Ford, D. J. Jonker, C. J. O'Callaghan, D. Tu, N. C. Tebbutt, R. J. Simes, H. Chalchal, J. D. Shapiro, S. Robitaille, T. J. Price, L. Shepherd, H. J. Au, C. Langer, M. J. Moore and J. R. Zalcberg, *N. Engl. J. Med.*, 2008, **359**, 1757–1765; doi:10.1056/NEJMoa0804385.
21. A. Lievre, J. B. Bachet, D. Le Corre, V. Boige, B. Landi, J. F. Emile, J. F. Cote, G. Tomasic, C. Penna, M. Ducreux, P. Rougier, F. Penault-Llorca

and P. Laurent-Puig, *Cancer Res.*, 2006, **66**, 3992–3995; doi:10.1158/0008-5472.CAN-06-0191.
22. C. Bokemeyer, I. Bondarenko, A. Makhson, J. T. Hartmann, J. Aparicio, F. de Braud, S. Donea, H. Ludwig, G. Schuch, C. Stroh, A. H. Loos, A. Zubel and P. Koralewski, *J. Clin. Oncol.*, 2009, **27**, 663–671; doi:10.1200/JCO.2008.20.8397.
23. E. Van Cutsem, M. Dicato, N. Arber, J. Berlin, A. Cervantes, F. Ciardiello, A. De Gramont, E. Diaz-Rubio, M. Ducreux, R. Geva, B. Glimelius, R. Glynne Jones, A. Grothey, T. Gruenberger, D. Haller, K. Haustermans, R. Labianca, H. J. Lenz, B. Minsky, B. Nordlinger, A. Ohtsu, N. Pavlidis, P. Rougier, W. Schmiegel, C. Van de Velde, H. J. Schmoll, A. Sobrero and J. Tabernero, *Ann. Oncol.*, 2010, **21**(Suppl. 6), vi1–10; doi:10.1093/annonc/mdq273.
24. J. Y. Douillard, S. Siena, J. Cassidy, J. Tabernero, R. Burkes, M. Barugel, Y. Humblet, G. Bodoky, D. Cunningham, J. Jassem, F. Rivera, I. Kocakova, P. Ruff, M. Blasinska-Morawiec, M. Smakal, J. L. Canon, M. Rother, K. S. Oliner, M. Wolf and J. Gansert, *J. Clin. Oncol.*, 2010, **28**, 4697–4705; doi:10.1200/JCO.2009.27.4860.
25. J. R. Hecht, E. Mitchell, T. Chidiac, C. Scroggin, C. Hagenstad, D. Spigel, J. Marshall, A. Cohn, D. McCollum, P. Stella, R. Deeter, S. Shahin and R. G. Amado, *J. Clin. Oncol.*, 2009, **27**, 672–680; doi:10.1200/JCO.2008.19.8135.
26. M. Peeters, T. J. Price, A. Cervantes, A. F. Sobrero, M. Ducreux, Y. Hotko, T. Andre, E. Chan, F. Lordick, C. J. Punt, A. H. Strickland, G. Wilson, T. E. Ciuleanu, L. Roman, E. Van Cutsem, V. Tzekova, S. Collins, K. S. Oliner, A. Rong and J. Gansert, *J. Clin. Oncol.*, 2010, **28**, 4706–4713; doi:10.1200/JCO.2009.27.6055.
27. D. B. Mendel, A. D. Laird, X. Xin, S. G. Louie, J. G. Christensen, G. Li, R. E. Schreck, T. J. Abrams, T. J. Ngai, L. B. Lee, L. J. Murray, J. Carver, E. Chan, K. G. Moss, J. O. Haznedar, J. Sukbuntherng, R. A. Blake, L. Sun, C. Tang, T. Miller, S. Shirazian, G. McMahon and J. M. Cherrington, *Clin. Cancer Res.*, 2003, **9**, 327–337.
28. R. J. Motzer, T. E. Hutson, P. Tomczak, M. D. Michaelson, R. M. Bukowski, S. Oudard, S. Negrier, C. Szczylik, R. Pili, G. A. Bjarnason, X. Garcia-del-Muro, J. A. Sosman, E. Solska, G. Wilding, J. A. Thompson, S. T. Kim, I. Chen, X. Huang and R. A. Figlin, *J. Clin. Oncol.*, 2009, **27**, 3584–3590; doi:10.1200/JCO.2008.20.1293.
29. G. D. Demetri, A. T. van Oosterom, C. R. Garrett, M. E. Blackstein, M. H. Shah, J. Verweij, G. McArthur, I. R. Judson, M. C. Heinrich, J. A. Morgan, J. Desai, C. D. Fletcher, S. George, C. L. Bello, X. Huang, C. M. Baum and P. G. Casali, *Lancet*, 2006, **368**, 1329–1338; doi:10.1016/S0140-6736(06)69446-4.
30. J. Goffin and E. Eisenhauer, *Ann. Oncol.*, 2002, **13**, 1699–1716.
31. M. A. Owens, B. C. Horten and M. M. Da Silva, *Clin. Breast Cancer*, 2004, **5**, 63–69.
32. D. J. Slamon, G. M. Clark, S. G. Wong, W. J. Levin, A. Ullrich and W. L. McGuire, *Science*, 1987, **235**, 177–182.

33. H. Yaziji, L. C. Goldstein, T. S. Barry, R. Werling, H. Hwang, G. K. Ellis, J. R. Gralow, R. B. Livingston and A. M. Gown, *JAMA*, 2004, **291**, 1972–1977; doi:10.1001/jama.291.16.1972.
34. M. J. Piccart-Gebhart, M. Procter, B. Leyland-Jones, A. Goldhirsch, M. Untch, I. Smith, L. Gianni, J. Baselga, R. Bell, C. Jackisch, D. Cameron, M. Dowsett, C. H. Barrios, G. Steger, C. S. Huang, M. Andersson, M. Inbar, M. Lichinitser, I. Lang, U. Nitz, H. Iwata, C. Thomssen, C. Lohrisch, T. M. Suter, J. Ruschoff, T. Suto, V. Greatorex, C. Ward, C. Straehle, E. McFadden, M. S. Dolci and R. D. Gelber, *N. Engl. J. Med.*, 2005, **353**, 1659–1672; doi:10.1056/NEJMoa052306.
35. E. H. Romond, E. A. Perez, J. Bryant, V. J. Suman, C. E. Geyer, Jr., N. E. Davidson, E. Tan-Chiu, S. Martino, S. Paik, P. A. Kaufman, S. M. Swain, T. M. Pisansky, L. Fehrenbacher, L. A. Kutteh, V. G. Vogel, D. W. Visscher, G. Yothers, R. B. Jenkins, A. M. Brown, S. R. Dakhil, E. P. Mamounas, W. L. Lingle, P. M. Klein, J. N. Ingle and N. Wolmark, *N. Engl. J. Med.*, 2005, **353**, 1673–1684; doi:10.1056/NEJMoa052122.
36. G. E. Konecny, M. D. Pegram, N. Venkatesan, R. Finn, G. Yang, M. Rahmeh, M. Untch, D. W. Rusnak, G. Spehar, R. J. Mullin, B. R. Keith, T. M. Gilmer, M. Berger, K. C. Podratz and D. J. Slamon, *Cancer Res.*, 2006, **66**, 1630–1639; doi:10.1158/0008-5472.CAN-05-1182.
37. R. Nahta, L. X. Yuan, Y. Du and F. J. Esteva, *Mol. Cancer Ther.*, 2007, **6**, 667–674; doi:10.1158/1535-7163.MCT-06-0423.
38. K. L. Blackwell, H. J. Burstein, A. M. Storniolo, H. Rugo, G. Sledge, M. Koehler, C. Ellis, M. Casey, S. Vukelja, J. Bischoff, J. Baselga and J. O'Shaughnessy, *J. Clin. Oncol.*, 2010, **28**, 1124–1130; doi:10.1200/JCO.2008.21.4437.
39. C. L. Arteaga and A. O'Neill, S. L. Moulder, M. Pins, J. A. Sparano, G. W. Sledge and N. E. Davidson, *Clin. Cancer Res.*, 2008, **14**, 6277–6283; doi:10.1158/1078-0432.CCR-08-0482.
40. A. A. Adjei and M. Hidalgo, *J. Clin. Oncol.*, 2005, **23**, 5386–5403; doi:10.1200/JCO.2005.23.648.
41. J. D. Hainsworth, D. R. Spigel, H. A. Burris, 3rd, D. Waterhouse, B. L. Clark and R. Whorf, *J. Clin. Oncol.*, 2010, **28**, 2131–2136; doi:10.1200/JCO.2009.26.3152.
42. J. C. Yang, L. Haworth, R. M. Sherry, P. Hwu, D. J. Schwartzentruber, S. L. Topalian, S. M. Steinberg, H. X. Chen and S. A. Rosenberg, *N. Engl. J. Med.*, 2003, **349**, 427–434; doi:10.1056/NEJMoa021491.
43. J. R. Merchan, G. Liu, T. Fitch, J. Picus, R. Qin, H. C. Pitot, W. Maples and C. Erlichman, *J. Clin. Oncol.*, 2007, **25**, 5034 (June 20 Supplement).
44. B. J. Escudier, S. Negrier, G. Gravis, C. Chevreau, R. Delva, J. Bay, L. Geoffrois, E. Legouffe, E. Blanc and C. Ferlay, *J. Clin. Oncol.*, 2010, **28**, 4516, Supplement 15.
45. A. Patnaik, A. Ricart, J. Cooper, K. Papadopoulos, M. Beeram, C. Mita, M. M. Mita, D. Hufnagel, E. Izbicka and A. W. Tolcher, *J. Clin. Oncol.*, 2007, **25**, 3512 (June 20 Supplement).

46. M. Reck, J. von Pawel, P. Zatloukal, R. Ramlau, V. Gorbounova, V. Hirsh, N. Leighl, J. Mezger, V. Archer, N. Moore and C. Manegold, *J. Clin. Oncol.*, 2009, **27**, 1227–1234; doi:10.1200/JCO.2007.14.5466.
47. M. Reck, J. von Pawel, P. Zatloukal, R. Ramlau, V. Gorbounova, V. Hirsh, N. Leighl, J. Mezger, V. Archer, N. Moore and C. Manegold, BO17704 Study Group, *Ann. Oncol.*, 2010, **21**, 1804–1809; doi:10.1093/annonc/mdq020.
48. A. Sandler, R. Gray, M. C. Perry, J. Brahmer, J. H. Schiller, A. Dowlati, R. Lilenbaum and D. H. Johnson, *N. Engl. J. Med.*, 2006, **355**, 2542–2550; doi:10.1056/NEJMoa061884.
49. F. A. Shepherd, J. Rodrigues Pereira, T. Ciuleanu, E. H. Tan, V. Hirsh, S. Thongprasert, D. Campos, S. Maoleekoonpiroj, M. Smylie, R. Martins, M. van Kooten, M. Dediu, B. Findlay, D. Tu, D. Johnston, A. Bezjak, G. Clark, P. Santabarbara and L. Seymour and National Cancer Institute of Canada Clinical Trials Group, *N. Engl. J. Med.*, 2005, **353**, 123–132; doi:10.1056/NEJMoa050753.
50. Y. D. Jung, P. F. Mansfield, M. Akagi, A. Takeda, W. Liu, C. D. Bucana, D. J. Hicklin and L. M. Ellis, *Eur. J. Cancer*, 2002, **38**, 1133–1140.
51. P. Perrotte, T. Matsumoto, K. Inoue, H. Kuniyasu, B. Y. Eve, D. J. Hicklin, R. Radinsky and C. P. Dinney, *Clin. Cancer Res.*, 1999, **5**, 257–265.
52. V. A. Miller, P. O'Connor, C. Soh and F. Kabbinavar, *J. Clin. Oncol.*, 2009, **27**, LBA8002, Supplemental 18.
53. M. N. Dickler, H. S. Rugo, C. A. Eberle, E. Brogi, J. F. Caravelli, K. S. Panageas, J. Boyd, B. Yeh, D. E. Lake, C. T. Dang, T. A. Gilewski, J. F. Bromberg, A. D. Seidman, G. M. D'Andrea, M. M. Moasser, M. Melisko, J. W. Park, J. Dancey, L. Norton and C. A. Hudis, *Clin. Cancer Res.*, 2008, **14**, 7878–7883; doi:10.1158/1078–0432.CCR-08-0141.
54. R. M. Bukowski, F. F. Kabbinavar, R. A. Figlin, K. Flaherty, S. Srinivas, U. Vaishampayan, H. A. Drabkin, J. Dutcher, S. Ryba, Q. Xia, F. A. Scappaticci and D. McDermott, *J. Clin. Oncol.*, 2007, **25**, 4536–4541; doi:10.1200/JCO.2007.11.5154.
55. J. A. Meyerhardt, K. Stuart, C. S. Fuchs, A. X. Zhu, C. C. Earle, P. Bhargava, L. Blaszkowsky, P. Enzinger, R. J. Mayer, S. Battu, C. Lawrence and D. P. Ryan, *Ann. Oncol.*, 2007, **18**, 1185–1189; doi:10.1093/annonc/mdm124.
56. H. S. Nimeiri, A. M. Oza, R. J. Morgan, G. Friberg, K. Kasza, L. Faoro, R. Salgia, W. M. Stadler, E. E. Vokes and G. F. Fleming, Chicago Phase II Consortium, PMH Phase II Consortium and California Phase II Consortium, *Gynecol. Oncol.*, 2008, **110**, 49–55; doi:10.1016/j.ygyno.2008.02.009.
57. D. M. Jackman, H. L. Kindler, B. Y. Yeap, P. Fidias, R. Salgia, J. Lucca, L. K. Morse, P. A. Ostler, B. E. Johnson and P. A. Janne, *Cancer*, 2008, **113**, 808–814; doi:10.1002/cncr.23617.
58. E. Van Cutsem, C. H. Kohne, E. Hitre, J. Zaluski, C. R. Chang Chien, A. Makhson, G. D'Haens, T. Pinter, R. Lim, G. Bodoky, J. K. Roh, G. Folprecht, P. Ruff, C. Stroh, S. Tejpar, M. Schlichting, J. Nippgen and

P. Rougier, *N. Engl. J. Med.*, 2009, **360**, 1408–1417; doi:10.1056/NEJMoa0805019.

59. E. Van Cutsem, W. L. Vervenne, J. Bennouna, Y. Humblet, S. Gill, J. L. Van Laethem, C. Verslype, W. Scheithauer, A. Shang, J. Cosaert and M. J. Moore, *J. Clin. Oncol.*, 2009, **27**, 2231–2237; doi:10.1200/JCO.2008.20.0238.
60. J. Tol, M. Koopman, A. Cats, C. J. Rodenburg, G. J. Creemers, J. G. Schrama, F. L. Erdkamp, A. H. Vos, C. J. van Groeningen, H. A. Sinnige, D. J. Richel, E. E. Voest, J. R. Dijkstra, M. E. Vink-Borger, N. F. Antonini, L. Mol, J. H. van Krieken, O. Dalesio and C. J. Punt, *N. Engl. J. Med.*, 2009, **360**, 563–572; doi:10.1056/NEJMoa0808268.
61. L. B. Saltz and H. J. Lenz, H. L. Kindler, H. S. Hochster, S. Wadler, P. M. Hoff, N. E. Kemeny, E. M. Hollywood, M. Gonen, M. Quinones, M. Morse and H. X. Chen, *J. Clin. Oncol.*, 2007, **25**, 4557–4561; doi:10.1200/JCO.2007.12.0949.
62. D. Cunningham, Y. Humblet, S. Siena, D. Khayat, H. Bleiberg, A. Santoro, D. Bets, M. Mueser, A. Harstrick, C. Verslype, I. Chau and E. Van Cutsem, *N. Engl. J. Med.*, 2004, **351**, 337–345; doi:10.1056/NEJMoa033025.
63. B. J. Giantonio, P. J. Catalano, N. J. Meropol, P. J. O'Dwyer, E. P. Mitchell, S. R. Alberts, M. A. Schwartz and A. B. Benson, 3rd and Eastern Cooperative Oncology Group Study E3200, *J. Clin. Oncol.*, 2007, **25**, 1539–1544; doi:10.1200/JCO.2006.09.6305.
64. H. Hurwitz, L. Fehrenbacher, W. Novotny, T. Cartwright, J. Hainsworth, W. Heim, J. Berlin, A. Baron, S. Griffing, E. Holmgren, N. Ferrara, G. Fyfe, B. Rogers, R. Ross and F. Kabbinavar, *N. Engl. J. Med.*, 2004, **350**, 2335–2342; doi:10.1056/NEJMoa032691.
65. D. J. Jonker, C. J. O'Callaghan, C. S. Karapetis, J. R. Zalcberg, D. Tu, H. J. Au, S. R. Berry, M. Krahn, T. Price, R. J. Simes, N. C. Tebbutt, G. van Hazel, R. Wierzbicki, C. Langer and M. J. Moore, *N. Engl. J. Med.*, 2007, **357**, 2040–2048; doi:10.1056/NEJMoa071834.
66. L. B. Saltz, S. Clarke, E. Diaz-Rubio, W. Scheithauer, A. Figer, R. Wong, S. Koski, M. Lichinitser, T. S. Yang, F. Rivera, F. Couture, F. Sirzen and J. Cassidy, *J. Clin. Oncol.*, 2008, **26**, 2013–2019; doi:10.1200/JCO.2007.14.9930.
67. A. F. Sobrero, J. Maurel, L. Fehrenbacher, W. Scheithauer, Y. A. Abubakr, M. P. Lutz, M. E. Vega-Villegas, C. Eng, E. U. Steinhauer, J. Prausova, H. J. Lenz, C. Borg, G. Middleton, H. Kroning, G. Luppi, O. Kisker, A. Zubel, C. Langer, J. Kopit and H. A. Burris, 3rd, *J. Clin. Oncol.*, 2008, **26**, 2311–2319; doi:10.1200/JCO.2007.13.1193.
68. E. Van Cutsem, M. Peeters, S. Siena, Y. Humblet, A. Hendlisz, B. Neyns, J. L. Canon, J. L. Van Laethem, J. Maurel, G. Richardson, M. Wolf and R. G. Amado, *J. Clin. Oncol.*, 2007, **25**, 1658–1664; doi:10.1200/JCO.2006.08.1620.
69. L. R. Kelland, *Eur. J. Cancer*, 2004, **40**, 827–836; doi:10.1016/j.ejca.2003.11.028.
70. S. A. Engelholm, L. L. Vindelov, M. Spang-Thomsen, N. Brunner, N. Tommerup, M. H. Nielsen and H. H. Hansen, *Eur. J. Cancer Clin. Oncol.*, 1985, **21**, 815–824.

71. (a) P. J. Ferguson and Y. C. Cheng, *Cancer Res.*, 1989, **49**, 1148–1153; (b) Food and Drug Administration (December 2010), Guidance for Industry Codevelopment of Two or More Unmarketed Investigational Drugs for Use in Combination.
72. A. Smith, C. van Haaften-Day and P. Russell, *Cancer Genet. Cytogenet.*, 1989, **38**, 13–24.
73. J. I. Johnson, S. Decker, D. Zaharevitz, L. V. Rubinstein, J. M. Venditti, S. Schepartz, S. Kalyandrug, M. Christian, S. Arbuck, M. Hollingshead and E. A. Sausville, *Br. J. Cancer*, 2001, **84**, 1424–1431; doi:10.1054/bjoc.2001.1796.
74. R. H. Shoemaker, *Nat. Rev. Cancer*, 2006, **6**, 813–823; doi:10.1038/nrc1951.
75. S. Kummar, H. X. Chen, J. Wright, S. Holbeck, M. D. Millin, J. Tomaszewski, J. Zweibel, J. Collins and J. H. Doroshow, *Nat. Rev. Drug Discovery*, 2010, **9**, 843–856; doi:10.1038/nrd3216.
76. N. Keshelava, T. Frgala, J. Krejsa, O. Kalous and C. P. Reynolds, *Methods Mol. Med.*, 2005, **110**, 139–153; doi:10.1385/1-59259-869-2:139.
77. M. S. Gitler, A. Monks and E. A. Sausville, *Mol. Cancer Ther.*, 2003, **2**, 929–932.
78. C. Shinn, D. Larsen and J. R. Suarez, *Blood*, 2000, **96**, 294b.
79. P. J. Houghton and C. F. Stewart, P. J. Cheshire, L. B. Richmond, M. N. Kirstein, C. A. Poquette, M. Tan, H. S. Friedman and T. P. Brent, *Clin. Cancer Res.*, 2000, **6**, 4110–4118.
80. D. Meco, T. Colombo, P. Ubezio, M. Zucchetti, M. Zaffaroni, A. Riccardi, G. Faircloth, J. Jose, M. D'Incalci and R. Riccardi, *Cancer Chemother. Pharmacol.*, 2003, **52**, 131–138; doi:10.1007/s00280-003-0636-6.
81. M. Tan, H. B. Fang, G. L. Tian and P. J. Houghton, *Stat. Med.*, 2003, **22**, 2091–2100; doi:10.1002/sim.1467.
82. M. Patlak, *J. Natl. Cancer Institute*, 2010, **102**, 841–843; doi:10.1093/jnci/djq227.
83. D. B. Solit, Y. She, J. Lobo, M. G. Kris, H. I. Scher, N. Rosen and F. M. Sirotnak, *Clin. Cancer Res.*, 2005, **11**, 1983–1989; doi:10.1158/1078-0432.CCR-04-1347.
84. D. R. Duckett and M. D. Cameron, *Expert Opin. Drug Metab. Toxicol.*, 2010, **6**, 1175–1193; doi:10.1517/17425255.2010.506873.
85. D. J. Slamon, B. Leyland-Jones, S. Shak, H. Fuchs, V. Paton, A. Bajamonde, T. Fleming, W. Eiermann, J. Wolter, M. Pegram, J. Baselga and L. Norton, *New Engl. J. Med.*, 2001, **344**, 783–792; doi:10.1056/NEJM200103153441101.

CHAPTER 12

The Discovery of Lapatinib

KAREN E. LACKEY

Hoffmann-La Roche, Discovery Chemistry, 340 Kingsland Street, Nutley, NJ 07110, USA
Email: karen_e.lackey@roche.com

12.1 Introduction to Inhibition of Kinases for Cancer Therapeutics

To truly understand the significance and impact of cell signal inhibition in cancer therapy, a high level view of the evolution of cancer treatments will offer a perspective for the rapid developments in the last few decades. The detection of cancer as the cause of death emerged from autopsies done in the mid-1700s. Within 25–30 years of this discovery, surgery was used for curing cancer that had not invaded tissues. Over 200 years later, the cure for breast cancer remained the same as in 1880, surgery (albeit under better conditions). Advances in cancer treatment were realized with radiation therapy (initiated a few years after 1900) and chemotherapy (post WWII), and had to be balanced with the debilitating, dose-limiting side effects from the cytotoxic mechanisms. The dream of all scientists in the field was to create effective medicines that targeted the tumor and tumor growth properties, and avoided interfering with normal healthy cells.

New technologies for manipulating and understanding genes led to a revolution in the understanding of the disease of cancer. In the mid-1970s, Michael Bishop and Harold Varmus (both virologists who won the 1989 Nobel Prize in Physiology or Medicine) discovered a gene in a chicken virus that transformed cells.[2] Two concepts emerged to describe this phenomenon:

RSC Drug Discovery Series No. 21
Designing Multi-Target Drugs
Edited by J. Richard Morphy and C. John Harris

Published by the Royal Society of Chemistry, www.rsc.org

oncogenes are a switch that transforms a cell from normal to cancerous, whereas *proto-oncogenes* perform functions critical to normal cell behavior, but 'mutate' to become oncogenes. Many oncogenes were discovered in human cancers over the next decade such as *ras* (colon, pancreatic), *abl* (chronic myelogenous leukemia), *myc* (neuroblastoma, lymphoma), *HER1* (lung, head, and neck), *HER2/neu* (breast, stomach), *src* (colon, breast), *b-raf* (melanoma), and *mdm2* (sarcomas). These oncogenes were found to encode proteins that were mutated, over-expressed, and/or constitutively active which led to aberrant cell growth. Signal transduction through these proteins is defined as the biochemical event that conducts a message (e.g. to proliferate) from the cell exterior, through the membrane, and into the cytoplasm. The simple concept of a signal transduction inhibitor was to create a drug that would selectively prevent the ability of the cancer cells to multiply and invade other tissues by targeting the aberrant proteins expressed, thus avoiding affects on normal cells.

More research was required to go from disease understanding to drug intervention. While the oncogenes and their connection to the cancer had been established, the selection of the biological target within the aberrant signal cascade was a challenge, given compensating pathways and links to normal cell viability. Each of the proteins within the signal cascade had unique opportunities for modulation with a small molecule or biological agent. For example, signal inhibition strategies included binding in SH2 and SH3 domains, peptide substrate mimics, antibodies to the extracellular domain, and kinase ATP binding inhibitors. Each strategy had its perceived advantages and disadvantages, and a tremendous amount of outstanding research efforts went into each of these approaches. It is notable that the discovery and subsequent launch of Herceptin in 1998 as the first antibody drug for signal inhibition by binding to the extracellular domain of erbB2 was ground breaking.[3] Early work on kinase inhibition was fraught with selectivity concerns that could lead to potential toxicity due to the similarity of the ATP binding sites for the >500 protein family members. The discovery and launch of Gleevec in 2003 was a turning point in the research field since it proved that effective medicines could be created by inhibiting kinase activity with a small molecule.[4]

The erbB family signaling network first described in the late 1980s led to many opportunities to select drug intervention points. It is beyond the scope of this chapter to elaborate on the details of the ligands, adaptor proteins, interconnected signaling pathways, transcription factors, and the resulting cellular outcomes such as apoptosis, migration, growth, adhesion, and differentiation.[5] Suffice to say, it is very complex, and questions remain even today. It is important for the context of the discovery of the dual kinase inhibitor to briefly review the general scheme for the initiation of signaling through type 1 receptors. There are four members of the erbB family tyrosine kinases: EGFR, erbB2, erbB3, and erbB4. While the overall protein structure is similar, the four members of the erbB family display differences in the autophosphorylation docking sites, in substrate specificity, and in the potency of the kinase activity. For example, erbB3 lacks kinase activity. ErbB2 has the highest transforming capability and has no known directly binding ligand. These erbB family receptors are membrane-spanning with a cysteine-rich extracellular ligand-binding domain, a hydrophobic membrane-spanning region, and an intracellular kinase domain. The homo- or hetero-dimerization of erbB receptors after extracellular ligand binding (e.g. epiregulin, B-cellulin, and EGF) leads to the activation of the intrinsic tyrosine kinase activity via autophosphorylation, or hetero-dimerization and transactivation-autophosphorylation of erbB2. The formation of the erbB family dimers activate downstream Erk1/2 MAP kinases and PI3K/AKT kinase survival pathways.[6] At the time of the project, the state of the art had conflicting biological and translational results with the role of erbB4 in signaling. There appeared to be evidence linking the progression of breast cancer to a shift in erbB4 over-expression to an over-expression of erbB2. Therefore, the aim of the drug discovery effort was to inhibit both EGFR and erbB2 via the kinase domain, and to avoid toxicities associated with non-selective kinase inhibition.

12.2 Developing a Lead Series for Dual Kinase Inhibition

The first ATP-competitive kinase inhibitors were found to be potent cell signaling inhibitors, albeit generally non-selective. Staurosporine analogs and flavone/isoflavones were among these early inhibitors. By 1994, a potent EGFR inhibitor, PD153035, was reported demonstrating an impressive selectivity profile thus offering the promise of discovering drugs in this protein class.[7] Figure 12.1 demonstrates the contrast in the inhibition profiles for staurosporine and PD153035 using representative enzyme data. In these representative 12 diverse kinase assays, staurosporine inhibited the kinase activity with IC_{50} values

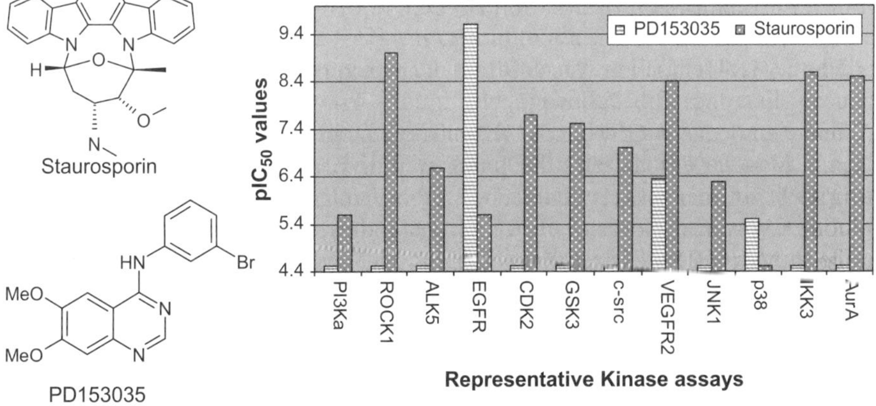

Figure 12.1 Structures and some representative kinase inhibition data for PD153035 and Staurosporine.

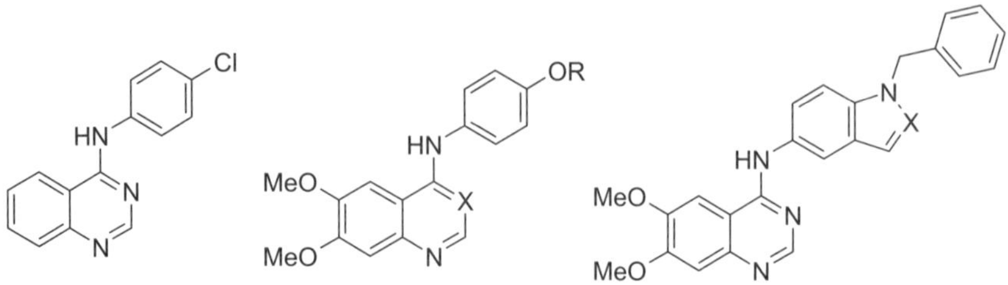

1 (CAQ)

2: R = Ph
3: R = Bn
X=N

4: X = C
5: X = N

Figure 12.2 Early potent ErbB family TK inhibitors.

less than 0.5 μM in 9 of them, whereas PD153035 only inhibited EGFR and VEGFR2.

Over 3500 compounds were synthesized on this drug discovery project, based on the quinazoline and quinoline substructure. Most likely due to ease of synthesis, many variants were generated with either an unsubstituted quinazoline (**1**) or the 6,7-dimethoxy variants (**2–5**), as shown in Figure 12.2. A very diverse, broad range of 4-substituents were synthesized covering anilines, amines, and heterocycles on the quinazoline and quinoline core. Many of the quinolines became important chemical series in a c-src inhibition project (cancer and osteoporosis disease areas).

At the time, the industry appeared to favor selective EGFR inhibitors, and several were discovered that were outstanding: Iressa and Tarceva for example.[8] However, our aim was to create a dual EGFR and erbB2 tyrosine kinase inhibitor. Substitutions off the 6- and 7-position affected the compound properties, potency, and pharmacokinetic properties. The aniline substitution pattern provided the key to achieving the erbB2 potency by the inclusion of a larger substituent such as the 4-benzyloxy as shown for **3** and the 'tied back' versions like 4-(*N*-benzylindole)quinazoline **4** and 4-(*N*-benzylindazole)quinazoline **5**.[9] Other similar substituents like 4-phenoxy aniline **2** provided good activity although lacking kinase selectivity.

By defining our project as a dual kinase inhibitor, in contrast to the dominant pharma industry preference for selective EGFR inhibition, a preclinical biological evaluation cascade had to be established that would not only identify a dual acting kinase inhibitor, but also validate the approach for key cancers such as breast and gastric. The tool compound, **3**, served a valuable purpose early in the program to build the cellular and *in vivo* models, and provide confidence to move forward into a lead optimization phase. Table 12.1 is a summary of the assays used; their importance in the discovery of lapatinib was so significant, that it requires a brief explanation to fully understand the subsequent analyses.

The *in vitro* assays were standard in the kinase field and grew in number over time from a panel of 10 kinases to well over 65 assays. The data were not used in the design of the dual inhibitor, but served as a great source of new lead

Table 12.1 Summary of primary assays used in the lead optimization phase.

In vitro *assays*	*Cell panel*	In vivo *assays*
Purified, catalytic EGFR, erbB2, and erbB4/peptide substrate	Normal control line: HFF (normal fibroblast)	Limited pharmacokinetic analyses
Mixed binding and catalytic kinase panel of assays	Tumor lines: BT474 (breast) erbB-2+++ NH5 (head/neck) EGFR+++ N87 (gastric) erbB-2+++/ EGFR+	Tumor xenograft assays: BT474 HN5
Variety of developability assays such as solubility, permeability, and p450	Transform lines: HB4a r4.1 (*Ha-ras*) HB4a c5.2 (*c-erbB2*)	*In vivo* kinase inhibition

compounds for other kinases important in cancer. The compound property assays included water solubility, log *P*, p450 enzymes, and protein binding. The efficacy of these dual inhibitors on human tumor cells was assessed in a cell-based proliferation assay using protein staining to estimate the relative cell number. The cell-based assay included the EGFR over-expressing cell line (HN5), the ErbB-2 over-expressing cell line (BT474), and a cell line over-expressing both receptors (N87).[10] Human foreskin fibroblasts (HFF) cells, from normal tissue and expressing low levels of EGFR and ErbB-2, were included as a control cell line for non-specific toxicity. Activity in tumor versus normal tissue was determined by comparing efficacy on tumor cells to efficacy on HFF cells, which is thought to be indicative of selectivity and thus, potentially, lower toxicity.

The type I receptors transduce a mitogenic signal after autophosphorylation by activating the downstream effector Ras, which then signals through Raf, MEK1, and ERK2, resulting in translation of response genes and then cell division. A cell line that is over-expressing activated Ras, therefore, should be resistant to the anti-proliferative effects of a selective EGFR or ErbB-2 inhibitor. A transfected cell system was created that contained either *erbB-2* (HB4a c5.2) or the valine 12 mutant of Ha-*ras* (HB4a r4.2) to determine if the anti-proliferative effects of the compounds were due to inhibiting the receptor kinase as opposed to targeting downstream effectors of cell proliferation.[10] Therefore, the ideal cellular profile was potent inhibition of the tumor lines (e.g. IC_{50} values <0.25 μM) N87, HN5, BT474, HB4a c5.2, and inactivity in the control lines (e.g. IC_{50} values >30 μM) HFF and HB4a r4.2. The *in vivo* efficacy models used were established by using the cell lines above in human tumor xenografts in immunocompromised mice and will be described later when the advanced compounds are discussed. A rigorous, limited pharmacokinetic protocol was used to assess plasma concentrations of drug after oral dosing.

The team generated many potent dual inhibitors and even nominated drug candidates from 1992 to 1998. One of the drug candidates related to **4** had the desired inhibition profile, but the cellular efficacy IC_{50} values on relevant tumor lines remained over 1 μM. The *in vivo* pharmacokinetic profile was not going to reach sufficient levels in the tumor to warrant taking the compound into the clinic. A second candidate (GW2974, discussed later) possessed a potent, and ideal, biological efficacy profile for cancer therapy, but did not achieve the desired therapeutic index for progression into clinical trials. The significant achievement of the research up to this point in the project was the thorough understanding of the disease biology, the link to the patient population, and the evolution of the preclinical disease models.

12.3 Performing Multi-Dimensional Data Analysis for Achieving the Target Profile

SAR investigations for dual (but selective) EGFR/erbB2 inhibition required following several IC_{50} values for each of the derivatives. A method to evaluate and drive the SAR of dual inhibition or pan-erbB family inhibition, with an objective measure of whether we were achieving the desired profile, was created. A simple calculation was used to reflect the combined measured IC_{50} values and was described as an index. This index combined multiple parameters into one value for rapid analysis of general trends and could be used for large amounts of screening data generated from *in vitro* assays such as kinase enzyme and cellular proliferation assays. For a more sophisticated quantitative data analysis method, BioMAP may be a more preferred method, but requires more expert users.[11]

$$\text{Index (I)} = \text{WM} - \text{RMS*}$$

where:

WM (weighted mean) $= \{x\text{EGFR}(\text{pIC}_{50}) + y\text{ERBB2}(\text{pIC}_{50}) + z\text{ERBB4}(\text{pIC}_{50})\}/(x+y+z)$

RMS* (modified root mean square) $= \sqrt{\{[x(\text{DP}_1 - \text{MP}_1)^2 + y(\text{DP}_2 - \text{MP}_2)^2 + z(\text{DP}_3 - \text{MP}_3)^2]/(x+y+z)\}}$

and DP = desired profile, MP = measured profile, and x, y, and z = relative weighting terms.

The index calculation formula reflects the desired profile more accurately than using the calculated average of the IC_{50} values. An index can be calculated for dual (I_d) or pan-family (I_p) inhibition and should reflect how well the compound's activity matches the desired profile. This value is simply the weighted mean average (WM) of the pIC_{50} determinations less the modified version of a root mean square term (RMS*). The RMS* is calculated by subtracting the measured pIC_{50} determination from the desired pIC_{50} value. This type of calculation can be done in an Excel™ spreadsheet (with easily constructed formulas) by importing large lists of screening data and doing the calculation on the fly. The weightings of the different parts of the profile can be adjusted, as can the enzymes included in the profile. The SAR can be understood by contrasting the features of the molecule that increase the dual or pan-family index, or avoid inhibiting certain kinases.

The average of a compound's pIC_{50} values can be misleading. For example, using just enzyme values to make the point, if compound A tested gave the results of 7.8, 8.0, and 4.5 as pIC_{50} values for EGFR, erbB2, and erbB4, it would be considered a dual inhibitor of EGFR and erbB2 with 6.8 as the average of the three values. If compound B had pIC_{50} values of 6.8, 7.2 and 6.5 for the same assays, it would likely be considered a moderate pan-family inhibitor, yet it has the same 6.8 average pIC_{50} value. Therefore the average value of 6.8 would indicate that these compounds are similar, and yet the index does distinguish their profiles. Using the index calculation for a dual inhibitor profile, compound A ($I_d = 6.6$) affords an I_d 1.5 log units higher than compound B ($I_d = 5.1$). By a similar analogy, I_p is better for compound B ($I_p = 5.1$) than for compound A ($I_P = 4.4$). The higher the index valuc is, the closer it approaches the desired profile. Setting the index for I_d, comparisons can be made with many other features such as physicochemical properties, cellular selectivity, and improved efficacy.

Since so many compounds had been generated in the quinazoline and quinazoline-like series, and two drug candidates had failed, it was questionable whether a suitable drug could be discovered. The data were analyzed to determine if a trend could be found to design dual inhibitors that were potent but also selective in the cellular assays. Figure 12.3 is an index plot comparing dual enzyme profile of inhibiting erbB-2 and EGFR on the x-axis and the formula applied to the cell data is on the y-axis using only compounds for which data are in the public domain. One twist to the cellular calculation is that four cell lines were used, with the following desired profile values: BT474 (8), N87 (8), HN5 (8), and HFF (4.5), taking into account the additional parameter of selectivity where 4.5 is the maximum concentration tested in the cellular assays.

Immediately, the lack of correlation between the enzyme profile and the desired cellular activity is apparent and the plot looked the same when all ~3000 compounds' data were included. The most obvious reason suggested for

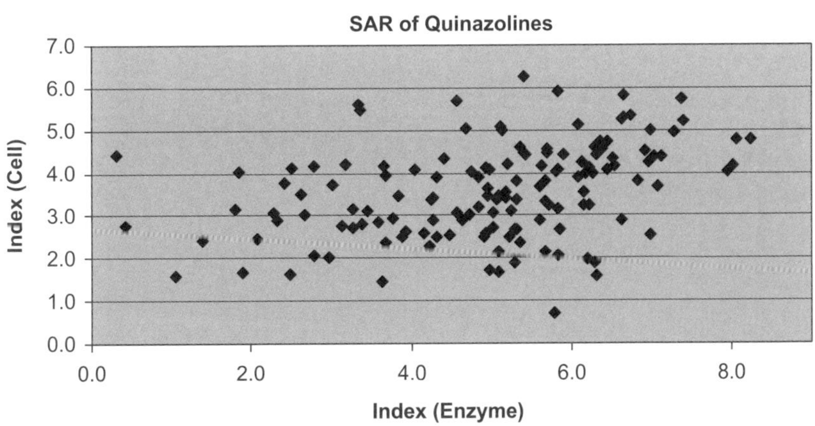

Figure 12.3 Plot of I_{dual} versus I_{cell}.

the lack of correlation was the physical chemical properties of the compounds. All compounds with a pIC_{50} value >7 in both the erbB2 and EGFR catalytic enzyme assays were evaluated in multiple developability assays (e.g. solubility in multiple solvents, cell permeability assays, protein binding measures, *in vitro* metabolic stability). The physical properties of the compounds could not explain the apparent lack of correlation between the cellular and enzyme activity. A line was drawn corresponding to the ideal enzyme/cell correlation, and only the compounds that fell along that line in the plot were included in the SAR analysis. A medical chemistry plan was built on the SAR of the index for dual inhibition and a diagram of the strategy is shown in Figure 12.4.

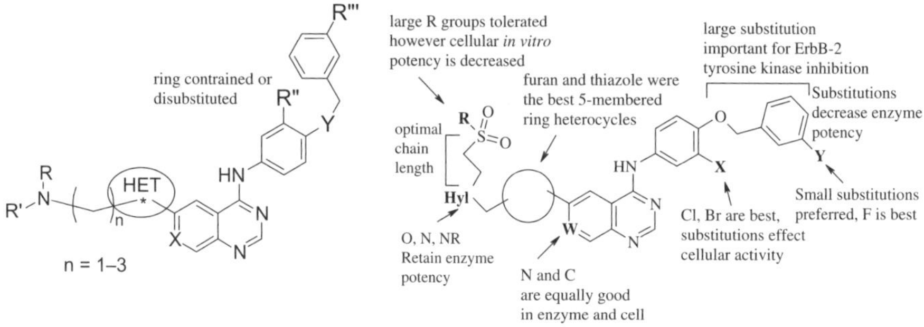

Figure 12.4 Medicinal chemistry plan and the resulting SAR.

Table 12.2 Data extracted from published reports on 4-anilinoquinazoline cell and enzyme activity converted into log scale before calculating an index for dual enzyme inhibition and cellular efficacy.

Example[a]	*erbB2, pIC_{50}*	*EGFR, pIC_{50}*	*BT474, pIC_{50}*	*HN5, pIC_{50}*	I_{enz}	I_{cell}	*Ref.*
GW2016	8.04	7.97	7.0	6.92	7.5	5.9	12
OSI-744	5.98	7.77	5.0	6.74	5.0	3.6	12
GW2974	7.8	8.15	6.4	6.47	7.4	4.9	9
GW9263	7.52	7.44	6.32	6.52	6.5	4.8	9
GW4263	7.68	7.64	6.23	6.37	6.8	4.6	9
GW0277	7.55	7.6	6.19	6.37	6.6	4.6	9
GW5289	7.49	7.59	5.94	6.08	6.6	4	9
GW5945	7.62	7.66	6.41	6.19	6.8	4.6	9
GW9525	7.52	7.54	6.23	5.95	6.6	4.2	9
1	6.7	7.46	5.62	6.24	5.6	3.8	13
2	7.32	7.09	6.07	5.8	5.9	3.9	13
3	7.55	7.15	7.22	6.96	6.2	6.2	13
4	6.26	6.27	4.82	4.74	4.0	1.6	13
5	7.85	8.1	6.55	6.64	7.4	5.2	13
6	7.85	8	6.96	7.1	7.3	6.1	13
8b	6.63	7.96	5.77	7.35	5.9	4.9	14
8c	6.68	7.7	5.86	7.15	5.8	4.9	14
8f	7.72	6.62	6.41	6.2	5.7	4.6	14

[a]Compound numbers correspond to those used in the references.

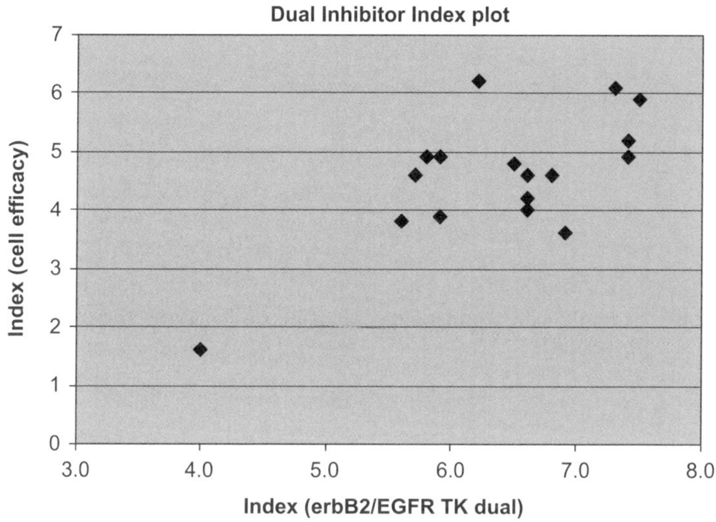

Figure 12.5 Calculated index plot of advanced quinazoline and pyridopyrimidine compounds.

A table for representative compounds that were designed based on the strategy is compiled in Table 12.2. The data from the papers were converted to pIC_{50} format, the index was calculated for dual inhibition, and plotted in Figure 12.5. Most compounds possessed the desired relationship between dual inhibition of EGFR/erbB2 and the cellular efficacy in relevant tumor lines, plus selectivity over normal cells. The data also suggested that there could be a mechanistic explanation for why some compounds were effective while other equipotent enzyme inhibitors in the same series were ineffective in the cellular context.

12.4 Optimizing Drug Properties whilst Retaining the Target Profile

12.4.1 Pyridopyrimidines

Modifications to the central core quinazoline became a feature of the medicinal chemistry campaign to search for potency and SAR, but also as an approach to reducing log *P*. A range of heteroatoms were investigated to provide alternatives to the phenyl ring of the quinazoline. Amongst these were thienopyrimidines, exemplified by **6**, and the pyrido[3,4-*d*]pyrimidine, exemplified by **7**, shown in Figure 12.6.

Initial data from the pyridopyrimidines was encouraging; **7** had a bioavailability of 37% in rats and represented an improvement over comparable quinazoline analogs. The cell assays were the key decision-making assays, dramatically overshadowing the kinase enzyme activities. The lead compound generated in this series was the 4-(*N*-benzyl)indazole **GW2974**, shown in

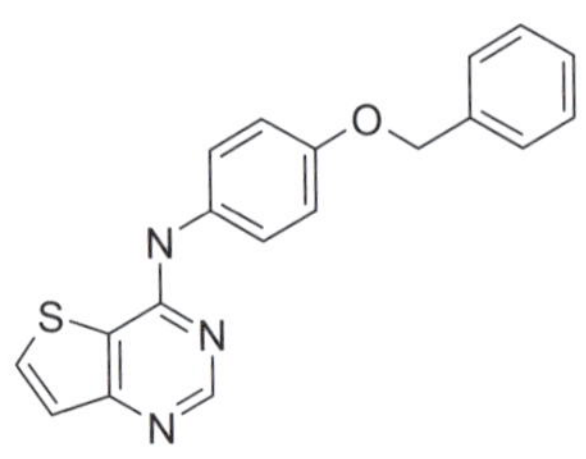

7

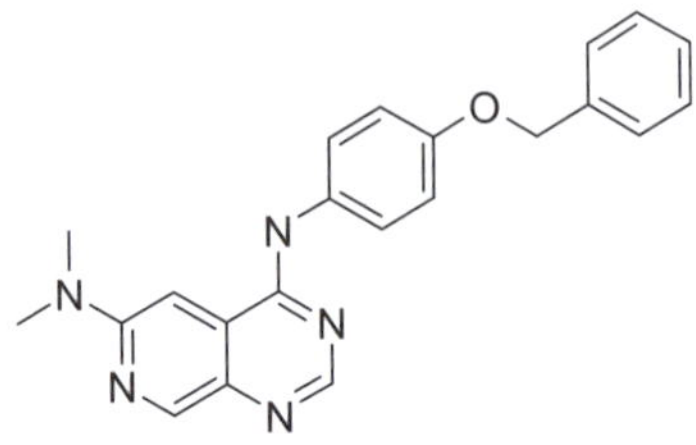

8

Figure 12.6 Example core quinazoline changes designed to improve compound properties.

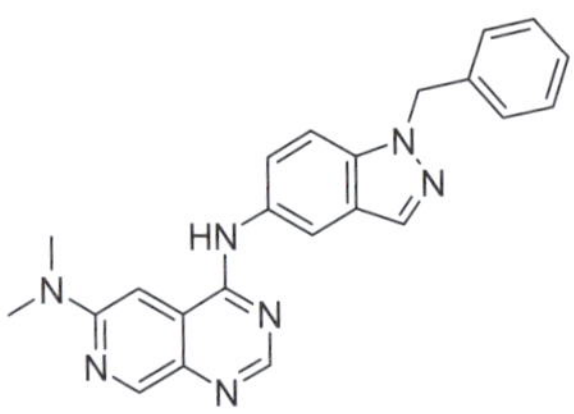

GW2974

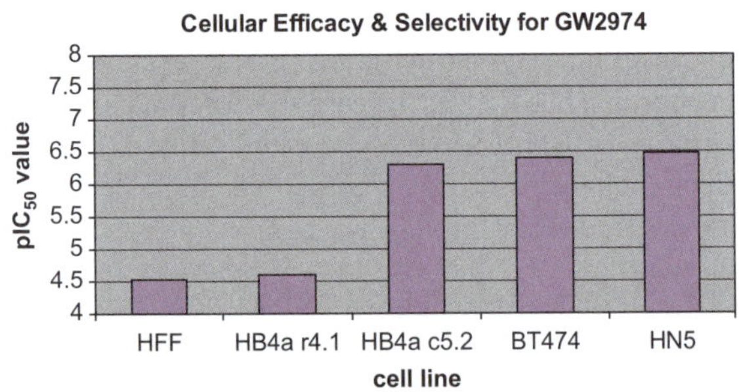

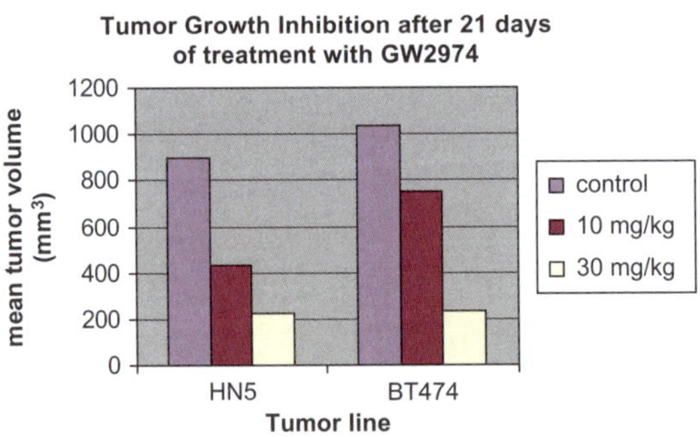

Figure 12.7 Advanced EGFR/erbB2 tyrosine kinase inhibitors.

Figure 12.7.[15] It was one of the first compounds to exhibit an IC_{50} value under 100 nM in the HB4a.c5.2, the c-erbB-2 over-expressing line cell line, and appeared potent (IC_{50} ~250 nM) in three important tumor cell lines: BT474 breast cancer cell line, CaLu3 lung line, and HN5 head and neck tumor line. Cell selectivity was good as measured by comparing the *erbB* driven versus the *ras* driven HB4a results, and are also shown in pIC_{50} format in Figure 12.7. **GW2974** had adequate pharmacokinetic parameters in rats, marmosets, and dogs. Mean plasma levels on repeat dosing in rats and mice showed that levels above the average tumor cell IC_{50} value could be maintained at doses of 20 mg/kg b.i.d. At a dose of 10 mg/kg, complete inhibition of tumor growth was observed for **GW2974** in the BT474 xenograft mouse model and >85% inhibition of tumor growth was observed in the Calu3 and HN5 model and when the dose was increased to 50 mg/kg, irreversible tumor regression was observed.[10] Based on the efficacy and mechanism of action of **GW2974**, it was progressed into preclinical development. Several issues were uncovered at this stage, including non-linear pharmacokinetics, the formation of variably active metabolites, and a number of toxicity findings that precluded progression into the clinic. The further development of **GW2974** was halted, as the strategy was taken to deliver a dual inhibitor that possessed a larger therapeutic index.

12.4.2 6-Ether Linked 4-Anilinoquinazolines

Quinazoline derivatives with a variety of C-4 anilines and C-6 ether linked substitutions were synthesized to improve dual EGFR/erbB2 activities.[14] Representative examples are highlighted in Table 12.3 and demonstrate the general SAR trends for the entire series.

In an homology model for erbB2, shown in Figure 12.8, the anilino portion of these derivatives is oriented deep in a hydrophobic region in the back of the active site not occupied by ATP, consistent with crystal structures acquired later in the drug discovery project. The 6-ether linker extends out of the ATP pocket into the solvent exposed region at the junction of the C- and N-terminal lobes of the kinase region, suggesting that modifications could be tolerated at the 6-position aimed at improving the pharmacokinetic properties and solubility. Comparison of compounds **8**–**11**, the methylsulfonylethylamino (optimized solubilizing group that was cell penetrable) linked with a four-carbon methylene group through a 6-position oxygen, **9** provided a dual inhibitor enzyme profile (IC_{50} EGFR: 0.074 μM, erbB2: 0.095 μM).

The target affinity could be modulated to a greater extent via the 4-anilino substitutions. Therefore, the optimal ether-linked side chain from **9** was combined with two of the best 6-anilino substituents that confer dual enzyme activity, **12** and **13**, and to achieve selective EGFR activity, via Tarceva SAR, 3-ethynylaniline was inserted into the 4-position to afford **14**. Representative examples of cellular activity are shown in Table 12.4. Consistent with the erbB family inhibition profile, **14** inhibited the proliferation of the HN5 cell line over-expressing EGFR (IC_{50} 0.045 μM), but was less active in a BT474 cell line over-expressing erbB2 (IC_{50}> 1.5 μM). **13** showed potent dual EGFR/erbB2 inhibitory activity as well as adequate efficacy (avg. IC_{50} = 0.54 μM) against tumor cell lines, and the cellular selectivity (normal cell versus tumor cell lines) exceeded 50-fold.

The cellular efficacy and selectivity correlated well with the enzyme inhibition profile in the 6-position side chain with an ether linkage combined with the optimal 4-anilino substituents that demonstrated the consistency of the SAR of dual inhibition and was able to block tumor growth through the desired

Table 12.3 EGFR and erbB2 inhibition data for representative 6-position ether linked substitutions on 4-anilinoquinazolines.

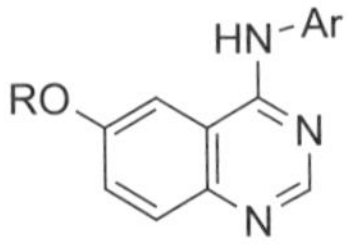

Example	*R*	*AR*	*EGFR, IC_{50} nM*[a]	*erbB2, IC_{50} nM*[a]
8	H	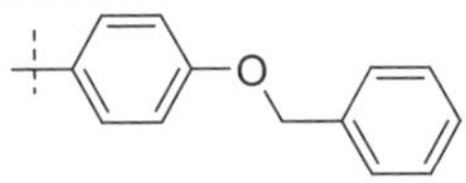	98	260
9	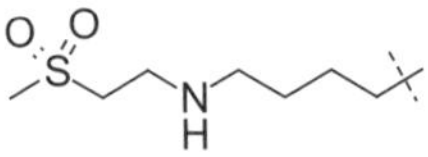	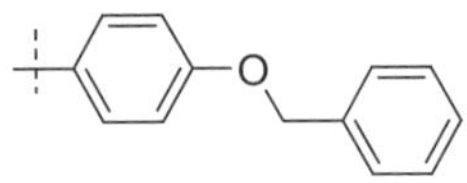	74	95
10	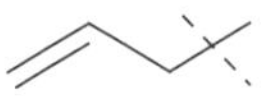	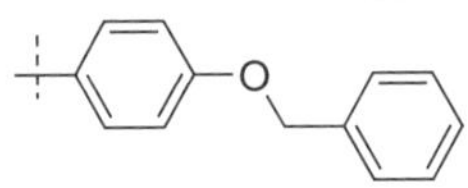	79	350
11	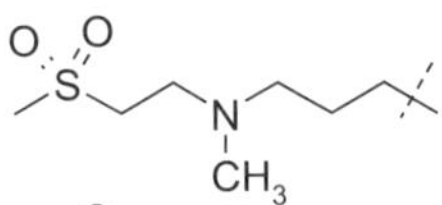	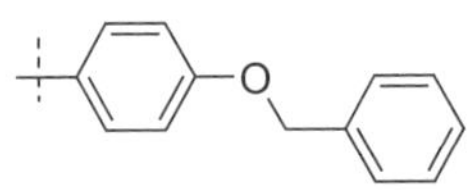	830	1800
12	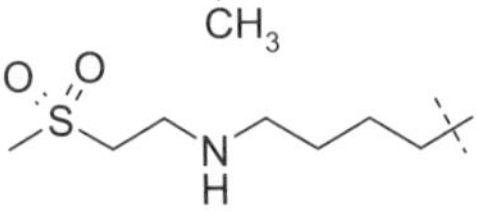	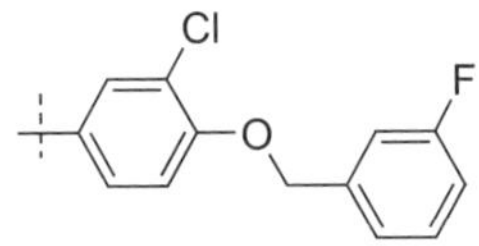	27	10
13	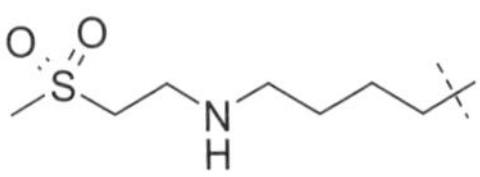	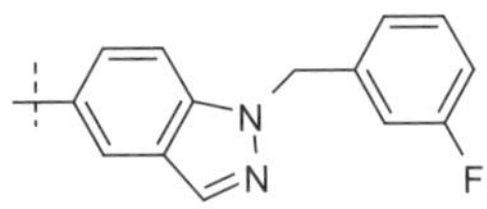	24	19
14	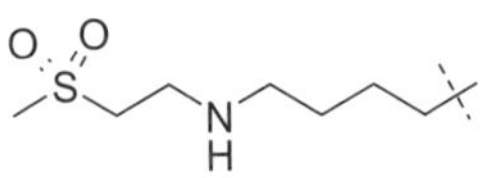	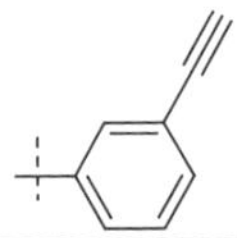	11	230

[a]IC_{50} values are generated by measuring inhibition of phosphorylation of a peptide substrate added to enzyme reaction.[14]

mechanism. However, there was no distinct property advantage over the 6 heteroaryl-linked series (discussed in Section 12.4.3) and the syntheses were more difficult. This series was not progressed further into clinical development.

12.4.3 6-Heteroaryl Linked 4-Anilinoquinazolines

12.4.3.1 6-Furanyl Quinazoline Series

Several trends were observed for 6-position side chains linked through a furan combined with 4-anilino modifications. For example, the orientation of the

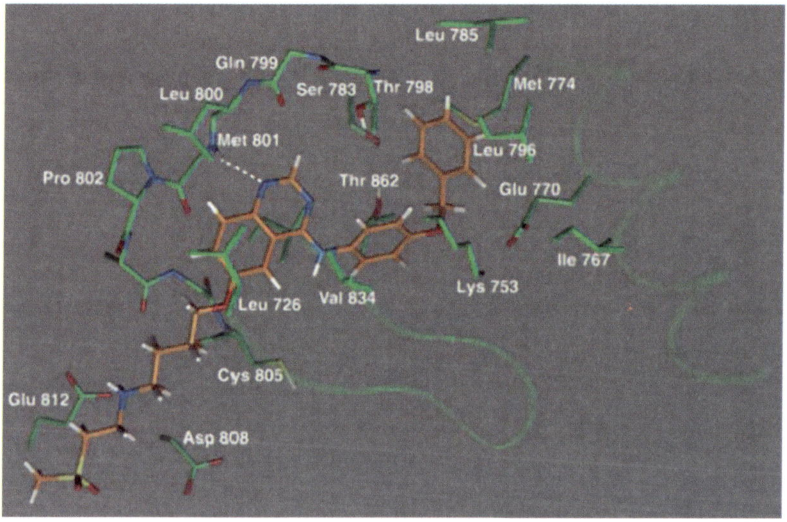

Figure 12.8 Homology model of erbB2 with **9** docked into the ATP-binding site.

Table 12.4 Cellular efficacy of EGFR/erbB2 dual inhibitors.

Example	*Tumor cell,* IC_{50}, μM *HN5/BT474*	*Cellular selectivity[a] HFF/avg. tumor*
14	0.045/1.68	>35×
13	0.63/0.39	>59×

[a]The value for cellular selectivity is calculated by dividing the IC_{50} value for the normal fibroblasts by the average tumor cell IC_{50} value.

substituents on the furan ring affected the cellular activity. Compounds **15–18** in Table 12.5 offer a comparison of each connection to the furan ring plus an example with no side chain appended. The substitution at the 5′ (**16**) and 4′ (**17**) position of the furan are equipotent dual EGFR/ErbB-2 inhibitors, yet **16** was 4–5 times more potent against the tumor cell lines. Further analogs synthesized in the series, therefore, retained the 5-substituted-2-furanyl quinazolines as the core scaffold.

Several areas of the 6-furanyl-4-anilinoquinazoline core structure were investigated to optimize the drug properties while retaining the biological efficacy profile.[16] For ease of comparing the compounds, the data presented has been simplified and includes just the erbB2 portion of the profile. In all cases included in the table and discussion, the EGFR data were equipotent. In the final section of the chapter, the dual inhibitory profile will be summarized with all relevant data included.

Table 12.6 shows data for a systematic assessment of each component of the core structure. Compounds **16**, **19**, and **20** highlight data that led to lapatinib's

Table 12.5 Type I receptor inhibition activity and cellular efficacy results for regiochemical furanylquinazoline analogs.

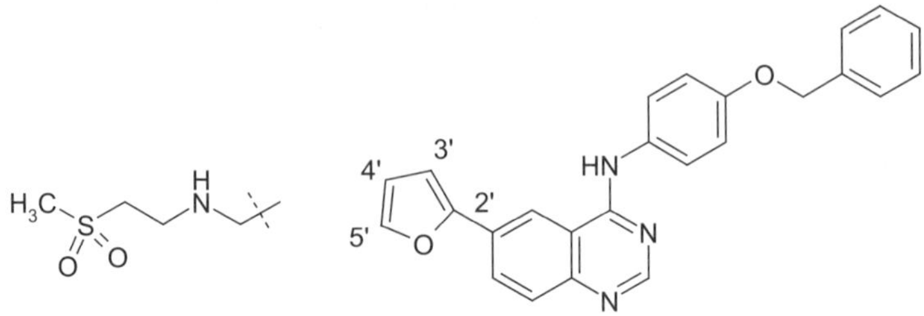

#	Position	*Average ErbB2/ EGFR IC_{50} μM*	*Average tumor HN5/BT474/ N87 IC_{50} μM*
15	none	0.074	2.98
16	5′	0.027	0.60
17	4′	0.023	2.68
18	3′	0.068	3.81

aniline optimized substitution as 4-(3-fluorobenzyloxy)-3-chloroaniline. Compounds **21** and **22** demonstrate that there was a large tolerance in the terminal substitution for enzyme inhibition, but no group was found superior to methyl. For cost of goods and synthetic ease, methyl was chosen even though there were many compounds equivalent in efficacy (data not shown). Because the amine in the side chain was metabolically labile, a variety of substitutions were tried to make a more stable compound, including oxygen (**23**), tertiary amines (**24**, **26**), and substituents with heteroatoms (**25**). Some were found equivalent in their biological profile, but none offered any drug property improvement. A thiomorpholino substituent linked through the furanyl group (**27**–**30**) showed great promise with potent enzyme and cellular results, improved water solubility, but it was highly variable in pharmacokinetic parameters.

Table 12.7 contains representative examples from each of the groups of modifications to the 6-furanyl-quinazoline template covered in the discussion thus far. The average of the tumor cell IC_{50} values was used to assess the dual EGFR and erbB-2 inhibitory properties in assays that are run under similar conditions. The therapeutic potential of this series was considered very high because greater than 100-fold selectivity was observed for all of these analogs.

The 6-furanyl-4-(4-benzyloxyanilino)-quinazoline scaffold afforded the necessary drug-like properties and dual ErbB-2 and EGFR tyrosine kinase inhibition to enable the discovery of a potential anti-cancer therapeutic agent. The halogen substitution on the benzyloxyanilino group was key to improving the enzyme/cell ratio of activity, with 4-(3-fluorobenzyloxy)-3-chloroanilino providing the most promising cellular efficacy. The substitutions on the furanyl ring were required to be in a 2,5-orientation for the desired cellular activity, while there was no apparent difference in the enzyme activity. Quite a diversity of amine substitutions was tolerated, due to the binding mode of these

Table 12.6 Type I receptor inhibition activity and cellular efficacy results for representative examples of the 6-furanyl-4anilinoquinazoline series.

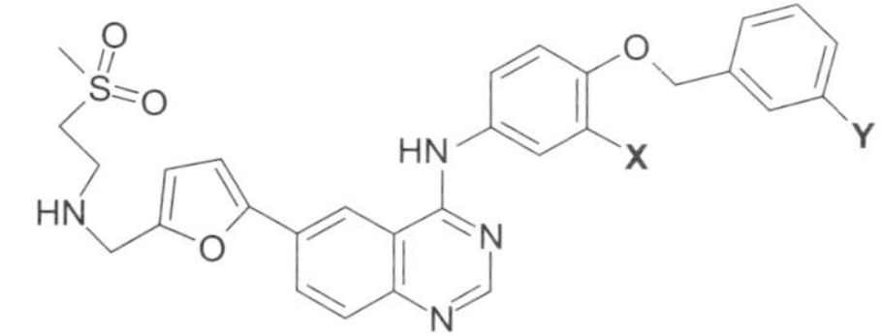

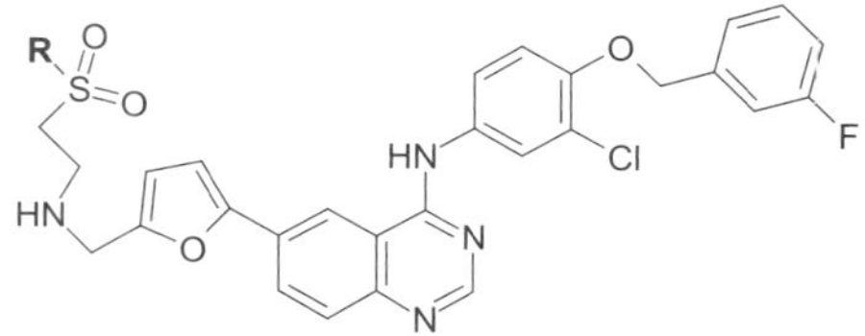

Example	X, Y	ErbB2 IC_{50} μM	BT474 IC_{50} μM	Example	R	ErbB2 IC_{50} μM	BT474 IC_{50} μM
1	Cl, F	0.010	0.08	**1**	$-CH_3$	0.010	0.08
2	H, H	0.026	0.28	**16**	-Ph	0.030	1.04
5	H, CF_3	0.36	4.2	**18**	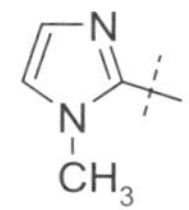	0.026	0.39
8	MeO, F	0.09	2.07				

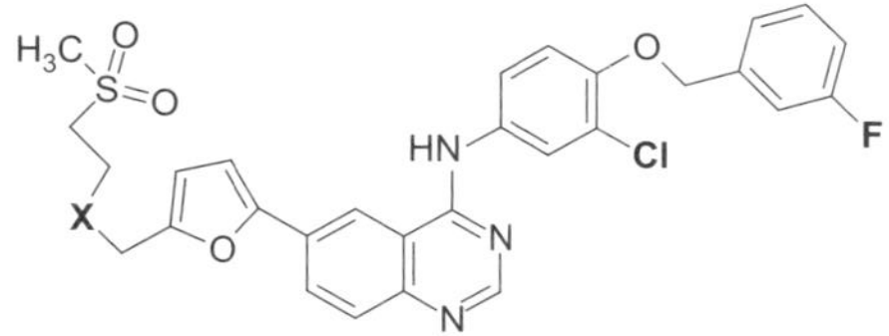

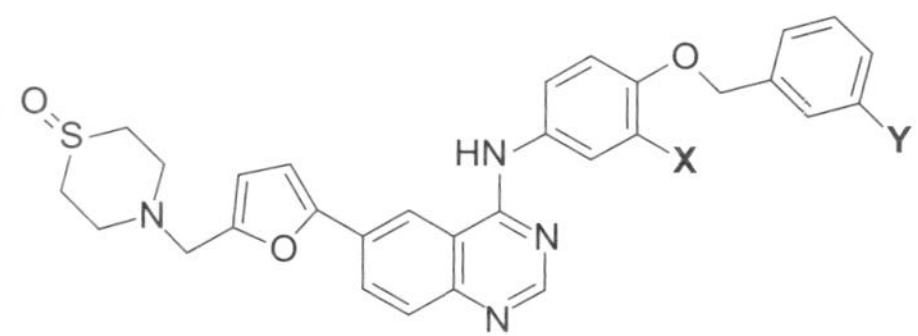

Example	X	ErbB2 IC_{50} μM	BT474 IC_{50} μM	Example	X, Y	ErbB2 IC_{50} μM	BT474 IC_{50} μM
19	-O-	0.017	0.17	**25**	CF_3, F	0.141	3.00
22	-N-(n-Pr)	0.02	0.25	**26**	Cl, F	0.018	0.10
23	-N-(CH_2CN)	0.022	0.067	**27**	F, H	0.035	1.20
24	-N-(Bn)	0.141	0.61	**28**	Cl, H	0.023	0.42

Table 12.7 Cellular activity and selectivity for representative substituted 6-furanyl-quinazolines.

Example	*Ave. tumor*[a] *IC_{50} μM*	*HFF* *IC_{50} μM*	*N/T selectivity*
lapatinib	0.09	9.9	111
23	0.16	>30	>188
25	0.11	>30	>273
28	0.08	8.18	102

inhibitors where the aniline is tucked into the back of the ATP-binding pocket, and the side chain on the furanyl portion extends out toward solvent.

12.4.3.2 6-Thiazolylquinazoline Series

A wide variety of substitutions in the 6-position of the quinazoline was tolerated with retention of potent enzyme inhibition; the SAR from the cellular activity suggested that 5-membered ring heterocycles were preferred over 6-membered ring heterocycles. For example, compounds having a 6-pyridyl or 6-phenyl ring with all other optimized substituents were 10–50-fold less efficacious in cellular assays. The 6-thiazole ring was investigated as a linking heterocycle by combining the 'best' anilines for dual inhibition with the optimized side chain linked via a thiazole ring.[13] A good correlation was observed between the enzyme catalytic activity profile and the cellular efficacy of these derivatives. Representative results are listed in Table 12.8, where the average IC_{50} value is used for the cellular efficacy in the three tumor lines: HN5, BT474, and N87. **30**, **32**, and **GW3340** are potent dual EGFR and erbB-2 tyrosine kinase inhibitors and show efficacy in all the appropriate cell lines. The best overall cellular activity is observed with compounds **32** (*N*-1-benzylindazole) and **GW3340** (4-[3-fluorobenzyloxy]-3-chloroaniline) and they exceed 100-fold selectivity for tumor cells over normal cells.

Pharmacokinetic parameters were generated in mice by treatment with a single i.v. or oral dose of compound at 10 mg/kg. The data for **32** and **GW3340** are compiled in Table 12.9 and demonstrate adequate plasma exposures for tumor studies with oral bioavailability (%F) of 11% and 40%, respectively. The compounds were administered orally at 100 mg/kg b.i.d. for 21 days in the subcutaneous xenograft studies for both BT474 (erbB-2+ + +) and NH5 (EGFR+ + +) human tumor cell lines. The tumor inhibition (%TI), which was recorded for the final day of the study, is also listed in Table 12.9. The control line represents a vehicle-treated group of animals and approximately two tumor size doublings occur during the treatment period. **GW3340** was the most efficacious compound in the 6-thiazoylquinazoline series, displaying approximately 80% tumor inhibition in both xenograft models. A dose response was seen for a 0, 30, 100 mg/kg b.i.d. orally of compound **GW3340** and the lack of significant body weight loss suggested a good therapeutic index.

Table 12.8 Catalytic enzyme assay results of 6-thiazolylquinazoline aniline derivatives.

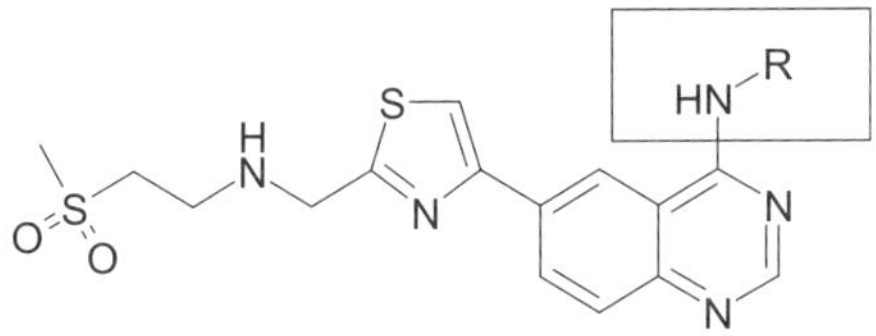

Entry	*Aniline group*	*ErbB-2 IC_{50}, μM^a*	*EGFR IC_{50}, μM*	*HFF IC_{50}, μM*	*Ave tumor IC_{50}, μM*	*N/T selectivity*
31	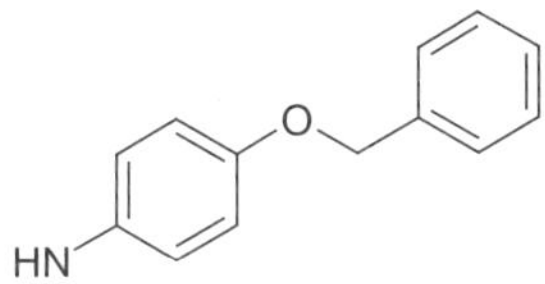	0.048	0.082	>18	1.3	>14
32	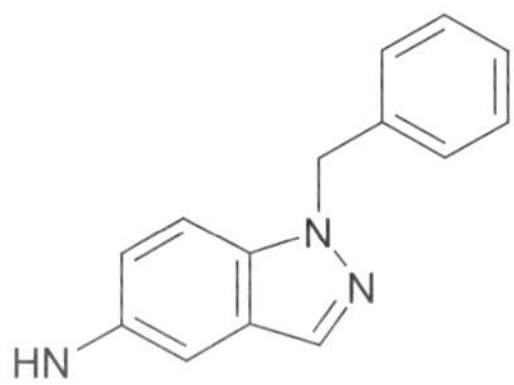	0.028	0.071	12	0.09	138
GW3340	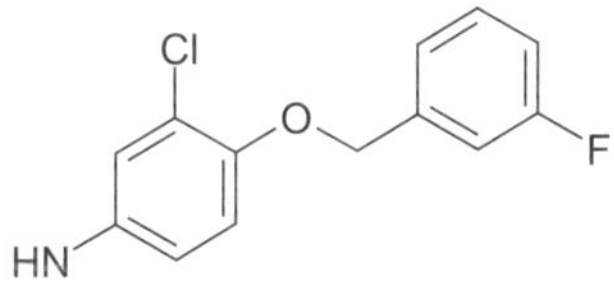	0.014	0.010	>23	0.11	>211

Table 12.9 Representative murine *in vivo* data.

Example	*i.v. AUC*	*PO AUC*	*F (%)*	*% TI[a] HN5*	*% TI[a] BT474*
32	2489	274	11	65	0
GW3340	19040	7520	39.5	81	83

12.4.4 Alkynylpyrimidine Series

Much effort was expended with the aim of replacing the quinazoline scaffold. One obvious example was the replacement of the quinazoline core with a pyrimidine whilst retaining the key 'large' aniline. The parent 5-phenylakynyl-4-anilinopyrimidine **33** had modest dual inhibition of EGFR and ErbB-2 (Table 12.10). Many analogs were synthesized with substituted phenyls and heterocycles linked through the acetylene.[17] Appropriately substituted ring

Table 12.10 Examples of SAR for alkynylpyrimidines.

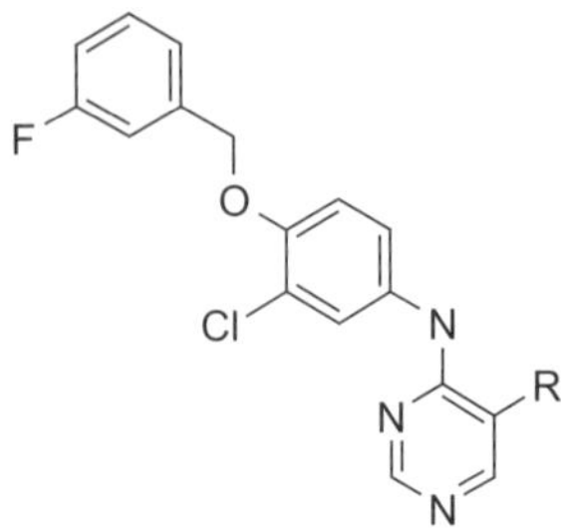

Compound	*R*	*EGFR IC_{50} (μM)*	*ErbB-2 IC_{50} (μM)*	*AveTumor IC_{50} (μM)*	*HFF IC_{50} (μM)*
33		0.079	0.78	—	—
34	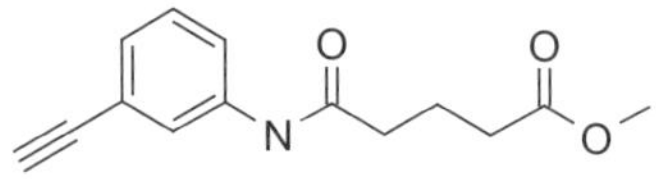	0.031	0.029	3.86	22.4
35	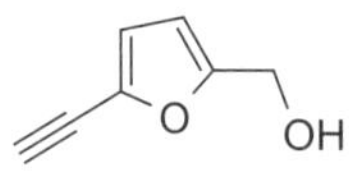	0.033	0.047	1.48	15.8
36	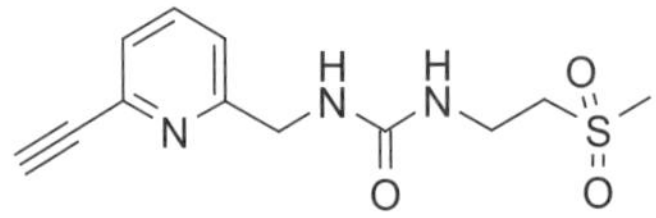	0.015	0.009	0.80	>30

systems appeared to occupy the same region of the ATP-binding pocket as the amine in the preferred quinazoline series 6-position side chain (methylsulfonylethylamine). A heteroatom in this position was important for successful dual kinase inhibition in these compounds, an interesting finding since most of the dual activity was derived from the aniline substitution in the quinazoline series. The potency of **34** with the extended substitution off the amide connection is an example of the tolerance in the binding site. The hydroxymethyl furan **35** retained dual enzyme activity. Heterocycles with a urea methylene linker, as in sulfone analog **36**, were quite potent with IC_{50} values below 0.015 μM in the EGFR and ErbB-2 enzyme assays.

Compounds **34**–**36** were greater than 100-fold selective against a panel of non-EGFR family kinases, including CDK2, GSK3, SRC, and VEGFR2 (data not shown). These compounds were also tested in cellular proliferation assays and the data are included in Table 12.10. The IC_{50} values were averaged for just the tumor cell lines BT474 (erbB-2 driven) and HN5 (EGFR driven), and the HFF line was used as the control comparison. Despite having a crystal structure, SAR from the quinazoline series, and robust assay systems, the conclusion of the research work was that a dual EGFR/erbB2 tyrosine kinase inhibitor drug candidate could not be achieved in this series.

12.5 Understanding the Mode of Inhibition That Makes Lapatinib Analogs Effective

As has been noted, plots of enzyme and cellular data, when taken all together, did not appear to correlate. There were compounds with good profiles, but it was difficult to determine how to interpret the SAR and use it to design better molecules. In the data analysis section, the index calculation was introduced as a way to compare the desired profiles rather than using a single parameter. A subset of the data was chosen along a correlation vector and only those compounds were included in the SAR studies. By examining the SAR of the compounds that were effective, and ignoring the ineffective ones, the desired properties were optimized without the full understanding of why there was a difference between the molecules that correlated and the ones that did not. The paper published by R. B. Lichtner and co-workers (on why quinazolines but not phthalimides worked in cellular assays) gave us some confidence that there could be a mechanistic reason why there might be a difference, but further studies were needed to understand the unique features of lapatinib.[18]

The kinase conformational state and the binding kinetics were found to significantly affect the receptor activity in tumor cells. The interaction of the bound inhibitor with the activated conformational state of the protein can provide added effectiveness in signal inhibition.[12] The off-rates observed for lapatinib and Tarceva are summarized in Table 12.11 and were evaluated using an enzyme reactivation procedure. The rate of product formation in the presence of Tarceva was virtually indistinguishable from the rate of product formation in the absence of inhibitor, indicating a rapid off-rate (half-life <10 minutes). After preincubation with lapatinib, the enzyme activity recovered slowly, suggesting a much slower off-rate (half-life = 300 minutes). A similar dissociation rate was observed with lapatinib using erbB-2 with no evidence for the covalent interaction that is observed with CI-1033, an irreversible kinase inhibitor.[19]

The general features of quinazolines bound in the ATP-binding site of kinases has been reported.[20] The crystal structure of EGFR bound to lapatinib revealed an inactive-like conformation in contrast to the published active-like structure with Tarceva.[21] The differences in the ligand bound structures included the shape of the ATP site (closed versus open conformation), the position of the C helix (large back pocket versus intact Glu738-Lys721 salt bridge), the conformation of the COOH-terminal tail (partially blocking the

Table 12.11 Binding kinetics of a selective EGFR TK inhibitor, Tarceva, compared with the dual EGFR/erbB2 TK inhibitor, lapatinib.

Compound	*erbB-2* K_i *(nM)*	*EGFR* K_i *(nM)*	*Dissociation rates*	*EGFR activity post compound washout*
lapatinib	13 ± 1	3.0 ± 0.2	$T_{1/2} = 300$ min	15% @ 72 h
Tarceva	870 ± 90	0.4 ± 0.1	$T_{1/2} < 10$ min	100% @ 24 h

ATP cleft versus poorly defined), the conformation of the activation loop (A-loop similar to ones found in inactive structures versus ones found in active structures), and the hydrogen bonding pattern with the quinazoline scaffold (water mediated interaction with Thr830 versus Thr766).

When HN5 tumor cells were treated with erlotinib and lapatinib, >85% reduction in tyrosine phosphorylation was observed without any reduction in total receptor content. HN5 tumor cells were treated for 4 hours with an inhibitor, and the receptor phosphorylation was analyzed at multiple time points after inhibitor washout to determine if the inhibitor dissociation properties affected the recovery of autophosphorylation. The slow off-rate found for lapatinib in the purified intracellular domain enzyme reaction correlated with the observed, prolonged signal inhibition in tumor cells based on receptor tyrosine phosphorylation measurements. Blencke and co-workers identified a critical mutation of Thr766 in the EGFR kinase domain as leading to a dramatically reduced sensitivity to the 4-anilinoquinazoline PD153035 and linked it to the mechanism of tumor resistance.[22] It is possible that lapatinib's lack of an interaction at this amino acid will lead to lower mutation rates or simply to a drug that retains inhibition despite the mutation. The slow binding kinetics may offer increased signal inhibition in the tumor, thus leading to greater potential to affect the tumor growth rates or effectiveness of co-dosing with other chemotherapeutic agents.

The exciting component of these ligand-bound kinase structures is the insightful links they may provide for enhancing efficacy via a parameter other than simple potency (i.e. binding to and stabilizing different conformations) or in patient selection for optimal therapeutic outcomes. Overcoming the expectation that there should be a direct correlation between simple enzyme inhibition and the cellular proliferation was key to the successful identification of lapatinib, since by focusing on the subset of quinazolines and pyridopyrimidines that possessed the desired correlation, the lessons learned could be applied to make highly effective erbB family signal inhibitors with a unique interaction with the protein.

12.6 Conclusion

Highlights of preclinical data for lapatinib as evaluated in a range of efficacy assays is summarized in Table 12.12, along with the references for methods and details of the assays. Lapatinib is potent on two of the erbB family members, with enzymatic IC_{50} values against erbB-2 and EGFR receptor tyrosine kinases of 9 and 10 nM respectively, with greater than an order of magnitude loss in activity for the other family member, erbB-4.[23] A small molecule-kinase interaction map was created for lapatinib (GW2016) by workers at Ambit using an ATP site-dependent competition binding assay in a panel of 119 kinases and this demonstrated a very clean profile.[1] An expanded panel of tumor cell lines, which incorporated varying levels of type I receptor expression, were examined for efficacy and lapatinib was equally effective in lines over-expressing either or

Table 12.12 Summary of lapatinib preclinical efficacy data.

Assay	*Efficacy*	*Comment*
Catalytic enzyme:	IC_{50} values	
erbB-2	9.2 nM	>10 000 nM for c-raf1, MEK,
EGFR	10.8 nM	ERK, c-fms, CDK1, CDK2, p38,
Cell growth inhibition:	IC_{50} values	
HFF (EGFR+)	12 000 nM	Expression levels are noted in parentheses on a relative scale. No detectable levels were found where the receptor is omitted
HN5 (EGFR+ + +, erbB-2+)	120 nM	
BT474 (EGFR+, erbB-2+ + +)	100 nM	
N87 (EGFR+ + +, erbB-2+)	90 nM	
CaLu-3 (erbB-2+ + +)	130 nM	
Cellular downstream phospho-signal inhibition:		Inhibition also occurred in the presence of EGF
pAKT & pErk levels (HN5)	>90% at 5 μM	
pAKT & pErk levels (HN5)	>90% at 1 μM	
In vivo xenograft downstream phospho-signal inhibition:		Same tumor implant biopsied pre- and post-treatment
pAKT & pErk levels (HN5)	≥30 mg/kg	
pAKT & pErk levels (HN5)	≥30 mg/kg	
In vivo human xenograft anti-tumor activity:		Values are growth inhibition percent relative to control animals receiving vehicle alone. Number of replicates in average is given in parentheses.
HN5, 30 mg/kg, b.i.d.	34% (32)	
HN5, 100 mg/kg, b.i.d.	101% (40)	
BT474, 30 mg/kg, b.i.d.	42% (32)	
BT474, 100 mg/kg, b.i.d.	94% (40)	

both EGFR and erbB-2. The average selectivity for the tumor cell lines tested over normal lines was 100-fold.

Lapatinib was shown to have *in vivo* anti-tumor activity in xenograft mouse models of BT474, the erbB-2 over-expressing human ductal carcinoma, and HN5, the EGFR over-expressing human squamous cell carcinoma. At doses of 30 and 100 mg/kg b.i.d., lapatinib demonstrated reproducible tumor growth inhibition of 34 ± 28% and 101 ± 20%, respectively, in the HN5 xenograft model with regression (defined as >25% reduction in tumor volume) in 33% of the treated animals.[16] In the BT474 model, inhibition of 42 ± 35% and 94 ± 18%, respectively, was observed with 10% of the treated animals with regressions. Because it was important to relate the anti-tumor activity to the inhibition of receptor phosphorylation, the level of erbB-2 phosphotyrosine in tumor tissue excised after therapy in the 100 mg/kg treatment, was reduced by 93% in the BT474 model and 85% in the HN5 model. Analysis of the tumor samples following therapy with lapatinib indicated that receptor phosphorylation was decreased in a dose responsive manner. Lapatinib had very little, if any, toxicity at this dose based on the general appearance of the study animals, lack of body weight loss, and their normal clinical chemistry. The activity observed in the xenograft mouse models suggested that lapatinib could be safely combined with other chemotherapies, potentially offering advantages in patient care where combination therapy is expected.

Research work was done in the preclinical setting to understand the effects of lapatinib on proliferation, survival, and downstream proteins in signaling cascades. These studies were important because, by understanding the mechanism of action, the selection of the patient population that would most benefit from a dual EGFR and erbB2 kinase inhibitor could be achieved. Also, it would be possible to assess potential biomarkers for determining effective doses (i.e. maximum biological effects without having to dose to the traditionally used maximum tolerated dose) and patient population (over-expressors of EGFR and/or erbB2) for use in chemotherapy. Lapatinib inhibits the activation of both erbB-2 and EGFR, and interrupts downstream activation of Erk1/2 MAP kinases and AKT.[23] The inhibition of AKT by lapatinib was associated with a 23-fold increase in apoptosis. The PI3K/AKT pathway is involved in regulating cell survival and activation of AKT in tumors has been shown to be involved in chemotherapy resistance. Lapatinib was found to inhibit signal transduction by measuring the p-Tyr, p-Erk, and p-AKT levels in the presence of saturating concentrations of epidermal growth factor (EGF) in tumor cell lines that over-express EGFR/erbB2 receptors as well as stimulated tumor lines that do not over-express EGFR. These studies have important implications in the clinical use of lapatinib, since EGF levels have been shown to be elevated in cancer patients.[24] The pre-clinical human xenograft *in vivo* inhibitory effects of lapatinib on receptor p-tyr expression and the same downstream signal cascade proteins were evaluated by biopsy of the same tumor implant prior to and following treatment with lapatinib. The experiment modeled the human clinical scenario where each patient's tumor served as their own control before and after treatment with drug using the same xenograft models for EGFR over-expression (HN5) and erbB-2 over-expression (BT474). Dose-dependent inhibition of EGFR p-tyr, erbB2 p-tyr p-AKT, and p-Erk was seen at 30 and 100 mg/kg without affecting steady state protein levels and was consistent with earlier findings of anti-tumor activity in these models. Figure 12.9 shows the graph of the anti-tumor activity in the BT474 xenograft model at two doses, with the p-tyr western blots showing the signal inhibition.[23]

The drug discovery effort for lapatinib was taking place just as personalized cancer treatments began to emerge from striking advances in oncogenes and their mechanisms of actions. The project required careful attention to link mechanism of action studies, dual inhibition of EGFR/erbB2, to all observed efficacy and selectivity. In the early work, the data for the inhibition of the targets did not correlate with the desired cellular efficacy, and thus called into question whether it was the right concept for the generation of an effective drug. By doing a multi-parameter analysis linking the enzyme profile to a cellular outcome (potent on tumor cells and inactive on normal or control lines), the data could be reviewed as index analysis for the ability of each series to reach the desired profile. A subset of the data that fits the desired correlation was used to drive the design of the final sets of synthetic compounds. Lapatinib was discovered as the most effective dual EGFR/erbB2 tyrosine kinase

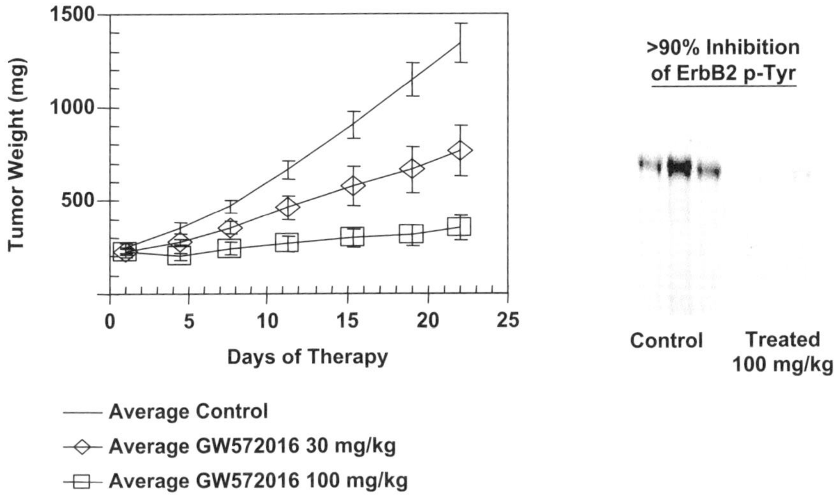

Figure 12.9 BT474 xenograft model showing how efficacy correlated to mechanism of action for lapatinib.

inhibitor with outstanding cellular efficacy and selectivity. When studied in multiple models of *in vivo* xenograft studies, tumor growth was inhibited in a dose-dependent manner, and the excised tumors showed blocked autophosphorylation consistent with the tumor inhibition. Subsequent crystal structure studies of lapatinib bound to the kinase domain of erbB2 showed a unique binding interaction as compared with the leading EGFR selective drug candidates (inactive versus active forms, respectively). Enzyme kinetic studies revealed a very slow off-rate for lapatinib, further supporting why this subset of quinazoline derivatives was effective. Lapatinib was selected for the full development process and was dosed to humans in January 2001 in the first clinical trial in healthy volunteers. In March of 2007, lapatinib was launched as an anti-cancer treatment for breast cancer in combination with Xeloda.[25]

Acknowledgements

The team of scientists responsible for the discovery of lapatinib include Karen Affleck, Patricia Allen, Krystal J. Alligood, Zongwei Cai, Malcolm C. Carter, G. Stuart Cockerill, Renae M. Crosby, Scott Dickerson, Stephen V. Frye, Micheal Gaul, Cassander Gauthier, Tona M. Gilmer, Kimberly Glennon, Robert Griffin, Stephen B. Guntrip, Yu Guo, Nelson W. Johnson, Barry R. Keith, Wilson B. Knight, Karen E. Lackey, Michael Luzzio, Robert A. Mook, Robert J. Mullin, Doris M. Murray, David W. Rusnak, Sarva M. Tadepalli, Achintya K. Sinhababu, Kathryn J. Smith, Edgar Wood, and Yue-mei Zhang.

References

1. M. A. Fabian, W. H. Biggs, D. K. Treiber, C. E. Atteridge, M. D. Azimioara, M. G. Benedetti, T. A. Carter, P. Ciceri, P. T. Edeen, M. Floyd, J. M. Ford, M. Galvin, J. L. Gerlach, R. M. Grotzfeld, S. Herrgard, D. E. Insko, M. A. Insko, A. G. Lai, J.-M. Lelias, S. A. Mehta, Z. V. Milanov, A. M. Velasco, L. M. Wodicka, H. K. Patel, P. P. Zarrinkar and D. J. Lockhart, *Nat. Biotechnol.*, 2005, **23**, 329–336.
2. H. Varmus, J. M. Bishop and P. Vogt, *J. Mol. Biol.*, 1973, **74**, 613–626.
3. R. Bazell, *Her-2: The Making of Herceptin, a Revolutionary Treatment for Breast Cancer*, Random House, New York, 1998, ISBN 0-679-45702-X.
4. *Gleevec: a Medical Dictionary, Bibliography, and Annotated Research Guide to Internet References*, ICON publications, 2004, ISBN 978-0-5978443-7-9.
5. Y. Yarden and M. Sliwkowski, *Nat. Rev. Mol. Cell Biol.*, 2001, **2**, 127–137.
6. J. Hancock, *Cell Signalling*, 3rd edn, Oxford University Press, USA, 2010, ISBN 978-0-1992321-0-9.
7. D. W. Fry, A. J. Kraker, A. McMichael, L. A. Ambroso, J. M. Nelson, W. R. Leopold, R. W. Connors and A. J. Bridges, *Science*, 1994, **265**, 1093–1095.
8. S. G. Cockerill and K. E. Lackey, *Curr. Top. Med. Chem.*, 2002, **2**, 1001–1010.
9. P. S. Brignola, K. Lackey, S. H. Kadwell, C. Hoffman, E. Horne, H. L. Carter, J. D. Stuart, K. Blackburn, M. B. Moyer, K. J. Alligood, W. B. Knight and E. R. Wood, *J. Biol. Chem.*, 2002, **277**, 1576–1585.
10. D. W. Rusnak, K. Affleck, S. G. Cockerill, C. Stubberfield, R. Harris, M. Page, K. J. Smith, S. B. Guntrip, M. C. Carter, R. J. Shaw, A. Jowett, J. Stables, P. Topley, E. R. Wood, P. S. Brignola, S. H. Kadwell, B. R. Reep, R. J. Mullin, K. J. Alligood, B. R. Keith, R. M. Crosby, D. M. Murray, W. B. Knight, T. M. Gilmer and K. E. Lackey, *Cancer Res.*, 2001, **61**, 7196–7203.
11. E. J. Kunkel, I. Plavec, I. Nguyen, J. Melrose, E. S. Rosler, L. T. Kao, Y. Wang, E. Hytopoulos, A. C. Bishop, R. Bateman, K. M. Shokat, E. C. Butcher and E. L. Berg, *Assay Drug Dev. Technol.*, 2004, **2**, 431–441.
12. E. R. Wood, A. T. Truesdale, O. B. McDonald, D. Yuan, A. Hassell, S. H. Dickerson, B. Ellis, C. Pennisi, E. Horne, K. Lackey, K. J. Alligood, D. W. Rusnak, T. M. Gilmer and L. A. Shewchuk, *Cancer Res.*, 2004, **64**, 6652–6659.
13. M. D. Gaul, Y. Guo, K. Affleck, G. S. Cockerill, T. M. Gilmer, R. J. Griffin, S. Guntrip, B. R. Keith, W. B. Knight, R. J. Mullin, D. M. Murray, D. W. Rusnak, K. Smith, S. Tadepalli, E. R. Wood and K. Lackey, *Bioorg. Med. Chem. Lett.*, 2003, **13**, 637–640.
14. Y.-M. Zhang, S. Cockerill, S. G. Guntrip, D. Rusnak, K. Smith, D. Vanderwall, E. R. Wood and K. E. Lackey, *Bioorg. Med. Chem. Lett.*, 2004, **14**, 111–114.
15. G. S. Cockerill, C. Stubberfield, J. Stables, M. Carter, S. Guntrip, K. Smith, S. McKeown, R. Shaw, P. Tapley, L. Thomsen, K. Affleck,

A. Jowett, D. Hayes, M. Willson, P. Woollard and D. Spalding, *Bioorg. Med. Chem. Lett.*, 2001, **11**, 1401–1405.
16. K. G. Petrov, Y.-M. Zhang, M. Carter, G. S. Cockerill, S. Dickerson, C. Gauthier, Y. Guo, D. W. Rusnak, R. A. Mook, A. L. Walker, E. R. Wood and K. E. Lackey, *Bioorg. Med. Chem. Lett.*, 2006, **16**, 4686–4691.
17. A. G. Waterson, K. L. Stevens and M. J. Reno, *Bioorg. Med. Chem. Lett.*, 2006, **16**, 2419–2422.
18. R. B. Lichtner, A. Menrad, A. Sommer, U. Klar and M. R. Schneider, *Cancer Res.*, 2001, **61**, 5790–5795.
19. P. A. Janne, J. von Pawel, R. B. Cohen, L. Crino, C. A. Butts, S. S. Olson, I. A. Eiseman, A. A. Chiappori, B. Y. Yeap, P. F. Lenehan, K. Dasse, M. Sheeran, P. D. Bonomi and D. Philip, *J. Clin. Oncol.*, 2007, **25**, 3936–3944.
20. L. Shewchuk, A. Hassell, B. Wisely, W. Rocque, W. Holmes, J. Veal and L. F. Kuyper, *J. Med. Chem.*, 2000, **43**, 133–138.
21. J. Stamos, M. X. Sliwkowski and C. Eigenbrot, *J. Biol. Chem.*, 2002, **277**, 46265–46272.
22. S. Blencke, A. Ulrich and H. Daub, *J. Biol. Chem.*, 2003, **278**, 15435–15440.
23. D. W. Rusnak, K. E. Lackey, K. Affleck, E. R. Wood, K. J. Alligood, N. Rhodes, B. Keith, D. M. Murray, R. J. Mullin, W. B. Knight and T. M. Gilmer, In Vitro *and* In Vivo *Mol. Cancer Ther.*, 2001, **1**, 85–94.
24. N. L. Spector, W. Xia, H. Burris, H. Hurwitz, E. C. Dees, A. Dowlati, B. O'Neil, B. Overmoyer, P. K. Marcom, K. L. Blackwell, D. A. Smith, K. M. Koch, A. Stead, S. Mangum, M. J. Ellis, L. Liu, A. K. Man, T. M. Bremer, J. Harris and S. Bacus, *J. Clin. Oncol.*, 2005, **23**, 2502–2512.
25. B. Corkery, N. O'Donovan and J. Crown, *OncoTargets Ther.*, 2008, **1**, 21–34.

CHAPTER 13

Identification and Optimization of Dual PI3K/mTOR Inhibitors

ANDREAS KARLSSON[a] AND
CARLOS GARCÍA-ECHEVERRÍA[b,*]

[a] Structure, Design, and Informatics; [b] Oncology Drug Discovery and Preclinical Research, Sanofi-Aventis, Vitry-sur-Seine, France
*Email: carlos.garcia-echeverria@sanofi-aventis.com

13.1 Introduction

Phosphoinositide-3-kinase (PI3K) and mammalian target of rapamycin (mTOR) are critical components of the so-called PI3K pathway (Figure 13.1). This signal transduction cascade plays an important role in controlling cell growth, proliferation, and survival. Through various mechanisms, the PI3K pathway is frequently constitutively activated in human cancers,[1–3] suggesting the use of pathway modulators as novel targeted anti-cancer agents. To this end, substantial drug discovery efforts have been devoted both in pharmaceutical companies and in academia to identify and develop therapeutic agents able to specifically down-regulate the kinase activity of some of its components, in particular PI3K and/or mTOR, in tumor cells.[4–7]

The PI3Ks are widely expressed lipid kinases that phosphorylate phosphoinositides at the D-3 position of the inositol ring.[8] The eight members of the PI3K family have been classified into three groups based on their primary sequence, *in vitro* substrate preference, domain structure, and mode of regulation. One of the subjects of this book chapter is class I PI3Ks, which is composed of two subgroups IA and IB. The class IA PI3K subgroup consists of

RSC Drug Discovery Series No. 21
Designing Multi-Target Drugs
Edited by J. Richard Morphy and C. John Harris

Published by the Royal Society of Chemistry, www.rsc.org

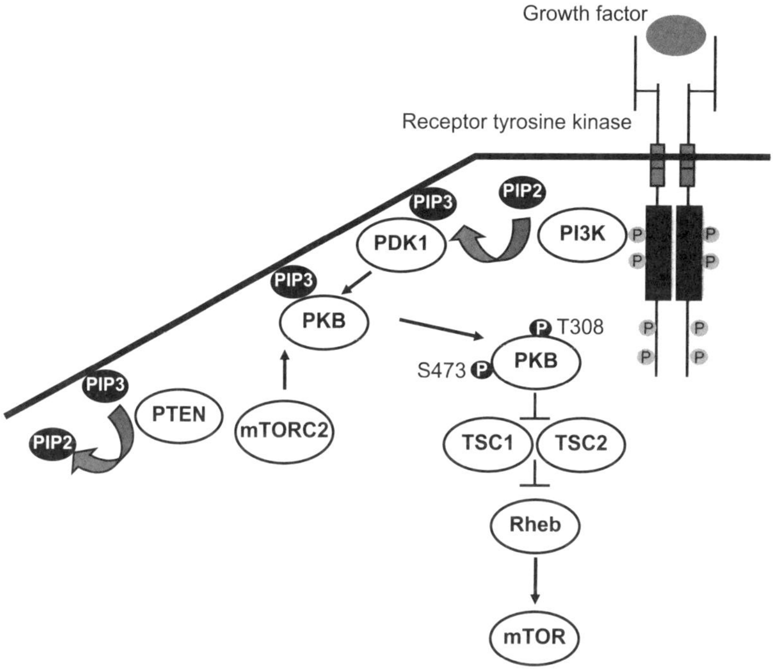

Figure 13.1 Schematic representation of the PI3K pathway and its main components.

three catalytic subunits, p110α, p110β, and p110δ, which form heterodimers with one of five regulatory domains: p85α, p85β, p85γ, p50α, and p55α. These PI3Ks are activated by cell-surface receptor tyrosine kinases, and in the case of p110β also by G-protein-coupled receptors (GPCRs). The class IB consists of one member, a heterodimer of p110γ and a regulatory p101 domain, which is exclusively activated by G-protein βγ subunits following the stimulation of GPCRs. PI3Ks IA and IB catalyze the formation of phosphatidylinositol-3,4,5-triphosphate (PtdIns(3,4,5)P_3; PIP3 in Figure 13.1), a process that is reversed by the action of the lipid phosphatase and tensin homolog deleted on chromosome 10 (PTEN). The other target that is covered herein is mTOR. This protein is a serine/threonine kinase that is a key intracellular component downstream of PI3K (Figure 13.1). mTOR is found in two structurally and functionally distinct multi-protein complexes known as mTOR complex 1 (mTORC1) and 2 (mTORC2), which have different subunit composition, downstream substrates, and biological effects. mTORC1 contains raptor, mLST8, PRAS40, and mTOR, while mTORC2 consists of rictor, mSIN1, mLST8, and mTOR.[9,10]

The pharmaceutical development of PI3K and mTOR modulators has made a great leap forward during the past few years, and some of these drugs (e.g. rapamycin and derivatives thereof)[11] have already provided proof-of-concept in oncology clinical settings and achieved marketing approval. This chapter is focused on the identification and development of PI3K inhibitors that concomitantly block the enzymatic activity of mTOR. The mTOR biological activity of the clinical compounds covered herein was found retrospectively, and their multi-target kinase profile was probably not part of the medicinal chemistry optimization strategy for these or related analogs. One reason behind the late identification of mTOR inhibitory activity of these dual inhibitors is that, contrary to the relatively early availability of a broad range of biochemical and cellular assays to assess the selectivity profile of preclinical candidates against protein kinases,[12,13] suitable biochemical or cellular systems to determine the activity profile of compounds against members of the lipid kinase family have been available only recently. The kinase selectivity profile of these dual PI3K/mTOR modulators is consistent with the high sequence homology and identity in the ATP-catalytic cleft of these kinases.

As briefly illustrated in the following sections, the available preclinical results demonstrate that the concomitant inhibition of PI3K and mTOR results in cell growth inhibition and induction of apoptosis in a variety of tumor cancer cells, particularly for those harboring mutations in the PIK3CA gene and/or overexpressing erbB2. In spite of their high lipophilicity and limited aqueous solubility, the pharmacological, biological, and preclinical safety profiles of some of the early dual PI3K/mTOR inhibitors supported their clinical development and a few of them are currently undergoing phase I/II clinical trials in cancer patients.

Although the polypharmacological activity of these compound was serendipitous, the identification of the structural features[14–16] that determine their inhibitory profile and extensive SAR data have been instrumental in guiding our continuing drug discovery efforts to target the PI3K pathway in tumor cells and understanding into the roles of PI3K and mTOR in human cancer biology.

13.2 Pyridofuropyrimidine Derivatives: From a Chemical Tool to a Development Candidate

Imidazopyridine (e.g. compound **1**, $IC_{50} = 0.67$ μM, p110α; Figure 13.2), pyridothienopyrimidine (e.g. compound **2**, $IC_{50} = 1.4$ μM, p110α; Figure 13.2) and quinazoline (e.g. compound **3**, $IC_{50} = 1.3$ μM, p110α; Figure 13.2) derivatives were early representative examples of class I PI3K inhibitors.[17–20] These compounds were identified by HTS using a scintillation proximity p110α assay, and then optimized to yield single digit nanomolar lead compounds. One of those lead compounds is PI-103 (compound **4**, Figure 13.3).[19,21] This pyridofuropyrimidine derivative is a PI3K inhibitor with enzymatic IC_{50} values in the 2 nM (p110α) to 15 nM range (p100δ), and a good selectivity profile against a panel of 70 other protein and lipid kinases, including class II and III PI3Ks. Despite its rapid *in vivo* metabolism (the phenolic hydroxyl group is rapidly glucuronidated) and its limited aqueous solubility (the planar tricyclic structure may significantly contribute to molecular packing), PI-103 allowed the demonstration of preclinical proof-of-concept for anti-tumor activity by delaying the growth of different subcutaneous human cancer xenografts, in

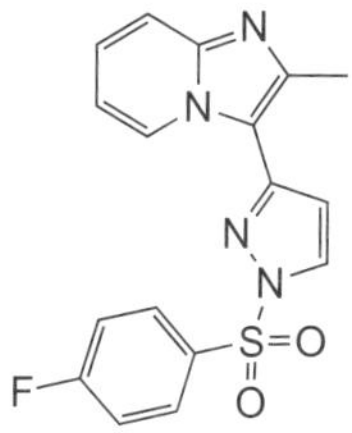

1

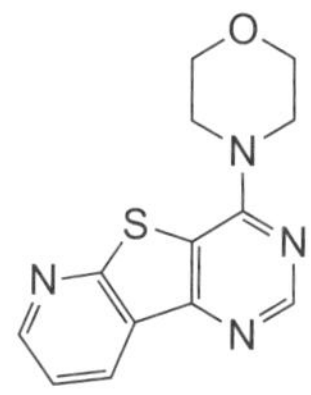

2

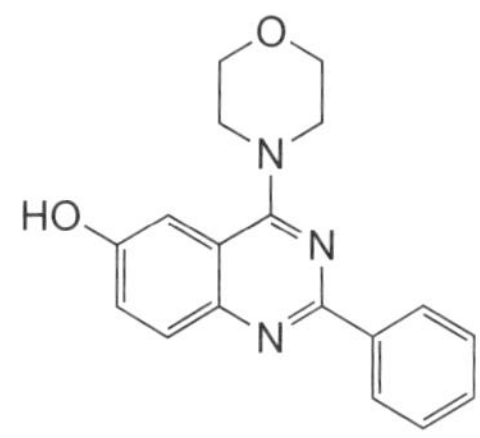

3

Figure 13.2 Early examples of class I PI3K inhibitors.

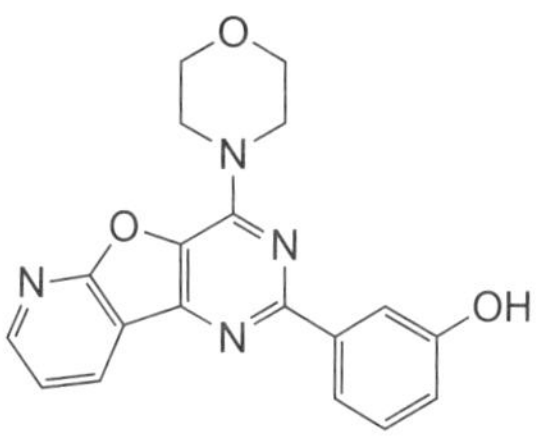

4

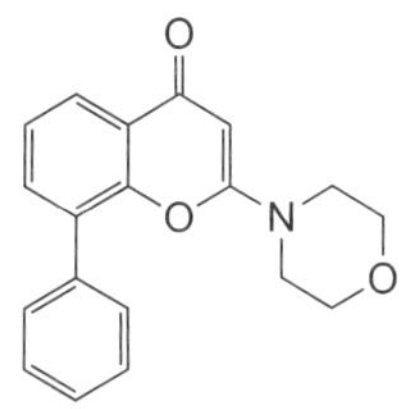

5

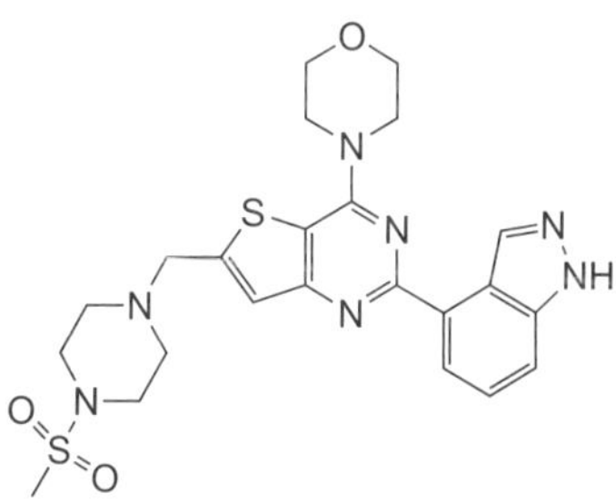

6

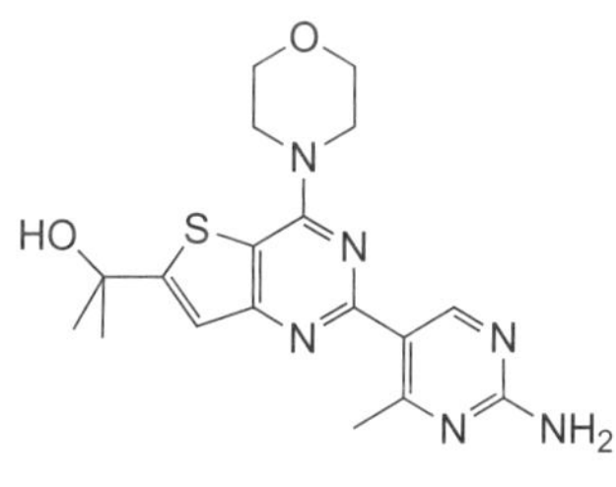

7

Figure 13.3 Representative examples of fused 4-morpholino-2-arylpyrimidine derivatives and structure of LY294002 (compound **5**).

particular cancer cells harboring PI3KCA mutations. The observed *in vivo* anti-cancer effects were consistent with the therapeutic mechanism being inhibition of the PI3K pathway as demonstrated by pharmacodynamic biomarkers (e.g. reduced phosphorylation of protein kinase B (PKB), Figure 13.1, or downstream effectors). It was later shown that PI-103 also inhibits both complexes of mTOR: the rapamycin-sensitive mTORC1 ($IC_{50} = 0.02$ μM) and the rapamycin-insensitive mTORC2 ($IC_{50} = 0.083$ μM). The mTOR biochemical activity was further confirmed in cells, where the compound blocks downstream markers of mTOR signaling (e.g. phosphorylation of p70 S6 kinase $IC_{50} < 0.1$ μM).

Docking studies performed in p110α and mTOR suggest that PI-103 has similar binding modes in the ATP-binding pockets of these proteins, where the compound makes key interactions mainly with residues that are conserved between class I PI3K family and mTOR (Figure 13.4).

The morpholino ring oxygen of PI-103 forms a pivotal hinge region hydrogen bond interaction with the backbone NH of Val851 in p110α and Val2240 in mTOR. In spite of its metabolic liability (e.g. via oxidation α to the morpholine ring oxygen), this pharmacophore group and H-bond interaction have been extensively exploited in PI3K/mTOR inhibitors, and it was already present in LY294002 (compound **5**, Figure 13.3),[22] the first synthetic, albeit unselective, PI3K inhibitor that has served as an important research tool for more than a decade in cancer biology.

The canonical H-bond of PI-103 in the hinge region is complemented by key interactions with other parts of the ATP-binding cleft (Figure 13.4). The planar tricyclic scaffold is sandwiched between hydrophobic residues in the central part of the ATP-binding side, while the phenol ring projects towards the inner part of the pocket. The hydroxyl displaces a conserved water molecule and makes hydrogen bond interactions with Tyr836 and Asp810, and Tyr2225 and Asp2195 in p110α and mTOR, respectively.

The versatility of fused 4-morpholino-2-arylpyrimidines to provide modulators of the PI3K pathway with varying activity toward mTOR has been recently demonstrated.[23,24] Modifications of the thieno[3,2-*d*]pyrimidine core of PI-103 and derivatives thereof resulted in the identification, by the Piramed group, of a selective PI3K modulator, GDC-0941 (compound **6**, Figure 13.3).[24–26] This ATP-competitive inhibitor is active against class I PI3K with biological activities

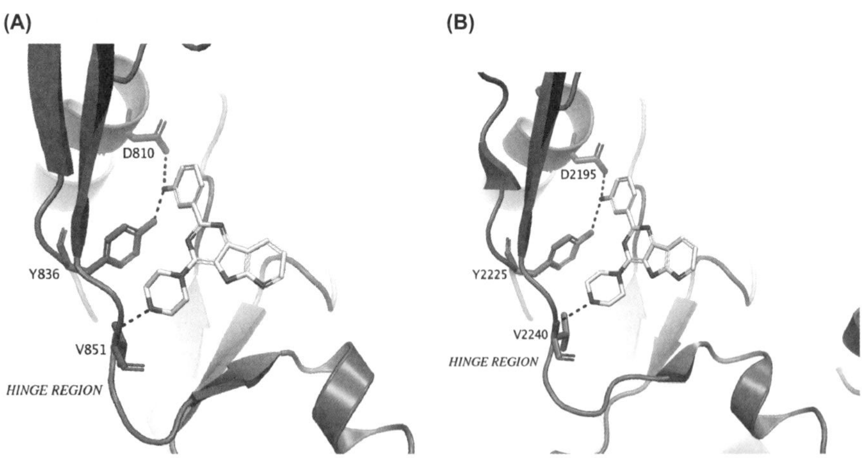

Figure 13.4 Model of PI-103 docked into (A) p110α (X-ray structure) and (B) mTOR (homology model). Potential hydrogen bonds are shown with hatched lines.

in the low nM range (IC_{50} = 3, 33, 3, and 75 nM for p110α, p110β, p110δ, and p110γ, respectively). In biochemical assays, GDC-0941 demonstrates selectivity over PI3K C2β (IC_{50} = 670 nM) and Vps34 (IC_{50} = 1 μM). In addition, GDC-0941 has a weak potency against DNA-PK (IC_{50} = 1.2 μM) and mTOR (IC_{50} = 580 nM). Key modifications were introduced to achieve the preceding activity and biological profile, and improve the pharmacological properties of the parent compound.

Although a comparison between the binding modes of PI-103 with GDC-0941 indicates structural similarities, evaluation of structure–activity relationship (SAR) data around this series suggests origins of their different inhibitory and selectivity profiles.[23] Closely related analogs of GDC-0941 in which the indazole moiety is replaced by an aminopyrimidine group showed that the addition of a single methyl group at position 4 (e.g. compound **7**, Figure 13.3) resulted in a significant decrease in mTOR binding affinity (IC_{50} = 740 nM versus 32 nM) without affecting potency against class I PI3K (IC_{50} = 3.5 nM).[23] The X-ray crystal structures of several of these analogs with p110γ showed that in contrast to the unsubstituted aminopyrimidine, the 4-methyl group slightly twists the aromatic ring out of the plane of the thienopyrimidine scaffold while steric clashes cause the 4-methyl group to point away from Tyr867 in p110γ and towards the upper surface ('ceiling') of the ATP-binding pocket. This region and the sequence around Tyr867 show differences in hydrophobic character and conformation/flexibility in PI3K (Met804 and Ile831, p110γ numbering) and mTOR (Ile2163 and Ley2185) may account for the observed differences in PI3K and mTOR selectivity within this compound class, although this is not really proven given that the X-ray structure of the kinase domain of mTOR has not been solved yet.

On the basis of its attractive molecular profile and preclinical anti-tumor activity and safety profile, GDC-0941 was selected by Genentech/Roche (Piramed was purchased by Roche in April 2008) for clinical development, and entered phase I dose escalation trial in patients with advanced solid tumors in April 2008.

This case study illustrates how initial hits identified in HTS p110α campaigns were improved to a dual PI3K/mTOR tool compound to interrogate cancer biology, then into a more advanced lead, and finally into a clinical development candidate.

13.3 Imidazoquinoline: NVP-BEZ235, the First Dual PI3K/mTOR Inhibitor to Enter Clinical Trials

NVP-BEZ235 (compound **8**, Figure 13.5) evolved by target-hopping from a previously described 3-phosphoinositide-dependent protein kinase-1 (PDK1) inhibitor.[27] As illustrated in this section, the imidazo[4,5-*c*]quinoline core of NVP-BEZ235 can be considered a privileged ATP site directed scaffold for protein and lipid kinase modulation. Depending on the substitution

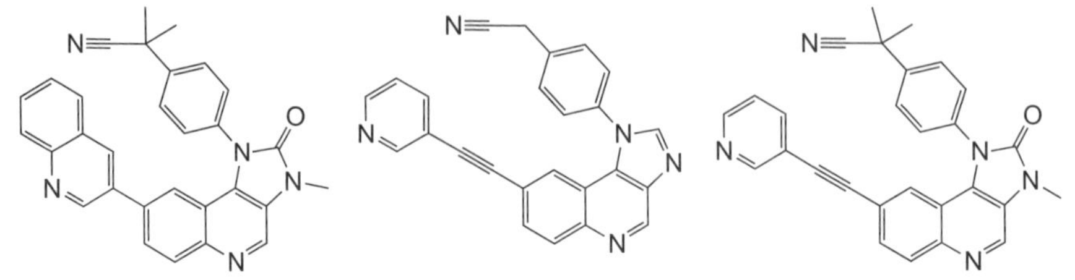

8 9 10

Figure 13.5 Representative examples of imidazo-quinoline derivatives.

pattern, it can adopt different and complementary binding modes in the ATP-binding cleft that can be exploited to achieve the desired activity and selectivity profile by reaching regions by the active site unique to the targeted kinase.

Starting from a previously identified hit, different structural elements and binding modes were exploited to reach the ATP-binding pocket of PDK1. This structure-based design optimization approach led to compound **9** (Figure 13.5), which inhibits PDK1 kinase activity with an IC_{50} of 34 nM.[27] PDK1 is a serine/threonine kinase belonging to the AGC kinase family that phosphorylates and partially activates PKB at Thr308 in a PI3K-dependent manner (Figure 13.1).[28] Compound **9** shows a good selectivity profile against other protein kinases, including members of AGC kinase family (e.g. PKA and PKB, 24% and 0% inhibition at 10 μM, respectively), but, unexpectedly, it also inhibits class I PI3K (IC_{50} = 64, 432, 98, and 67 nM for p110α, p110β, p110δ, and p110γ, respectively). Changing the central imidazo ring to *N*-methyl imidazolinone (e.g. compound **10**, Figure 13.5) resulted in a complete loss in PDK1 inhibitor activity (IC_{50} > 25 μM), while potency against PI3K was preserved (IC_{50} = 72, 2336, 201, and 383 for p110α, p110β, p110δ, and p110γ, respectively). Analysis of the docking of compound **9** in the PDK1 active site (data not shown) reveals a short distance (3.4 Å) between the carbon in position 2 of the imidazole ring and the backbone carbonyl group of Leu88 in the P-loop of this protein kinase. It was postulated that the *N*-methyl imidazolinone moiety could create an electrostatic mismatch with Leu88 in PDK1 while such a strong repulsive interaction could not happen in the PI3K active site because the carbonyl group is facing the protein–water solvent interface.

Starting from compound **10** and derivatives thereof, a focused medicinal chemistry approach guided by biochemical and cellular assays led to the identification of NVP-BEZ235.[29] On the basis of docking and energy minimization studies in p110α, it is proposed that NVP-BEZ235 forms three-H-bond interactions in the ATP-binding cleft of PI3K (Figure 13.6). The nitrogen atom of the core quinoline ring is involved in a canonical H-bond interaction with the backbone of the conserved p110α Val851 (Val2240 in mTOR). As shown previously, the H-bond interaction with the hinge region of the ATP-binding pocket of PI3K is observed for all the inhibitors for which X-ray

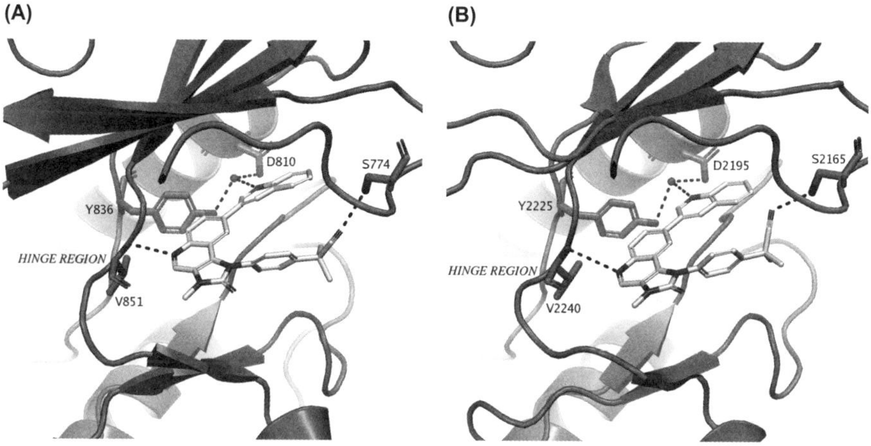

Figure 13.6 Model of NVP-BEZ235 docked into (A) p110α (X-ray structure) and (B) mTOR (homology model). Potential hydrogen bonds are shown with hatched lines.

co-crystal structures exist. The second and third H-bond interactions involve other parts of the binding cleft. The quinoline substituent at C-6 of NVP-BEZ235 is hydrogen bonded with a water molecule that interacts with Tyr836 and Asp810 in p110α (Tyr2225 and Asp2195 in mTOR) at the back end of the ATP-binding site. Towards the entrance of this binding cleft, the cyano group could reach the side chain of p110α Ser774 (Ser2165 in mTOR). These H-bond interactions are complemented by favorable van der Waals contacts with conserved hydrophobic residues of the ATP catalytic site of both enzymes.

NVP-BEZ235 is a potent inhibitor of PI3K (IC_{50} = 4, 76, 7, 5, and 21 nM against p110α, β, γ, δ) and mTOR (IC_{50} = 21 nM) kinase activity,[29] and exhibits anti-proliferative activity against a broad panel of tumor cell lines by specifically blocking the biological function of PI3K signaling components. The anti-proliferative activity of this imidazoquinoline derivative in cellular settings translates well in *in vivo* models of human cancer. Thus, compound treatment results in disease stasis or tumor regression when administered orally and enhanced efficacy when used in *in vivo* combination studies with targeted therapies or standard-of-care agents.[30–32] *Ex vivo* PK/PD analyses of tumor tissues upon acute dose or after termination of *in vivo* efficacy studies showed a time-dependent correlation between compound concentration and inhibition of PKB phosphorylation. In all preclinical *in vivo* models, NVP-BEZ235 was well tolerated at the efficacious doses when compared with vehicle-treated animals, with no significant difference seen in the body weight. Unlike other modulators of the PI3K pathway, no statistically significant changes in blood glucose levels were observed in the animals treated with NVP-BEZ235 after *in vivo* efficacy experiments in mice or rats. Moreover, no effect on insulin or glucose levels was reported for NVP-BEZ235 at efficacious doses at the end of a 13-week rat toxicity study.[33] These preclinical data demonstrated for the first time the feasibility of effectively blocking the PI3K pathway *in vivo* with manageable effects on glucose regulation.

One of the recurrent clinical problems of molecularly targeted kinase inhibitors in oncology has been the inevitable occurrence of resistance on account of point mutations that impair drug binding and hence effective enzymatic blockade.[34–36] Although it is too early to determine if PI3K drug-resistance mutations will emerge in cancer patients as is the case for protein kinase modulators, preclinical studies have shown that NVP-BEZ235 and other dual PI3K/mTOR inhibitors may be less susceptible to PI3K drug resistance than selective PI3K-targeted agents, owing to their preserved activity against mTOR.[37] This early preclinical finding provides evidence that the simultaneous engagement of multiple targets, in this case PI3K and mTOR, may represent a suitable strategy to delay the appearance of resistance mechanisms within a pathway.

Some of the structural features of NVP-BEZ235 and derivatives thereof have been exploited in the identification of compounds that exclusively inhibit mTOR kinase activity.[38,39] Originally identified by a medium-throughput biochemical assay utilizing purified mTORC1 from mammalian cells, compound **11** (Figure 13.7) was optimized using a structure-based strategy to produce Torin 1, which exhibits significant selectivity for mTOR over p110α (IC50 = 4.32 nM and 250 nM, respectively).[38] The introduction of the six-membered lactam ring and modifications of the piperazine acyl moiety contributed to achieve the desirable mTOR cellular activity and selectivity profile. A recent article has described the identification of another mTOR modulator (Torin 2, compound **12**, Figure 13.7) that has significantly improved pharmacological properties relative to Torin 1.[39]

NVP-BEZ235 entered phase I clinical trials at the end of 2006 for dose escalation in patients with advanced solid tumors, particularly enriched for patients with advanced breast cancers. It was the first dual PI3K/mTOR inhibitor to be tested in cancer patients.

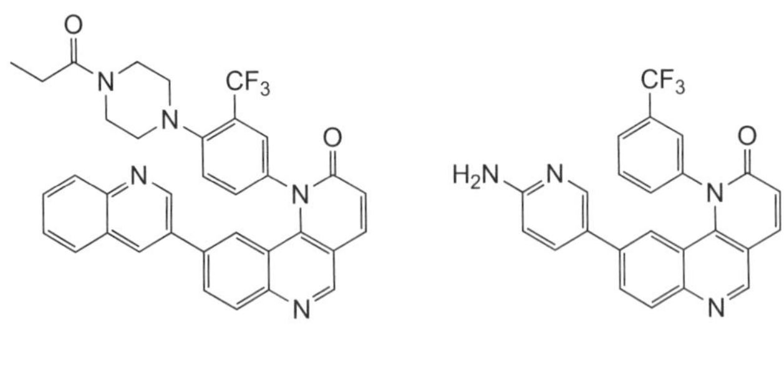

11 **12**

Figure 13.7 Structures of ATP-competitive mTOR inhibitors that do not inhibit PI3K.

13.4 Quinoline Derivatives: GSK1059615 and GSK2126458

Two quinoline derivatives (GSK1059615, compound **13**; and GSK2126458, compound **14**, Figure 13.8) with outstanding biochemical activity against PI3K and mTOR have been described. GSK1059615 (IC_{50} = 0.4, 0.6, 5, 2, and 12 nM for p110α, p110β, p110γ, p110δ, and mTOR, respectively) was the first compound from this series to enter phase I dose escalation trials in patients with solid tumors and lymphoma in September 2008, but its development has been discontinued recently. The X-ray structure of GSK1059615 bound into p110γ (data not shown) reveals that, as previously reported for the central core of NVP-BEZ235, the quinoline is involved in a H-bond interaction with the hinge region, while the thiazolinedione ring forms an interaction with the catalytic lysine (Lys883) within the ATP-binding pocket.[40] The structure also shows that a larger group could potentially replace the thiazolinedione synthon providing a suitable approach to identify inhibitors with improved potency and pharmacological properties. In this context, a recent publication has provided the key molecular modifications that allowed the identification of the second PI3K/mTOR development candidate, GSK2126458.[40] Thus, the X-ray co-crystal structure of p110γ and a synthetic precursor containing an azaindazole group at C-7 (compound **15**, Figure 13.8) confirmed that the nitrogen at the 2-position of this synthon is involved in a H-bond interaction with Lys833 and that the pyridyl nitrogen interacts with a conserved active site water molecule. Although these interactions positively impacted on its biological activity against PI3K, they did not address the limited aqueous solubility and poor pharmacokinetic profile of this compound series. A major breakthrough was achieved by exploiting a series of arylsulfonamide derivatives that allowed the pharmacological limitations of the parent to be addressed with no loss of PI3K/mTOR inhibitory activity. Further efforts to improve the CYP inhibition profile by the incorporation of a pyridazine at the 4-position of the quinoline scaffold led to the identification of GSK2126458 (K_i = 0.019, 0.13, 0.024, 0.06, and 0.18 nM for p110α, p110β, p110δ, p110γ, and mTOR, respectively). Of special note is the boost in potency obtained by exploring small substituents (e.g. methoxy) at C-2 of the pyridine ring. The authors reasoned that this increase in potency was

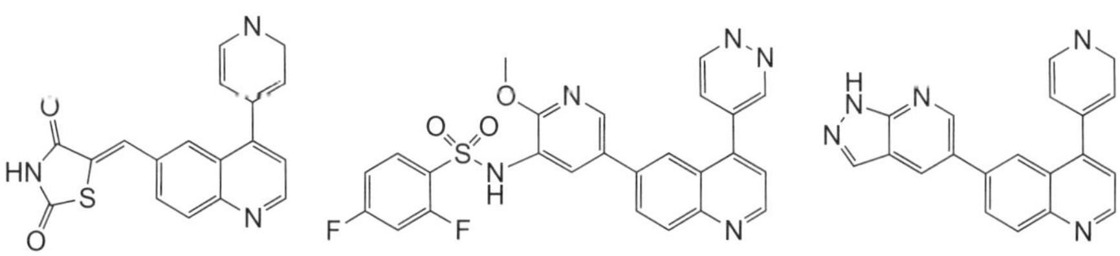

13 **14** **15**

Figure 13.8 Representative examples of quinoline derivatives.

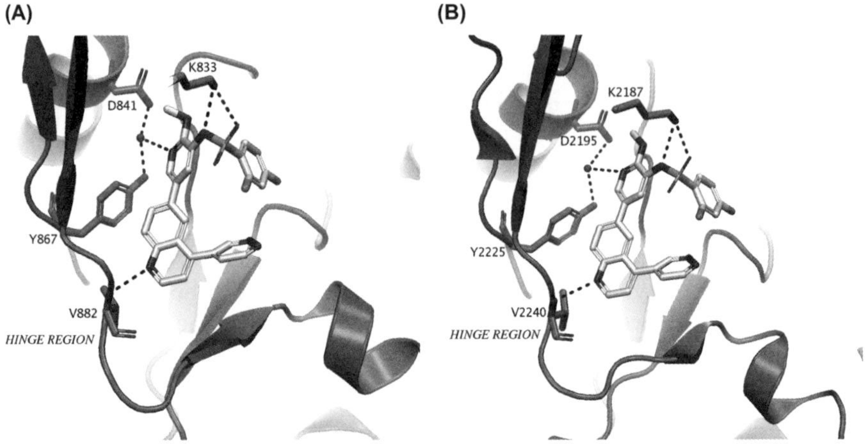

Figure 13.9 (A) X-ray co-crystal structure of p110γ with GSK2126458 and (B) model of the same compound docked into a homology model of mTOR. Potential hydrogen bonds are shown with hatched lines.

due to the substituent filling an unoccupied cavity within the enzyme, as well as a potential conformational effect on the orientation of the neighboring sulfonamide moiety (Figure 13.9). The methoxy group of GSK2126458 projects into an unoccupied space located in vicinity of the DFG motif and the catalytic lysine. This pocket was previously identified as a key region for target modulation.[41] The so-called affinity pocket is lined with residues conserved within the class I PI3K and, as illustrated in previous sections, it can harbor a wide range of substituents from chlorine (e.g. PIK-93) to larger groups such as phenol (e.g. PI-103) or indazole (e.g. GDC-0941). The pronounced mTOR inhibitory potency of GSK2126458 is in line with mTOR docking studies (Figure 13.9) that show similar interactions to the ones previously described for PI3K.

GSK2126458 entered phase I clinical trials in August 2009 in patients with refractory solid tumors or lymphoma.

13.5 Outlook

The development of modulators of the PI3K pathway either as a single agent or in combination with inhibitors of other oncogenic pathways (e.g. Mek allosteric kinase inhibitors)[31] is progressing. As shown in previous sections, most of the first PI3K inhibitors to enter clinical trials also block the enzymatic activity of mTOR, although they have exquisite selectivity over protein kinases, and, to a lesser extent, over other lipid kinases (e.g. DNA-PK or Vps34). This fortuitous biological profile may provide a more profound clinical benefit than by exclusively targeting one of the two kinases by maximizing efficacy and delay the onset of resistance mechanisms within this pathway. The identification by

X-ray crystallography or molecular modeling of the structural features responsible of the biological activity of the early dual PI3K/mTOR and extensive SAR data have been instrumental in helping medicinal chemists to carry out the optimization of new compounds with a PI3K/mTOR multi-dimensional target profile or selective PI3K or mTOR inhibitors.[42–49]

The possibility to target tumor cells at multiple nodes with poly-pharmacological inhibitors is an interesting area of drug discovery, but it is expected to be even more challenging than for the most selective inhibitors. The current dual PI3K/mTOR inhibitors should offer insight to this drug discovery strategy and serve to validate the use of polypharmacological modulators in oncology to achieve sustained pathway inhibition and block compensation mechanisms. If these compounds provide clinical benefit at well-tolerated doses and delay the appearance of resistance, this will spur the development of other polypharmacological inhibitors and stimulate interest in broader application of this strategy for targeted anti-cancer therapies.

Acknowledgements

We thank our present and past colleagues for their outstanding efforts to identify and develop inhibitors of the PI3K/mTOR pathway.

References

1. A. Carnero, C. Blanco-Aparicio, O. Renner, W. Link and J. F. M. Leal, *Curr. Cancer Drug Targets*, 2008, **8**, 187.
2. E. Tokunaga, E. Oki, A. Egashira, N. Sadanaga, M. Morito, Y. Kakeji and Y. Maehara, *Curr. Cancer Drug Targets*, 2008, **8**, 27.
3. S.-M. Maira, C. Voliva and C. Garcia-Echeverria, *Expert Opin. Ther. Targets*, 2008, **12**, 223.
4. C. Garcia-Echeverria and W. R. Sellers, *Oncogene*, 2008, **27**, 5511.
5. S. Brachmann, C. Fritsch, S.-M. Maira and C. Garcia-Echeverria, *Curr. Opin. Cell Biol.*, 2009, **21**, 1.
6. S.-M. Maira, P. Finan and C. Garcia-Echeverria, *Curr. Top. Microbiol. Immunol.*, 2010, **347**, 209.
7. A. Carnero, *Expert Opin. Investig. Drugs*, 2009, **18**, 1265.
8. B. Vanhaesebroeck, P. K. Vogt and C. Rommel, *Curr. Top. Microbiol. Immunol.*, 2010, **347**, 1.
9. D. A. Guertin and D. M. Sabatini, *Sci. Signaling*, 2009, **2**, 1.
10. D. M. Sabatini, *Nat. Rev. Cancer*, 2006, **6**, 729.
11. C. Garcia-Echeverria, *Bioorg. Med. Chem. Lett.*, 2010, **20**, 4308.
12. M. A. Fabian, W. H. Biggs III, D. K. Treiber, C. E. Atteridge, M. D. Azimiora, M. G. Benedetti, T. A. Carter, P. Civeri, P. T. Edeen, M. Floyd, J. M. Fors, M. Galvin, J. L. Gerlach, R. M. Grotzfeld, S. Herrgard, D. E. Insko, M. A. Insko, A. G. Lai, J.-M. Léilas, S. A. Mehta, Z. V. Milanov, A.

M. Velasco, L. M. Wodicka, H. K. Patel, P. P. Zarrinkar and D. J. Lockhart, *Nature Biotech.*, 2005, **3**, 336.
13. M. W. Karaman, S. Herrgard, D. K. Treiber, P. Gallant, C. E. Atteridge, B. T. Campbell, K. W. Chan, P. Cicieri, M. I. Davis, P. T. Edeen, R. Faraoni, M. Floyd, J. P. Hunt, D. J. Lockhart, Z. V. Milanov, M. J. Morrison, G. Pallares, H. K. Patel, S. Pritchard, L. M. Wodicka and P. P. Zarrinkar, *Nature Biotech.*, 2008, **26**, 127.
14. E. H. Walker, O. Perisic, C. Ried, L. Stephens and R. L. Williams, *Nature*, 1999, **402**, 313.
15. E. H. Walker, M. E. Pacold, O. Perisic, L. Stephens, P. T. Hawkins, M. P. Wymann and R. L. Williams, *Mol. Cell*, 2000, **6**, 909.
16. Z. A. Knight, B. Gonzalez, M. E. Feldman, E. R. Zunder, D. D. Goldenberg, O. Williams, R. Loewith, D. Stokoe, A. Balla, B. Toth, T. Balla, W. A. Weiss, R. L. Williams and K. M. Shokat, *Cell*, 2006, **125**, 733.
17. M. Hayakawa, H. Kaizawa, H. Morimoto, T. Koizumi, T. Ohishi, M. Okada, M. Ohta, S. Tsukamoto, P. Parker, P. Workman and M. Waterfield, *Bioorg. Med. Chem.*, 2006, **14**, 6847.
18. M. Hayakawa, H. Kaizawa, K. Kawaguchi, N. Ishikawa, T. Koizumi, T. Ohishi, M. Yamano, M. Okada, M. Ohta, S. Tsukamoto, F. I. Raynaud, M. Waterfield, P. Parker and P. Workman, *Bioorg. Med. Chem.*, 2007, **15**, 403.
19. M. Hayakawa, H. Kaizawa, H. Moritomo, T. Koizumi, T. Ohisi, M. Yamano, M. Okada, M. Ohta, S. Tsukamoto, F. I. Raynaud, P. Workman, M. D. Waterfield and P. Parker, *Bioorg. Med. Chem.*, 2007, **17**, 2438.
20. M. Hayakawa, K. Kawaguchi, H. Kaizawa, T. Koizumi, T. Ohishi, M. Yamano, M. Okada, M. Ohta, S. Tsukamoto, F. I. Raynaud, P. Parker, P. Workman and M. D. Waterfield, *Bioorg. Med. Chem.*, 2007, **15**, 5833.
21. F. I. Raynaud, S. Eccles and P. A. Clarke, *et al., Cancer Res.*, 2007, **67**, 5840.
22. C. J. Vlahos, W. F. Matter, K. Y. Hui and R. F. Brown, *J. Biol. Chem.*, 1994, **269**, 5241.
23. D. P. Sutherlin, D. Sampath, M. Berry, G. Castanedo, Z. Chang, I. Chuckowree, J. Dotson, A. Folkes, L. Friedman, R. Goldsmith, T. Heffron, L. Lee, J. Lesnick, C. Lewis, S. Mathieu, J. Nonomiya, A. Olivero, J. Pang, W. W. Prior, L. Salphati, S. Sideris, Q. Tian, V. Tsui, N. C. Wan, S. Wang, C. Wiesmann, S. Wong and B. Y. Zhu, *J. Med. Chem.*, 2010, **53**, 1086.
24. A. Folkes, K. Ahmadi, W. K. Alderton, S. Alix, S. J. Baker, G. Box, I. S. Chuckowree, P. A. Clarke, P. Depledge, S. A. Eccles, L. S. Friedman and A. Hayes, *et al., J. Med. Chem.*, 2008, **51**, 5522.
25. P. Workman, P. A. Clarke, F. I. Raynaud and R. L. M. van Montfort, *Cancer Res.*, 2010, **70**, 2146.
26. F. I. Raynaud, S. A. Eccles, S. Patel, S. Alix, G. Box, I. Chuckowree, A. Folkes, S. Gowan, B. A. De Haven, S. F. Di, A. Hayes, A. T. Henley, L. Lensun, G. Pergl-Wilson, A. Robson, N. Saghir, A. Zhyvoloup, E. McDonald, P. Sheldrake, S. Shuttleworth, M. Valenti, N. C. Wan, P. A. Clarke and P. Workman, *Mol. Cancer Ther.*, 2009, **8**, 1725.

27. F. Stauffer, S.-M. Maira, P. Furet and C. Garcia-Echeverria, *Bioorg. Med. Chem. Lett.*, 2008, **18**, 1027.
28. D. R. Alessi, M. Deak, A. Casamayor, F. B. Caudwell, N. Morrice, D. G. Norman, P. Gaffney, C. B. Reese, C. N. Macdougall, D. Harbison, A. Ashworth and M. Bownes, *Curr. Biol.*, 1997, **7**, 776.
29. S.-M. Maira, F. Stauffer, J. Brueggen, P. Furet, C. Schnell, C. Fritsch, S. Brachmann, P. Chène and A. De Pover, *et al., Mol. Cancer. Ther.*, 2008, **7**, 1851.
30. S. M. Brachmann, I. Hofmann, C. Schnell, C. Fritsch, S. Wee, H. Lane, S. Wang, C. Garcia-Echeverria and S. M. Maira, *Proc. Natl. Acad. Sci. USA*, 2009, **106**, 22299.
31. J. A. Engelman, L. Chen, X. Tan, K. Crosby, A. R. Guimares, R. Upadhyay, K. McNamara, S. A. Perera, Y. Song, L. R. Chirieac, R. Kaur, *et al., Nat. Med.*, 2008, **14**, 1351.
32. V. Serra, B. Markman, M. Scaltriti, P. Eichhorn, V. Valero, M. Guzman, M. L. Botero, E. Llonch, F. Atzori, S. Di Cosimo, S.-M. Maira, C. Garcia-Echeverria, J. L. Parra, J. Arribas and J. Baselga, *Cancer Res.*, 2008, **68**, 8022.
33. E. Marrer, S.-M. Maira, C. Schnell and C. Garcia-Echeverria, *Proc. Am. Assoc. Cancer Res.*, 2008, Abstract 215.
34. M. E. Gorre, M. Mohammed, K. Ellwood, N. Hsu, R. Paquette, P. N. Rao and C. L. Sawyers, *Science*, 2001, **293**, 876.
35. Y. L. Choi, M. Soda, Y. Yamashita, T. Ueno, J. Takashima, T. Nakajima, Y. Yatabe, K. Takeuchi, T. Hamada, H. Haruta, Y. Ishikawa, H. Kimura, T. Mitsudomi, Y. Tanio and H. Mano, *N. Engl. J. Med.*, 2010, **363**, 1734.
36. A. Pallis, E. Briasoulis, H. Linardou, C. Papadimitriou, D. Bafaloukos, P. Kosmidis and S. Murray, *Curr. Med. Chem.*, 2011, **18**, 1613.
37. E. R. Zunder, Z. A. Knight, B. T. Houseman, B. Apsel and K. M. Shokat, *Cancer Cell*, 2008, **14**, 107.
38. Q. Liu, J. W. Chang, J. Wang, S. A. Kang, C. C. Thoreen, A. Markhard, W. Hur, J. Zhang, T. Sim, D. M. Sabatini and N. S. Gray, *J. Med. Chem.*, 2010, **53**, 7146.
39. Q. Liu, J. Wang, S. A. Kang, C. C. Thoreen, W. Hur, T. Ahmed, D. M. Sabatini and N. S. Gray, *J. Med. Chem.*, 2011, **54**, 1473.
40. S. D. Knight, N. D. Adams, J. L. Burgess, A. M. Chaudhari, M. G. Darcy, C. A. Donatelli, J. I. Luengo, K. A. Newlander, C. A. Parrish, L. H. Ridgers, M. A. Sarpong, S. J. Schmidt, G. S. Van Aller, J. D. Carson, M. A. Diamond, P. A. Elkins, C. M. Gardiner, E. Garver, S. A. Gilbert, R. R. Gontarek, J. R. Jackson, K. L. Kershner, L. Luo, K. Raha, C. S. Sherk, C.-M. Sung, D. Sutton, P. T. Tummino, R. J. Wegrzyn, K. R. Auger and D. Dhanak, *ACS Med. Chem. Lett.*, 2010, **1**, 39.
41. Z. A. Knight, B. Gonzalez and M. E. Feldman, *et al., Cell*, 2006, **125**, 733.
42. Z. Chen, A. M. Venkatesan, C. M. Dehnhardt, S. Ayral-Kaloustian, N. Brooijmans, R. Mallon, L. Feldberg, I. Hollander, J. Lucas, K. Yu, F. Kong and T. S. Mansour, *J. Med. Chem.*, 2010, **53**, 3169.

43. C. M. Dehnhardt, A. M. Venkatesan, S. E. Delos, Z. Chen, O. Santos, S. Ayral-Kaloustian, N. Brooijmans, R. Mallon, I. Hollander, L. Feldberg, J. Lucas, I. Chaudhary, K. Yu, J. Gibbons, R. Abraham and T. S. Mansour, *J. Med. Chem.*, 2010, **53**, 798.
44. A. M. Venkatesan, C. M. Dehnhardt, Z. Chen, E. Delos Santos, O. dos Santos, M. Bursavich, A. M. Gilbert, J. W. Ellingboe, S. Ayral-Kaloustian, G. Khfizova, N. Brooijmans, R. Mallon, L. Feldberg, J. Lucas, K. Yu, J. Gibbons, R. Abraham and T. S. Mansour, *Bioorg. Med. Chem. Lett.*, 2010, **20**, 653.
45. A. M. Venkatesan, Z. Chen, S. O. dos, C. Dehnhardt, E. D. Santos, S. Ayral-Kaloustian, R. Mallon, I. Hollander, L. Feldberg, J. Lucas, K. Yu, I. Chaudhary and T. S. Mansour, *Bioorg. Med. Chem. Lett.*, 2010, **20**, 5869.
46. A. M. Venkatesan, C. M. Dehnhardt, S. E. Delos, Z. Chen, S. O. dos, S. Ayral-Kaloustian, G. Khafizova, N. Brooijmans, R. Mallon, I. Hollander, L. Feldberg, J. Lucas, K. Yu, J. Gibbons, R. T. Abraham, I. Chaudhary and T. S. Mansour, *J. Med. Chem.*, 2010, **53**, 2636.
47. A. M. Venkatesan, C. M. Dehnhardt, Z. Chen, E. D. Santos, S. O. dos, M. Bursavich, A. M. Gilbert, J. W. Ellingboe, S. Ayral-Kaloustian, G. Khafizova, N. Brooijmans, R. Mallon, I. Hollander, L. Feldberg, J. Lucas, K. Yu, J. Gibbons, R. Abraham and T. S. Mansour, *Bioorg. Med. Chem. Lett.*, 2010, **20**, 653.
48. H. R. Tsou, G. MacEwan, G. Birnberg, G. Grosu, M. Bursavich, J. Bard, N. Brooijmans, L. Toral-Barza, I. Hollander, T. Mansour, S. Ayrai-Kaloustain and K. Yu, *Bioorg. Med. Chem. Lett.*, 2011, **20**, 2321.
49. H. R. Tsou, G. MacEwan, G. Birnberg, N. Zhang, N. Brooiymans, L. Toral-Barza, I. Hollander, S. Ayral-Kaloustian and K. Yu, *Bioorg. Med. Chem. Lett.*, 2010, **20**, 2258.

CHAPTER 14

Discovery of HDAC-Inhibiting Multi-Target Inhibitors

XIONG CAI* AND CHANGGENG QIAN

Curis, Inc., Lexington, MA, USA
*Email: xcai@curis.com

14.1 Introduction

Histone deacetylases (HDACs) comprise a family of 18 genes in humans and are divided into four classes.[1] Among them, Class I (HDACs 1, 2, 3, and 8), Class II (HDACs 4, 5, 6, 7, 9, and 10) and Class IV (HDAC 11) HDACs are zinc dependent, while Class III HDACs (sirtuins 1–7) are not zinc dependent and generally not inhibited by classes of inhibitors that depend on zinc chelation for their activity. HDACs play an important role in regulating expression of tumor suppressor genes and activities of transcriptional factors involved in cancer and have been clinically validated as a target for cancer. Two HDAC inhibitors, vorinostat (Figure 14.1)[2] and romidepsin,[3] have been approved by FDA for the treatment of cutaneous T-cell lymphoma (CTCL). In addition, there are several others, including panobinostat (LBH-589), belinostat (PXD-101), entinostat (MS-275), mocetinostat (MGCD-0103), PCI-24781, SB-939, and givinostat (ITF-2357), at various stages of clinical trials for cancer.[4] These agents have been used either as monotherapy or in combination with other anti-cancer drugs in both hematological and solid malignancies. In many reports, combining HDAC inhibitors with other anti-cancer agents achieves synergistic effects to suppress proliferation and induces apoptosis in tumor cells via signaling network disruption modulated by HDAC inhibition.[5–12] The synergistic effects of HDAC inhibitors

RSC Drug Discovery Series No. 21
Designing Multi-Target Drugs
Edited by J. Richard Morphy and C. John Harris

Published by the Royal Society of Chemistry, www.rsc.org

with other agents can be used as a strategy to overcome the common limitations of a single anti-cancer agent, such as low response rate due to molecular heterogeneity among and within tumors,[13,14] as well as acquired resistance that frequently emerges following treatment.[15,16] There are a number of approaches to execute this strategy on inhibition of multiple pathogenic pathways, including combination therapies and HDAC-inhibiting multi-target inhibitors.

The combination therapy approach involves two or more drugs. The challenges with such combination therapy include the dosing optimization of two or more drugs to achieve effective therapeutic pharmacokinetics (PK) and pharmacodynamics (PD) in clinical practice, the potential of exacerbated toxicities due to drug–drug interactions, on-target and off-target adverse effects, the complex regulatory requirements to demonstrate safety and efficacy of combination therapies,[17] as well as other issues such as high cost to the patient or third party payer and intellectual property complications.

In contrast, the HDAC-inhibiting multi-target inhibitor approach involves only a single drug with multi-target inhibitory activities. This approach has several key advantages, including aligned pharmacokinetics and pharmacodynamics, no inherited adverse effects from separate agents, reduced drug cost, and high patient compliance. This latter approach has shown some promising results in preclinical and clinical studies with the first drug candidate CUDC-101 (Curis, Inc., Figure 14.1) currently being in phase I/Ib clinical trials. This compound is further discussed in this chapter as the case study, along with a brief overview of other HDAC-inhibiting multi-target inhibitors.

14.2 CUDC-101: a Potent Multi-Target EGFR, HER2, and HDAC Inhibitor

14.2.1 Compound Design and Synthesis

CUDC-101 was derived from Curis' HDAC-inhibiting multi-target cancer programs. This compound was designed to introduce an HDAC inhibitory functionality into an epidermal growth factor receptor (EGFR) and human epidermal growth factor 2 (HER2) inhibitor scaffold. Known HDAC inhibitors can be classified into several classes, such as hydroxamic acid, benzamide, short-chain fatty acid, and cyclic tetrapeptide.[18] Among these classes of inhibitors, hydroxamic acids are considered the broadest class of HDAC inhibitors, with high affinity for HDACs which depends on zinc chelation for activity.[19] One representative of this class is vorinostat (Figure 14.1), which is FDA approved for the treatment of CTCL. The quinazoline structure, on the other hand, is a well-characterized scaffold of kinase inhibitors, with examples of marketed drugs being erlotinib (an EGFR inhibitor, Figure 14.1) and lapatinib (an EGFR and HER2 inhibitor, Figure 14.1). Our HDAC-inhibiting multi-target inhibitor design strategy was to introduce an HDAC-inhibiting hydroxamic acid functionality into the quinazoline pharmacophore of EGFR/HER2 inhibitors through a structure-based rational drug design approach

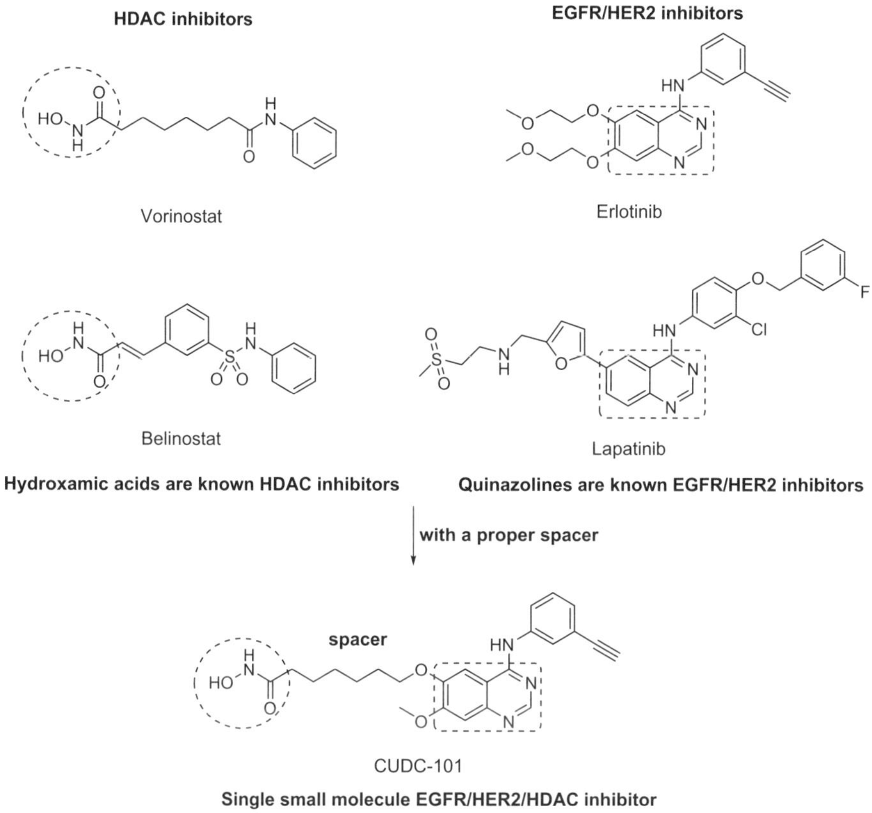

Figure 14.1 Design of multi-target EGFR, HER2, and HDAC inhibitors.

(Figure 14.1) based on, amongst other observations, structure requirements of prototype drugs with their respective target proteins. The X-ray co-crystal structure of EGFR with erlotinib shows that erlotinib binds to the EGFR-binding domain with the quinazoline occupying the adenine region of the ATP-binding site of the receptor and forming two hydrogen bonds at N1 and N3, while the phenyl-amino group occupies the hydrophobic pocket and the two methoxy-ethoxy groups at C-6 and C-7 of the quinazoline ring stick out of the protein.[20,21] The X-ray co-crystal structure of HDAC inhibitor trichostatin A (TSA) or vorinostat with HDLP (an HDAC-like protein) reveals that both compounds bind to HDLP with their large hydrophobic (phenyl) group contacting the hydrophobic part of the enzyme, the aliphatic chain sitting in the 11 Å tube-like channel, through multiple contacts with hydrophobic residues, and with the terminal hydroxamic acid interacting with zinc at the active site to disrupt enzyme activity.[22,23] Since the methoxy-ethoxy groups extend outside of the EGFR-binding pocket and are flexible, we reasoned that introducing a hydroxamic acid functionality onto the terminus of these side chains with a proper spacer should not dramatically affect its binding to EGFR, and molecules formed in this manner should also be able to bind to HDAC because the hydrophobic pocket within the HDAC dimeric interface should be large enough to accommodate the phenylaminoquinazoline backbone of the molecules. The spacer serves as a bridge to connect the hydroxamic acid zinc-binding group to the quinazoline backbone. A proper spacer should be one that allows the hydroxamic acid to reach the active site so as to interact with zinc, which is required to disrupt the HDAC enzyme activity, and yet still maintain the molecule's ability to potently inhibit EGFR and HER2. Based on the above design, we synthesized a number of analogs with various substituents on the quinazoline backbone and a wide variety of spacers between the terminal hydroxamic acid and the quinazoline backbone, and discovered a number of analogs that display potent inhibitory activity against HDAC, EGFR, and HER2.[24] After an extensive structure–activity relationship (SAR) study and lead optimization, CUDC-101 was identified as the development candidate for further clinical development.[10,24]

The synthesis of CUDC-101 is illustrated in Scheme 14.1.[24] This eight-step process involves no column chromatography separation and is suitable for scale-up preparation. Compound **1** is treated with iodomethane to give compound **2**. Alkylation of **2** with ethyl 7-bromoheptanoate yields intermediate **3**, which is then treated with fuming nitric acid to introduce a nitro group onto the phenyl ring to yield **4**. Reduction of nitro compound **4** with iron/hydrochloric acid gives the corresponding amine **5**. Cyclization of **5** using formamide and ammonium formate results in the quinazoline intermediate **6**. Chlorination of **6** using phosphoryl chloride gives the desired intermediate **7**. The subsequent coupling of **7** with 3-ethynylaniline results in the corresponding ester **8** which is finally reacted with freshly prepared hydroxylamine to form the final product CUDC-101.[24]

14.2.2 *In Vitro* Potency and Mechanism of Action

In an *in vitro* assay, CUDC-101 potently inhibits EGFR and HER2 kinase and HDAC enzyme activities with IC_{50} values of 2.4, 15.7, and 4.4 nM, respectively, representing 10-fold more potency on HDAC inhibition than the HDAC inhibitor vorinostat, 10-fold more potency on EGFR than the EGFR inhibitor erlotinib, and similar potency on HER2 to lapatinib, a HER2 inhibitor (Table 14.1). CUDC-101 inhibits both Class I and Class II HDACs, but not Class III, Sir-type HDACs (Table 14.2).[24] CUDC-101 is a selective HDAC, EGFR, and HER2 inhibitor with only weak inhibition of the following protein kinases (IC_{50}): KDR (VEGFR2) (849 nM), Src (11 000 nM), Lyn (840 nM), Lck (5910 nM), Abl-1 (2890 nM), FGFR-2 (3430 nM), Flt-3 (1500 nM), and Ret (3200 nM) among 69 other kinases tested.[10,24]

In treated cancer cells, CUDC-101 dose-dependently increases the acetylation of histones H3 and H4 (Figure 14.2a and 14.2b) and inhibits EGFR autophosphorylation (Figure 14.2d and 14.2e), as well as HER2 phosphorylation in cultured cancer cells at a concentration of 1 μM (Figure 14.2f).[10]

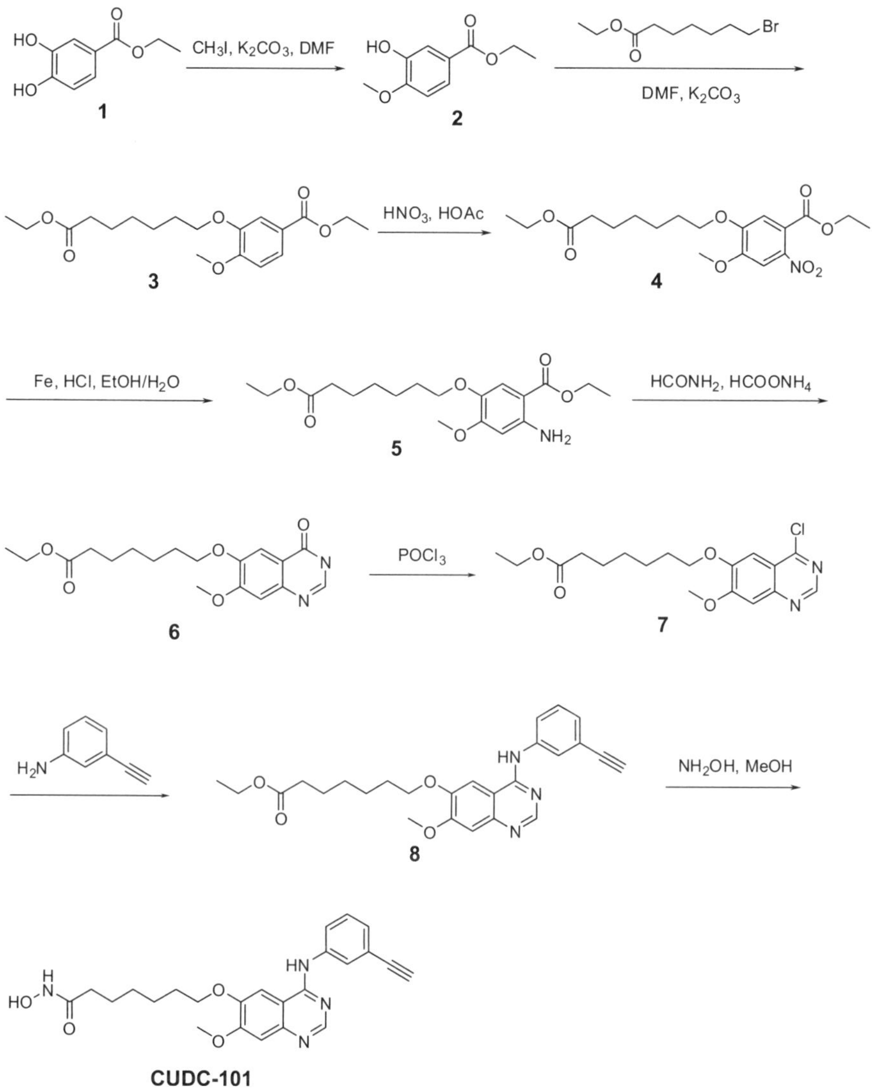

Scheme 14.1 Synthesis of CUDC-101.[24]

In addition, we observed that CUDC-101 increases the acetylation of p53 (Figure 14.2c) and α-tubulin (Figure 14.3b), nonhistone substrates of HDAC, in treated cancer cells.[10] These results suggest that CUDC-101 effectively inhibits all three targets in cancer cells.

To assess the synergistic effects between the HDAC and receptor tyrosine kinase (RTK) inhibition mediated by CUDC-101, we carried out a cancer cell growth inhibition study.[10,24] In this study, we used two reference compounds, vorinostat and erlotinib, to achieve HDAC and EGFR inhibition, respectively.

Table 14.1 *In vitro* activity of CUDC-101 and reference compounds.

	IC_{50} (nM) in enzyme assays		
Compound	*HDAC*	*EGFR*	*HER2*
CUDC-101	4.4	2.4	15.7
vorinostat	40.0	N/A*	N/A*
erlotinib	N/A*	48.0	134.5
lapatinib	N/A*	11.2	10.2

*N/A, not active.

Table 14.2 Inhibition of HDAC Class I and Class II family members by CUDC-101[24] (reproduced with permission from ref. 24, copyright 2010 American Chemical Society).

Class I	*HDAC1*	*HDAC2*	*HDAC3*	*HDAC8*
IC_{50} (nM)	4.5	12.6	9.1	79.8

Class II	*HDAC4*	*HDAC5*	*HDAC6*	*HDAC7*	*HDAC9*	*HDAC10*
IC_{50} (nM)	13.2	11.4	5.1	373	67.2	26.1

We applied a well-established mathematical model for analyzing multi-drug interactions in these studies.[25] This experiment allowed us to assess whether there is a synergistic effect between HDAC and RTK inhibition. IC_{50} values of proliferation inhibition of several cancer cells incubated with vorinostat, erlotinib, or their combination at various concentrations were generated, and combination indices were calculated to interpret the effect of concurrent inhibition of the HDAC and RTK targets. Our results show that co-treatment with vorinostat and erlotinib (1:1 ratio) results in greater than additive growth inhibition as calculated combination indices are in the range of 0.22 to 0.50 μM, indicating considerable synergy between the two targets (Table 14.3).[24] We further investigated whether this synergistic effect of cancer cell growth inhibition requires combination of a specific ratio of vorinostat and erlotinib. We performed similar studies by applying varying ratios of vorinostat and erlotinib. In all cases, the combination indices were calculated to be well below 1, indicating that the inhibition of a combination of these two targets is synergistic with no requirement of a specific dosing ratio.[10] These results indicate that one single molecule with the ability to inhibit multiple targets, such as CUDC-101, can achieve improved therapeutic outcomes.

In cultured human cancer cells, CUDC-101 displays broad anti-proliferative activity in many cancer cell types, including lung, pancreas, liver, colon, breast, prostate, and head and neck cancer.[10,24] CUDC-101 also displays equal or greater potency in these assays than vorinostat, erlotinib, lapatinib, and combinations of vorinostat with either erlotinib or lapatinib, respectively. These results are summarized in Table 14.4. Noticeably, CUDC-101 suppresses the

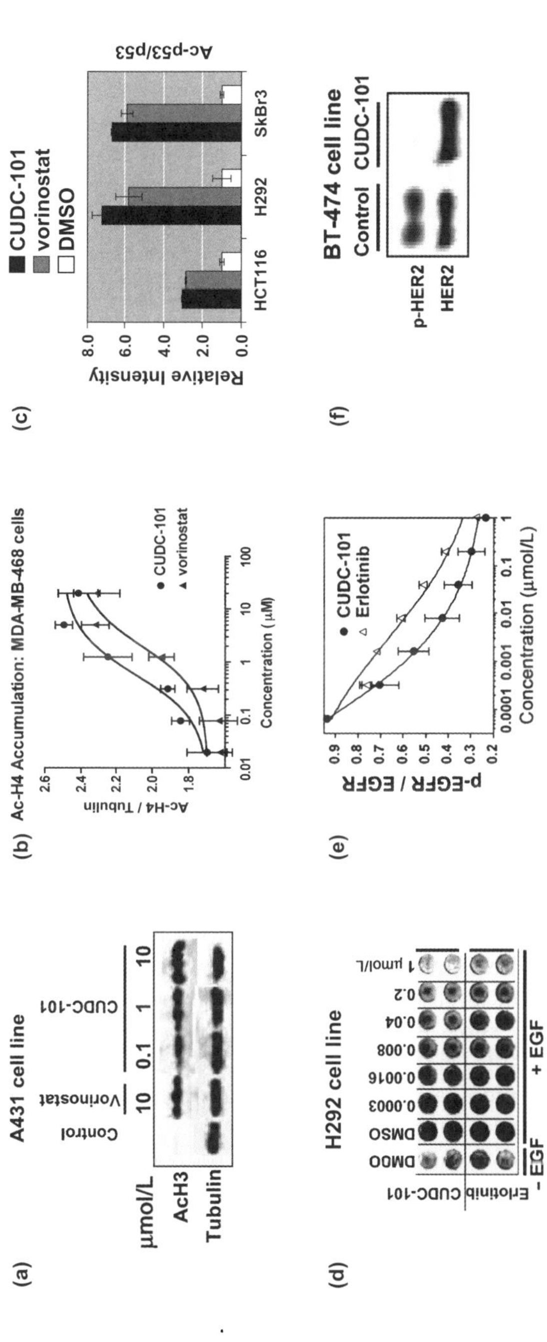

Figure 14.2 CUDC-101 induces the accumulation of acetylated histone H3, H4 (a and b) and p53 (c), inhibits EGFR (d and e) and HER2 phosphorylation (f).[10]

Table 14.3 Synergy between HDAC and PTK pathway inhibition (IC_{50} μM)[24] (reproduced with permission from ref. 24, copyright 2010 American Chemical Society).

Cancer type	*Cell line*	*Vorinostat*	*Erlotinib*	*Vorinostat + erlotinib*	*Combination index*
NSCLC	H2122	7.50	1.00	0.40	0.45
pancreatic	BxPC3	2.70	7.60	1.00	0.50
breast	MDA-MB-468	5.00	11.40	1.60	0.46
prostate	LN-Cap	0.58	18.35	0.15	0.27
HNSCC	SCC-15	2.49	1.94	0.3	0.28
HNSCC	SCC-25	1.93	3.34	0.61	0.50
HNSCC	CAL-27	3.72	10.15	0.6	0.22

growth of cancer cell lines that are not sensitive to lapatinib or erlotinib, for example, the lapatinib-insensitive, triple-negative breast cancer cell line MDA-MB-231 and the erlotinib-insensitive NSCLC cell line H1975.[10,24] In addition, CUDC-101 induces the activities of caspases 3 and 7 in HCT-116 colon cancer cells (Figure 14.3a), reduces the levels of the anti-apoptotic proteins survivin and Bcl-xL, and induces the expression of p21 in MDA-MB-231 breast cancer cells (Figure 14.3b).[10] These data suggest that CUDC-101, with its simultaneous and synergistic inhibition of multiple targets, may improve the response rates to traditional kinase inhibitors.

Further mechanism of action studies of CUDC-101 reveal that CUDC-101 not only directly and potently inhibits all three targets, but also indirectly disrupts multiple cancer network signaling pathways (listed below). These combined effects may provide enhanced clinical efficacy, response rates, and the ability to overcome resistance to RTK inhibitors.

- **CUDC-101 modulates RTK activity and expression**

Using breast cancer cell lines as a model, we found that CUDC-101 not only inhibits HER2 kinase activity (Figure 14.2f) but also suppresses *Her2* transcription and protein levels, while both lapatinib and erlotinib inhibit HER2 activity, yet neither show effects on *Her2* transcription (Figures 14.4a and 14.4b).[10] These results suggest that CUDC-101 can achieve its synergistic effects through the modulation of multiple variables affecting RTK signaling, including activity and protein levels of RTKs.

- **CUDC-101 suppresses HER3 expression and AKT reactivation**

One of the proposed underlying mechanisms of developing drug resistance against RTK inhibitors is a compensatory shift in the HER3 phosphorylation-dephosphorylation equilibrium that leads to reactivation of the downstream PI3K/Akt survival pathway.[26–28] To confirm this hypothesis, we used erlotinib to explore this effect. Within 8 hours of erlotinib treatment, Akt reactivation

Table 14.4 Inhibition of proliferation in selected human cancer cell lines by CUDC-101 and reference compounds.

		IC_{50} (μM)					
Cancer type	*Cell line*	*CUDC-101*	*Vorinostat*	*Erlotinib*	*Vorinostat + erlotinib*	*Lapatinib*	*Vorinostat + lapatinib*
NSCLC	H1975	0.50	4.70	>10	4.0		
NSCLC	HCC827	0.60	1.80	7.50	2.3		
NSCLC	H1993	0.36	2.06	17.70		4.15	1.11
NSCLC	H358	0.40	2.50	6.00	1.10		
NSCLC	H460	0.70	1.70	8.20	1.40		
liver	HepG2	0.13	1.66	>20	1.95	6.27	1.23
liver	Hep3B2	0.23	2.44	>20	2.64	5.49	2.36
liver	Sk-Hep-1	0.22	3.46	10.37	3.97	5.30	1.50
pancreatic	Capan1	0.80	7.30	>20	4.40		
pancreatic	HPAC	0.08	1.57	10.87	0.77		
pancreatic	BxPc3	0.27	2.70	7.60	1.00		
breast	MCF-7	0.55	2.80	>20	2.70	6.60	2.60
breast	MDA-MB-231	0.10	2.11	>20	2.00	5.40	1.80
breast	Sk-Br-3	0.04	1.19	1.56	0.90	0.04	0.05
breast	MDA-MB-468	0.21	5.00	11.40	1.60	3.60	1.10
prostate	22RV1	0.06	1.70	>20	1.20		
prostate	PC3	0.61	4.80	>20	3.00		
colon	HCT-116	0.10	1.15	18.60	1.37	6.40	1.18
colon	WiDr	0.07	0.61	16.10	0.55	1.64	0.58
HNC	CAL-27	0.06	3.72	10.15	0.60	0.98	0.46
HNC	SCC-4	0.08	2.25	>20	0.63	1.39	0.39
HNC	SCC-9	0.19	4.81	5.69	2.53	2.64	1.28

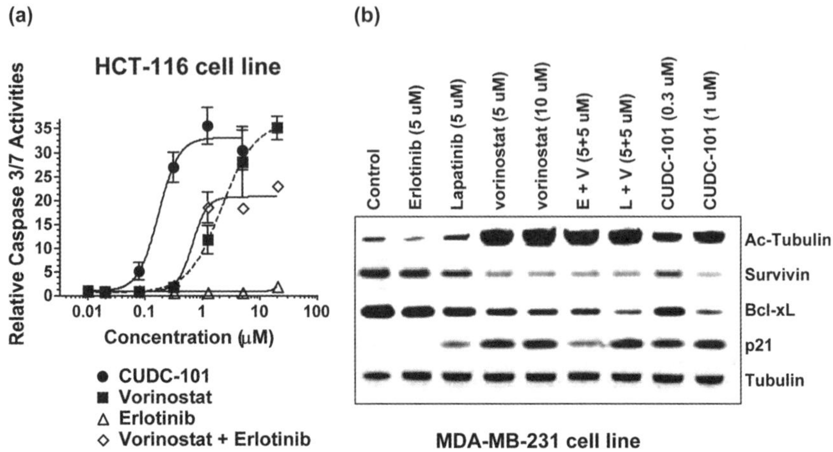

Figure 14.3 CUDC-101 induces the activities of caspases 3 and 7 in HCT-116 colon cancer cells (a), increases acetylation of α-tubulin, induces the expression of p21, and reduces survivin and Bcl-xL in MDA-MD-231 breast cancer cells (b).[10]

and increased levels of phosphorylated HER3 can be observed despite an initial down-regulation following exposure of cultured cancer cells to erlotinib (Figure 14.4c).[10] In contrast, CUDC-101 shows rapid and prolonged inhibition of both Akt and HER3 signaling in cultured cancer cells, with no subsequent reactivation during 24 hours of continuous exposure (Figure 14.4c and 14.4d).[10] Similar to CUDC-101, vorinostat shows inhibition of Akt and HER3 signaling, but this effect is observed only after 7 or 8 hours of treatment (Figure 14.4c and 14.4d).[10] These data suggest that CUDC-101, with its multi-target EGFR, HER2, and HDAC inhibition, can rapidly and sustainably suppress RTK and Akt signaling and may overcome the ability of heterogeneous tumors to acquire resistance to RTK inhibitor therapy.

- **CUDC-101 prevents drug resistance due to MET amplification and EGFR-T790M secondary mutation**

MET amplification and secondary EGFR mutation are potential mechanisms responsible for EGFR inhibitor resistance in patients.[29–31] In our study, we found that the NSCLC cell line H1993, which harbors an amplification of MET, is sensitive to CUDC-101 with an IC_{50} of ~300 nM, but fails to respond to erlotinib treatment (IC_{50} of ~20 μM) in a proliferation assay (Table 14.4).[10] We further observed that CUDC-101, but not erlotinib, can down-regulate the levels of phosphorylated and total MET and inhibit downstream Akt signaling in treated cancer cells (Figure 14.4e).[10] Similarly, we observed that H1975, which harbors a drug-resistant secondary EGFR mutation, T790M, is sensitive to CUDC-101 (IC_{50} of 500 nM), but not erlotinib treatment (IC_{50} of >10 μM) (Table 14.4). Similar to its effect on reduction of total HER2 and MET protein

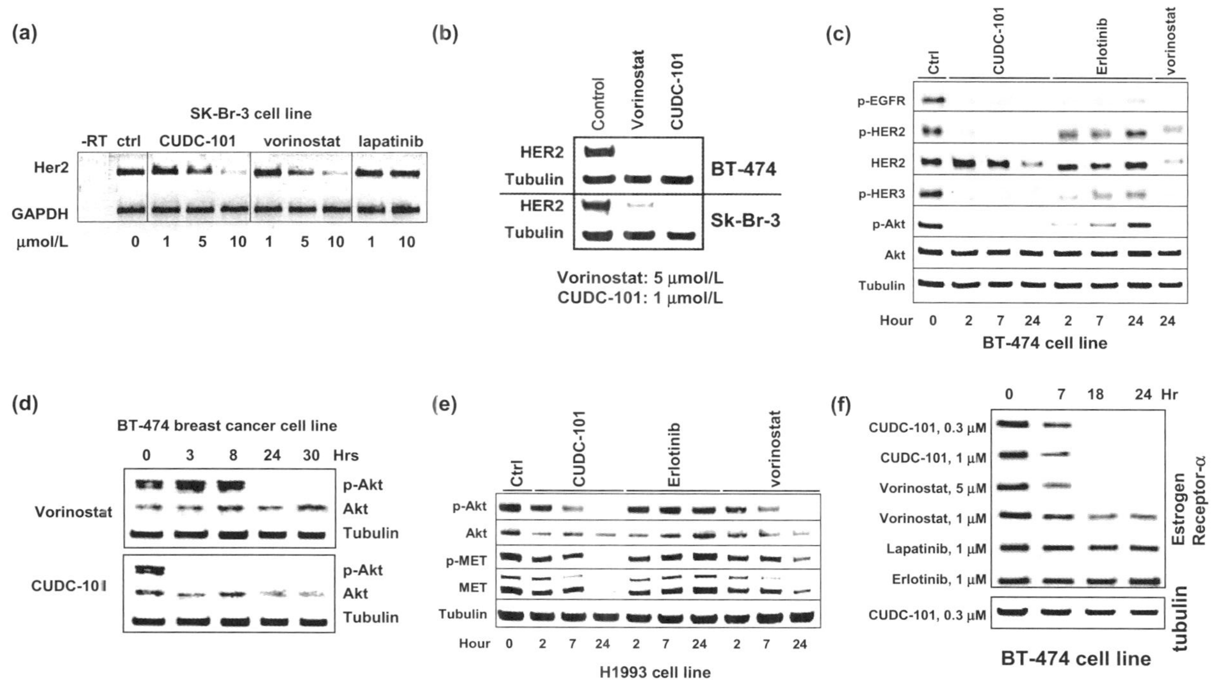

Figure 14.4 CUDC-101 suppresses Her2 transcription (a), reduces HER2 protein level (b), inhibits Akt signaling with no subsequent reactivation (c and d), down-regulates phosphorylated and total MET (e) and suppresses the expression of estrogen receptor α protein levels (f).[10]

levels, we detected that CUDC-101 down-regulates the EGFR protein level in H1975 cells. Interestingly, we also note that CUDC-101 inhibits the purified EGFR-T790M mutant, which can no longer bind to erlotinib, to an A_{max} of ~60% of peak enzyme activity in a cell-free enzyme assay. These results further suggest that CUDC-101 may overcome EGFR inhibitor resistance in patients with secondary EGFR-mutation and/or MET amplification.

- **CUDC-101 reduces estrogen receptor α protein levels**

It has been shown that estrogen receptor signaling may underlie lapatinib resistance in HER2-positive breast cancers.[32] Our study of the effects of CUDC-101 on estrogen receptor α expression revealed that CUDC-101, but not lapatinib, can effectively suppress estrogen receptor α protein levels (Figure 14.4f).[10] This finding suggests that CUDC-101 has the potential to overcome drug resistance in HER2-positive breast cancers.

14.2.3 *In Vivo* Efficacy and Pharmacodynamics

Consistent with its potent *in vitro* inhibitory activity and broad anti-proliferative activity in a range of human cancer cell types, CUDC-101 has demonstrated robust *in vivo* efficacy in a wide array of human tumor xenograft models including those resistant to erlotinib and lapatinib. In general, CUDC-101 was administrated daily i.v. (intravenous) and i.p. (intraperitoneal) as a solution in 30% Captisol, and anti-tumor activity was determined at its maximum tolerated dose (MTD) (120 mg/kg).

CUDC-101 induces 30% tumor regression in the Hep-G2 liver cancer model and is more efficacious than erlotinib at its MTD as determined in-house by ascending dose toxicity studies (25 mg/kg) and vorinostat at an equimolar dose (72 mg/kg) (Figure 14.5a).[10] In the erlotinib-resistant A549 NSCLC xenograft model, CUDC-101 shows potent inhibition of tumor growth (Figure 14.5b).[10] CUDC-101 causes significant tumor regression in the lapatinib-resistant, HER2-negative, EGFR-overexpressing MDA-MB-468 breast cancer model compared to control or lapatinib treated animals (Figure 14.5c).[10] Similarly, significant tumor regression can be observed in the EGFR-overexpressing CAL-27 head and neck squamous cell carcinoma (HNSCC) model (Figure 14.5d).[10] CUDC-101 also inhibits tumor growth in the K-ras mutant HCT116 colorectal and in EGFR/HER2 (neu)-expressing HPAC pancreatic cancer models.[10,24]

In addition to the robust *in vivo* efficacy as a multi-target single agent, CUDC-101 demonstrated enhanced anti-tumor effects when combined with chemotherapeutic agents. In the MDA-MB-468 breast cancer xenograft model, CUDC-101 alone induces tumor regression similar to that of paclitaxel, but the combination of CUDC-101 and paclitaxel causes greater tumor regression (Figure 14.5e).[10]

Pharmacodynamic studies confirm that anti-tumor activity of CUDC-101 is target specific. For example, in the H292 NSCLC xenograft model, CUDC-101

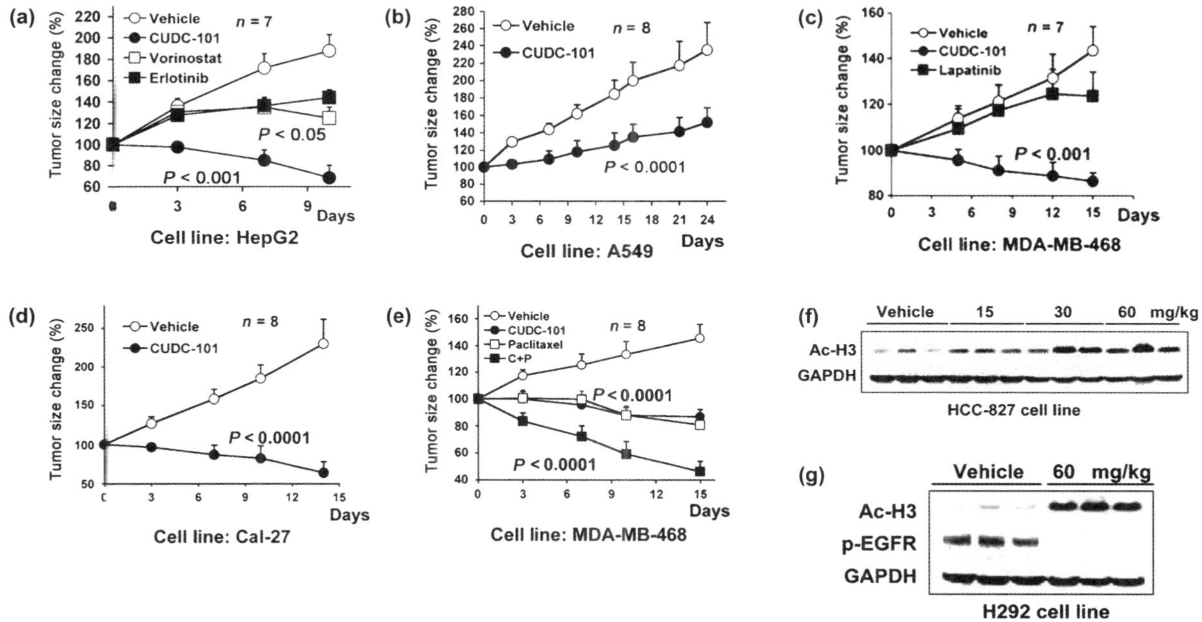

Figure 14.5 CUDC-101 induces tumor regression in the Hep-G2 liver cancer xenograft model (a), inhibits tumor growth in A549 NSCLC xenograft model (b), causes significant tumor regression in MDA-MB-468 breast cancer (c) and CAL-27 HNSCC xenograft models (d), acts alone and in combination with paclitaxel in MDA-MB-468 breast cancer xenograft model (e). CUDC-101 induces the accumulation of acetylated histone H3 level (f, g) and inhibits the activation of EGFR (g).[10]

causes an increase of acetylated histone H3 level and a reduction of phosphorylated EGFR and HER3 in tumor tissues isolated from CUDC-101 treated animals (Figure 14.5f and 14.5g) after administration of one i.v. dose at 60 mg/kg.[10] A decrease in the cell proliferation marker Ki-67 and an increase in the apoptosis marker activated caspase-3 are also observed in these tumors.

14.2.4 Pharmacokinetics and Safety Profile

Pharmacokinetic studies of CUDC-101 in several nude mice xenograft models reveal good penetration and persistence in tumor tissues. For example, in the H292 NSCLC mouse xenograft model, after a single i.v. dose of CUDC-101 at 30 mg/ml, CUDC-101 shows 2.5-fold higher exposure (AUC) and 5-fold longer half-life in the tumor than in plasma (Figure 14.6). Similarly, in the BxPC3 pancreatic cancer model, it displays 1.5-fold higher exposure (AUC) and a 5.4-fold longer half-time in tumor than in plasma after a single i.v. dose of CUDC-101 at 60 mg/ml (Figure 14.6). Further pharmacokinetic analysis of CUDC-101 in mice, rats, and monkeys indicates that CUDC-101 is most stable in monkeys and least stable in mice. Metabolic stability analysis of CUDC-101 in hepatocytes of five species (mouse, rat, dog, monkey, and human) indicates that its metabolite profile is similar across all species tested.

In *in vitro* safety profiling assays, including for 71 receptors, 16 non-kinase enzymes, 5 CYP450 isotypes (1A2, 2C19, 2C9, 2D6, and 3A4), and the hERG

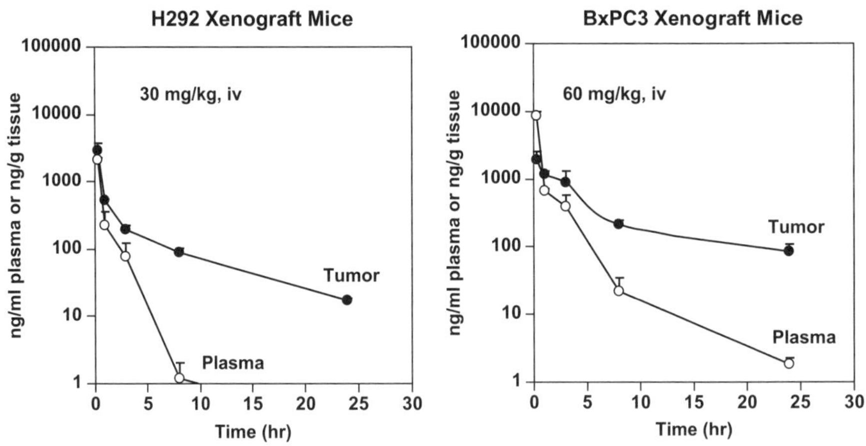

	H292 Xenograft (Mice)		BxPC3 Xenograft (Mice)	
	Plasma	Tumor	Plasma	Tumor
T1/2 (hr)	0.79	4.02	1.05	5.65
Cmax (ng/g)	2138	2983	8755	1905
AUC0-24 (hr*ng/g)	1397	3590	5848	8381

Figure 14.6 PK of CUDC-101 in mice xenograft models.

channel, CUDC-101 displays a favorable safety profile (data not shown). CUDC-101 was further evaluated in several GLP toxicology studies, including 28-day repeat dose studies in rats and dogs, and a cardiovascular telemetry study in dogs. The results indicate that CUDC-101 displays an overall favorable safety profile with high MTD/highest nonseverely toxic dose (HNSTD) of 75 mg/kg [450 mg/m^2/day] for the rat and 40 mg/kg [800 mg/m^2/day] for the dog, has minimal effects on bone marrow, and displays generally minimal-to-mild clinical and histopathology findings with good reversibility. It also has a favorable cardiovascular profile and is unlikely to be a mutagen, as assessed by AMES test, which is negative.

Based on its high *in vitro* potency, impressive *in vivo* efficacy, and favorable overall profiling, CUDC-101 was selected for clinical development and is now in phase Ib clinical trials.

14.2.5 Clinical Progress and Results

Based on the potent *in vitro* and *in vivo* activity and excellent safety profile, CUDC-101 entered the clinic in a phase I trial designed as an open-label, dose escalation study of CUDC-101 in patients with advanced, refractory solid tumors. The primary objectives were to evaluate the safety and tolerability of escalating doses of CUDC-101 and to establish the MTD and dose-limiting toxicities. Secondary objectives were to study the pharmacokinetics, evaluate pharmacodynamic biomarkers, and assess efficacy and ability of CUDC-101 to inhibit HDAC, EGFR, and HER2 in this patient population. The study was conducted at two clinical sites within the United States and enrolled 25 patients across several dose-escalating cohorts. The phase 1 clinical trial data were presented at the 22nd EORTC-NCI-AACR meeting.[33] CUDC-101 was generally well tolerated when administered at doses of 75-275 mg/m^2 as i.v. infusion over 1 hour on 5 consecutive days every 2 weeks cycle. The MTD was determined to be 275 mg/m^2, with the most common dose-limiting toxicity (DLT) being transient increases in serum creatinine. The most common CUDC-101 treatment-related adverse events were mild to moderate and included fatigue, nausea, dry skin, rash, transient increase in hepatic enzymes, and increases in creatinine. Preliminary evidence of anti-tumor activity was observed, indicating CUDC-101, through its combined inhibition of HDAC and RTK signaling (EGFR, HER2), has therapeutic potential. The observed anti-tumor activity of CUDC-101 includes one confirmed partial response that was achieved in a gastric cancer patient at 275 mg/m^2, with concomitant decrease in alkaline phosphatase (Figure 14.7),[33] and a stable disease of greater than three months observed in a refractory breast cancer patient at 150 mg/m^2 (data not shown). Two head and neck cancer patients exhibited shrinkage of greater than 20% in their respective target lesions (Figures 14.8 and 14.9).[33] Pharmacokinetic analysis demonstrated dose-proportional increases in C_{max} and AUC (Table 14.5).[33] Pharmacodynamic analysis including EGFR, HER2 inhibition in paired skin and tumor biopsies, and histone acetylation in

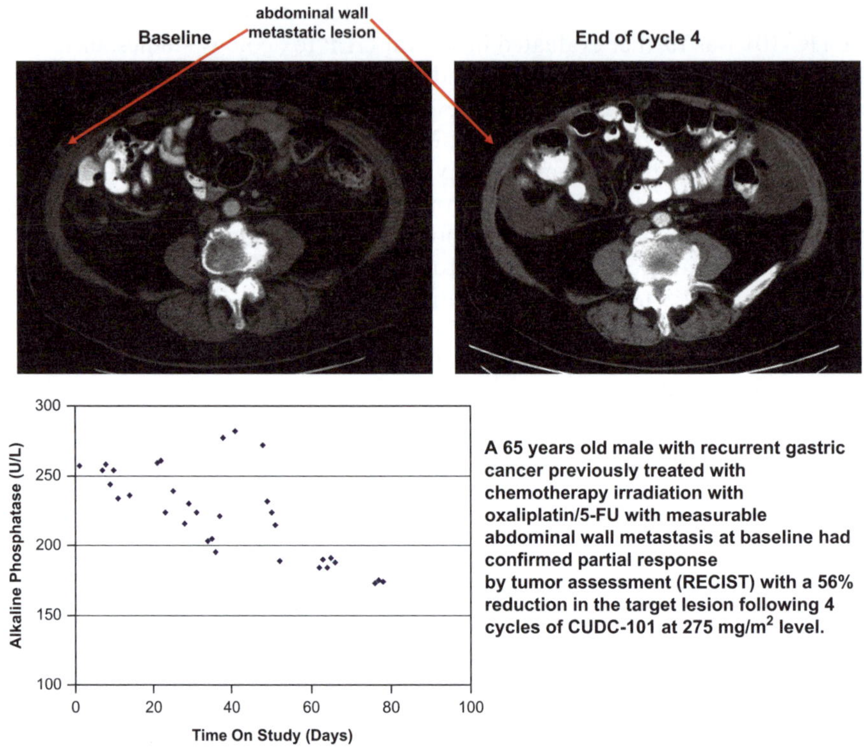

Figure 14.7 One confirmed partial response in a gastric cancer patient by CUDC-101.

PBMCs is ongoing. A phase Ib expansion study exploring alternative dosing schedules of CUDC-101 in specific tumor types (head and neck, gastric, breast, liver, and non-small cell lung cancer) is ongoing at the time of writing of this review.

14.3 Overview of Multi-Target HDAC Inhibitors

In addition to CUDC-101, Curis is also developing other multi-target HDAC inhibitors using its proprietary multi-target inhibitor platform.

In January 2011, Curis announced the selection of its second cancer network-targeted development candidate, CUDC-907, an orally bioavailable, multi-target phosphatidylinositol-3-kinase (PI3K) and HDAC inhibitor, from its HDAC-inhibiting multi-target inhibitor portfolio, which is scheduled to undergo IND-enabling studies.[34] CUDC-907 inhibits Class I PI3Ks (PI3Kα $IC_{50} = 37.7$ nM) and Class I and II HDACs ($IC_{50} = 14.84$ nM), and displays potent anti-proliferative activity against a broad range of cultured cancer cells including cell lines that are insensitive to single-target PI3K inhibitors. CUDC-907 not only inhibits the PI3K pathway, but also up-regulates the tumor

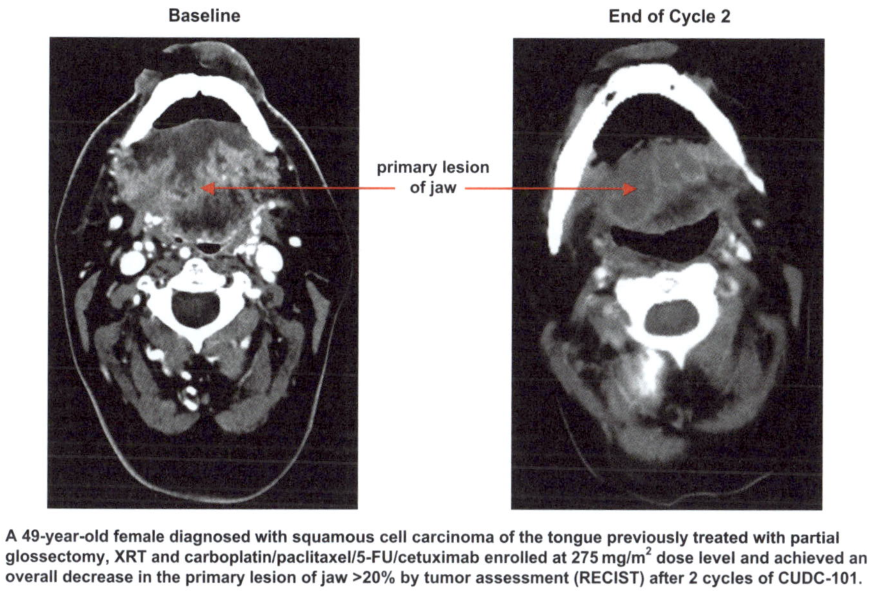

Figure 14.8 Decrease of $>$ 20% in primary lesion of jaw in a head and neck patient by CUDC-101.

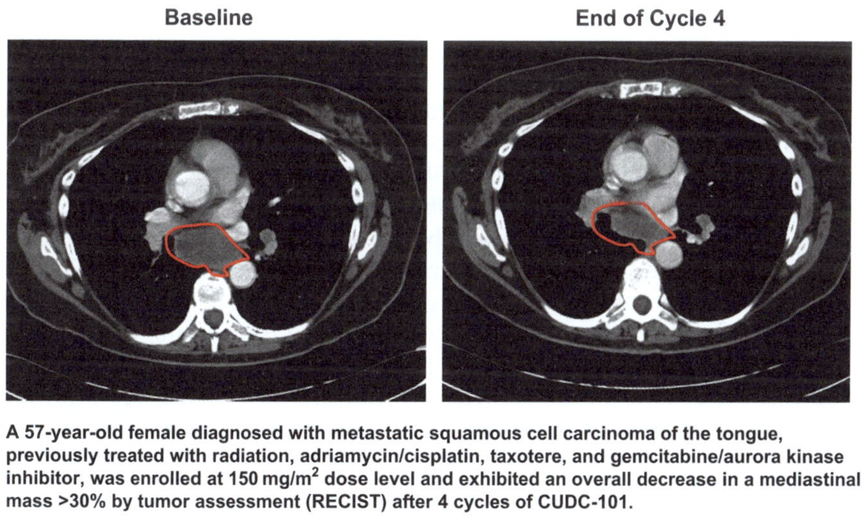

Figure 14.9 Decrease of $>$ 30% in a mediastinal mass in a head and neck patient by CUDC-101.

suppressors p53 and p21, and suppresses the MEK pathway via epigenetic modifications. Furthermore, CUDC-907 inhibits tumor growth in preclinical xenograft models of hematology cancers and solid tumors with K-RAS mutations that are insensitive to PI3K inhibitors, indicating that this

Table 14.5 Pharmacokinetic assessment of CUDC-101 in cancer patients.

Dose	*N*	*Day*	C_{max} *(ng/mL)*	$AUC_{(0\text{-}24)}$ *(ng.h/mL)*	$T_{1/2}$ *(hr)*	*Vss* (L/m^2)	*CL* $(L/hr/m^2)$
75 mg/m^2	4	1	2253.3 ± 416.5	2479.7 ± 492.2	1.6 ± 0.3	12.4 ± 0.7	33.1 ± 4.7
	4	5	1690.1 ± 355.9	2036.1 ± 362.2	4.5 ± 1.6	110 ± 85	41.9 ± 9.8
150 mg/m^2	7	1	4929.3 ± 695.1	5624.3 ± 958.4	2.1 ± 0.3	16.2 ± 0.9	29.8 ± 3.2
	6	5	4458.8 ± 868.6	5288.7 ± 1275.0	2.9 ± 0.1	25.3 ± 3.3	34.3 ± 5.4
225 mg/m^2	4	1	8547.5 ± 1090.3	9460.0 ± 1319.8	2.0 ± 0.1	11.8 ± 1.9	25.7 ± 4.5
	4	5	6947.8 ± 905.9	8010.8 ± 1178.1	2.3 ± 0.2	19.7 ± 2.3	30.3 ± 5.1
275 mg/m^2	6	1	8534.9 ± 731.6	9954.0 ± 925.8	2.6 ± 0.5	20.3 ± 3.0	29.1 ± 3.2
	5	5	8777.1 ± 528.9	10242.4 ± 675.3	2.4 ± 0.7	20.9 ± 4.7	27.4 ± 2.2
300 mg/m^2	4	1	8873.5 ± 456.4	10191.9 ± 784.2	1.9 ± 0.3	17.6 ± 0.6	30.0 ± 2.5
	2	5	9081.3 ± 1603.4	10137.1 ± 1783.5	2.8 ± 0.0	20.7 ± 3.7	30.5 ± 5.4

Mean values (± SEM).

compound may be more efficacious than other leading PI3K inhibitors currently in clinical development (data not shown).[34] CU-906, an analog of CUDC-907, also displays potent inhibition to both PI3K and HDAC. Its preclinical data were presented in the AACR 101st Annual Meeting.[35]

CU-201, a potent inhibitor of Abl (IC_{50} = 25.8 nM), Src (IC_{50} = 2.4 nM) and HDAC (IC_{50} = 6.8 nM), is a leading compound in Curis' Abl/Src/HDAC inhibitor program. CU-201 displays greater anti-proliferation and apoptosis-inducing potency against a broad range of cancer cell lines than vorinostat and/or dasatinib. CU-201 inhibits cancer cell growth, survival, angiogenesis, migration, and invasion and retains its potency in tumor cells resistant to Brc-Abl or HDAC inhibitors.[36,37]

Similar approaches of generating multi-target HDAC inhibitors have also been carried out by other organizations. The University of Regensburg in Germany reported a series of hybrid molecules combining structural features of imatinib, an inhibitor of Abl, platelet derived growth factor receptor (PDGFRβ), and Kit, with HDAC inhibitory benzamide or hydroxamate head groups.[38] In general, the selected compounds described display HDAC inhibition profiles and a potency similar to vorinostat, and maintain Abl kinase inhibition with selectivity profiles different from that of imatinib. Cytotoxicity was evaluated by using a broad panel of tumor cell lines, with selected analogs displaying mean IC_{50} values between 3.6 and 7.1 μM.[38] Most recently, the University of Regensburg published study results on lapatinib-based hybrid molecules, and some selected molecules from this series showed selective and potent inhibition of EGFR/HER2 as well as HDAC enzymatic activity.[39] McGill University worked on bifunctional 1α,25-dihydroxyvitamin D3 (1,25D) analogs that combine nuclear vitamin D receptor (VDR) agonism with inhibition of HDAC activity. Triciferol, the lead compound from this class, displays enhanced cytostatic and cytotoxic activities relative to 1,25D.[40] In extending the structural scope beyond triciferol, saturated hydroxamic acid side chains and other zinc-binding groups were also incorporated into the side chain of 1,25D, and such derived bifunctional hybrid molecules display *in vitro* anti-proliferative activity against the AT84 squamous cell carcinoma cell line, while lacking the *in vivo* hypercalcemic effects of 1,25D.[41] Furthermore, researchers from the University of Minnesota reported dual inhibitors of inosine monophosphate dehydrogenase (IMPDH) and HDAC for cancer treatment. These dual inhibitors, which show potent inhibition of both targets,[42,43] were constructed by combining structural features of the known inhibitors IMPDH and HDAC.

14.4 Conclusions

Low response rates and acquired resistance of single-target inhibitor treatment has promoted a shift in the treatment of cancer to combination and multi-target inhibition approaches. A single molecule possessing HDAC-inhibiting multi-target inhibitory activity is a novel approach to achieving improved response rates and reduced drug resistance. Compared with combination therapy, this HDAC-inhibiting multi-target approach offers the potential advantages of aligned pharmacokinetics, single agent toxicity liabilities, reduced drug cost, and high patient compliance. CUDC-101 is the first and, to the best of the authors' knowledge, only HDAC-inhibiting multi-target inhibitor in clinical trials to date. Its high potency in inhibiting EGFR, HER2, and HDAC, broad anti-proliferative and pro-apoptotic activities in cultured tumor cells, including RTK-resistant cell lines, effective network disruption in survival signaling pathways, high efficacy in *in vivo* xenograft animal models, and favorable preclinical safety profile resulted in it being advanced to clinical trials. Phase I data indicate that CUDC-101 is well tolerated and displays dose-proportional pharmacokinetics. Preliminary evidence of anti-tumor activity was observed in this heavily pre-treated patient population. These results support the network disruption approach and indicate that a single compound possessing multi-target EGFR, HER2, and HDAC inhibitory activities has therapeutic potential. The ongoing phase 1b expansion study will further examine the dosing schedule and efficacy in specific tumor types. Ongoing and future clinical development and regulatory activities are designed to verify this multi-target single compound approach clinically and commercially.

Acknowledgement

The authors are grateful to the Curis research and development teams for their contributions to the CUDC-101 and other HDAC-inhibiting multi-target programs, and would also like to thank Drs Cheng-Jung Lai, Rudi Bao, Carmen Pepicelli, Mitchell Keegan, Hai-Xiao Zhai. and Mr Mark Noel for helpful discussions and their review of this chapter.

References

1. P. A. Marks and W.-S. Xu, *J. Cell. Biochem.*, 2009, **107**, 600.
2. B. S. Mann, J. R. Johnson, M. H. Cohen, R. Justice and R. Pazdur, *Oncologist*, 2007, **12**, 1247.

3. S. Jain and N. B. Agarwal, *Pharma Rev.*, 2010, **8**, 112.
4. J. Tan, S. Cang, Y. Ma, R. L. Petrillo and D. Liu, *J. Hematol. Oncol.*, 2010, **3**, 5.
5. D. Z. Qian, X. Wang, S. K. Kachhap, Y. Kato, Y. Wei, L. Zhang, P. Atadja and R. Pili, *Cancer Res.*, 2004, **64**, 6626.
6. P. Bali, M. Pranpat, R. Swaby, W. Fiskus, H. Yamaguchi, M. Balasis, K. Rocha, H.-G. Wang, V. Richon and K. Bhalla, *Clin. Cancer Res.*, 2005, **11**, 6382.
7. P. Bali, P. George, P. Cohen, J. Tao, F. Guo, C. Sigua, A. Vishvanath, A. Scuto, S. Annavarapu, W. Fiskus, L. Moscinski, P. Atadja and K. Bhalla, *Clin. Cancer Res.*, 2004, **10**, 4991.
8. A. Edwards, J. Li, P. Atadja, K. Bhalla and E. B. Haura, *Mol. Cancer Ther.*, 2007, **6**, 2515.
9. L. Fuino, P. Bali, S. Wittmann, S. Donapaty, F. Guo, H. Yamaguchi, H.-G. Wang, P. Atadja and K. Bhalla, *Mol. Cancer Ther.*, 2003, **2**, 971.
10. C.-J. Lai, R. Bao, X. Tao, J. Wang, R. Atoyan, H. Qu, D.-G. Wang, L. Yin, M. Samson, J. Forrester, B. Zifcak, G.-X. Xu, S. DellaRocca, H.-X. Zhai, X. Cai, W. E. Munger, M. Keegan, C. V. Pepicelli and C. Qian, *Cancer Res.*, 2010, **70**, 3647.
11. C. Yu, B. B. Friday, J.-P. Lai, A. McCollum, P. Atadja, L. R. Roberts and A. A. Adjei, *Clin. Cancer Res.*, 2007, **13**, 1140.
12. L. Paoluzzi, L. Scotto, E. Marchi, J. Zain, V. E. Seshan and O. A. O'Connor, *Clin. Cancer Res.*, 2010, **16**, 554.
13. S. V. Sharma, D. W. Bell, J. Settleman and D. A. Haber, *Nat. Rev. Cancer*, 2007, **7**, 169.
14. H. Zhang, A. Berezov, Q. Wang, G. Zhang, J. Drebin, R. Murali and M. I. Greene, *J. Clin. Invest.*, 2007, **117**, 2051.
15. W. Pao, V. A. Miller, K. A. Politi, G. J. Riely, R. Somwar, M. F. Zakowski, M. G. Kris and H. Varmus, *PLoS Med.*, 2005, **2**, e73.
16. E. Avizienyte, R. A. Ward and A. P. Garner, *Biochem. J.*, 2008, **415**, 197.
17. FDA draft guidance for industry entitled 'Codevelopment of Two or More Unmarketed Investigational Drugs for Use in Combination', December 2010.
18. J. E. Bolden, M. J. Peart and R. E. Johnstone, *Nat. Rev. Drug Discovery*, 2006, **5**, 769.
19. M. R. Acharya, A. Sparreboom, J. Venitz and W. D. Figg, *Mol. Pharmacol.*, 2005, **68**, 917.
20. G. Keri, L. Orfi, D. Eros, B. Hegymegi-Barakonyi, C. Szantai-Kis, Z. Horvath, F. Waczek, J. Marosfalvi, I. Szabadkai, J. Pato, Z. Greff, D. Hafenbradl, H. Daub, G. Muller, B. Klebl and A. Ullrich, *Curr. Signal Transduction Ther.*, 2006, **1**, 67.
21. J. Stamos, M. X. Sliwkowski and C. Eigenbrot, *J. Biol. Chem.*, 2002, **277**, 46265.
22. D.-F. Wang, P. Helquist, N. L. Wiech and O. Wiest, *J. Med. Chem.*, 2005, **48**, 6936.

23. M. S. Finnin, J. R. Donigian, A. Cohen, V. M. Richon, R. A. Rifkind, P. A. Marks, R. Breslow and N. P. Pavletich, *Nature*, 1999, **401**, 188.
24. X. Cai, H.-X. Zhai, J. Wang, J. Forrester, H. Qu, L. Yin, C.-J. Lai, R. Bao and C. Qian, *J. Med. Chem.*, 2010, **53**, 2000.
25. T. C. Chou and P. Talalay, *Adv. Enzyme Regul.*, 1984, **22**, 27.
26. A. C. Hsieh and M. M. Moasser, *Br. J. Cancer*, 2007, **97**, 453.
27. N. V. Sergina, M. Rausch, D. Wang, D. J. Blair, B. Hann, K. M. Shokat and M. M. Moasser, *Nature*, 2007, **445**, 437.
28. D. L. Wheeler, S. Huang, T. J. Kruser, M. M. Nechrebecki, E. A. Armstrong, S. Benavente, V. Gondi, K.-T. Hsu and P. M. Harari, *Oncogene*, 2008, **27**, 3944.
29. J. A. Engelman, K. Zejnullahu, T. Mitsudomi, Y. Song, C. Hyland, J. O. Park, N. Lindeman, C.-M. Gale, X. Zhao, J. Christensen, T. Kosaka, A. J. Holmes, A. M. Rogers, F. Cappuzzo, T. Mok, C. Lee, B. E. Johnson, L. C. Cantley and P. A. Jaenne, *Science*, 2007, **316**, 1039.
30. J. A. Engelman and P. A. Janne, *Clin. Cancer Res.*, 2008, **14**, 2895.
31. C. L. Arteaga, *Nat. Med.*, 2007, **13**, 675.
32. W. Xia, S. Bacus, P. Hegde, I. Husain, J. Strum, L. Liu, G. Paulazzo, L. Lyass, P. Trusk, J. Hill, J. Harris and N. L. Spector, *Proc. Natl. Acad. Sci. USA*, 2006, **103**, 7795.
33. T. Shimizu, A. W. Tolcher, K. Papadopoulos, A. Patnaik, P. LoRusso, L. S. Smith, M. Beeram, D. Rasco, G. Chambers and M. Keegan, 22nd EORTC-NCI-AACR Symposium (November 16–19, Berlin, Germany), 2010, Abstract #364.
34. Curis, Inc. Press Release, January 6, 2011, http://www.curis.com/
35. R. Bao, C.-J. Lai, H. Qu, D. Wang, Y. Lin, B. Zifcak, R. Atoyan, J. Wang, M. Samson, S. DellaRocca, J. Forrester, M. Borek, G.-X. Xu, T. Xu, H.-X. Zhai, X. Cai and C. Qian, AACR 101st Annual Meeting (April 17–21, Washington, DC), 2010, Abstract #4469.
36. C.-J. Lai, X. Tao, Y. Liu, R. Atoyan, J. Wang, J. Forrester, S. Liu, X. Cai, H.-X. Zhai, R. Bao and C. Qian, AACR Annual Meeting 2008 (April 12–16, San Diego), 2008, Abstract #5717.
37. R. Bao, C.-J. Lai, H. Qu, D. Wang, Y. Lin, B. Zifcak, R. Atoyan, J. Wang, M. Samson, S. DellaRocca, J. Forrester, M. Borek, G.-X. Xu, H.-X. Zhai, X. Cai and C. Qian, 22nd EORTC-NCI-AACR Symposium (November 16–19, Berlin, Germany), 2010, Abstract #185.
38. S. Mahboobi, S. Dove, A. Sellmer, M. Winkler, E. Eichhorn, H. Pongratz, T. Ciossek, T. Baer, T. Maier and T. Beckers, *J. Med. Chem.*, 2009, **52**, 2265.
39. S. Mahboobi, A. Sellmer, M. Winkler, E. Eichhorn, H. Pongratz, T. Ciossek, T. Baer, T. Maier and T. Beckers, *J. Med. Chem.*, 2010, **53**, 8546.
40. L. E. Tavera-Mendoza, T. D. Quach, B. Dabbas, J. Hudon, X. Liao, A. Palijan, J. L. Gleason and J. H. White, *Proc. Natl. Acad. Sci. USA*, 2008, **105**, 8250.

41. M. Lamblin, B. Dabbas, R. Spingarn, R. Mendoza-Sanchez, T.-T. Wang, B.-S. An, D. C. Huang, R. Kremer, J. H. White and J. L. Gleason, *Bioorg. Med. Chem.*, 2010, **18**, 4119.
42. L. Chen, D. Wilson, H. N. Jayaram and K. W. Pankiewicz, *J. Med. Chem.*, 2007, **50**, 6685.
43. L. Chen, R. Petrelli, G. Gao, D. J. Wilson, G. T. McLean, H. N. Jayaram, Y. Y. Sham and K. W. Pankiewicz, *Bioorg. Med. Chem.*, 2010, **18**, 5950.

CHAPTER 15

Targeting Protein–Protein Interactions: Dual Inhibitors of Bcl-2 and Bcl-xL

MICHAEL D. WENDT

Oncology Research, Global Pharmaceutical Research and Development, Abbott Laboratories, 100 Abbott Park Road, Abbott Park, IL 60064, USA
Email: mike.d.wendt@abbott.com

15.1 Introduction

Apoptosis, or programmed cell death, is a highly regulated process used to eliminate defective and unneeded cells, and is important in development, tissue remodeling, and immune response.[1,2] Evasion of apoptosis is a hallmark of cancer. Cancer cells persist in part by overcoming cellular stresses such as hypoxia, genetic damage, abnormal cell cycle progression, and most importantly, chemotherapeutics and radiation.[3] Cytotoxic therapies kill cells by creating aberrations in cellular physiology sufficient to induce apoptosis, while resistance develops when apoptotic pathways become blocked. Agents able to specifically restore pathways for apoptosis could not only block tumor progression, but might also potentiate existing therapies to overcome resistance.[4,5]

The Bcl-2 family of proteins is composed of both pro-apoptotic (pro-death) and anti-apoptotic (pro-survival) members that act through a complex series of protein–protein interactions to mediate the intrinsic apoptotic pathway.[6–8] There is a wealth of evidence that the anti-apoptotic family members (Bcl-xL, Bcl-2, Mcl-1, Bcl-w, A1, or Bfl-1), especially Bcl-2 and Bcl-xL, are associated

RSC Drug Discovery Series No. 21
Designing Multi-Target Drugs
Edited by J. Richard Morphy and C. John Harris

Published by the Royal Society of Chemistry, www.rsc.org

with tumor progression, poor prognosis, and drug resistance. This group is characterized by four α-helical Bcl-2 homology (BH) domains. Pro-death proteins are further broken into those that contain three BH domains, (Bax, Bak, Bok) and those containing a single BH3 domain (BH3-only: Bad, Bik, Bid, Bim, Hrk, Bmf, Noxa, and Puma). The BH3-only proteins are mobilized and activated to initiate apoptosis in response to cellular stresses.[9,10] Bax and Bak subsequently mediate apoptosis by oligomerizing and forming pores on the mitochondrial outer membrane, resulting in release of apoptotic factors and caspase-dependent cell death.[11,12] Anti-apoptotic Bcl family proteins sequester their pro-apoptotic counterparts and stop the apoptotic signal. Cells relying on this check of the apoptotic signal can be considered 'primed for death,' the degree of priming dependent on the relative amounts of pro- and anti-apoptotic proteins at the mitochondria.[13] Thus, a small molecule antagonist of the anti-apoptotic proteins may restore apoptotic signaling, and may enhance the effects of other chemotherapeutics, or work as a single agent. Additionally, a significant therapeutic window should exist between 'primed' tumor cells and normal, 'unprimed', cells.

15.2 Rationale

At the outset of this work, Bcl-2 and in particular Bcl-xL were far more thoroughly studied and implicated in cancer than other anti-apoptotic family members. In particular, Bcl-xL overexpression had shown a more consistent correlation with intractability of cancer cell lines than had Bcl-2. As an illustration, an informatics study on the NCI 60 tumor cell line panel showed that Bcl-xL expression had a strong negative correlation with sensitivity to both 122 standard chemotherapeutic agents and a larger set of 1200 cytotoxic agents.[14] This correlation was not observed for either Bcl-2 or Bax, and was more significant than that between cytotoxicity and p53 status. Internally, we had successfully generated an NMR structure of a Bcl-xL/Bak peptide complex using a biologically active version of Bcl-xL with deletions of both the COOH terminal transmembrane helix and the majority of an unstructured loop.[15] An analogous structure of this version of Bcl-xL complexed to a Bad-derived BH3 peptide followed,[16] and a fluorescent polarization binding assay (FPA) using labeled Bax peptide as a ligand was developed.[17] Bcl-xL activity was then initially targeted as a strategy for developing a small molecule that would probably act primarily as a potentiator in conjunction with standard cancer chemotherapies, while at the same time trying to develop structural and biological support for Bcl-2 parallel to that already present for the Bcl-xL program. It was known that structural homology between the two proteins was very high, particularly with regard to the binding site. Only four active site amino acid pairs vary between the two proteins.[18] Bcl-w also has a high degree of homology to both Bcl-xL and Bcl-2, whereas Mcl-1 and A1 possess much lower homology to Bcl-2/xL. Phylogenetic analysis has shown that those latter two members of the multidomain family are more distantly related to Bcl-2, Bcl-xL, and Bcl-w.[19] Additionally, a recent paper reports major structural differentiation of Mcl-1.[20] In the course of this work, compounds consistently had Bcl-w affinity that closely tracked Bcl-2, while having essentially no affinity to Mcl-1 or A1.

15.3 Discovery Program

15.3.1 Screening and Hit to Lead

Anti-apoptotic Bcl-2 family proteins show common structures consisting of two hydrophobic α-helices surrounded by six or seven amphipathic α-helices.[18,20–22] A large (620 Å^2) hydrophobic groove approximately 20 Å long serves as the binding site for BH3 domains of their pro-apoptotic partners. As with many protein–protein interaction targets, the nature of the binding surface greatly impacted a number of issues throughout the project.[23] The effort to find chemical matter for Bcl-xL was first among them. High throughput screening (HTS) against Bcl-xL produced six hits with $IC_{50} < 10$ μM from a collection of > 100 000 compounds, and several others with slightly lower affinities. Hit compounds were extremely large and, though many contained an acid group, were also very hydrophobic. These compounds thus had low binding efficiencies, as affinity was driven by sheer size. Also, an inability to obtain NMR structures of compounds bound to Bcl-xL indicated a lack of specific binding modes. Thus, these compounds were unworkable for hit-to-lead chemistry. This result was duplicated in subsequent top-off screens of both Bcl-xL and of Bcl-2. In this case, then, screening for a target that binds almost solely via hydrophobic interactions, and absent any kind of recognition element or active site, failed to provide hit compounds suitable for further exploration.

Parallel to this effort, Bcl-xL was also subjected to fragment screening.[24,25] Fragment library compounds are small enough to bind efficiently to 'hot spot,' regions of a protein surface that generate a disproportionately large contribution to binding energy when paired with an appropriate ligand.[26,27] Ideally, fragments binding at different sites can be linked together, forming a more tightly bound lead compound. At Abbott, fragment identification was normally performed through NMR screening.[28–30] For Bcl-xL, screening a library of 10 000 fragments (210 Da average) produced a group of hits, led by the fluorobiphenyl acid **1**.[31,32] Screening again, this time in the presence of **1**, found a structurally tight cluster of second-site ligands such as the tetrahydronaphthol **2** (Figure 15.1C). The binding sites for **1** and **2** coincided with two hydrophobic hot spots determined through alanine scanning of a Bak peptide bound to Bcl-xL (Figure 15.1A).[15] Alanine substitution of Leu78 caused a 900-fold drop in affinity to Bcl-xL, while Ile85 replacement caused a 310-fold drop. These two residues point directly into the hydrophobic groove, and Bcl-xL forms a more locally concave surface in response (Figure 15.1B). In the absence of a typical enzyme–substrate interaction and nomenclature system, the four hydrophobic residues directed into the surface and their corresponding interaction sites are

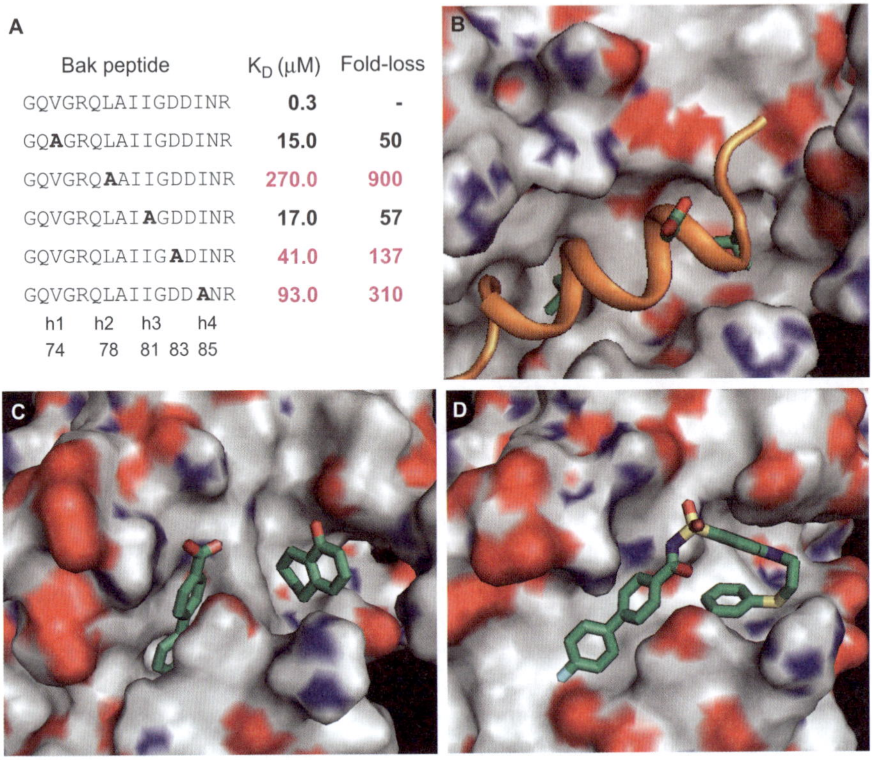

Figure 15.1 (A). Alanine scan of Bak peptide bound to Bcl-xL. (B)–(D). Aligned views of NMR structures. (B) Bcl-xL/Bak peptide complex with key residues determined from alanine scan highlighted. (C) Bcl-xL and fragment screening ligands **1** and **2**. Compound **1** mimics Leu78 and Asp83, while **2** occupies the same space as Ile85. The Phe97 side chain is situated between the two ligands. (D) Bcl-xL/**3** complex. Phe97 has rotated below the –SPh group of **3**, expanding the P4 pocket. Reproduced from M. D. Wendt, *Exp. Opin. Drug Discovery*, 2008, **3**, 1123.

designated h1–h4 and P1–P4, respectively, starting from the N-terminal end of the BH3 domain. As determined above, the h2/P2 and h4/P4 pairs were of primary importance. A third key residue, Asp83, was identified based on a 137-fold drop in affinity. Asp83 clearly interacts with Arg139 of Bcl-xL, as does the acid group of **1**. Based on the above information, it seemed that although **1** and **2** bound Bcl-xL with only 300 and 4300 μM affinities, respectively, they constituted an efficient point from which to begin chemistry work.

Attempts to join **1** and **2** via the position *ortho* to the carboxylate of **1** failed with a variety of linkers. It was ascertained that Phe97 blocked this direct path to the second site (Figure 15.1C). Fortunately, an acylsulfonamide isostere of **1**, with similar acidity, had near equal binding affinity. With acylsulfonamides providing a new vector toward the second site, two rounds of directed parallel synthesis quickly led to compound **3** (Figure 15.1D, Table 15.1).[31,32] An NMR

Table 15.1 Discovery of lead molecule **3** via consecutive library synthesis.

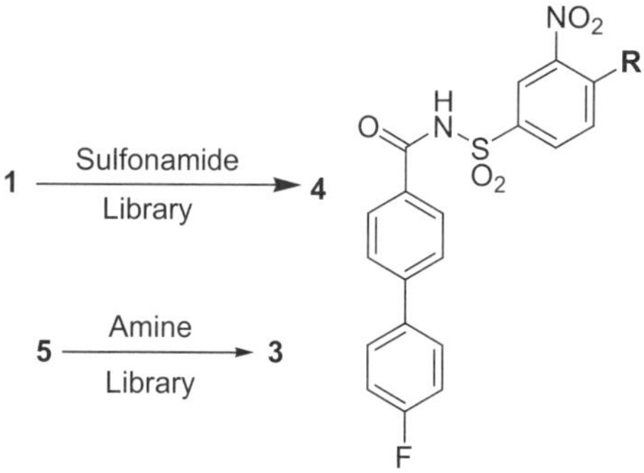

Compound	*R*	*Bcl-xL FPA IC_{50} (μM)*
4	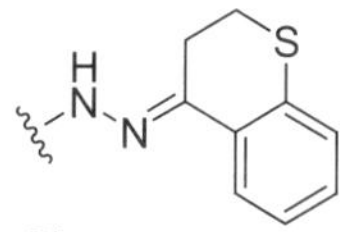	3.5
5	Cl	>10
6	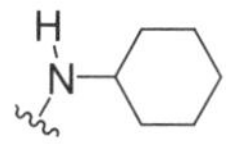	0.56
7	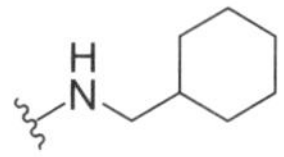	0.33
8	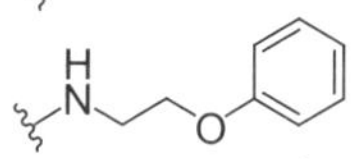	1.20
3	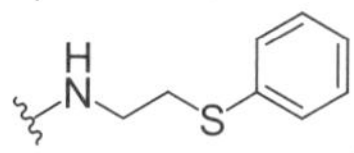	0.093
9	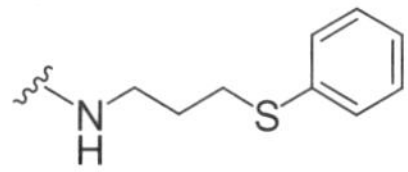	0.70

structure of Bcl-xL-bound **3** shows the acylsulfonamide changing the angle of the biphenyl group, and pulling the charge-carrying part of the acylsulfonamide farther from Arg139 than the carboxylate of **1**. The nature of the interaction of Arg139 with **3** and later acylsulfonamides is not entirely clear; nevertheless, eliminating the anionic charge in any way consistently abrogated both affinity and cellular activity.

The serendipitous discovery of **3** was crucial to the success of the project. The first acylsulfonamide library produced the nitroaryl **4** (Table 15.1), and the following library kept the nitro group, and appended different 'tail' groups *ortho* to it, via **5**. Many compounds from the second library, such as **6** and **7**, with hydrophobic tails of varying size, had affinities in the sub-μM range. Several compounds very similar to **3** (**8**, **9**, and others) were in this library as well, yet were distinctly poorer binders than **3**. It was not clear why **3** should stand out, and it was fortunate that a Bcl-xL-complexed NMR structure of its 'bent-back' conformation became available so quickly.

Importantly, movement of the Phe97 side chain expands the P4 hot spot, accommodating the hydrophobic collapse of **3** and forming an extended π-network with the Tyr194 side chain (Figure 15.1D). The phenylthio group occupies the canonical h4 position, with the nitroaryl group also filling the hydrophobic pocket, as do the h4 + *4* residues of Bak and Bad. It is known from compounds such as **5** that a single aromatic ring in P4 does little to improve affinity, and that most of the improved binding of compounds such as **4**, **6**, or **7** derives from their hydrophobic tails, which occupy regions outside of P4. Compound **3** is not only more potent still, but that potency derives solely from interaction with P4, so the potential remained to grow compounds back into the 'tail' region, and presumably pick up still more affinity. The stacked phenyl arrangement of **3** bears little resemblance to fragment **2** in the original tertiary structure; still, finding a second-site ligand helped validate the results of the alanine scan, and thereby set a strategy for the elaboration of **1**, which ultimately led to an excellent ligand for the expanded P4 site.

15.3.2 Bcl-xL-Selective Compounds

Though **3** bound to Bcl-xL with a K_i of 36 nM in the FPA assay, the same assay run with 1% serum resulted in 69-fold lower affinity. As lead compounds were very likely to continue becoming more hydrophobic during subsequent optimization, it seemed necessary to begin trying to address the serum deactivation issue immediately.[33] The majority of the serum binding was traced to domain III of human serum albumin (HSA-III). This domain, which is known to bind medium- and long-chain fatty acids and aromatic anions, was an obvious candidate to bind **3**.[34,35] Bcl-xL/**3** affinity in the presence of the 1% serum equivalent of HSA-III showed a 68-fold deactivation.

NMR-derived structures again drove a rational plan to approach this problem. Polar groups such as amines, amides, and carbamates, known to reduce HSA-III affinity,[36] were appended to two sites on **3** where they were expected to be accommodated by Bcl-xL, but by not HSA-III. These sites were identified by comparison of NMR structures of a close analog of **3** bound to HSA-III, and **3** itself bound to Bcl-xL. Both the Bcl-xL region occupied by the majority of hydrophobic tails from the second library synthesis, and the terminal biphenyl end of **3** are partly solvent-exposed. The tail region had already accepted hydrophobic groups, and the BH3 proteins fill the binding groove well beyond P2. If polar groups were tolerated at those sites, they were expected to also clash with deeper hydrophobic regions of the HSA-III binding site.[33]

For this effort affinity was tracked with and without serum in both the FPA and cellular assay (Table 15.2). As was common for this program, it proved difficult to effectively situate polar groups in the binding groove, near P2. Many amines or amides lowered serum deactivation, but affinity also suffered unless groups were located very far down the groove (**10**), which in turn reduced the improvement in the serum effect. Piperidine **11** appeared to have slightly better affinity in serum, and it also appeared capable of further elaboration via different vectors than **3**.[33]

The other region of emphasis, lying outside the binding groove, did accommodate polarity. Amines provided not only an affinity boost, similar to what had been seen with non-polar groups, but maintained more of this improvement in the presence of serum in both assays. A two-atom linker proved optimal for cellular potency (**12**, **14**, **15**), and the (*R*) enantiomer was preferred (**12**, **13**). Less basic (**16**) and non-basic groups such as amides had similar, though smaller, effects.[33]

Many compounds in addition to those in Table 15.2 began to show mechanism-based levels of cellular activity, with EC_{50} values distinct from those of enantiomers or other compounds with weaker binding. Cellular data were obtained using Bcl-xL-transfected FL5.12 cells in both the absence and presence of 3% fetal bovine serum. FL5.12 is an IL-3-dependent murine pro-B lymphoid cell line characterized by low to moderate Bcl-xL and Bcl-2 expression levels. Transfection with human Bcl-xL produced cells that expressed roughly tenfold higher protein levels as quantitated by Western blot analysis, allowing cells to survive in the absence of IL-3. It was also during this stage of the discovery effort that a stable, soluble Bcl-2 mutant began to become available. One mutant had the same COOH terminal deletion as the Bcl-xL construct, along with deletion of most of the unstructured loop. However, the remaining 15 residues of this loop are in Bcl-2 very hydrophobic, and this

Table 15.2 Modifications designed to reduce serum deactivation.

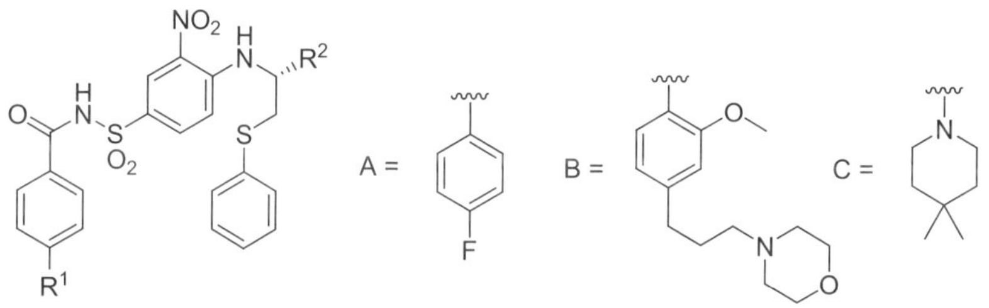

			Bcl-xL K_i *(nM)*[a]			*FL5.12 Bcl-xL* EC_{50} *(μM)*[a]	
	R[1]	*R*[2]	*0%*	*1% serum*	*10% serum*	*gelatin*	*3% FBS*
3	A	H	36 ± 9*	2500 ± 580*	>10000	1.35	24.9
10	B	H	10.4 ± 4.0	580 ± 300	4330 ± 1680*	n.d.	n.d.
11	C	H	19	652	>10000	1.2	18.4
12	C	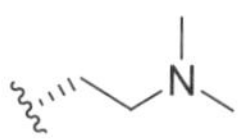	0.8 ± 0.2*	n.d.	360 ± 67*	0.47 ± 0.05*	5.1 ± 0.5*
13	C	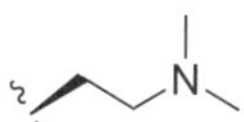	252 ± 16*	n.d.	3850 ± 660*	9.5 ± 0.7	16 ± 1
14	C	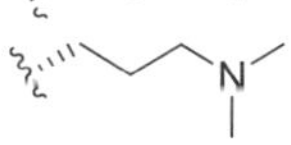	26	n.d.	728	2.0 ± 0.6*	14 ± 3*
15	C	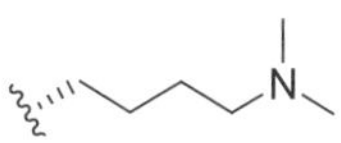	12	n.d.	174	1.1 ± 0.4*	3.9 ± 0.7*
16	C	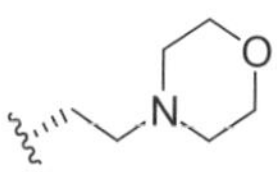	11 ± 0.8	n.d.	>10000	0.37 ± 0.09*	7.0 ± 1.6*

[a]Values with standard deviation for two experiments run in duplicate; with asterisk and standard error for three or more experiments run in duplicate. n.d. – no data.

mutant had solubility and stability difficulties. A solution was to use the corresponding residues of Bcl-xL for the truncated loop.[18] Using this new construct, analogous FPA and FL5.12 assays were introduced for Bcl-2, and complexed structures of Bcl-2 began to be generated, just as with Bcl-xL.

Lead compound **12** was evaluated for efficacy in a number of experiments. Compound **12** enhanced the activity of UV radiation and multiple cytotoxic agents against human tumor cell lines *in vitro* and *in vivo*; however, **12** had little or no single agent effect.[37] This provided confidence in the value of inhibiting Bcl-xL going forward, but **12** still was only just below 1 μM in cellular assays, a single order of magnitude better than background, off-target activity of inactive compounds, and it seemed clear that much more potency would be required for robust activity. Also, though **12** had good affinity toward Bcl-xL (K_i = 0.8 nM), it bound much less tightly to Bcl-2 (K_i = 67 nM), which was not surprising as compounds had been optimized solely against Bcl-xL for most of this time. Given the overexpression of Bcl-2 in many cancers, the t(14;18) chromosomal translocation driving Bcl-2 overexpression in non-Hodgkin's lymphoma,[38] and the possibility of redundant function of Bcl-2 and Bcl-xL, it was clearly a necessity to improve the potency of this series against both targets, and in particular against Bcl-2.

15.3.3 Dual Inhibitors of Bcl-xL and Bcl-2

Testing previously made compounds against Bcl-2 showed that many compounds were, like **12**, roughly 100-fold selective for Bcl-xL, even though the tertiary structures of the proteins are very similar.[18,39] Figure 15.2 compares X-ray structures of a **12**/Bcl-xL complex and a Bcl-2-complex of a close analog of **12** without the amine tail, with the surfaces deriving from residues that differ

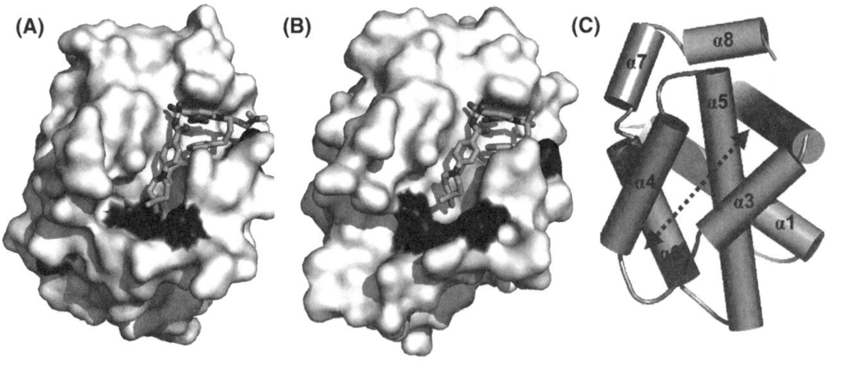

Figure 15.2 (A), (B). X-ray structures of **12**/Bcl-xL and analog/Bcl-2 complexes. Connolly surfaces from amino acid changes are in black and are, on α3, Ala104Asp, and Leu108Met, on α4, Ser122Arg, and outside P4, Glu96Asp. (C) Schematic of 3D structure of Bcl-xL and Bcl-2 with labeled helices. The dotted line is the axis of the binding groove.

between the two proteins in black. The surfaces involved in binding to project compounds up to this point were essentially identical, particularly in the P4 area. It was later determined that the Glu96Asp change from Bcl-xL to Bcl-2 was responsible for some of the Bcl-xL selectivity of compounds with an amine tail, such as **12**. The longer glutamic acid side chain of Bcl-xL makes a closer interaction with amines than the Asp of Bcl-2. The presence of an amino group was in any case required, and was maintained going forward. The primary difference arising from two changes near the bottom of the α3 helix is that the Bcl-2 groove appears to be a bit wider, and more open in the direction of P1. Growing compounds in that direction was expected to preferentially increase Bcl-2 affinity.

Extension of compounds further down the groove was accomplished by first simply growing the 4,4-disubstituted piperidine of **13**.[40] One methyl group was replaced with larger groups to better interact with the protein surface while for the other, replacement with a polar group to improve solubility and decrease protein binding was planned. In practice, increasing polarity in this way reduced cell activity. Nevertheless, facile synthesis from a 4-piperidone intermediate produced a large series of methyl ethers (Table 15.3). This series showed improved Bcl-2 affinity with increasing bulk, eventually achieving parity between Bcl-2 and Bcl-xL (**17-20**). A series of benzyl compounds showed that *ortho* substitution produced still better activity, and eventually led to the *ortho*-phenyl **21**.

Replacement of the piperidine ring with a piperazine created a synthetically tractable isostere with a slightly different vector for a hydrophobic group. Because of the basicity of piperazines, a small potency drop was expected. Earlier, amino groups had caused a loss of affinity unless they were situated much farther down the groove, as with **10**. There was indeed a small cost, as the benzyl **22** was somewhat less potent than the corresponding piperidine **19**. However, with further elaboration this series seemed to catch up, and the *ortho* phenyl **23** proved equal to or better than **21**. At this point neither the FPA nor the FL5.12 assay in gelatin could distinguish between the best compounds, and the FL5.12 assays with 3% serum (FBS) drove further work. EC_{50} values for **23** in that assay were 1.3 and 1.2 μM for Bcl-2 and Bcl-xL, respectively. However, SAR of the terminal phenyl ring showed that nonpolar *para* groups had superior efficacy in the presence of serum.[40] Though **23** and the chloro-substituted **24** appear similar. based on Table 15.2, in the FL5.12 3% serum assay **24** had values of 0.050 and 0.22 μM for Bcl-2 and Bcl-xL, respectively. Compound **24** thus achieved >250 fold (Bcl-2) and >20 fold (Bcl-xL) better potency in this assay than the piperidine **12**, further illustrating the success of the strategy for introducing more Bcl-2 affinity, and **24** was identified as our first development candidate, ABT-737.

These later improvements derive almost completely from better interaction with the P2 hot spot, and while P4 was efficiently liganded early in the drug discovery process, P2 optimization clearly did not occur until synthesis of the last group of *para*-phenyl-substituted piperazine series compounds, such as **24**. Moreover, similar substitution patterns in the other series did not result in

Table 15.3 SAR of P2 site modifications.

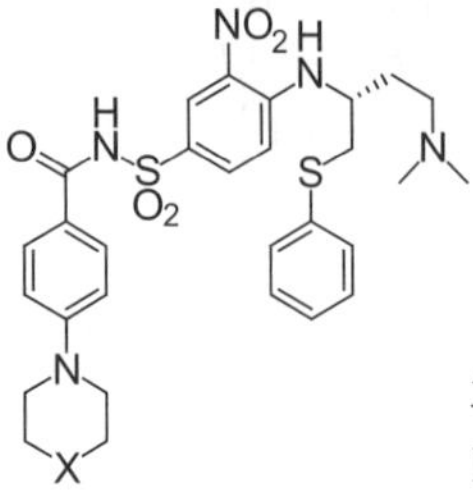

		FPA (nM)[a]			*FL5.12 cells (μM)*[a]	
		K_i		*10%HS*	EC_{50}	
	X	*Bcl-2*	*Bcl-xL*	*Bcl-xL*	*Bcl-2*	*Bcl-xL*
12	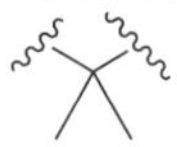	67 ± 6*	0.8 ± 0.2*	360 ± 67*	2.2 ± 0.15*	0.47 ± 0.05*
17	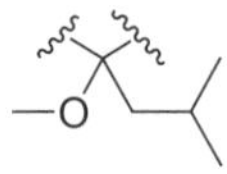	57 ± 6*	5.7 ± 1.7*	360/75	0.70	0.35
18	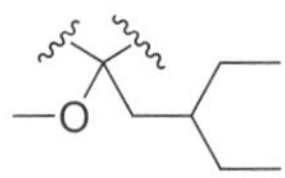	2.2	< 0.5	142/48	0.40	0.33
19	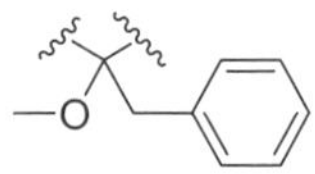	8.1	1.8 ± 0.3	150/21	0.93 ± 0.37*	0.68 ± 0.10*
20	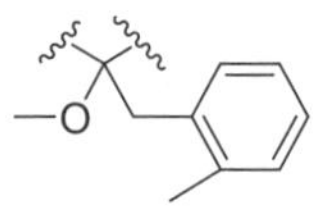	1.4 ± 0.3*	< 0.5	71/21	0.20	0.15 ± 0.05
21	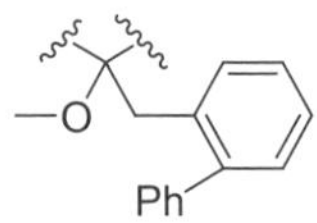	< 1	< 0.5	34.5/4.0	0.02 ± 0.00	0.035 ± 0.005
22	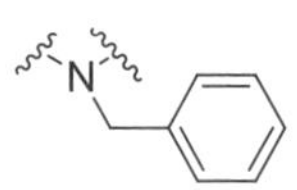	40	2.6	300/53	1.7 ± 0.1	1.1 ± 0.0
23	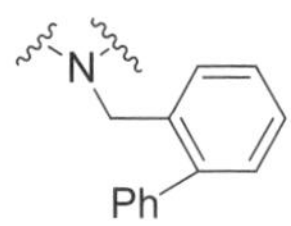	< 1	< 0.5	36.7/1.5	0.016 ± 0.004	0.018 ± 0.004
24	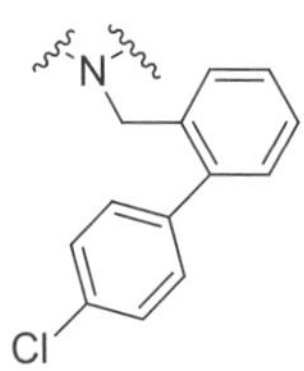	< 1	< 0.5	< 60	0.008 ± 0.002*	0.030 ± 0.0088*

[a]Values with standard deviation for two experiments run in duplicate; with asterisk and standard error for three or more experiments run in duplicate.

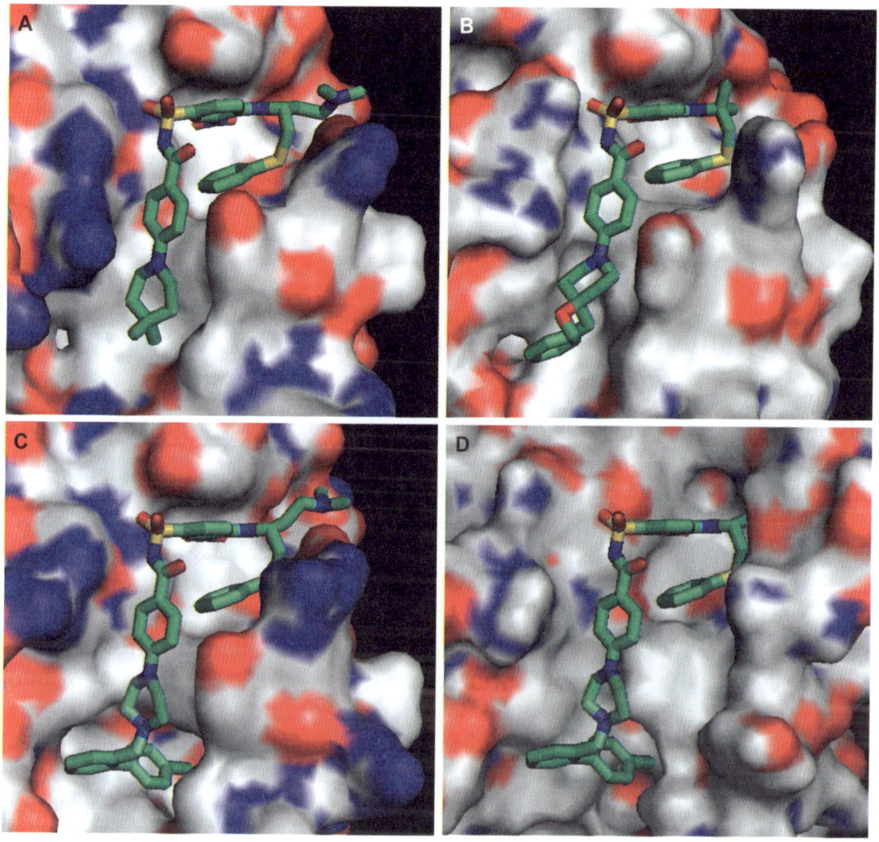

Figure 15.3 Aligned Connolly surfaces of inhibitors bound to Bcl-xL. (A) X-ray structure of **12**. (B) NMR structure of derivative of **19**. (C) X-ray crystal structure of **24**. (D) Structure of **24** from (C) overlaid onto Bcl-xL surface derived from NMR structure of Bcl-xL/Bad peptide complex. Reproduced from M. D. Wendt, *Exp. Opin. Drug Discovery*, 2008, **3**, 1123.

corresponding late improvements. In Figures 13.3B and C, an NMR structure of a tail-modified analog of **19** bound to Bcl-xL shows a version of P2 similar to that of **12**, while **24** achieves much better penetration into a better-defined P2.[41] The deep P2 and P4 pockets induced by **24** are similar to those complementing the Bad peptide, even though the gross surfaces are quite different, and P2 and P4 actually appear better defined in the Bcl-xL/**24** structure (Figure 15.3C and D).[16] Bcl-2 complexes respond similarly to changes in ligand. It is not clear from the above compound development if efficient P2 occupancy is intrinsically Bcl-2-selective, but Bcl-2-selective compounds have been reported that bind only to P2 and the lower part of the groove.[42–44]

ABT-737 (**24**) has excellent activity against primary patient-derived samples of follicular lymphoma,[32] and primary chronic lymphocytic leukemia (CLL) cells are extremely sensitive and undergo rapid apoptosis upon ABT-737 treatment ($EC_{50} = 4.5\,nM$, $EC_{50} = 7.0\,nM$).[45,46] Many laboratories have shown ABT-737 to be very effective ($EC_{50} < 100\,nM$) in primary patient samples of multiple types of blood-borne cancers.[47–53] Overexpression of Bcl-2 occurs in 55–90% of small-cell lung cancer (SCLC) cases, and may be important in the initiation, growth, and apoptotic resistance of SCLC.[54–58] ABT-737 has single-agent activity against SCLC cell lines,[32,59] and has shown efficacy with etoposide in SCLC xenograft models.[60]

15.3.4 Orally Bioavailable Compounds and ABT-263

The impressive efficacy of ABT-737 was mitigated by its extremely low water solubility (<1 μg/mL) and high lipophilicity, limiting options for i.v. formulation. Also, preclinical work had determined that ABT-737 was most efficacious when dosed on a continuous daily basis, so while ABT-737 was still in preclinical development, work began toward an orally dosed compound within the preexisting series.[61] Already in hand were examples of small modifications to ABT-737 (**24**) that greatly improved oral plasma levels. Table 15.4 shows data from a first-pass pharmacokinetic (PK) screen using Sprague-Dawley rats. The morpholino derivative **25** had four-fold higher plasma levels, with a commensurate decrease in potency both *in vitro* and *in vivo*. The functional result of these independent effects was tracked using the ratio AUC/EC_{50}, which derived its inputs from the rat PK model and a cell assay using H146 cells,[60] a small-cell lung cancer (SCLC) line expressing high levels of Bcl-xL and some Bcl-2, in 10% serum, thus taking into account both plasma exposure and cellular potency. More encouragingly, a nitro-for-trifluoromethyl change gave **26**, which had outstanding plasma levels and a long half-life (12 h). This also came with a commensurate drop in potency; nevertheless it was clear that PK parameters of ABT-737 could be meaningfully improved despite its size and hydrophobicity. Also, the dimethylamino group was metabolically labile, while the nitro group, though minimally metabolized in our assays, remained a toxicology concern. Much of the subsequent work, then, focused on changes to these two moieties. The h2 fragment was also targeted, as some modifications of the terminal chlorobiphenyl group had not yet been explored.

The possibility of nitro group replacement had been periodically explored throughout the project, but the nitro group had remained necessary for cellular efficacy. At this point, over two dozen ABT-737 analogs were tested. Many substitutions out-performed the nitro group in the rat PK model, though to a lesser extent than the trifluoromethyl group. Lower acylsulfonamide acidity correlated strongly with higher oral exposure, indicating that a reduction in charged species improved oral absorption. However, lower acidity also correlated with reduced potency.

Fortunately, in seeking to combine electron-withdrawing capacity with the hydrophobicity the binding pocket required, a trifluoromethylsulfonyl, or

Table 15.4 PK parameters after 5 mpk oral dose in rat.

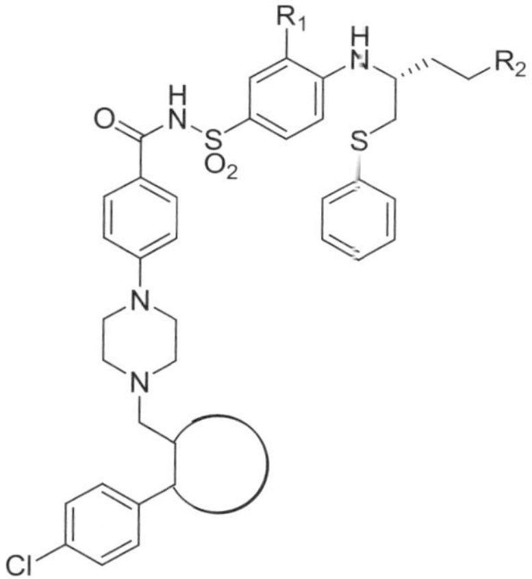

	ring	R_1	R_2	AUC^a	$F\%$	$EC_{50}{}^b$	AUC/EC_{50}
24	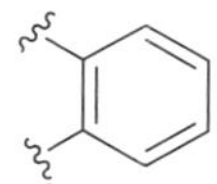	NO_2	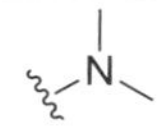	0.23	5.9	87	2.6
25	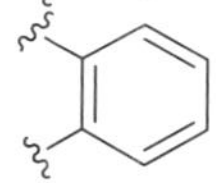	NO_2	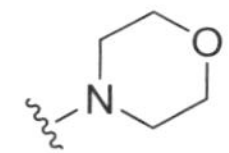	0.99	16	570	1.7
26	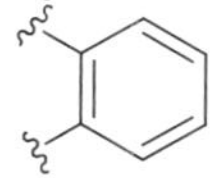	CF_3	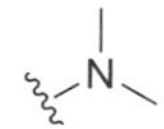	3.7	24	2200	1.7
27	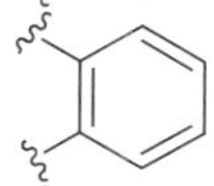	SO_2CF_3	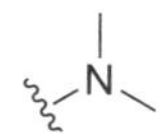	0.87	8.7	39	19
28	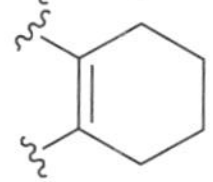	SO_2CF_3	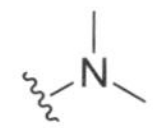	0.91	12.5	21	43
29	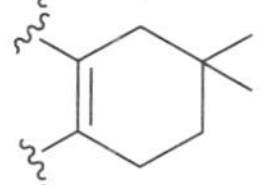	SO_2CF_3	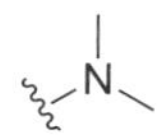	2.9	14.4	28	104
30	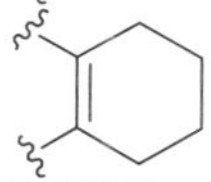	SO_2CF_3	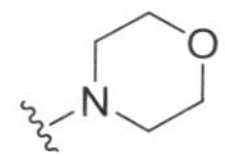	3.66	13.5	59	62

[a]0–8 h, μg/ml.
[b]nM, H146 SCLC cells, 10% serum.

trifluoryl group was hit upon (**27**). This change produced small improvements in both potency and oral exposure over **24**, resulting in an eight-fold increase in the AUC/EC_{50} ratio. The trifluoryl and nitro groups have similar electron-withdrawing effects, but the trifluoryl group is much larger, requiring a modification of either the protein surface, or of the bound conformation of **27**. As expected, this change did not seem to impact Bcl-2/Bcl-xL potency ratios. Trifluoryl compounds routinely out-performed nitro compounds going forward, and the latter group were eventually dropped.

Chlorobiphenyl SAR concerned changes to the non-terminal phenyl group. Ring substitution failed to improve compounds; however, cyclohexenes such as **28** (Table 15.4) were roughly equipotent, and exhibited higher oral plasma levels. Dimethylcyclohexenes (**29**) typically had still higher plasma levels, and importantly improved tissue/plasma distributions in pharmacokinetic/pharmacodynamic (PK/PD) models.

With regard to morpholines, their consistent jump in AUC values led to the hypothesis that, as with the acylsulfonamide, amino groups decreasing the amount of charged species present at a given moment and pH in the digestive tract would increase absorption. Several other oxygen-containing mono- and bicyclic amines were investigated, as well as bulkier amino groups such as isopropylmethyl or pyrrolidino. Most of these improved absorption relative to the dimethylamine, albeit to a small degree, but the morpholino group, with its roughly three-pK_b unit drop in basicity, typically provided the highest oral exposures, albeit accompanied by a small loss of potency (**28**, **30**).

As work on the three modification sites neared conclusion, a select few compounds with combinations of the best functional groups were compared in various animal efficacy and PK models. Finally, **31** (ABT-263) was chosen for development (Figure 15.4).[61] Summarizing the changes from ABT-737, the former nitro and phenyl positions incrementally improved potency and various PK inputs, while the morpholine sacrificed some of the gained potency in return for a further improvement in oral absorption. ABT-263 is even larger than ABT-737 (mol. wt. 974), and is still highly protein bound ($\geq$99%) with extremely low water solubility. However, ABT-263 has fewer routes of metabolism, is less charged over the relevant pH range, and has better tissue distribution. With intravenous dosing, ABT-263 has low volumes of distribution and low clearance.[62] ABT-263 also has a low solubility-driven, long dissolution rate-limited oral absorption in all species (Table 15.5). Alternate formulations increase absorption and significantly enhance PK parameters.

ABT-263, like ABT-737, is very Bad-like in its binding profile, with high affinity for Bcl-2, Bcl-xL, and Bcl-w, but not Mcl-1 or A1 (Table 15.6).[62] ABT-263 is also very comparable to ABT-737 across a range of *in vivo* studies. ABT-263 induces complete tumor regressions in SCLC and ALL xenograft models, and potentiates clinically relevant regimens in lymphoma and multiple myeloma models (Figure 15.5).[62–65] ABT-263 entered phase I clinical studies in late 2007, and is currently in phase II studies as a single agent for leukemia/lymphoma and in combination therapies for leukemia/lymphoma and solid tumors.

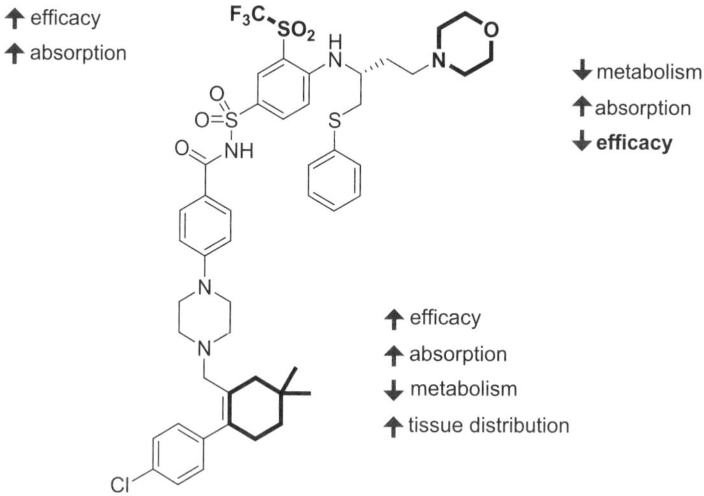

Figure 15.4 Summary of changes made to ABT-737 (**24**) to produce ABT-263 (**31**). Reproduced from M. D. Wendt, *Exp. Opin. Drug Discovery*, 2008, **3**, 1123.

Table 15.5 Multispecies PK of **31** (ABT-263). Reproduced from M. D. Wendt, *Exp. Opin. Drug Discovery*, 2008, **3**, 1123.

Species	*Oral dose (mg/kg)*	$t_{1/2}$ *(h)*	C_{max} *(μg/mL)*	*AUC (μg hr/ml)*	*F(%)*
Mouse[a]	50	5.7	5.46	55	21
Rat[b]	5	6.5	0.67	7.5	22
Monkey[b]	2.5	n.d.	0.43	6.4	13
Dog[b]	2.5	5.6	3.67	27	22
Dog[c]	5	8.9			48

[a]Delivered in 2% DMSO, 5% Tween80, 20% propylene glycol, 73% D5W solution;
[b]Delivered in 10% DMSO in PEG-400;
[c]Delivered in 10% EtOH, 30% PEG-400, 60% Phosal 50-PG solution;
n.d. = unable to determine.

Table 15.6 Comparison of binding of hBad, **24** and **31**.

	FPA serum-free K_i (nM)				
	Bcl-xL	*Bcl-2*	*Bcl-w*	*Mcl-1*	*A1*
hBad	0.5	15	33	4800	>10000
24 ABT-737	<0.5	<1	0.9	>1000	>1000
31 ABT-263	<0.5	<1	<1	550	>3000

*103-127: NLWAAQRYGRELRRMSDEFVDSFK

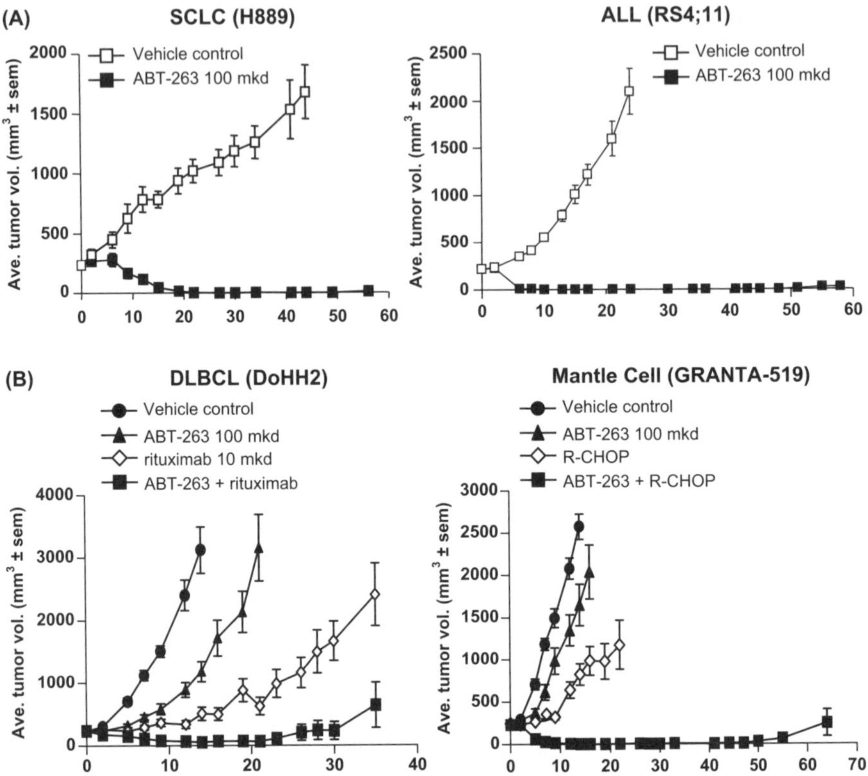

Figure 15.5 *In vivo* activity of ABT-263 (**31**) in xenograft models. Graphs show days after tumor staging versus tumor volume. ABT-263 given p.o. once daily for 21 d. (A) Xenograft models of SCLC and ALL. (B) ABT-263 in combination with other agents. Rituximab given on day 1. R-CHOP: rituximab at 10 mg/kg, i.v. qd ×1; cyclophosphamide at 25 mg/kg, i.p., qd ×1; doxorubicin at 3 mg/kg, i.v., qd ×1, vincristine at 0.25 mg/kg, i.v., qd ×1, prednisone at 0.5 mg/kg, p.o., qd ×1.

15.4 Conclusion

The discovery of dual inhibitors of anti-apoptotic proteins Bcl-2 and Bcl-xL was particularly challenging in that the targets act via protein–protein interactions with other Bcl family members.[23] Synthesis of stable, soluble versions of the target proteins, and of suitable excised α-helical BH3 peptides, facilitated parallel assay development and structural biology support in the form of fragment screening and structure-based drug design. The target proteins are similar enough that early work, performed on Bcl-xL alone, produced lead compounds that were sufficient as starting points for further elaboration into dual inhibitors. Subtle structural differences between the two proteins were later identified, which aided understanding of binding modes. Later compounds were more potent against Bcl-xL and much more potent against Bcl-2, leading to compounds with robust potentiation of standard chemotherapeutics and single agent activity in a variety of hematological cancers. This work ultimately resulted in the discovery of ABT-263, a rationally designed Bcl-2/Bcl-xL inhibitor in phase II clinical trials for cancer.

Acknowledgements

Tables 15.1–15.4 are adapted by permission from Informa Healthcare: ref. 23. Figure 15.5 and Table 15.6 are adapted from ref. 62 and used with permission of *AACR publications*. Table 15.2 is also adapted with permission from ref. 33, and Table 15.3 is also adapted with permission from ref. 40, copyright 2006 and 2007 American Chemical Society.

References

1. N. N. Danial and S. J. Korsmeyer, *Cell*, 2004, **116**, 205.
2. J. C. Reed, *Am. J. Pathol.*, 2000, **157**, 1415.
3. A. Strasser, D. C. Huang and D. L. Vaux, *Biochim. Biophys. Acta*, 1997, **1333**, F151.
4. J. C. Reed, *Toxicol. Lett.*, 1995, **82/83**, 155.
5. H. J. Guchelaar, A. Vermes, I. Vermes and C. Haanen, *Pharm. World Sci.*, 1997, **19**, 119.
6. R. J. Youle and A. Strasser, *Nat. Rev. Mol. Cell. Biol.*, 2008, **9**, 47.
7. D. R. Green and G. I. Evan, *Cancer Cell*, 2002, **1**, 19.
8. M. F. van Delft and D. C. Huang, *Cell. Res.*, 2006, **16**, 203.
9. D. C. S. Huang and A. Strasser, *Cell*, 2000, **103**, 839.
10. T. Chittenden, *Cancer Cell*, 2002, **2**, 165.
11. E. H. Cheng, M. C. Wei, S. Weiler, R. A. Flavell, T. W. Mak, T. Lindsten and S. J. Korsmeyer, *Mol. Cell*, 2001, **8**, 705.
12. K. Degenhardt, R. Sundararajan, T. Lindsten, C. Thompson and E. White, *J. Biol. Chem.*, 2002, **277**, 14127.
13. M. Certo, V. Del Gaizo Moore, M. Nishino, G. Wei, S. Korsmeyer, S. A. Armstrong and A. Letai, *Cancer Cell*, 2006, **9**, 351.
14. S. A. Amundson, T. G. Myers, D. Scudiero, S. Kitada, J. C. Reed and A. J. Fornace, *Cancer Res.*, 2000, **60**, 6101.
15. M. Sattler, H. Liang, D. Nettesheim, R. P. Meadows, J. E. Harlan, M Eberstadt, H. S. Yoon, S. B. Shuker, B. S. Chang, A. J. Minn, C. B. Thompson and S. W. Fesik, *Science*, 1997, **275**, 983.
16. A. M. Petros, D. G. Nettesheim, Y. Wang, E. T. Olejniczak, R. P. Meadows, J. Mack, K. Swift, E. D. Matayoshi, H. Zhang, C. B. Thompson and S. W. Fesik, *Protein Sci.*, 2000, **9**, 2528.
17. H. Zhang, P. Nimmer, S. H. Rosenberg, S. C. Ng and M. Joseph, *Anal. Biochem.*, 2002, **307**, 70.

18. A. M. Petros, A. Medek, D. G. Nettesheim, D. H. Kim, H. S. Yoon, K. Swift, E. D. Matayoshi, T. Oltersdorf and S. W. Fesik, *Proc. Natl. Acad. Sci. USA*, 2001, **98**, 3012.
19. A. Aouacheria, F. Brunet and M. Gouy, *Mol. Biol. Evol.*, 2005, **22**, 2395.
20. C. L. Day, L. Chen, S. J. Richardson, P. J. Harrison, D. C. S. Huang and M. G. Hinds, *J. Biol. Chem.*, 2005, **280**, 4738.
21. S. W. Muchmore, M. Sattler, H. Liang, R. P. Meadows, J. E. Harlan, H. S. Yoon, D. Nettesheim, B. S. Chang, C. B. Thompson, S. L. Wong, S. L. Ng and S. W. Fesik, *Nature*, 1996, **381**, 335.
22. A. Y. Denisov, M. S. R. Madiraju, G. Chen, A. Khadir, P. Beauparlant, G. Attardo, G. C. Shore and K. Gehring, *J. Biol. Chem.*, 2003, **278**, 21124.
23. M. D. Wendt, *Exp. Opin. Drug Discovery*, 2008, **3**, 1123.
24. P. J. Hajduk and J. Greer, *Nat. Rev. Drug Discovery*, 2007, **6**, 211.
25. M. Congreve, G. Chessari, D. Tisi and A. J. Woodhead, *J. Med. Chem.*, 2008, **51**, 3661.
26. B. C. Cuningham and J. A. Wells, *Science*, 1989, **244**, 1081.
27. T. Clackson and J. A. Wells, *Science*, 1995, **267**, 383.
28. S. B. Shuker, P. J. Hajduk, R. P. Meadows and S. W. Fesik, *Science*, 1996, **274**, 1531.
29. P. J. Hajduk, R. P. Meadows and S. W. Fesik, *Science*, 1997, **278**, 497.
30. P. J. Hajduk, *Modern Mag. Res.*, 2006, **2**, 1383.
31. A. M. Petros, J. Dinges, D. J. Augeri, S. A. Baumeister, D. A. Betebenner, M. G. Bures, S. W. Elmore, P. J. Hajduk, M. K. Joseph, S. K. Landis, D. G. Nettesheim, S. H. Rosenberg, W. Shen, S. Thomas, X. Wang, I. Zanze, H. Zhang and S. W. Fesik, *J. Med. Chem.*, 2006, **49**, 656.
32. T. Oltersdorf, S. W. Elmore, A. R. Shoemaker, R. C. Armstrong, D. J. Augeri, B. A. Belli, M. Bruncko, T. L. Deckwerth, J. Dinges, P. J. Hajduk, M. K. Joseph, S. Kitada, S. J. Korsmeyer, A. R. Kunzer, A. Letai, C. Li, M. J. Mitten, D. G. Nettesheim, S. Ng, P. M. Nimmer, J. M. O'Connor, A. Oleksijew, A. M. Petros, J. C. Reed, W. Shen, S. K. Tahir, C. B. Thompson, K. J. Tomaselli, B. Wang, M. D. Wendt, H. Zhang, S. W. Fesik and S. H. Rosenberg, *Nature*, 2005, **435**, 677.
33. M. D. Wendt, W. Shen, A. Kunzer, W. J. McClellan, M. Bruncko, T. K. Oost, H. Ding, M. K. Joseph, H. Zhang, P. M. Nimmer, S. C. Ng, A. R. Shoemaker, A. M. Petros, A. Oleksijew, K. Marsh, J. Bauch, T. Oltersdorf, B. A. Belli, D. Martineau, S. W. Fesik, S. H. Rosenberg and S. W. Elmore, *J. Med. Chem.*, 2006, **49**, 1165.
34. H. Mao, P. J. Hajduk, R. Craig, R. Bell, T. Borre and S. W. Fesik, *J. Am. Chem. Soc.*, 2001, **123**, 10429.
35. S. Curry, H. Mandelkow, P. Brick and N. Franks, *Nat. Struct. Biol.*, 1998, **5**, 827.
36. P. J. Hajduk, R. Mendoza, A. M. Petros, J. R. Huth, M. Bures, S. W. Fesik and Y. C. Martin, *J. Comp.-Aided Mol. Des.*, 2003, **17**, 93.
37. A. R. Shoemaker, A. Oleksijew, J. Bauch, B. A. Belli, T. Borre, M. Bruncko, T. Deckwirth, D. J. Frost, K. Jarvis, M. K. Joseph, K. Marsh, W. McClellan, H. Nellans, S. Ng, P. Nimmer, J. M. O'Connor, T. Oltersdorf,

W. Qing, W. Shen, J. Stavropoulos, S. K. Tahir, B. Wang, R. Warner, H. Zhang, S. W. Fesik, S. H. Rosenberg and S. W. Elmore, *Cancer Res.*, 2006, **66**, 8731.
38. Y. Tsujimoto, L. R. Finger, J. Yunis, P. C. Nowell and C. M. Croce, *Science*, 1984, **226**, 1097.
39. A. M. Petros, E. T. Olejniczak and S. W. Fesik, *Biochim. Biophys. Acta*, 2004, **1644**, 83.
40. M. Bruncko, T. K. Oost, B. A. Belli, H. Ding, M. K. Joseph, A. Kunzer, D. Martineau, W. J. McClellan, M. Mitten, S. C. Ng, P. M. Nimmer, T. Oltersdorf, C.-M. Park, A. M. Petros, A. R. Shoemaker, X. Song, X. Wang, M. D. Wendt, H. Zhang, S. W. Fesik, S. H. Rosenberg and S. W. Elmore, *J. Med. Chem.*, 2007, **50**, 641.
41. E. F. Lee, P. F. Czabotar, B. J. Smith, K. Deshayes, K. Zobel, P. M. Colman and W. D. Fairlie, *Cell Death Differ.*, 2007, **14**, 1711.
42. J. Porter, A. Payne, B. de Candole, D. Ford, B. Hutchinson, G. Trevitt, J. Turner, C. Edwards, C. Watkins, I. Whitcombe, J. Davis and C. Stubblefield, *Bioorg. Med. Chem. Lett.*, 2009, **19**, 230.
43. A. C. Castro, M. J. Grogan, E. B. Holson, B. T. Hopkins, N. O. Koney, S. Peluso and D. A. Snyder, *US Patent App. Publ.*, 2007/0155705 A1, July 5, 2007.
44. A. M. Petros, J. R. Huth, T. Oost, C.-M. Park, H. Ding, X. Wang, H. Zhang, P. Nimmer, R. Mendoza, C. Sun, J. Mack, K. Walter, S. Dorwin, E. Gramling, U. Ladror, S. H. Rosenberg, S. W. Elmore, S. W. Fesik and P. J. Hajduk, *Bioorg. Med. Chem. Lett.*, 2010, **20**, 6587.
45. V. Del Gaizo Moore, J. R. Brown, M. Certo, T. M. Love, C. D. Novina and A. Letai, *J. Clin. Invest.*, 2007, **117**, 112.
46. M. Vogler, D. Dinsdale, X.-M. Sun, K. W. Young, M. Butterworth, P. Nicotera, M. J. S. Dyer and G. M. Cohen, *Cell Death Differ.*, 2008, **15**, 820.
47. S. Trudel, A. K. Stewart, Z. Li, Y. Shu, S.-B. Liang, Y. Trieu, D. Reece, J. Paterson, D. Wang and X.-Y. Wen, *Clin. Cancer Res.*, 2007, **13**, 621.
48. D. Chauhan, M. Velankar, M. Brahmandam, T. Hideshima, K. Podar, P. Richardson, R. Schlossman, I. Ghobrial, N. Raje, N. Munshi and K. C. Anderson, *Oncogene*, 2007, **26**, 2374.
49. M. P. Kline, S. V. Rajkumar, M. M. Timm, T. K. Kimlinger, J. L. Haug, J. A. Lust, P. R. Greipp and S. Kumar, *Leukemia*, 2007, **21**, 1549.
50. T. M. Kohl, C. Hellinger, F. Ahmed, C. Buske, W. Hiddemann, S. K. Bohlander and K. Spiekermann, *Leukemia*, 2007, **21**, 1763.
51. M. Konopleva, R. Contractor, T. Tsao, I. Samudio, P. P. Ruvolo, S. Kitada, X. Deng, D. Zhai, Y.-X. Shi, T. Sneed, M. Verhaegen, M. Soengas, V. R. Ruvolo, T. McQueen, W. D. Schober, J. C. Watt, T. Jiffar, X. Ling, F. C. Marini, D. Harris, M. Dietrich, Z. Estrov, J. McCubrey, W. S. May, J. C. Reed and M. Andreeff, *Cancer Cell*, 2006, **10**, 375.
52. J. Kuroda, H. Puthalakath, M. S. Cragg, P. N. Kelly, P. Bouillet, D. C. Huang, S. Kimura, O. G. Ottmann, B. J. Druker, A. Villunger, A. W. Roberts and A. Strasser, *Proc. Natl. Acad. Sci. USA*, 2006, **103**, 14907.

53. V. Del Gaizo Moore, K. D. Schlis, S. E. Sallan, S. A. Armstrong and A. Letai, *Blood*, 2008, **111**, 2300.
54. M. Higashiyama, O. Doi, K. Kodama, H. Yokouchi and R. Tateishi, *Tumor Biol.*, 1996, **17**, 341.
55. U. Kaiser, M. Schilli, U. Haag, K. Neumann, H. Kreipe, E. Kogan and K. Havemann, *Lung Cancer*, 1996, **15**, 31.
56. U. A. Sartorius and P. H. Krammer, *Int. J. Cancer*, 2002, **97**, 584.
57. Y. K. Zhang, N. Fujita and T. Tsuruo, *Int. J. Cancer*, 1999, **83**, 790.
58. U. Zangemeister-Wittke, T. Schenker, G. H. Luedke and R. A. Stahel, *Br. J. Cancer*, 1998, **78**, 1035.
59. S. K. Tahir, X. Yang, M. G. Anderson, S. E. Morgan-Lappe, A. V. Sarthy, J. Chen, R. B. Warner, S.-C. Ng, S. W. Fesik, S. W. Elmore, S. H. Rosenberg and C. Tse, *Cancer Res.*, 2007, **67**, 1176.
60. C. L. Hann, V. C. Daniel, E. A. Sugar, I. Dobromilskaya, S. C. Murphy, L. Cope, X. Lin, J. S. Hierman, D. L. Wilburn, D. N. Watkins and C. M. Rudin, *Cancer Res.*, 2008, **68**, 2321.
61. C.-M. Park, M. Bruncko, J. Adickes, J. Bauch, H. Ding, A. Kunzer, K. C. Marsh, P. Nimmer, A. R. Shoemaker, X. Song, S. K. Tahir, C. Tse, X. Wang, M. D. Wendt, X. Yang, H. Zhang, S. W. Fesik, S. H. Rosenberg and S. W. Elmore, *J. Med. Chem.*, 2008, **51**, 6902.
62. C. Tse, A. R. Shoemaker, J. Adickes, M. Anderson, J. Chen, S. Jin, E. Johnson, K. Marsh, M. Mitten, P. Nimmer, L. Roberts, S. Tahir, Y. Xiao, X. Yang, H. Zhang, S. Fesik, S. Rosenberg and S. Elmore, *Cancer Res.*, 2008, **68**, 3421.
63. A. R. Shoemaker, M. J. Mitten, J. Adickes, S. Ackler, M. Refici, D. Ferguson, A. Oleksijew, J. M. O'Connor, B. Wang, D. J. Frost, J. Bauch, K. Marsh, S. K. Tahir, X. Yang, C. Tse, S. W. Fesik, S. H. Rosenberg and S. W. Elmore, *Clin. Cancer Res.*, 2008, **14**, 3268.
64. S. Ackler, M. J. Mitten, K. Foster, A. Oleksijew, M. Refici, S. K. Tahir, Yu Zhao, C. Tse, D. J. Frost, S. W. Fesik, S. H. Rosenberg, S. W. Elmore and A. R. Shoemaker, *Cancer Chemother. Pharm.*, 2010, **66**, 869.
65. S. Ackler, Y. Xiao, M. J. Mitten, K. Foster, A. Oleksijew, M. Refici, S. Schlessinger, B. Wang, S. R. Chemburkar, J. Bauch, C. Tse, D. J. Frost, S. W. Fesik, S. H. Rosenberg, S. W. Elmore and A. R. Shoemaker, *Mol. Cancer Ther.*, 2008, **7**, 3265.

CHAPTER 16

Discovery of the Anti-Psychotic Drug, Ziprasidone

JOHN A. LOWE, III

Pfizer Global Research & Development, Groton CT 06340, USA
Email: jal3rd@gmail.com

16.1 Introduction: Multi-Target Drug Design by Serendipity

The antipsychotic drugs offer many lessons for the design of multi-target drugs.[1] The first antipsychotic drug, chlorpromazine, **1** (Figure 16.1) was originally prepared at Rhone-Poulenc in 1950 and was intended as an antagonist for the histamine H_1 receptor with sedating properties. Jean Delay investigated this sedative effect as a possible treatment for one of his schizophrenic patients, and discovered its profound calming effects and eventually its efficacy in treating the positive symptoms of the disease.[2] Its efficacy, however, stems not from its H_1 receptor affinity, but rather from its blockade of the D_2 receptor subtype for the neurotransmitter dopamine.[3] Additional receptors that may contribute to its behavioral effects include various subtypes of the alpha adrenergic, muscarinic, and serotonin receptors.[4] It thus became a poster child for the multi-target drug, though not by design.

The early success of chlorpromazine stimulated the development of numerous phenothiazine drugs that were eventually marketed as antipsychotic agents.[5] Despite their success, however, these drugs were plagued by motor side effects, such as extra-pyramidal symptoms (EPS) and tardive dyskinesia, which

RSC Drug Discovery Series No. 21
Designing Multi-Target Drugs
Edited by J. Richard Morphy and C. John Harris

Published by the Royal Society of Chemistry, www.rsc.org

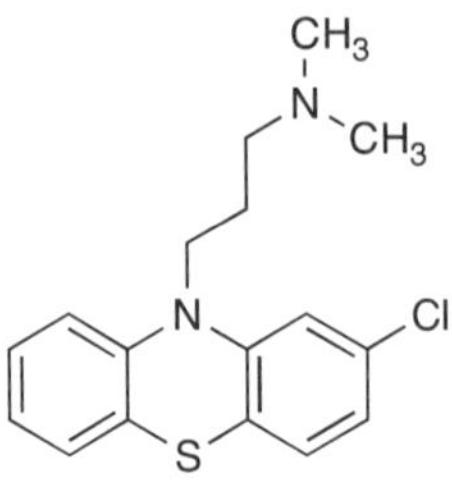

1

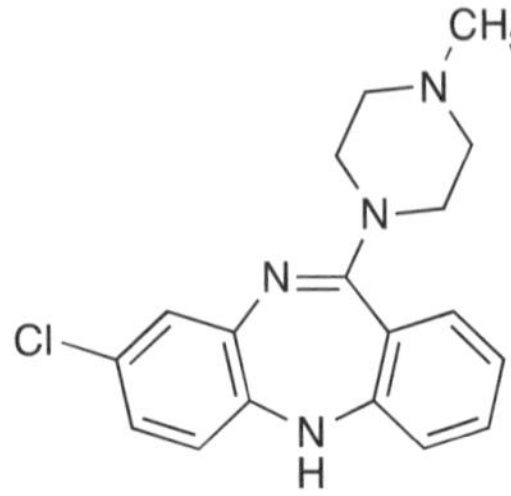

2

Figure 16.1 The typical antipsychotic drug chlorpromazine, **1**, and the atypical antipsychotic drug clozapine, **2**.

became more pronounced with time. Another serendipitous discovery came to the rescue, clozapine, **2** (Figure 16.1), an efficacious antipsychotic drug without the motor side effects of the phenothiazines.[6] Many theories have been advanced for clozapine's 'atypicality' – efficacy without side effects. Clozapine, like chlorpromazine, binds to a multitude of CNS receptors, with affinities throughout the nanomolar to micromolar range,[7] thus stimulating many interesting hypotheses as to the exact receptor combination responsible for its 'atypicality'. It retains the D_2 receptor affinity required for antipsychotic efficacy, but apparently somehow ameliorates the side effect potential of this activity by adding another receptor activity or activities. Efforts to replicate its favorable profile by finding the right receptor profile began soon after its introduction to the market, and were given greater impetus when clozapine had to be withdrawn from the market due to a side effect, the potentially fatal blood disorder agranulocytosis.[8]

16.2 Initial Efforts Towards an Atypical Antipsychotic Drug

Our initial efforts at Pfizer to find a replacement for clozapine began by investigating how it differs from the typical antipsychotic drugs. Preclinical animal models established a number of important differences. We considered the most significant of these findings to be the lack of catalepsy induction by clozapine, which was postulated to reflect its lack of EPS induction in humans. Hence our first approach was to combine efficacy, achieved with potent D_2 receptor blockade, with the ability to reverse catalepsy induced by the typical antipsychotic drug haloperidol. The anxiolytic agent buspirone was reported to reverse haloperidol-induced catalepsy, presumably by virtue of its potent agonism at the serotonin $5HT_{1a}$ receptor.[9] So our initial target was a combination of D_2 receptor blockade for efficacy with $5HT_{1a}$ agonist activity to alleviate motor side effects. There were no agents reported with this combination of activity, so we decided to take advantage of the chemical similarity in the endogenous ligands for the two receptors, which are arylethyl amines

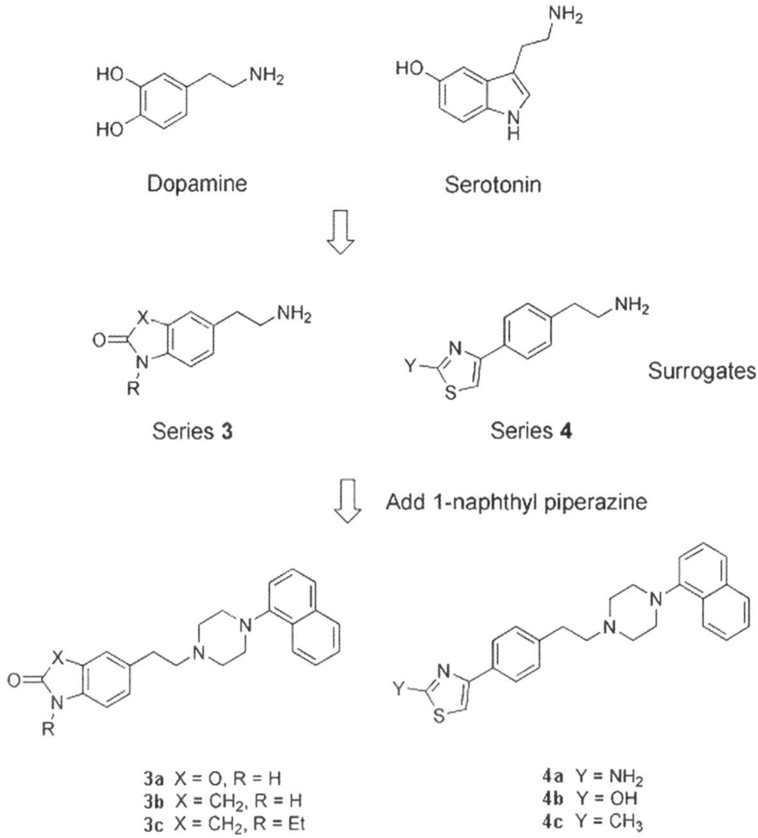

Figure 16.2 Design of combined $D_2/5HT_{1a}$ agents, starting with the structures of the endogenous ligands dopamine and serotonin.

containing a phenol function on the aryl group. We combined surrogates designed to mimic this pharmacophore with aryl piperazines to provide receptor blockade, settling eventually on 1-naphthylpiperazine as the best choice for the aryl piperazine moeity.[10] The strategy and resulting activity profiles are shown in Figure 16.2 and Table 16.1.

Representatives from both series **3** and series **4** afforded potent affinity for D_2 and $5HT_{1a}$ receptors, as well as inhibition of amphetamine-induced hypermotility (a preclinical indication of antipsychotic efficacy) and catalepsy reversal. Series **3**, however, added potent activity at another serotonin receptor, $5HT_2$, which would prove important for our next series of compounds. The hypothesized requirement for a phenol mimetic, as in the N-H group in compound **3a**, was not supported by the potent binding profile in compound **3c**. Compound **3b** was felt to embody the optimal profile, and was subjected to further pharmacological testing. As this testing progressed, however, its activity was found to be inconsistent with our goal of an atypical antipsychotic drug, and this series was not pursued further.

Table 16.1 Activity of series **3** and **4**.

Compound	D_2	$5HT_{1a}$	$5HT_2$	*AMPH*	*CAT REV*
3a	44	6.2	10	40.5	>10 (s.c.)
3b	38 ± 11	4.87 ± 0.6	20 ± 2	13.27	10
3c	81.2	2.9	ND	ND	ND
4a	72 ± 13	11	385	25.34	3.2
4b	80 ± 37	ND	249	ND	10 (s.c.)
4c	108	ND	ND	ND	ND

D_2, $5HT_{1a}$, and $5HT_2$ receptor affinity values are given in nM units. AMPH (ED_{50} for inhibition of amphetamine-induced hypermotility) and CAT REV (minimal effective dose for reversal of haloperidol-induced catalepsy) are given in mg/kg units after oral administration in rats, except where indicated by s.c., subcutaneous administration. ND – not determined. Data from ref. 10.

16.3 A New Direction and the Discovery of Ziprasidone

The failure of **3b** to provide the desired atypical antipsychotic profile forced us to reconsider our strategy. Instead of relying on $5HT_{1a}$ activity to alleviate EPS side effects, we decided to pursue activity at the $5HT_2$ receptor. At the time, one hypothesis for the lack of EPS in clozapine was its potent affinity for the $5HT_2$ receptor compared to its D_2 receptor affinity. In the case of clozapine, the $D_2/5HT_2$ ratio is 5.5.[11] This ratio also seemed to correlate with EPS propensity for a number of other antipsychotic drugs.[12] For example, the atypical antipsychotic drug risperidone, **5** (Figure 16.3), has a $D_2/5HT_2$ ratio of 6.8 and a low propensity to cause EPS.[13]

While compounds **3a** and **3b** above showed a favorable $D_2/5HT_2$ ratio, both this ratio and the D_2 receptor affinity did not meet our requirements. The benzisothiazolyl piperazine found in the experimental atypical antipsychotic agent tiospirone, **6** (Figure 16.3), was designed to mimic part of the structure of clozapine.[14] This aryl piperazine proved a suitable replacement for the 1-napthylpiperazine in our series, leading to series **7** as shown in Figure 16.4 and Table 16.2.

The multi-target analysis required for these atypical antipsychotic drugs is apparent from Table 16.2. The D_2 and $5HT_2$ receptor affinities and ratios are similar for compounds **7a–c** and **7e**, and the favorable $D_2/5HT_2$ ratios translate to a favorable ratio *in vivo* for inhibition of amphetamine-induced hypermotility as compared with induction of catalepsy.[15] This latter ratio indicates an onset of efficacy at a much lower dose than the onset of motor side effects modeled by the induction of catalepsy. Compound **7c**, ziprasidone, also has weaker affinity for the alpha-1 adrenergic receptor, which would predict a lower propensity to cause orthostatic hypotension, an adrenergic effect shown by clozapine, **2**, presumably because its alpha-1 receptor affinity exceeds its D_2 receptor affinity. Along with the outcome of advanced testing, these results led to the selection of **7c**, ziprasidone, for advancement to clinical trials and ultimately to marketing as a novel atypical antipsychotic drug.

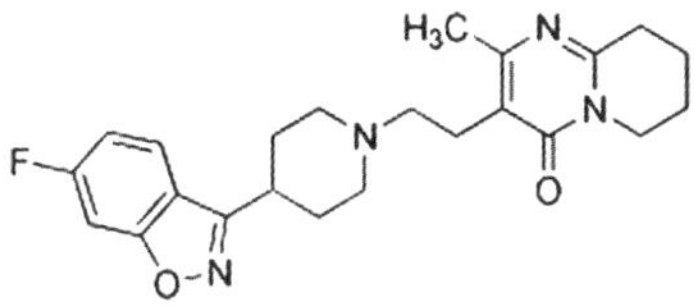
5

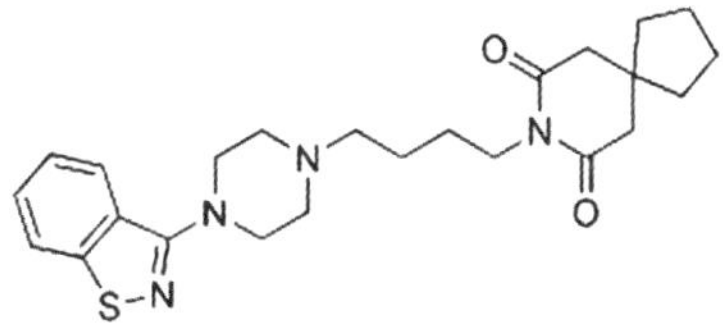
6

Figure 16.3 Compounds **5**, risperidone, and **6**, tiospirone.

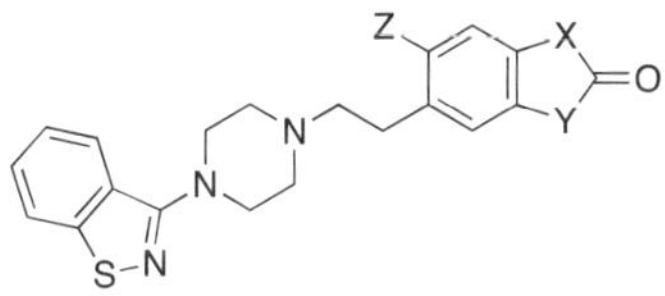

Figure 16.4 Compounds in series **7**.

Table 16.2 Activity of series **7**.

CPD	*D_2*	*$5HT_2$*	*Alpha-1*	*$D_2/5HT_2$*	*AMPH*	*CAT*	*CAT/AMPH*
7a	6.3	0.15	4.2	42	4.2	>32	7.6
7b	3.8	0.30	3.0	13	1.4	18.8	21
7c	4.8	0.42	11	11	1.5	12.1	8.1
7d	3.0	1.4	4.5	2.1	1.1	10	6.2
7e	3.1	0.33	2.1	9.4	2.3	17.8	8.1
2	83	15	5.6	5.5	12.8	>32	>32
5	3.7	0.54	0.74	6.8	0.44	5.7	13
6	2.1	0.58	1.5	3.6	4.8	20.1	6.7

D_2, $5HT_2$, and alpha-1 receptor affinity values are given in nM units. AMPH (ED_{50} for inhibition of amphetamine-induced hypermotility) and CAT (ED_{50} for induction of catalepsy in rats) are given in mg/kg units after oral administration in rats. Data from ref. 14.

In addition to the *in vitro* activity shown in Table 16.2, ziprasidone binds to the $5HT_{1a}$, $5HT_{1d}$, and $5HT_{2c}$ receptors for serotonin, and the D_1, D_3, and D_4 receptors for dopamine, with high affinity.[11] It also binds to the transporters for serotonin, dopamine, and norepinephrine with high affinity.[16] While these receptor affinities may result in an improved efficacy and side effect profile for ziprasidone, not all multi-target activity is necessarily beneficial. For example, recent work of Kim *et al.*, showed that antipsychotic drug-induced weight gain is mediated by histamine H_1 receptor blockade, which activates the hypothalamic enzyme AMP-kinase (AMPK).[17] AMPK in turn is thought to regulate food intake. The H_1 receptor affinities reported in this paper for clozapine (9 nM), risperidone (80 nM), and ziprasidone (150 nM), correlate with their weight-gain potential, from very significant for clozapine, to somewhat significant for risperidone, to weight-neutral for ziprasidone. Because weight gain can be a serious side effect with some of the atypical antipsychotic drugs, ziprasidone's weight neutrality may benefit some schizophrenic patients. Finally, ziprasidone's lack of significant affinity for M_1 muscarinic receptors, compared with the potent M_1 receptor affinity of clozapine, removes another potential side effect liability.

In addition to its efficacy in numerous clinical trials in controlling the positive symptoms of schizophrenia,[18] ziprasidone's binding profile affords it potential activity as an anxiolytic and antidepressant.[19] Its potential for causing EPS and weight gain is minimal. Following its approval by the FDA in 2001,[20] as well as approvals worldwide, ziprasidone is now marketed in over 80 countries and achieved sales of over $1 billion in 2009.[21] Thus ziprasidone has been a valuable addition to the repertoire of antipsychotic drugs.

Ziprasidone illustrates many of the challenges of multi-target drug design. The relevant receptor for antipsychotic efficacy, the D_2 dopamine receptor, shares structural homology with receptors for several other neurotransmitters, including serotonin ($5HT_1$ and $5HT_2$), adrenaline (alpha-1), histamine (H_1), and acetylcholine (M_1). While this homology proved to be a benefit in the case of the $5HT_2$ receptor, it could have caused side effect issues in the case of the alpha-1, H_1, or M_1 receptors. In some cases, the consequences of these secondary activities were not appreciated until many years after the medicinal chemistry design was completed, as for example in the case of the liability of H_1 receptor affinity in weight gain. Successful multi-target drug design often involves sculpting a compound's activity profile given the available data and what is known at the time, and then being fortunate when that profile is still beneficial once the unknowns are subsequently discovered.

References

1. B. L. Roth, D. J. Sheffler and W. K. Kroeze, *Nat. Rev. Drug Discovery*, 2004, **3**, 353.
2. J. Delay, P. Deniker and J. Harl, *Ann. Med.-Psycholog.*, 1952, **110**, 112.
3. S. Kapur and D. Mamo, *Prog. Neuro-Psychopharmacol. Biol. Psychiatr.*, 2003, **27**, 1081.
4. O. Blin, *Can. J. Psychiatr.*, 1999, **44**, 235.
5. L. Rees, *Br. Med. J.*, 1960, **2**, 522.
6. A. Fitton and R. Heel, *Drugs*, 1990, **40**, 722.
7. H. A. Nasrallah, *Mol. Psychiatr.*, 2007, **13**, 27.
8. G. Grunder, H. Hippius and A. Carlsson, *Nat. Rev. Drug Discovery*, 2009, **8**, 197.
9. B. McMillen and L. Mattiace, *J. Neural Transm.*, 1983, **57**, 255.
10. J. A. Lowe, III, T. F. Seeger, A. A. Nagel, H. R. Howard, P. A. Seymour, J. H. Heym, F. E. Ewing, M. E. Newman, A. W. Schmidt, J. S. Furman, L. A. Vincent, P. R. Maloney, G. L. Robinson, L. S. Reynolds and F. J. Vinick, *J. Med. Chem.*, 1991, **34**, 1860.
11. T. F. Seeger, P. A. Seymour, A. W. Schmidt, S. H. Zorn, D. W. Schulz, L. A. Lebel, S. McLean, V. Guanowsky, H. R. Howard and J. A Lowe, III, *J. Pharmacol. Exp. Ther.*, 1995, **275**, 101.

12. A. C. Altar, A. Wasley, R. Neale and G. Stone, *Brain Res. Bull.*, 1986, **16**, 517.
13. J. Leysen, W. Gommeren, A. Eens, D. De Chaffoy de Courcelles, J. Stoof and P. Janssen, *J. Pharmacol. Exp. Ther.*, 1988, **247**, 661.
14. J. P. Yevich, J. S. New, D. W. Smith, W. G. Lobeck, J. D. Catt, J. L. Minielli, M. S. Eison, D. P. Taylor, L. A. Riblet and D. L. Temple, Jr., *J. Med. Chem.*, 1986, **29**, 359.
15. H. R. Howard, J. A. Lowe, III, T. F. Seeger, P. A. Seymour, S. H. Zorn, P. R. Maloney, F. E. Ewing, M. E. Newman, A. W. Schmidt, J. S. Furman, G. L. Robinson, E. Jackson, C. Johnson and J. Morrone, *J. Med. Chem.*, 1996, **39**, 143.
16. A. W. Schmidt, L. A. Lebel, H. R. Howard and S. H. Zorn, *Eur. J. Pharmacol.*, 2001, **425**, 197.
17. S. Kim, A. Huang, A. Snowman, C. Teuscher and S. Snyder, *Proc. Nat. Acad. Sci. USA*, 2007, **104**, 3456.
18. D. Addington, C. Pantelis, M. Dineen, I. Benattia and S. Romano, *J. Clin. Psychiatr.*, 2004, **65**, 1624.
19. G. L. Stimmel, M. A. Gutierrez and V. Lee, *Clin. Ther.*, 2002, **24**, 21.
20. C. Ukens, *Drug Topics*, 2001, **2**, 19.
21. Pfizer Inc., *2009 Financial Report*, 2010.

CHAPTER 17

The Rational Design of Triple Reuptake Inhibitors for the Treatment of Depression

ROBERT J. WEIKERT

Hoffmann-La Roche, Discovery Chemistry, 340 Kingsland Street, Nutley, NJ 07110, USA
Email: robert.weikert@roche.com

17.1 Introduction

Despite advances in treatment, major depression continues to be a disabling condition that exacts an enormous toll on society, both socially and economically. It is estimated that as much as 17% of the US population age 18 and older will suffer from depression at some point in their lifetime, with the loss of productivity estimated at more than $31 billion per year.[1,2] Current pharmacological treatments are derived from the monoamine hypothesis which states that the molecular basis for depression is a deficiency in one or more of the three major neurotransmitters – serotonin, norepinephrine, and dopamine. This theory has been clinically supported by the fact that the underlying mechanism of most effective treatments for depression are through the modulation of one or more of these transmitters. Early therapies included the tricyclic antidepressants (TCAs) such as amitriptyline and clomipramine, which essentially block the reuptake of both serotonin and norepinephrine, resulting in increased synaptic concentrations of these two neurotransmitters. While the TCAs are effective, they are associated with severe, and often life threatening,

RSC Drug Discovery Series No. 21
Designing Multi-Target Drugs
Edited by J. Richard Morphy and C. John Harris

Published by the Royal Society of Chemistry, www.rsc.org

side effects due to their concomitant interactions at histaminergic, adrenergic, and muscarinic receptors.

A significant advance in the safety of antidepressant treatment was associated with the discovery of the selective serotonin reuptake inhibitors (SSRIs) such as paroxetine and fluoxetine, which inhibit the serotonin transporter. These transporters are 12 trans-membrane polytopic proteins located peri-synaptically on the pre-synaptic terminals of serotonergic neurons. They are responsible for the reuptake and storage of serotonin, resulting in termination of signaling. While the efficacy of this class of drugs is not optimal, likely due to their preferential effects only on serotonin neurotransmission, their relatively benign side effect profile has led to wide acceptance within the medical community. Although hailed as a breakthrough for the treatment of depression, the SSRIs still leave considerable unmet medical need. In terms of efficacy, only one-third of the patients respond to treatment and there is a considerable delay in the onset of action, with most patients not responding before 4 to 6 weeks of treatment. This delay in onset suggests antidepressant efficacy is driven by complex adaptations of downstream events of chronic treatment, including gene expression changes (up- and down-regulation of pre-and post-synaptic receptors) as well as alterations in the endocrine and immune systems.[3–6] In terms of side effects, the SSRIs are associated with sexual dysfunction including erectile dysfunction, anorgasmia, and decreased libido. The combination of low remission rates, delayed onset of action, and side effects leads to a low compliance rate, making treatment even more difficult.

A more recently discovered class of antidepressants are the selective serotonin and norepinephrine reuptake inhibitors (SNRIs) exemplified by venlafaxine and duloxetine, which exert their actions on both serotonergic and noradrenergic neurotransmission. This class inhibits both the serotonin transporter (SERT) and the norepinephrine transporter (NET), resulting in increased synaptic concentrations of both neurotransmitters. While arguably more efficacious than the SSRIs, there is still a low rate of response accompanied by a similar side effect profile and delay in onset. One clear benefit of this class of compounds though is their efficacy for the treatment of pain symptoms which are often co morbid with depression. This is exemplified with duloxetine which has attained regulatory approval for both neuropathic pain, commonly associated with diabetic neuropathy, and fibromyalgia.

In order to improve upon the remission rates for antidepressant treatment, many psychiatrists have employed an augmentation strategy, the practice of adding another drug to a patient's SSRI or SNRI regimen. This is a common strategy for addressing patients with treatment-resistant depression (TRD) or atypical depression. Patients are classified as having TRD if they have failed adequate courses of two first-line antidepressants, whereas atypical depression is characterized by mood reactivity accompanied by ancillary symptoms including weight gain, hypersomnia, leaden paralysis, and interpersonal rejection sensitivity.[7,8] Two common drugs for augmentation treatment are pindolol, a modulator of serotonin (5-HT) 1A receptors, and bupropion, a norepinephrine and dopamine reuptake inhibitor (NDRI).[9–11] This strategy of combining drugs for their polypharmacology is an attempt to regain the improved efficacy observed with the earlier non-selective TCAs while minimizing their off target activities. The FDA has recognized the benefits of this multi-target approach and has recently approved Symbyax, a fixed dose combination of olanzapine, an atypical antipsychotic, and fluoxetine, an SSRI.[12] While combining compounds in order to achieve their mixed pharmacologies has proven benefits, it also has many drawbacks. Clearance rates and metabolic profiles vary for each compound between individuals making it difficult to deliver the same proportion of components to any two patients over time. This differing proportion of drugs between individuals not only modulates efficacy through differential receptor occupancies, but can alter the rate at which other concomitant medications are cleared, resulting in complex circulating mixtures of pharmacologically active compounds and potential drug–drug interactions (DDIs). There is also the inconvenience of taking multiple pills affecting compliance rates. For these reasons, it is beneficial to design single agents that have multiple desired pharmacologies with minimal off target activity. There is currently a large effort among multiple drug companies to discover compounds that combine additional pharmacologies with the SSRI/SNRI mechanism in order to improve efficacy, shorten onset rates, and/or address side effects. Some of these additional pharmacological targets include 5-HT_{1A} (pre- and post-synaptic), 5-HT_{2A}, 5HT_{2C}, α-2 adrenergic, D_2/D_3 agonism, dopamine release, and dopamine reuptake.

17.2 Role of Dopamine in Depression

While current antidepressant treatments primarily focus on increasing synaptic concentrations of serotonin and/or norepinephrine, there is considerable evidence, both preclinical and clinical, suggesting that decreased dopamine levels also contribute to depressive states. Figure 17.1 outlines some of the physiological behaviors associated with these three neurotransmitters and aligns some of the treatment options accordingly.

The STAR*D trial (sequenced treatment alternatives to relieve depression) was one of the largest trials evaluating current antidepressants and specifically assessed treatment options for patients who do not respond initially to an SSRI.[13] This study demonstrated that the combination of an SSRI plus bupropion was more effective than switching patients to a different SSRI, SNRI, or bupropion alone. Since bupropion is primarily considered an NDRI, the observed beneficial effects on remission rates in conjunction with SSRI treatment can be linked to the additional inhibition of the dopamine transporter (DAT) and increased synaptic dopamine levels. In addition, anhedonia, which is an inability to experience pleasurable emotions from normally pleasurable life events, is a core symptom of depression and has been linked to deficits in mesocortical and mesolimbic dopamine function.[14,15] Several clinical studies have evaluated the effects of increasing dopamine function on the onset of action of antidepressant (SSRI/SNRI) treatment. Amisulpride, which increases dopamine signaling at low doses, has been shown to decrease the time

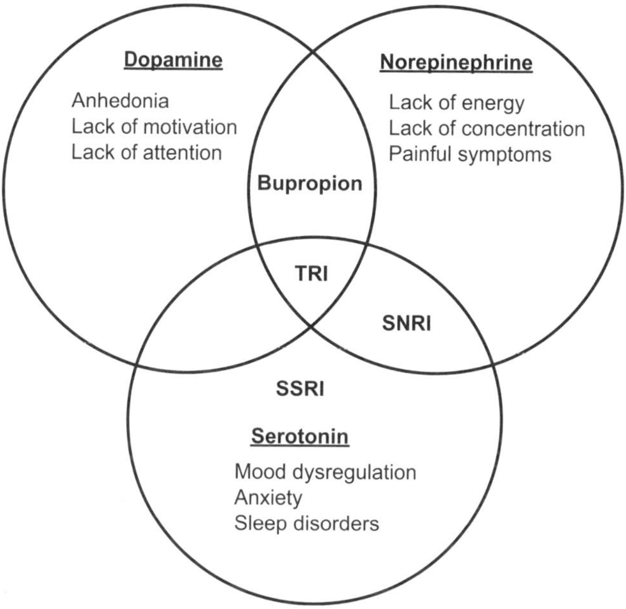

Figure 17.1 Physiological behaviors associated with serotonin, norepinephrine, and dopamine combined with corresponding treatment options.

to onset of antidepressant efficacy, both preclinically and clinically.[16,17] Studies have also shown that the addition of methylphenidate (Ritalin), a dopamine releaser, to citalopram (an SSRI treatment) also leads to accelerated response rates.[18] Along with the efficacy and response rate benefits associated with increased dopamine function, are data to support the effectiveness of bupropion in reversing the sexual dysfunction in patients treated with SSRI/SNRIs.[19]

17.3 Challenges Associated with Increased Dopamine Function

While significant evidence points to a beneficial role of increasing dopamine function within depressed patients, there are also considerable concerns. Dopamine systems, especially within the ventral tegmental region of the brain, have been directly linked to reward and craving behaviors.[20] Stimulants and drugs of abuse such as amphetamine, methamphetamine, and cocaine are believed to derive their addictive properties through modulation of these dopamine pathways. Preclinical data suggest that dopamine related abuse liability through DAT inhibition is associated with several factors including both the degree and rate of occupancy.[21] These studies, utilizing DAT positronic emission topography (PET), suggest that abuse liability correlates with a greater than 50% DAT occupancy achieved in a short period of time (less than

15 minutes). While these data suggest that the addition of DAT inhibition to an SNRI could result in drug-seeking behavior, the monoamine oxidase inhibitors (MAOIs), which increase serotonin, norepinephrine, and dopamine through inhibiting their metabolic degradation, are not associated with abuse liability nor is bupropion. Nonetheless, these properties will need to be evaluated in preclinical and clinical settings to better define the abuse liability potential and therapeutic index for the Triple Reuptake Inhibitor (TRI) class of compounds.

17.4 Compound Target Profiles

While it was apparent that the incorporation of dopamine reuptake inhibition to an SNRI to provide a TRI could lead to a more efficacious antidepressant, the question remained as to the optimal balance of inhibition between the three targets. Insufficient increases in dopamine function could lead to a compound which would not clinically differentiate from an SNRI while excessive increases in dopamine could potentially lead to self administration, abuse, and subsequent regulatory scheduling. We wanted to maintain efficacy for the treatment of pain conditions so we chose to pursue a relatively balanced SET:NET profile similar to duloxetine. PET studies in human have shown that SSRIs, at effective doses, result in ~80% or greater occupancy of SERT while bupropion treatment results in ~25% occupancy at DAT.[22,23] Since bupropion augmentation is effective and reasonably well tolerated, we chose 25% DAT occupancy as our minimum requirement and set out to discover compounds with a range of DAT potencies. Our strategy was to evaluate these varying profiles for their potential benefits in both preclinical and clinical settings using endpoints such as sexual dysfunction, time to onset of action, abuse liability, and efficacy in patients with TRD. We also laid out an initial set of desired preclinical properties based on our previous work with central nervous system (CNS) targets. Compounds needed to cross the blood/brain barrier (BBB), which required high permeability as measured in Caco-2 cells (A to B diffusion $>5 \times 10^{-6}$ cm/s) with no P-glycoprotein (P-gp) mediated efflux. We also wanted a good cardiovascular safety window so our initial hERG requirement, as measured by electrophysiology, was an $IC_{20} > 2\ \mu M$, although flexible based on transporter potency and other ADME properties. Our primary DDI concerns were focused on the cytochrome P450s (CYP) 2D6 and 3A4. Since 3A4 is involved in the clearance of several drug classes, inhibition of this isozyme is generally to be avoided. The CYP2D6 isozyme is polymorphic in humans, with both high and low metabolizers, so our goal was to minimize clearance through this enzyme, as well as inhibition, in order to improve plasma exposure predictability. A further target was for a moderate clearance rate (5–30 μL/min/mg) as measured in human liver microsomes (HLM). While our target was once daily dosing, previous experience revealed that liver-stable compounds can lead to very long half-lives with accumulation upon repeat dosing. Finally, we wanted to have pharmacologically inactive metabolites to avoid complex PK/PD profiles through multiple circulating active species.

17.5 Genesis of the TRI Series

Our research into reuptake inhibitors began with an SNRI program with the goal of differentiating from the yet to be approved duloxetine. We took a multi-pronged approach to discover our initial leads, including a high throughput campaign at both SERT and NET, an informational approach based on literature compounds as well as the construction of a pharmacophore model from literature inhibitors. Two indole series derived from the duloxetine template were identified and entered into lead optimization. These originated by substituting an electron-rich indole ring for the naphthyloxy group present in duloxetine, as shown in Figure 17.2. Our primary screening assay was a scintillation proximity assay measuring the binding of compounds to human SERT and NET, with the data reported as pK_i values. These two novel indole scaffolds were quite active at both transporters and provided rapid access to novel and potent SNRI compounds.[24]

With the successful launch of duloxetine, it was apparent that the business case for another SNRI molecule was diminishing. A new, successful entry into the antidepressant market would need to clearly differentiate from existing treatment options, so we began to focus our efforts on the discovery of triple reuptake inhibitors.[25] We had made all the regioisomers of the indole propylamine series, as shown in Table 17.1, and the 5-substituted analog was moderately active at DAT (pK_i of 6.7) with a reasonable clearance and CYP profile.

Attempts to optimize potency, clearance, and hERG on this series proved difficult. Potency was generally improved through lipophilic substitutions, which led to increased clearance and more potent hERG inhibition. Metabolite identification for this series consistently showed a rapid *N*-demethylation pathway leading to the primary amines which were not as active as the parent compounds at the transporters but remained potent hERG inhibitors. In addition, we had yet to develop an efficient chiral synthesis and were separating enantiomers by HPLC which made obtaining larger quantities of material

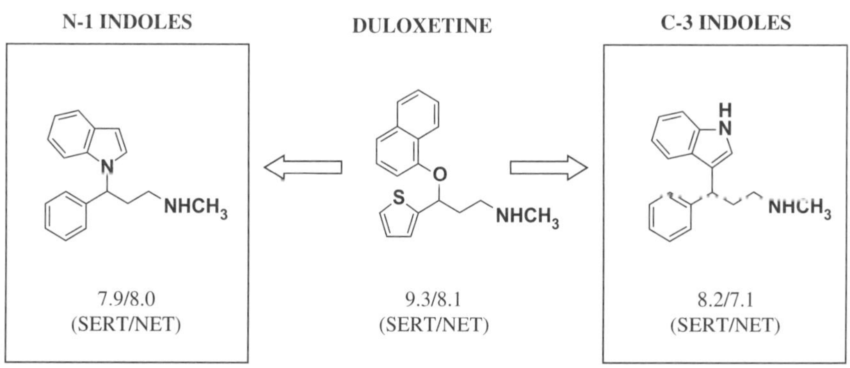

Figure 17.2 Potency (pK_i) of novel SNRI templates derived from duloxetine.

Table 17.1 Transporter potency, microsomal clearance, CYP, and hERG data for the regioisomers of the SNRI indole propylamine series.

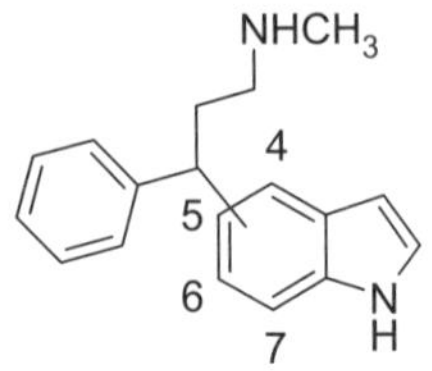

Substitution	*SERT/NET/DAT* pK_i	*HLM* *(μl/min/mg)*	*2D6/3A4* *(IC_{50} μM)*	*hERG (1 μM)* *% remaining*
4	7.4/8.0/6.1	11	13/10.6	90
5	7.5/8.0/6.7	17	2.8/5.1	53
6	8.1/7.9/5.9	33	7.8/37.5	31
7	7.6/7.4/6.0	57	4.7/38.3	71

laborious. With many issues to resolve, we sought out new templates that could provide better starting points.

17.6 Scaffold Hopping

17.6.1 Aminopiperidine Series

Our next template was specifically designed to avoid the chiral center as well as the *N*-demethylated metabolites, as illustrated in Table 17.2.[26] Replacing the chiral carbon with a nitrogen and cyclizing the amine to a piperidine ring provided compound **1**, which proved to be inactive. However, replacing the phenyl group with a benzyl resulted in our first achiral, potent triple reuptake inhibitor **2**. Replacing the metabolically labile 2,3-unsubstituted indole with an indazole to give compound **3** boosted SERT potency while diminishing NET. Changing the connectivity on the piperidine ring to provide analog **4** significantly decreased NET potency while the pyrrolidine compound **5** had substantially decreased SERT potency.

While varying potency with these changes suggested the ability to fine tune transporter balance through SAR, all of these templates remained potent hERG inhibitors with IC_{20} values at 1 μM or below. In addition, they remained potent inhibitors of both CYP 2D6 and 3A4. Several examples from this series were also tested *in vivo*, utilizing the tail suspension test (TST) for antidepressant-like activity and a locomotor assay (LMA) for the evaluation of psychomotor stimulation, and even with significant plasma exposures, these compounds displayed little to no activity in these behavioral models, suggesting limited BBB penetration.[27,28] In order to address the CYP and hERG liabilities, our optimization efforts focused on decreasing lipophilicity through the addition of heteroatoms and/or polar functionality. The resulting compounds from

Table 17.2 Potency, hERG, and CYP data for the aminopiperidines.

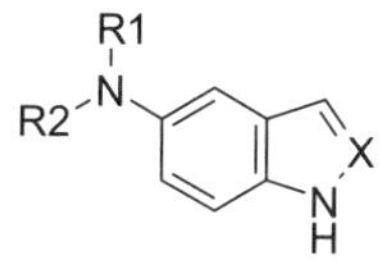

Cmpd	*X*	*R1*	*R2*	*SERT/NET/DAT* pK_i	*hERG* IC_{20}(μM)	*2D6/3A4* IC_{50} (μM)
1	C	Ph	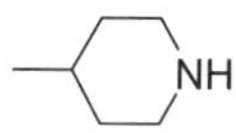	6.0/5.8/5.8	—	—
2	C	Bn	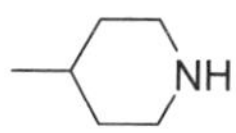	8.5/8.3/7.3	1.0	0.1/3.0
3	N	Bn	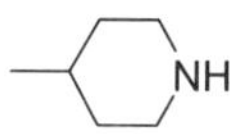	8.9/7.8/7.3	<1	0.5/0.4
4	C	Bn	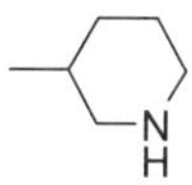	8.2/7.2/7.4	<1	0.3/5.0
5	C	Bn	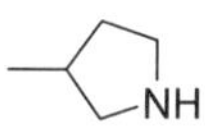	7.7/8.0/7.5	<1	2.0/1.9

this effort had further reduced permeability in the Caco-2 assay as well as moderate P-gp mediated efflux while not significantly improving the therapeutic window over hERG or CYP inhibition.[29]

17.6.2 Aminopiperazine Series

Since hERG interactions are often a function of both lipophilicity and basicity, we envisioned lowering the pK_a of the piperidine amine by replacement with a piperazine ring.[30,31] As illustrated in Figure 17.3, the first piperazine analog we prepared (**6**) was very potent and balanced at the three transporters.

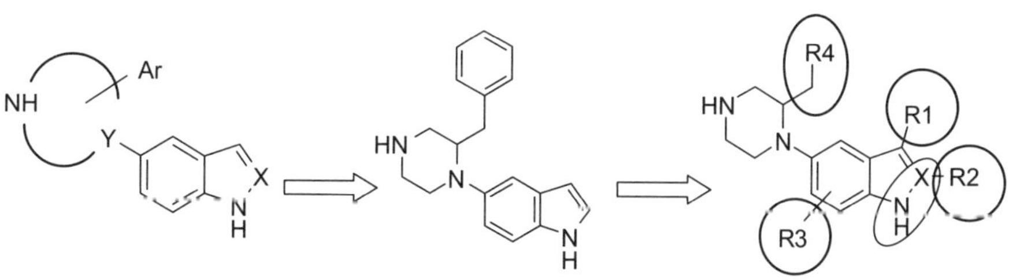

6

SERT/NET/DAT (pK_i):
8.0/8.0/8.2

Figure 17.3 Cyclization strategy leading to the piperazine series and sites for SAR exploration.

Table 17.3 Potency, microsomal clearance, CYP, and hERG data for the aminopiperazines.

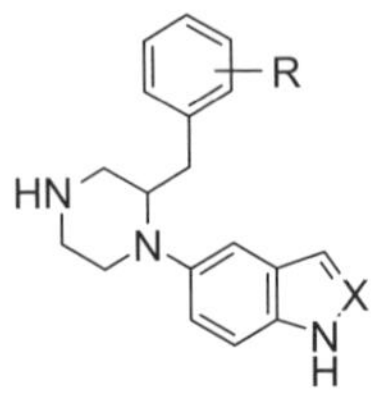

Cmpd	*X*	*R*	*SERT/NET/DAT* pK_i	*HLM* (*μL/min/mg*)	*2D6/3A4* IC_{50} (*μM*)	*hERG* IC_{20} (*μM*)
6	C	H	8.0/8.0/8.2	43	0.03/3.3	2.1
7	N	H	8.9/7.6/7.8	3.8	0.04/0.7	3.4
8	C	3-F	8.4/7.5/8.0	52	0.1/2.7	—
9	N	3-F	8.9/7.3/7.4	2.5	0.04/0.23	—
10	C	3-OCH_3	8.0/7.7/8.1	58	0.04/1.6	—
11	N	3-OCH_3	8.7/7.6/7.7	21	0.03/0.09	—

We assessed this compound *in vivo* in the rat and determined it had reasonable oral bioavailability but rapid clearance. We also evaluated it in the TST and found it to be very active at 10 mg/kg. At this point we began to develop the SAR, focusing on improving the clearance rate and decreasing the DDI potential. *In vitro* metabolite identification pointed to oxidation of the indole ring, in either the 2- or 3-position, as responsible for the rapid clearance. Our subsequent search for other heterocyclic replacements or bioisosteres for the indole were unsuccessful; only the indole and indazole rings retained sufficient potency. As illustrated in Table 17.3, the parent indole **6** has rapid clearance in human microsomes and is a potent inhibitor of 2D6. Fortunately, the reduced pK_a of the piperazine ring over the piperidine ring did have the desired effect on hERG inhibition, improving the IC_{20} to 2.1 μM. While clearance rate and 2D6 inhibition needed to be addressed, we felt this was a reasonable starting place given the high transporter potency and *in vivo* activity.

The indazole analogs **7**, **9** and **11** also had very good potency with a consistent trend of increasing SERT activity while slightly decreasing DAT. The indazoles also had considerably slower clearance rates, supporting our *in vitro* data implicating the indole oxidation pathway, as well as decreased hERG interaction. Despite these improvements, they were still very potent 2D6 and 3A4 inhibitors. Extensive SAR around the benzyl ring to address this revealed the 3-position as being the most tolerable to substitution; the 3-fluoro and 3-methoxy analogs **8**, **9** and **10**, **11** respectively, were tolerated for potency but had little effect on CYP inhibition. A further concern beyond the DDI liability was the potential to generate reactive metabolites of these 5-amino systems through metabolic oxidation to provide diiminoquinones, as illustrated in Figure 17.4.

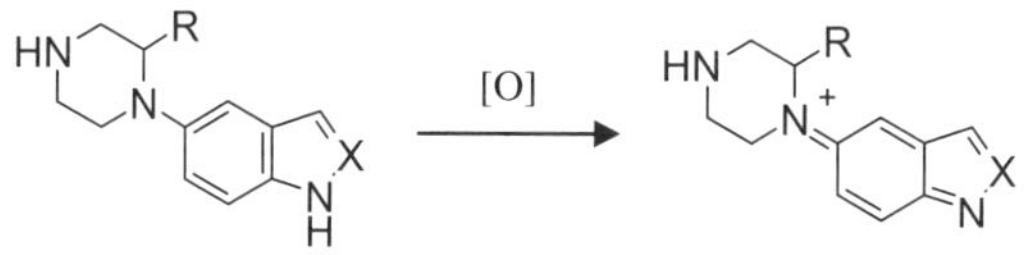

Figure 17.4 Potential structure of reactive metabolite generated via metabolic oxidation.

In fact, when both the parent indole and indazole compounds were evaluated *in vitro* for 3A4 time-dependent inactivation (TDI), one measure of the generation of reactive metabolites, they both gave a strong signal.[32] This was of considerable concern as even small amounts of reactive metabolites in humans have the potential to lead to lethal idiosyncratic toxicity.[33] Since we required either the indole or indazole core for potency, our remaining two strategies for avoiding the diiminoquinone formation were to, first, lower overall lipophilicity to decrease CYP involvement and, second, to decrease the electron density of the ring system to disfavor oxidative metabolic pathways. Both of these strategies were successful to a degree: changing the lipophilic benzyl group to a propyl retained most of the transporter potency and decreased 3A4 TDI; while the addition of heteroatoms into the side chain reduced TDI, they also diminished transporter potency.[34] Initial modeling and simulation with these compounds to project efficacious plasma exposure, coupled with the measurement of their 3A4 K_{inact} and K_i, suggested we might have a window to allow further progression for this series. From previous experience with an aromatic piperazine that entered the clinic, we knew that the piperazine nitrogen could be metabolically acetylated via *N*-acetyltransferase, which is well known.[35] Unsubstituted piperazines are typically protonated under physiological conditions and this charged species may somewhat disfavor metabolic oxidation of the indole/indazole ring. If the piperazine were acetylated it would no longer be charged and could then pose a higher risk of diiminoquinone formation. To assess this, we acetylated an analog which had a low risk of TDI *in vitro* and discovered that the acetylated version was associated with a much stronger TDI signal than the parent. Although we did not observe the acetylated metabolite *in vitro* with hepatocyte incubation, we felt the potential risk once in the clinic was too great and we therefore abandoned this series.

17.6.3 Pyrrolidine Series

Another series that developed from our cyclization strategy was the 3-pyrrolidines, as shown in Table 17.4.[36] Although this template removed the diiminoquinone liability associated with the piperazine series, it is also more lipophilic resulting again in high clearance and potent CYP inhibition. Surprisingly, the hERG inhibition for this initial compound **12** was moderate, with an IC_{20} of around 3 μM.

Table 17.4 Properties of the 3-pyrrolidinyl indole enantiomers.

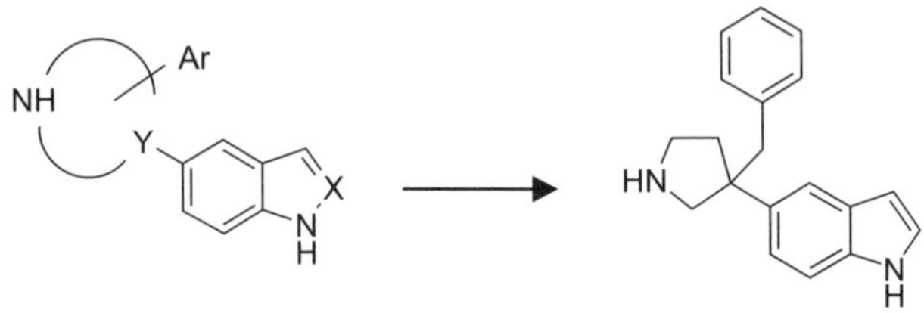

Cmpd	Chirality	*SERT/NET/DAT* pK$_i$	*HLM* *μL/min/mg*	*2D6/3A4* *IC$_{50}$ (μM)*	*hERG* *IC$_{20}$ (μM)*
12	Racemic	8.0/9.1/8.0	100	0.5/1.8	~3.0
13	"R"	8.3/9.1/8.0	153	0.3/0.8	—
14	"S"	7.3/7.8/7.1	—	0.2/4.5	—

While separation of the enantiomers by chiral HPLC revealed one isomer (**13**) as being more potent at all three transporters, there were no significant differences in CYP inhibition or clearance rates. As with previous efforts, we set out to decrease the lipophilicity through changes in the benzyl region, shown in Table 17.5, as well as heteroatom and polar substitutions to the core ring.

In order to probe for the optimal distance between the hydrophobes and the basic amine, we prepared the phenethyl homolog **15**. This helped to balance the SERT and NET potencies but was less potent at DAT. Furthermore, there were no improvements in microsomal clearance or CYP inhibition. Replacing the phenyl group with a pyridyl (**16**) decreased potency at all three transporters by 10-fold while significantly improving the CYP profile. Unfortunately, there was only a marginal improvement in clearance, with metabolism likely shifted to the core ring. In the piperazine series, replacement of the benzyl group with a propyl only slightly diminished potency while decreasing CYP interactions. In contrast, the propyl analog (**17**) in this series significantly reduced transporter potency while having the desired effect on both the microsomal rate and CYP interactions. The butyl analog (**18**) regained some of the transporter potency

Table 17.5 Potency, microsomal clearance, and CYP properties of the 3-pyrrolidinyl indoles.

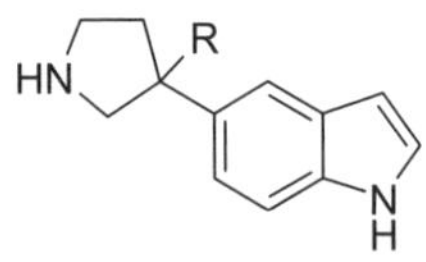

Cmpd	R	*SERT/NET/DAT* pK$_i$	*HLM* *(μL/min/mg)*	*2D6/3A4* *IC$_{50}$ (μM)*
15	CH_2CH_2Ph	7.9/7.8/7.2	105	0.3/0.7
16	Pyridin-2-ylmethyl	7.3/8.3/7.2	86.1	3.5/7.0
17	n-Propyl	7.2/7.2/6.4	2.3	3.3/1.7
18	n-Butyl	7.8/8.1/7.1	38.3	7.5/11.6

but the increased lipophilicity resulted in a higher clearance rate. It was clear that replacement of the benzyl group could address many of the undesired properties.

We next turned our attention to reducing the metabolic liability of the 2,3-unsubstituted indole core. Replacing the indole with indazole or benzothiophene was tolerated for potency but did not offer any other significant benefits.[37] Table 17.6 outlines some of the work attempting to block oxidation at the 2- or 3-position of the indole with polar functional groups to concomitantly lower lipophilicity.

Compared to the parent compound, the 2-carbomethoxy analog **19** boosted SERT potency while lowering both NET and DAT. Although esters are not generally considered to be metabolically stable groups, there was some improvement in the microsomal rate reinforcing our belief that the unsubstituted indole was at least partially responsible for the observed clearance. Placement of a nitrile in the 2-position provided compound **20**, one of the most potent SERT inhibitors to date, but was accompanied by a decrease in DAT potency. This nitrile stabilized the molecule towards human liver microsomes but the compound still carried a considerable 2D6 liability. Moving the nitrile to the 3-position to provide **21** significantly reduced DAT inhibition while having the desired effect of slowing clearance relative to the parent. We routinely measured permeability in Caco-2 cells and this analog was the first example to show a potential for P-gp mediated efflux with an efflux ratio (ER) of 11. Hoping to improve on the properties of the 2-cyano analog (**20**), we prepared the 2-carboxamide analog **22**. This compound not only had good potency at the transporters but we felt the balance between the potencies was also desirable. Consistent with the other 2-substituted compounds, the 2-carboxamide is more metabolically stable with a reduced CYP-related DDI potential. Unfortunately, the additional hydrogen bond donors of the amide, a

Table 17.6 Potency, microsomal clearance, and CYP properties of the 3-pyrrolidinyl indoles.

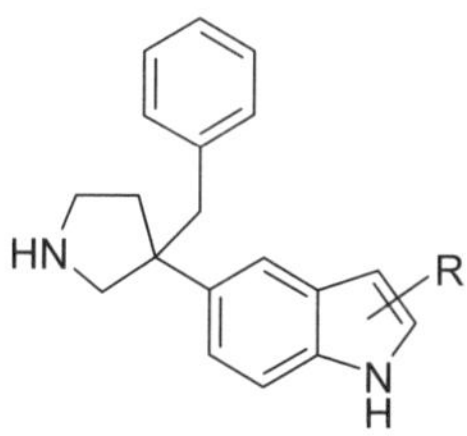

Cmpd	*R*	*SERT/NET/DAT* pK_i	*HLM* *(μL/min/mg)*	*2D6/3A4* IC_{50} *(μM)*	*BA/AB (ER)* $\times 10^{-6}$ *cm/sec*
12	H	8.0/9.1/8.0	100	0.5/1.8	13.7/16.7 (1)
19	2-CO_2CH_3	8.9/7.8/7.4	43.1	1.5/4.7	5.6/5.5 (1)
20	2-CN	9.6/8.3/7.3	0	0.9/4.8	7.8/7.9 (1)
21	3-CN	7.2/8.3/6.1	18.8	0.8/2.4	24.8/2.2 (11)
22	2-$CONH_2$	8.7/8.1/7.7	1.4	4.1/3.5	16.7/0.6 (28)

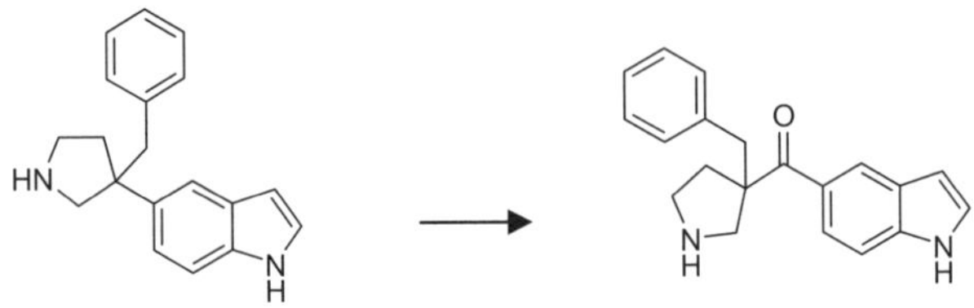

12

pK_i: 8.0/9.1/8.0
HLM: 100 μL/min/mg
hERG IC_{20}: ~3.0 μM

23

pK_i: 8.3/7.8/8.1
HLM: 30 μL/min/mg
hERG IC_{20}: ~6.1 μM

Figure 17.5 Potency, microsomal clearance, and hERG data associated with ketone insertion into the 3-pyrrolidinyl indole.

well-known contributor to P-gp mediated efflux, led to a decrease in overall permeability and an efflux ratio of 28, properties which would limit CNS exposure.[38]

17.6.4 Ketone Series

Since our attempts to mitigate high clearance and CYP inhibition through the addition of peripheral polar groups was limiting our ability to cross the BBB, we decided to incorporate polar functionality within the molecule. Towards this end, we inserted a carbonyl group between the indole and pyrrolidine ring to provide the ketone compound **23**, as shown in Figure 17.5.[39]

The potency and initial properties of this compound were very encouraging. While SERT and DAT activity were similar for both compounds, the NET activity associated with the ketone was decreased to provide an overall balance between the transporters. There was an improvement in microsomal clearance, as well as hERG, with no change in permeability. We took a multi-faceted approach to optimizing this series by addressing the hydrophobic benzyl region, the indole core, and the amine-containing ring simultaneously.[40] As shown in Figure 17.6, we contracted and expanded the pyrrolidine ring to evaluate the 4-, 5-, 6-, and 7-membered rings.

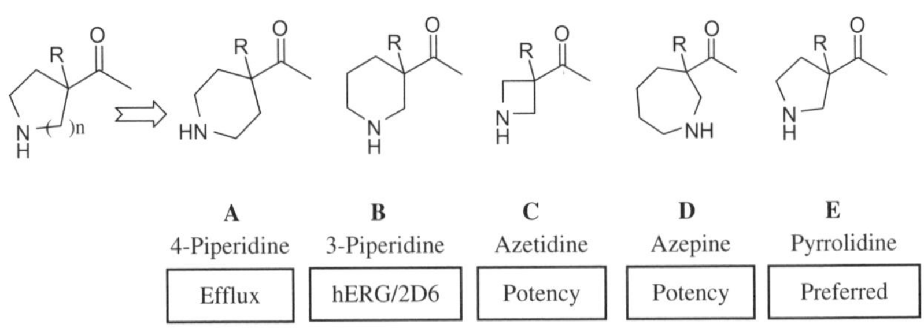

Figure 17.6 Overview of issues associated with the ring expansion and contraction of the 3-pyrrolidinyl ketone system.

While an achiral template was desirable, the 4-piperidine compounds (**A**) had P-gp mediated efflux which limited their BBB permeability and the azetidine analogs (**C**) lost considerable potency. The 3-piperidines (**B**), while potent, were strong inhibitors of both 2D6 and hERG and the 7-membered azepines (**D**) lost transporter potency. The 3-substituted pyrrolidines, template **E**, had the best overall profile to move forward into lead optimization. Traditional SAR around the benzyl group utilizing the Topliss tree approach was generally tolerated for potency but with no significant improvement to clearance or CYP properties.[41] As shown in Table 17.7, replacement of the benzyl group with alkyl groups provided very potent and balanced compounds.

The propyl analog **24** had slightly improved potency at SERT and NET with a decrease at DAT. More importantly, the microsomal clearance was lowered considerably along with significant improvement in the 2D6 and 3A4 liability. Increasing the size of the alkyl chain to the butyl analog **25** provided a boost in potency at all three transporters and maintained a good clearance rate but resulted in greater CYP inhibition. Further increasing the alkyl chain to provide the methylbutyl (**26**) and dimethylbutyl (**27**) analogs continued to improve transporter potency but was accompanied with greater CYP liability. We also addressed the indole liability through replacement with bioisosteres and alternate heterocycles, and in contrast to previous series, several different ring systems were tolerated. Utilizing the dimethylbutyl group as the side chain for its inherent DAT potency, we examined the properties for many of these, as shown in Table 17.8.

The dichloro fluorophenyl analog **28** showed very good transporter potency along with moderate clearance and a good CYP profile. There was also a slight improvement with respect to hERG inhibition with an IC_{20} of 2 μM. The isosteric thiophene **29** recovered the DAT potency of the indole and had improved clearance and CYP values. Also of particular interest were the chloroaniline (**31**) and benzothiophene (**32**) analogs due to their improved

Table 17.7 Potency, microsomal clearance, CYP, and hERG properties of the alkyl-substituted indole ketones.

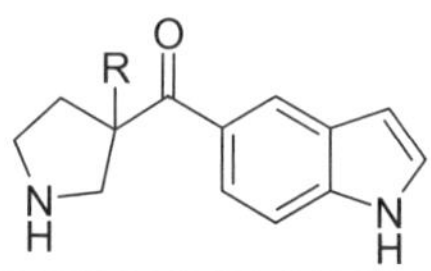

Cmpd	*R*	*SERT/NET/DAT* pK_i	*HLM* (*μL/min/mg*)	*2D6/3A4* IC_{50} (*μM*)	*hERG* IC_{20} (*μM*)
23	Benzyl	8.3/7.8/8.1	30	0.2/0.7	6.1
24	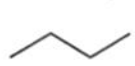	8.5/8.0/7.6	8.6	8.9/7.5	3.7
25	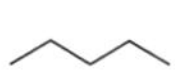	8.9/8.4/8.4	5.1	1.2/5.7	1.9
26	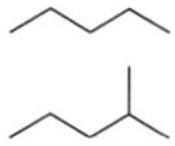	8.9/8.6/8.7	7.6	0.4/2.6	2.6
27	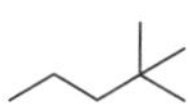	8.9/8.6/8.7	27.7	1.4/0.8	1.5

Table 17.8 Potency, microsomal clearance, CYP, and hERG properties associated with indole replacement in the ketone series.

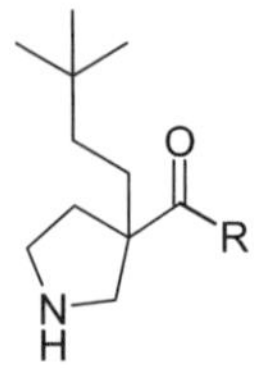

Cmpd	*R*	*SERT/NET/DAT* pK_i	*HLM* *(μL/min/mg)*	*2D6/3A4* *IC_{50} (μM)*	*hERG* *IC_{20} (μM)*
27	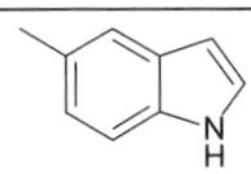	8.9/8.6/8.7	27.7	1.4/0.8	1.5
28	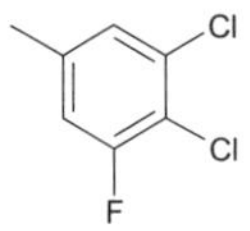	8.5/8.0/8.0	15.3	12.5/4.3	2.0
29	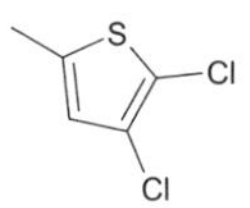	8.3/8.1/8.7	17	14.5/4.4	2.0
30	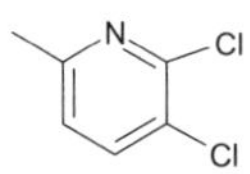	7.9/8.0/8.0	9	3.4/11.5	1.7
31	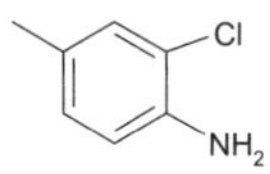	9.0/8.3/8.5	12	4.3/0.8	2.8
32	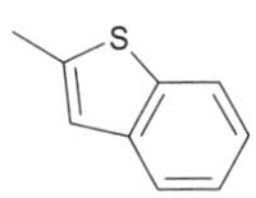	9.0/8.6/8.9	22	12.1/2.0	3.7
33	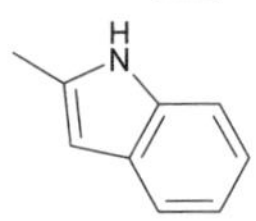	8.2/8.6/8.6	8	23.2/16.9	1.1

hERG profiles. Several of these compounds were scaled up and their enantiomers separated, as shown in Table 17.9.

In all cases examined, the transporter potency was associated with the (+) rotating isomer. While this isomer generally had slightly higher clearance rates than the racemate, there was an improvement in the CYP profile, especially with respect to 3A4 inhibition. The dichloro fluoro analog (+)**28** improved from an IC_{50} of 4.3 μM to 30 μM, the chloro aniline, (+)**31**, from 0.8 μM to 4.3 μM, and the benzothiophene enantiomer (+)**32** from 2 μM to 6.3 μM. With several novel and potent core rings identified in the 3-pyrrolidine series (Table 17.8), we reinvestigated the achiral 4-piperidines due to their ease of synthesis and low cost of goods. As shown in Table 17.10, several heterocyclic

Table 17.9 Potency, microsomal clearance, CYP, and hERG properties of single enantiomers in the ketone series.

Cmpd	*SERT/NET/DAT* pK_i	*HLM* *(μL/min/mg)*	*2D6/3A4* IC_{50} *(μM)*	*hERG* IC_{20} *(μM)*
(+) 28	8.8/8.3/8.4	25.1	16/30	1.4
(+) 31	9.4/8.7/8.8	12	5.5/4.3	3.3
(+) 32	9.1/8.6/8.8	25	15/6.3	4.0

Table 17.10 Potency, microsomal clearance, CYP, and hERG properties associated with heterocycle variations of the 4-piperidines.

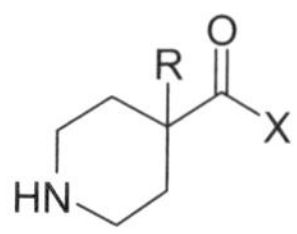

Cmpd	*X*	*R*	*SERT/NET/ DAT* pK_i	*HLM* *(μL/min/mg)*	*2D6/3A4* IC_{50} *(μM)*	*BA/AB (ER)* $\times 10^{-6}$ *cm/sec*
34	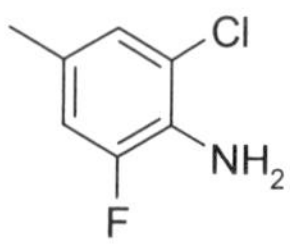		8.1/8.2/8.7	7.7	11.6/14.1	13.7/9.1 (1.5)
35	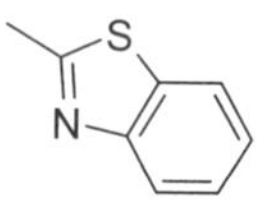	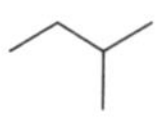	9.8/9.2/8.7	30	4.9/18	3.6/3.5 (1)
36	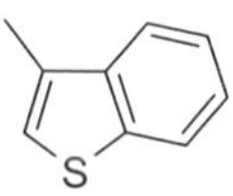	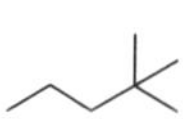	8.8/8.6/8.3	9.3	16/22	14.5/12.5 (1.)
37	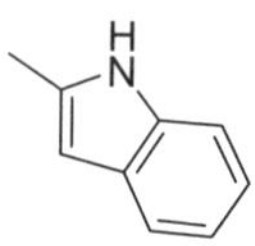	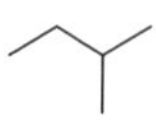	9.0/7.7/8.0	10.6	>50/>50	7/9.5 (0.7)
38	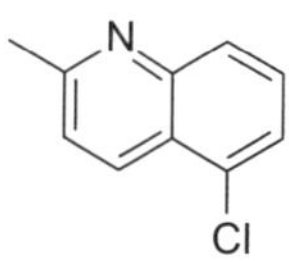	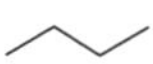	9.9/8.2/8.7	11.7	10.2/21.6	—

replacements in the 4-piperidine series were potent and displayed good *in vitro* drug-like properties.

All of these variations displayed high transporter potencies, low to moderate microsomal clearance, and significantly reduced CYP interactions compared to their indole progenitors. Of note is the fact that these heterocycles were no longer efflux substrates, as demonstrated by their apical and basal permeability

Table 17.11 Receptor occupancies of selected 3-pyrrolidines and 4-piperidines.

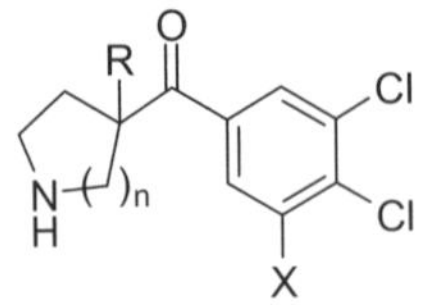

Cmpd	*X*	*n*	*R*	*SERT/NET/DAT* pK_i	*SERT/DAT* *% Occupancy*
(+) **39**	F	1	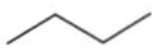	9.0/7.4/7.4	80:26
(+) **40**	H	1		9.4/8.6/8.9	80:34
41	H	2		9.3/7.5/8.5	80:40
(+) **42**	F	1		9.1/8.1/7.8	80:52
43	F	2	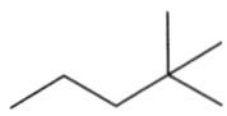	8.3/8.0/8.0	80:80

in Caco-2 cells. When the 4-piperidines were first evaluated in the indole series, template **A** (Figure 17.6), they were associated with a significant P-gp liability with efflux ratios generally greater than 10. While removing the hydrogen bond donor of the indole to provide compounds **35** and **36** provides a reasonable explanation for this observation, the aniline compound **34** maintains a similar capacity to hydrogen bond but has modest efflux potential as measured in Caco-2 cells.

At this point in the program, we were measuring free brain concentrations and had established robust *in vivo* occupancy assays for both SERT and DAT in the rat and mouse.[42] Through dose response, we established the dose that would result in 80% SERT occupancy and then with this dose, determined the DAT occupancy in a separate experiment. As shown in Table 17.11, we were able to identify several potent compounds that exhibit varying SERT:DAT occupancy ratios *in vivo* that may lead to differentiated antidepressant treatments. As a more functional measure of transporter occupancy, we also established microdialysis in the rat which allowed for the direct measurement of brain neurotransmitter levels. Several of these compounds displayed dose--dependent increases of serotonin in the prefrontal cortex and dopamine in the striatum, providing increased confidence to proceed with these compounds into the clinic.

17.7 Conclusion

Starting from two novel series of SNRIs derived from the duloxetine template and utilizing scaffold hopping techniques, we were able to discover several new series of potent TRIs. These successive templates evolved from the need to address specific drug-like issues such as microsomal clearance, CYP-related DDI potential, BBB penetration, hERG interactions, and chiral complexity. The pharmacophores responsible for these properties were modified to provide potent TRIs with developable *in vitro* profiles. Several of these compounds were evaluated *in vivo* (rat, mouse) for their ability to block SERT and DAT using radiolabelled PET ligands. Utilizing doses that provided 80% SERT occupancy, consistent with clinically efficacious SSRIs, we identified compounds with a range of DAT occupancies allowing us to evaluate the ability of these differentiated profiles to address some of the current unmet medical needs of antidepressant treatment such as low remission rates, delayed time to onset, and sexual dysfunction.

Acknowledgements

I would like to thank Luca Santarelli and Linda Bristow for their outstanding biology support, Xingrong Liu and Olga Vilenski for DMPK support, Cyril DeColle and Subbu Apparsundaram for *in vitro* pharmacology, the entire chemistry group (as exemplified in references 24, 26, 31, and 36), and finally Hoffman-La Roche for the opportunity to lead this effort.

References

1. (a) M. E. Thase, *J. Clin. Psychiatry*, 2006, **61**, 17; (b) R. C. Keesler, P. Berglung, O. Demler, R. Jin, D. Loretz, K. R. Merikangas, A. J. Rush and E. E. Walters, *J. Am. Med. Assoc.*, 2003, **289**, 3095.
2. W. F. Stewart, A. J. Ricci, E. Chee and S. R. Hahn, *J. Am. Med. Assoc.*, 2003, **289**, 3135.
3. (a) G. J. Marek, *Neurosci. Lett.*, 2008, **438**, 312–316; (b) E. Syvalahti, J. Penttila, H. Majasuo, E. P. Palvimaki, A. Laakso and J. Hietala, *J. Pharmacopsychiatry*, 2006, **39**, 1–8.
4. (a) I. Sillaber, M. Panhuysen, M. S. H. Henniger, F. Ohl, C. Kuehne, B. Puetz, T. Pohl, J. M. Deussing, M. Paez-Pereda and F. Holsboer, *Psychopharmacology*, 2008, **200**, 557–572; (b) B. Conti, R. Maier, A. M. Barr, C. M. Morale, X. Lu, P. P. Sanna, G. Bilbe, D. Hoyer and T. Bartfai, *Mol. Psychiatry*, 2007, **12**, 167–189.
5. (a) J. Savitz, I. Lucki and W. C. Drevets, *Prog. Neurobiol.*, 2009, **88**, 17–31; (b) J. H. Broadbear, T. Nguyen, I. J. Clarke and B. J. Canny, *Psychopharmacology*, 2004, **175**, 247–255; (c) S. Hesketh, D. S. Jessop, S. Hogg and M. S. Harbuz, *J. Endocrinol.*, 2005, **185**, 373–382.
6. D. V. West and M. Maes, *Neuroendocrinol. Lett.*, 1999, **20**, 11–17.
7. (a) R. J. Seo, H. MacPherson and A. H. Young, *Pharmaceuticals*, 2010, **3**, 3522–3542; (b) G. I. Papakostas, J. J. Worthington, III, D. V. Iosifescu, G. Kinrys, A. M. Burns, L. B. Fisher, C. H. Homberger, D. Mischoulon and M. Fava, *Depression Anxiety*, 2006, **23**, 178–181.
8. C. Pae, H. Tharwani, D. M. Marks, P. S. Masand and A. A. Patkar, *CNS Drugs*, 2009, **23**, 1023–1037.
9. K. N. Sokolski, J. C. Conney, B. J. Brown and E. M. DeMet, *Psychiatry Res.*, 2004, **125**, 81–86.

10. M. Jaronczyk, Z. Chilmonczyk, A. P. Mazurek, G. Nowak, A. W. Ravna, K. Kristiansen and I. Sylte, *Bioorg. Med. Chem.*, 2008, **16**, 9283–9294.
11. (a) M. H. Trivedi, M. Fava, S. R. Wisniewski, M. E. Thase, F. Quitkin, D. Warden, L. Ritz, A. A. Nierenberg, B. D. Lebowitz and M. M. Biggs, *N. Engl. J. Med.*, 2006, **354**, 1243–1252; (b) C. DeBattista, H. B. Solvason, J. Poirier, E. Kendrick and A. F. Schatzberg, *J. Clin. Psychopharmacology*, 2003, **23**, 27–30.
12. J. D. Croxtall and L. J. Scott, *CNS Drugs*, 2010, **24**, 245–262.
13. E. S. Friedman, S. R. Wisniewski, W. Gilmer, A. A. Nierenberg, A. J. Rush, M. Fava, S. Zisook, G. K. Balasubramani and M. H. Trivedi, *Depression Anxiety*, 2009, **26**, 612–621.
14. M. T. Treadway and D. H. Zald, *Neurosci. Biobehav. Rev.*, 2011, **35**, 537–55.
15. E. J. Nestler and W. A. Carlezon, *Biol. Psychiatry*, 2006, **59**, 1151–1159.
16. P. Willner, *Int. Clin. Psychopharmacology*, 1997, **12**, Suppl. 3, S7–14.
17. (a) R. Torta, C. Berra, L. Binaschi and R. Borio, *Supportive Care in Cancer*, 2007, **15**, 539–546; (b) M. C. Hardoy and M. G. Carta, *Clinical Practice and Epidemiology in Mental Health*, 2010, **6**, 1–3.
18. H. Lavretsky, M. Kim, A. Kumar and C. F. Reynolds, III, *J. Clin. Psychiatry*, 2003, **64**, 1410–1414.
19. (a) C. DeBattista, B. Solvason, J. Poirier, E. Kendrick and E. Loraas, *J. Clin. Psychiatry*, 2005, **66**, 844–848; (b) A. H. Clayton, J. K. Warnock, S. G. Kornstein, R. Pinkerton, A. Sheldon-Keller and E. L. McGarvey, *J. Clin. Psychiatry*, 2004, **65**, 62–67.
20. T. A. Zhang, A. N. Placzek and J. A. Dani, *Neuropharmacology*, 2010, **59**, 431–436.
21. (a) L. L. Howell, F. I. Carroll, J. R. Votaw, M. M. Goodman and H. L. Kimmel, *J. Pharmacol. Exp. Ther.*, 2007, **320**, 757–765; (b) R. I. Desai, T. A. Kopajtic, D. French, A. H. Newman and J. L. Katz, *J. Pharmacol. Exp. Ther.*, 2005, **315**, 397–404.
22. J. H. Meyer, A. A. Wilson, S. Sagrati, D. Hussey, A. Carella, W. Z. Potter, N. Ginovart, E. P. Spencer, A. Cheok and S. Houle, *Am. J. Psychiatry*, 2004, **161**, 826–835.
23. (a) S. M. Learned-Coughlin, M. Bergstrom, I. Savitcheva, J. Ascher, V. D. Schmith and L. Bengt, *Biol. Psychiatry*, 2003, **54**, 800–805; (b) J. H. Meyer, V. S. Goulding, A. A. Wilson, D. Hussey, B. K. Christensen and S. Houle, *Psychopharmacology*, 2002, **163**, 102–105.
24. R. Greenhouse, S. Jaime-Figueroa, L. Raptova, D. C. Reuter, K. A. Stein and R. J. Weikert, 2005, WO 2005118539.
25. Z. Chen and P. Skolnick, *Exp. Opin. Invest. Drugs*, 2007, **16**, 1365–1377.
26. P. Iyer, M. C. Lucas, R. C. Schoenfeld and R. J. Weikert, 2009, WO 2009118254.
27. J. F. Cryan, C. Mombereau and A. Vassout, *Neurosci. Biobehav. Rev.*, 2005, **29**, 571–625.
28. T. Romeas, M. C. Morissette, O. Mnie-Filali, G. Piñeyro and S. M. Boye, *Psychopharmacology*, 2009, **205**, 293–303.

29. M. C. Lucas, D. S. Carter, H. Cai, E. K. Lee, R. C. Schoenfeld, S. Steiner, M. Villa, R. J. Weikert and P. S. Iyer, *Bioorg. Med. Chem. Lett.*, 2009, **19**, 4630–4633.
30. (a) Z. Rankovic, *Abstracts of Papers*, 233rd ACS National Meeting, Chicago, IL, United States, March 25–29, 2007, MEDI-463; (b) D. J. Diller, *Curr. Comput.-Aided Drug Des.*, 2009, **5**, 106–121.
31. D. S. Carter, M. C. Lucas, R. C. Schoenfeld and R. J. Weikert, 2009, WO 2009101018.
32. (a) S. Zhou, S. Y. Chan, B. C. Goh and E. Chan, *Clin. Pharmacokinet.*, 2005, **44**, 279; (b) A. S. Kalgutkar and J. R. Soglia, *Exp. Opin. Drug Metab. Toxicol.*, 2005, **1**, 91.
33. (a) D. A. Smith, R. S. Obach, D. P. Williams and B. K. Park, *Chem.-Biol. Interact.*, 2009, **179**, 60–67; (b) J. C. Erve, *Expert Opin. Drug Metab. Toxicol.*, 2006, **2**, 923–946.
34. D. S. Carter, H. Cai, E. K. Lee, P. S. Iyer, M. C. Lucas, R. Roetz, R. C. Schoenfeld and R. J. Weikert, *Bioorg. Med. Chem. Lett.*, 2010, **20**, 3941–3945.
35. J. Rawal, R. Jones, A. Payne and I. Gardner, *Xenobiotica*, 2008, **38**(9), 1219–1239.
36. M. C. Lucas, R. C. Schoenfeld and R. J. Weikert, 2009, WO 2009115427.
37. L. M. Bannwart, D. S. Carter, H. Cai, J. C. Choy, R. Greenhouse, S. Jaime-Figueroa, P. S. Iyer, C. J. Lin, E. K. Lee, M. C. Lucas, S. M. Lynch, A. M. Madera, A. Moore, K. O. Ozboya, L. Raptova, R. Roetz, R. C. Schoenfeld, K. A. Stein, S. Steiner and M. Villa, *Bioorg. Med. Chem. Lett.*, 2008, **18**, 6062–6066.
38. T. J. Raub, *Mol. Pharm.*, 2006, **3**, 3–25.
39. P. Iyer, C. J. Lin, M. C. Lucas, S. M. Lynch, A. M. Madera, K. E. Ozboya, R. C. Schoenfeld and R. J. Weikert, 2009, WO 2009153178.
40. M. C. Lucas, R. J. Weikert, D. S. Carter, H. Cai, R. Greenhouse, P. S. Iyer, C. J. Lin, E. K. Lee, A. M. Madera, A. Moore, K. Ozboya, R. C. Schoenfeld, S. Steiner, Y. Zhai and S. M. Lynch, *Bioorg. Med. Chem. Lett.*, 2010, **20**, 5559–5566.
41. J. G. Topliss, *J. Med. Chem.*, 1972, **15**, 1006–1111.
42. X. Liu, O. Vilenski, J. Kwan, S. Apparsundaram and R. Weikert, *Drug Metab. Dispos.*, 2009, **37**, 1548.

CHAPTER 18

Discovery of Multi-Target Agents for Neurological Diseases via Ligand Design

MARIA LAURA BOLOGNESI,[a,*] CARLO MELCHIORRE,[a] CORNELIS J. VAN DER SCHYF[b] AND MOUSSA YOUDIM[c]

[a] University of Bologna, Italy; [b] Northeast Ohio Medical University (NEOMED), Rootstown, OH, USA; [c] Technion Israel Institute of Technology, Haifa, Israel
*Email: marialaura.bolognesi@unibo.it

18.1 The Rationale for Multi-Target Ligands in Alzheimer's and Parkinson's Diseases

Alzheimer's disease (AD) and Parkinson's disease (PD) are degenerative disorders that seriously and progressively impair the functions of the nervous system of specific brain regions.[1] They are the two most common age-related neurodegenerative disorders (with AD being much more prevalent), and represent a growing public health concern.[2] In fact, it has been estimated that, as the population continues to age, treatment of AD and PD patients will consume most of the gross national product of Western countries by the latter decades of this century. Although the two diseases are very distinctive in terms of pathology, they share a similar history in terms of strategies for their treatment. Despite massive ongoing attempts to produce disease-modifying medicines, current treatments still rely on neurotransmitter replacement

RSC Drug Discovery Series No. 21
Designing Multi-Target Drugs
Edited by J. Richard Morphy and C. John Harris

Published by the Royal Society of Chemistry, www.rsc.org

therapy.[3] This approach is based on an over-simplistic view of brain neurochemistry, which dominated drug discovery scenarios in the 1980s.[4] The observed relationship between cholinergic and dopaminergic dysfunctions and the neuropathological features of AD and PD symptomatology provided the rationale for developing agents aimed at improving cholinergic or dopaminergic neurotransmission, respectively. Thus, the AD and PD drugs available today exert cholinergic or dopaminergic effects that only impact the disease symptoms and do not blunt the progression of the disease.[5] Therefore, although initially helpful in improving cognitive, behavioral, and functional impairments, they are often ineffective for ongoing administration and in some cases produce severe side effects. In recent years, huge gains in basic science knowledge have broadened the horizons of conventional drug discovery, inspiring more mechanism-based therapies. Intriguingly, attention has been drawn to the fact that AD and PD neurodegenerative pathways are characterized by many common pathogenic molecular denominators. These include protein misfolding and aggregation, phosphorylation impairment, oxidative stress and free radical formation, metal dysregulation, and mitochondrial dysfunction, all occurring concurrently.[6] From time to time, these pathophysiological events have been the targets of various pharmaceutical efforts; however, none of them has so far been translated into a new disease-modifying treatment. As a consequence of the numerous late-stage trial failures in the last 10 years of drug candidates targeted towards sole elements of these diseases' hypothesized etiologies, AD and PD have emerged as two of the most frustrating and uncertain disease states for current drug discovery. This has forced a further reconsideration of the strategies adopted thus far. We conclude that both diseases are global brain network diseases, multi-factorial in nature, and resulting from diverse molecular abnormalities, instead of single target/gene defect. This view explains why the traditional 'one protein, one target, one drug' strategies have failed and suggest the development of drugs that simultaneously affect more of the multiple targets in the neurodegenerative cascade.[7–10] This approach has led us and others to exploit a new paradigm in medicinal chemistry, the 'multi-target directed ligand' (MTDL) design strategy.[11] The successful examples of many marketed multi-target drugs for cancer and depression have increasingly convinced the pharmaceutical community that modulating a multiplicity of targets can be an asset in treating complex disorders. But we believe that certain peculiarities of AD and PD pathology further strengthen the therapeutic validity of multi-target agents for their treatment. An important issue in AD and PD drug discovery is that aged patients are susceptible to a wide range of concomitant medical conditions (co-morbidity), including depression, hypertension, vascular diseases, or diabetes. This means that they take several drugs at the same time, with a greater risk of drug–drug interaction and toxicities. The mean number of drugs taken daily in such a polypharmaceutical approach has been estimated at 4.0 in men and 4.7 in women, with an expected increase in these numbers in the future.[12] The risks associated with polypharmacy in the geriatric population have thus been recognized as particularly critical.

There is therefore a need for therapeutic tools to be tailored to their specific conditions. An MTDL could be one such tool. Furthermore, simplifying the therapeutic regimen with a single multi-target agent could add significant value for elderly patients who have difficulty with compliance and therapy adherence.[13,14]

In the last 10 years, by following this strategy and exploiting our expertise and experience in the field, we and others have developed several ligands able to hit multiple AD and PD biological targets simultaneously.[15]

18.2 Discovery of Memoquin, a Multi-Target Lead Candidate for AD

18.2.1 The Polyamine Scaffold: A Universal Template for Designing Multi-Target Ligands

As set out above, MTDLs are likely to be more beneficial than single target agents in treating AD and PD. But the question remains: how best to develop them? Two conceptually different strategies have been exploited to date: a serendipitous screening approach and a rational approach.[16] The first is the most common, especially in the industrial settings, and involves compound libraries known to be active against one given target being profiled against a set of targets of interest. The second approach has been elegantly discussed by Morphy, and is also referred to as a 'framework combination'. It involves combining two molecular frameworks into a new single chemical entity.[17,18] Its main drawback is that the resulting molecules are more likely to have a high molecular weight and less likely to have oral drug-like properties.[19]

The rational approach would be facilitated if there were a single molecule able to modulate any target, thus functioning as a pharmaceutical skeleton key. Any lead molecule endowed with such properties can be dubbed a 'universal template' because it can provide, through proper manipulation, new molecules with high affinity and selectivity for any target. Polyamines are considered to be universal templates (or 'passepartout') because, thanks to their flexibility and charges, they can assume different conformations and interact with the anionic sites of various macromolecules (nucleic acids, structural proteins, or enzymes). From a drug design perspective, we see polyamines as the skeleton on which it is possible to build different functionalities in order to obtain pharmacological agents with a novel mechanism of action, as schematically shown in Figure 18.1.[20]

In support of this paradigm, we can consider the cases of benextramine, methoctramine, and caproctamine. The first two share a tetramine skeleton carrying terminal *o*-methoxybenzyl groups. Benextramine (**1**),[21] is an alpha-adrenegic, neuropeptide Y,[22] and H_2-histaminergic receptors antagonist,[23] while methoctramine (**2**),[24] which lacks the disulfide bridge, shows a multi-target profile towards cholinergic metabotropic[25] and ionotropic receptors.[26] Furthermore, it interacts directly with G proteins.[27] Caproctamine (**3**) is a dual

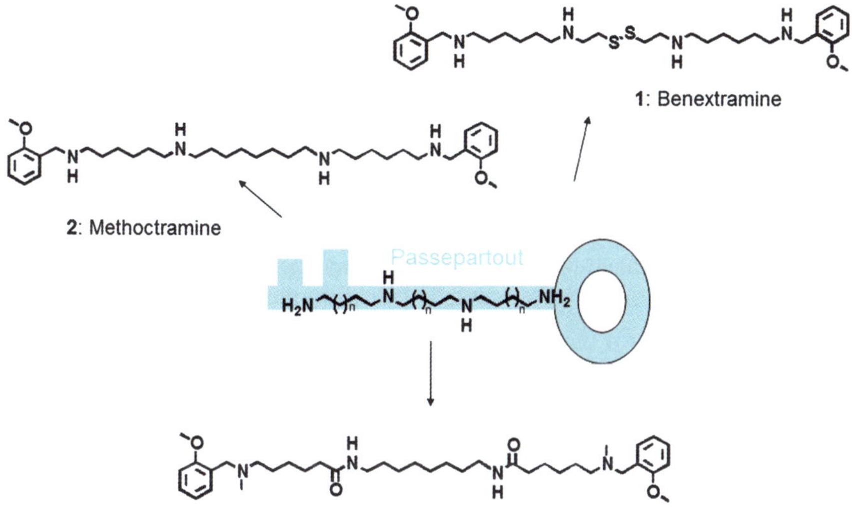

Figure 18.1 Polyamines as universal or 'passepartout' templates. Due to their flexibility and charges of polyamines, these scaffolds can assume different conformations and interact with anionic sites of nucleic acids, proteins, or enzymes.

cholinergic ligand for AD, which acts by inhibiting acetylcholinesterase (AChE) and antagonizes the M_2 muscarinic autoreceptors.

A polyamine cannot be considered an orthodox drug-like molecule when analyzed under Lipinski's Rule of Five (Ro5). But, as already noted by Lipinski, there are compound classes with structural features that allow them to act as substrates for naturally occurring transporters. Thus, carrier-mediated uptake of a drug is an exception to which the Ro5 does not apply.[28] Most of these compounds are natural products or derivatives thereof and this could also be true for polyamines. To support the drug-likeness of polyamines, it is worth mentioning that several derivatives have been developed as anti-proliferative agents,[29] neurotransmitter receptor ligands,[30] neuroprotectants,[31] anti-prion chemotherapeutics, and anti-parasitic compounds.[32]

18.2.2 Memoquin's *In Vitro* and *In Vivo* Profile

Building on the universal template approach, we initiated a drug discovery program aimed at the identification of ligands with a predefined multi-target mechanism of action against AD. The targets selected were AChE, Aβ, and oxidative stress, which are interrelated in AD pathogenesis. A well-established connection involves Aβ and synaptic AChE, shown to accelerate amyloid fibril formation.[33,34] Aβ and oxidative stress are linked in a vicious cycle, since not only can Aβ induce oxidative stress, but its activity is also increased as a consequence of oxidative stress.[35]

We reasoned we could convert the AChE inhibitor caproctamine into such a rationally designed MTDL, by incorporating into its backbone the benzoquinone fragment of the mitochondrial antioxidant coenzyme Q (CoQ). CoQ has demonstrated numerous beneficial properties against AD.[36,37] Its benzoquinone moiety, carrier of its radical-scavenging activity, was thus inserted to replace the internal amide groups of **3**. This modification afforded a 2,5-bis-diaminobenzoquinone nucleus, where the two nitrogen atoms at positions 2 and 5 have amide-like character, resembling the amide groups of **3**. Furthermore, the same nucleus due to a resonance effect affords a hydrophobic and planar system able, in principle, to disrupt the fibrillogenetic processes involving Aβ.

From a lead generation campaign, we identified **4** (memoquin) as our lead candidate for further investigation (Figure 18.2). Numerous modifications to the quinone scaffold, substitution of the tertiary amine, shortening or elongation of the polymethylene linkers, and variation of the benzene to different heteroaromatic rings were all found to compromise activity regardless of an α-methyl benzyl substituent.[38,39] Methodologies for the synthesis, separation,

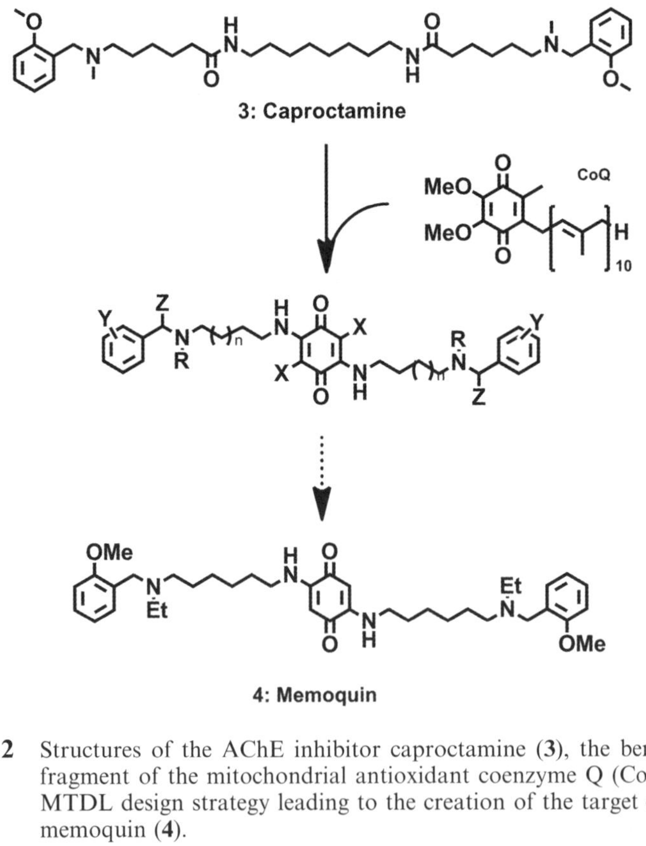

Figure 18.2 Structures of the AChE inhibitor caproctamine (**3**), the benzoquinone fragment of the mitochondrial antioxidant coenzyme Q (CoQ) and the MTDL design strategy leading to the creation of the target compound, memoquin (**4**).

Table 18.1 Activities of **4** at investigated molecular targets.

Activity	*Assay*	*4 (memoquin)*
AChE inhibition	Ellman's test	$IC_{50} = 1.55 \pm 0.11$ nM; $K_i = 2.60 \pm 0.48$ nM
Aβ aggregation (AChE-induced) inhibition	Thioflavin T fluorescence assay	$IC_{50} = 28.3 \pm 0.30$ μM
Aβ spontaneous aggregation inhibition	Thioflavin T fluorescence assay	$IC_{50} = 5.93 \pm 0.33$ μM
Neuroprotective activity against Aβ oligomer-induced toxicity	MTT	70% of protection at 1 μM
Reduction by NQO1		$V_{max} = 3.48$ mM/min/mg; $K_M = 12.7$ μM
Antioxidant profile in SHSY5Y, expressing high levels of NQO1	Intracellular ROS measurement	42% of reduction at 3 μM

purification, and analysis of **4**, which achieve excellent versatility, yield, purity, and HPLC resolution were developed.[40] Furthermore, its photostability profile was investigated, with encouraging results.[41] Table 18.1 shows a summary of all memoquin's (**4**) *in vitro* activities.

18.2.2.1 AChE Inhibition

The inhibitory potency of **4** against human AChE was remarkable ($IC_{50} =$ 1.55 nM), and significantly improved with respect to **3** (170 nM). This value is also 10 times lower than that displayed by the AChE inhibitor donepezil,[42] which is the most potent among the marketed AD drugs. Purposely addressed kinetic studies established that **4** acts as a mixed-type inhibitor, with a K_i value of 2.60 nM. These results perfectly matched the docking and molecular dynamics analysis, which showed that the benzylammonium ends of **4** were able to contact Trp86 and Trp286 belonging to the catalytic and peripheral sites of the enzyme, respectively.[43] This structural feature could be responsible for the experimentally observed capability of **4** to inhibit AChE-induced Aβ aggregation (see below).

18.2.2.2 Aβ-Aggregation Inhibition

In vitro studies[42] revealed that Aβ self-assembly into amyloid oligomers and fibrils is facilitated by the presence of AChE and it has been confirmed that this mechanism may operate also *in vivo*.[44] It has also been demonstrated that AChE inhibitors binding at the PAS can interfere with the enzyme pro-aggregating action.[34] We thus evaluated the inhibitory profile of **4** against AChE-induced amyloid aggregation by a thioflavin T-based fluorometric assay.[42] As shown in Table 18.1, compound **4** at a concentration of 100 μM blocked Aβ aggregation at a percentage (87%) very similar to that displayed by propidium (82%), which is a specific inhibitor of PAS.[45] Notably, the calculated IC_{50} values for the two inhibitors were 28.3 ± 0.3 μM and 12.6 ± 0.3 μM, respectively.

Based on the anti-aggregating properties of similar planar molecules with aromatic end groups,[46–48] the ability of **4** to blunt Aβ(1–42) spontaneous aggregation was also studied.[38,49] Data in Table 18.1 show that **4** halts Aβ aggregation with a 75% inhibition at 10 μM. Importantly, the effect was concentration-dependent with an IC_{50} value of 5.93 ± 0.33 μM, which makes **4** among the most effective small molecules that inhibit protein–protein interactions.[50]

To better clarify the effects of **4** on Aβ(1–42) polymerization, a circular dichroism (CD) assay was performed. At a concentration of 10 μM (1:5 inhibitor/Aβ) **4** caused a concentration-dependent decrease in the rate of the Aβ-transitional exponential phase, suggesting an inhibiting effect on the adoption of the β-sheet secondary structure. We interpreted this behavior as evidence of the ability of **4** to disrupt hydrophobic and hydrophilic interactions by hydrogen bonding either at the C or N terminus of Aβ.[51] The neuroprotective effects of **4** against Aβ(1–42)-induced toxicity were also determined in human neuronal-like SH-SY5Y cells. We used a range of concentrations of tested compounds (0.1–1 μM) that did not affect neuronal viability. As shown in Table 18.1, compound **4** at 1 μM exerted a strong inhibition (70%) of oligomer-induced neurotoxicity.[49]

Altogether, these results suggest a possible interaction of **4** with Aβ oligomers that would mask their toxic conformation or induce subtle structural changes, resulting in their decreased interaction with cellular substrates that mediate Aβ toxicity. Although future studies are necessary to verify this mechanism, the demonstration that **4** delays fibril formation indicates at least that it interacts with Aβ oligomers, in agreement with our initial hypothesis.

18.2.2.3 Antioxidant Activity

Before further discussion of compound **4**, some general considerations regarding quinone antioxidant properties should be pointed out. For several natural and synthetic quinones, it has been demonstrated that their antioxidant properties are associated mainly with their hydroquinone forms, since the quinones, in principle, cannot scavenge radicals.[52] NAD(P)H:quinone oxidoreductase 1 (NQO1) is the enzyme that physiologically reduces quinones to hydroquinones, in a single two-electron step. In addition to yielding substrates for phase II conjugation reactions and promoting excretion, this process bypasses the potentially toxic semiquinone radical intermediates. What was particularly attractive to us was that NQO1 is up-regulated in AD[53] and specifically co-localizes with AD pathological hallmarks.[54] We hypothesized that, being reduced by NQO1 into the corresponding hydroquinone, **4** might exert its beneficial antioxidant activity particularly in those brain regions affected by AD pathology.[55] On this basis, we tested whether **4** could be a substrate for NQO1. The apparent V_{max} and K_m values of 12.7 μM and 3.48 μmol/min/mg, respectively, showed that **4** acts similarly to menadione, a typical substrate of NQO1.[38]

As expected, the antioxidant cellular activity of **4** is markedly influenced by the overexpression of NQO1. In fact, in two different human neuronal cell lines (SH-SY5Y and T67) **4** in its oxidized form shows a low basal antioxidant activity.[38,56] Conversely, when the experiment was repeated by pre-treating cells with sulforaphane, which increases NQO1 activity by 50%,[57] a significant and dose-dependent reduction of ROS formation (1–3 μM) with respect to the untreated cells was observed. This result clearly confirms that the ability of **4** to prevent free radical toxicity is dependent upon its metabolic transformation into the hydroquinone form via NQO1.[38]

18.2.2.4 In Vivo *Effects*

The promising *in vitro* results showed by **4** encouraged us to perform an *in vivo* investigation. Being aware that selection of the correct mouse model in AD is a very critical issue, we turned our attention to an animal model which we considered suitable to profile the multi-target activities of **4**, the AD11 mouse. These mice, which express antibodies directed against NGF, are characterized by a progressive neurodegenerative phenotype defined by the simultaneous deposition of Aβ, intracellular neurofibrillary tangles, and by a marked cholinergic depletion.[58–60] We evaluated the effects of treatment with **4** on cholinergic deficit, hyperphoshorylated tau levels, and cognitive impairment.

To determine whether **4** can rescue the cholinergic deficit by reducing the loss of choline acetyltransferase positive neurons in the basal forebrain, the compound was administered first to AD11 mice at a very early stage of neurodegeneration (from 1.5 months to 2 months of age). The treatment fully reverts cholinergic impairment. More interestingly, the same results were obtained in mice aged 15 months, where the treatment led to an analogous improvement.[61]

In 15-month-old mice, where extracellular deposition of Aβ-reactive material is evident, the oral administration of **4** resulted in a significant attenuation of plaques in comparison with transgenic controls. Although the number of plaques did not diminish to the level (equal to zero) shown by age-matched wild-type animals, the morphology of plaques residual in mice treated with **4** was different from that observed in untreated AD11 mice.[61]

Hyperphosphorylated tau is present as early as two months of age in some brain regions in AD11 mice, whereas in 15–17-month-old mice nearly 90% of cortical neurons display phosphorylated tau. At all ages, the overall increase in tau hyperphoshorylation, was either completely (at 2 months of age) or partially (at 15 months of age) prevented by treatment with **4**.

Furthermore, the efficacy of **4** in reverting cognitive deficits was assessed in AD11 mice after oral administration by means of the object recognition test.

The same test was applied to non-transgenic mice treated with **4** 20 minutes before the administration of scopolamine. After administration of **4**, neither AD11 nor scopolamine-treated mice showed signs of amnesia, being able to distinguish familiar objects from new ones.[61]

Notably, memoquin is currently tested using the triple transgenic 3xTg-AD models, which could be suitably adopted to further characterize the *in vivo* profile of this promising multi-target drug candidate.

18.3 Alkylxanthines as Dual-Target-Directed Drug Candidates for PD

18.3.1 Coffee, Caffeine, Adenosine Receptors, and Monoamine Oxidase: The PD Connection

18.3.1.1 PD's Multifactorial Nature

Current pharmacotherapies for PD are mostly focused on symptom mitigation, and are inadequate to modify disease progression. This fact alone, amongst others, have energized efforts to develop agents that address newly discovered drug targets in the pathophysiological pathway of this disease. The multifactorial nature of PD offers an array of molecular drug targets and several treatment strategies have been suggested in recent reviews.[62] Some of these newly developed therapies offer novel approaches to provide relief of PD symptoms while others are aimed at protecting against the underlying neurodegenerative processes, therefore offering the potential to modify the disease and blunt its progression. Therapies that act at multiple targets, and offer the option to provide both symptomatic and neuroprotective benefits, may significantly advance the treatment options of complex neurodegenerative diseases such as PD.[63]

18.3.1.2 Monoamine Oxidase-B Inhibition

Among the most common adjunctive drugs to standard levodopa treatment of PD are inhibitors of the monoamine oxidase B enzyme (MAO-B).[64] Oxidative deamination of many endogenous amines in the brain is catalyzed by MAO-B (and to a lesser extent by MAO-A) and this reaction has been identified as a major catabolic pathway for dopamine in the striatum.[65–67] Notably, inhibition of these enzymes in the brain slows the depletion of dopamine stores and elevates the levels of endogenously produced synaptic dopamine, as well as dopamine derived from exogenously administered levodopa.[67,68] In addition, inhibitors of MAO also may exert a neuroprotective effect by decreasing the production of potentially hazardous catalytic by-products (including reactive oxygen species) of dopamine metabolism in the brain.[69] Considering research data that clearly suggest that MAO-B activity in the human brain increases with age[70–72] while MAO-A activity remains relatively constant throughout life,[73] it is notable that inhibition of MAO-B (rather than -A) is especially relevant in PD.

18.3.1.3 Adenosine A_{2A} Receptor Inhibition

In addition to the rather direct link of MAO-B and dopamine with PD, antagonists of the adenosine A_{2A} receptor are emerging as a contender class of promising anti-parkinsonian agents and may evolve into a leading candidate class for the non-dopaminergic treatment of PD.[74–76] In order to understand the relationship of A_{2A} antagonism and mitigation of PD symptoms, it needs to be understood that A_{2A} agonists decrease the binding affinity of D_2 agonists for the D_2 receptor in the striatum and also reduce the G-protein coupling activity of the D_2 receptor.[77,78] The antagonism of D_2 receptors by activation of A_{2A} receptors may also occur at the second messenger level or further downstream in the signal transduction pathway since both D_2 and A_{2A} receptors couple to adenylate cyclase and other messenger systems.[74] Blockade of the A_{2A} receptor in striatopallidal neurons potentiates D_2 receptor-mediated neurotransmission and therefore mitigates the effects of striatal dopamine depletion in PD.[75,79] As a result, antagonism of the A_{2A} receptor partially but reliably restores motor activity in animal models of PD.[80] Notably, it has been observed that motor stimulation afforded by A_{2A} antagonists is still present in D_2 knock-out mice, suggesting that A_{2A} receptor antagonism also exerts anti-parkinsonian effects that are independent of an interaction with D_2 receptors.[81] The antagonistic relationship between these receptors as a first point of departure, and the distinctive CNS distribution of A_{2A} receptors in brain, provide the basis for targeting the A_{2A} receptor as a treatment strategy for PD. Since the therapeutic benefits of A_{2A} antagonists are additive to that of dopamine replacement therapy (including levodopa therapy), amalgamated use of these two modalities may offer the option for dose reduction of dopaminergic drugs and the associated occurrence of dopaminergic side effects.

Of significant importance is emerging evidence that suggests that A_{2A} antagonists may also possess neuroprotective properties and may prevent the development of dyskinesias that are often associated with levodopa treatment.[82,83] Based on these observations, dual-target-directed drugs, i.e. compounds that simultaneously inhibit MAO-B and antagonize A_{2A} receptors, may have significant value as disease-modifying agents in the management of PD.

As a prototype for this approach, caffeine (**5**) and the consumption of caffeinated coffee have been shown to correlate with a reduced risk of developing PD in men and in women who have not taken postmenopausal estrogens.[84] Similarly, *trans*-8-(3-chlorostyryl)caffeine (CSC) (**6**, Figure 18.3), a well-known A_{2A} antagonist,[85,86] has been shown to be a potent, reversible inhibitor of MAO-B, and protect against striatal dopamine depletion in the *N*-methyl-4-phenyl-1,2,3,6-tetrahydropyridine (MPTP) mouse model.[87–89] In addition to being a potent and selective A_{2A} antagonist, CSC was also discovered to be a highly potent, competitive, and reversible inhibitor of MAO-B.[87]

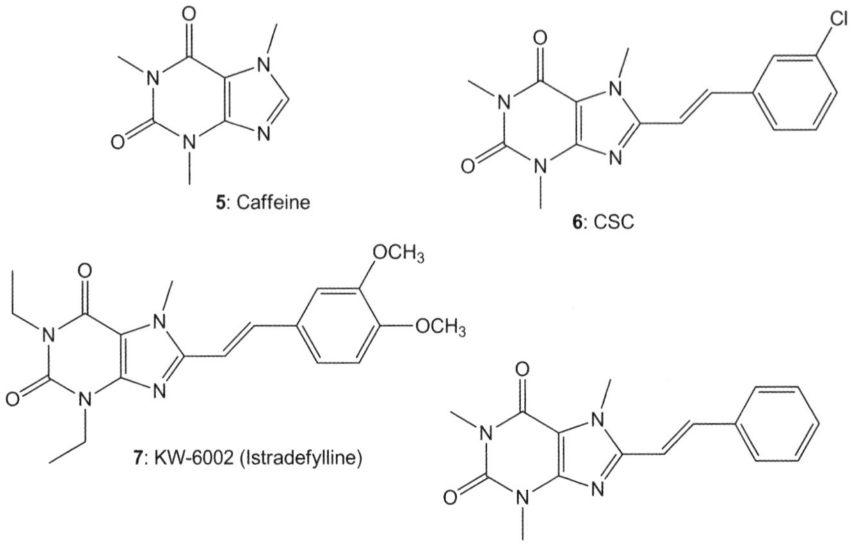

Figure 18.3 Chemical structures of caffeine and representative *trans*-8-styrylalkylxanthines described in the text: all compounds possess varying degrees of MAO-B inhibitory and A_{2A} antagonistic activity.

The latter finding has raised the possibility of utilizing the methylxanthine (caffeine-derived, **5**, Figure 18.3) system as a design scaffold in an MTDL paradigm for dual-target-directed drugs that block at both A_{2A} receptors and at MAO-B. Based on these initial findings, such compounds may offer enhanced symptomatic relief while also slowing the progression of PD by protecting against further neurodegeneration.

18.3.2 *Trans*-8-Styryl and (*E*,*E*)-8-(4-Phenylbutadien-1-yl) Alkylxanthines

18.3.2.1 Trans-*8-Styryl Alkylxanthines: Structure–Activity Relationships*

In order to determine the structural requirements necessary for alkylxanthines to act as MAO-B inhibitors, various substituted methyl- and ethylxanthines were evaluated as MAO-B inhibitors.[88–91] Analogous to what has been observed with A_{2A} antagonists, substitution of the xanthinyl core at C-8 with a variety of groups yielded compounds endowed with more potent MAO-B inhibition activities than caffeine. Also, analogous to A_{2A} antagonists, the styryl side chain was found to be especially efficient in enhancing the MAO-B inhibitory potency of caffeine-derived inhibitors. For example, **6** inhibited baboon liver MAO-B with a K_i value of 80.6 nM, approximately 45 000 times more potently than caffeine ($K_i = 3.6$ mM).[92] The K_i value for the inhibition of MAO-B by **6** is comparable to that reported for its antagonism of A_{2A} receptors ($K_i = 36–54$ nM).[85,86] SAR studies (summarized in Figure 18.4) further revealed that an electron-deficient styryl side chain was more effective in enhancing MAO-B inhibitory potency, and that the *trans* (or *E*) geometry about the styryl double bond is a requirement for MAO-B inhibition,[88,89] and saturation of this bond has a negative effect on MAO inhibitory potency.[88] The latter observation supports data that suggest that many MAO-B inhibitors contain planar conjugated heterocyclic systems. Of significance was the finding that ethyl substitution at positions 1 and 3 of the xanthinyl core has a negative effect on the potency of MAO-B inhibition compared to methyl substitution.[88,92] This substitution pattern requirement represents a limitation in the development of caffeine-derived MTDLs since, in general, 1,3-diethyl substitution of the xanthinyl core leads to enhanced A_{2A} antagonism.[86,92,93] While 1-, 3-, and 7-trimethyl substitution is probably optimal for the design of xanthine-based reversible MAO-B inhibitors, ethyl or propyl functional groups at C-1 and C-3 are optimal for A_{2A} antagonism. Accordingly, the potent A_{2A} antagonist KW-6002 (istradefylline, **7**, Figure 18.3) was found to be a relatively weak MAO-B inhibitor with a K_i value of 28 μM.[88] As for A_{2A} antagonists,[85] alkylation at N-7 of the xanthinyl core is also a requirement for potent MAO-B inhibition (Figure 18.4).[88]

18.3.2.2 Exploring MAO-B Activity

The potency of MAO-B inhibition by *trans*-8-styrylcaffeinyl analogs may possibly be explained by comparative inspection of the crystal structure of

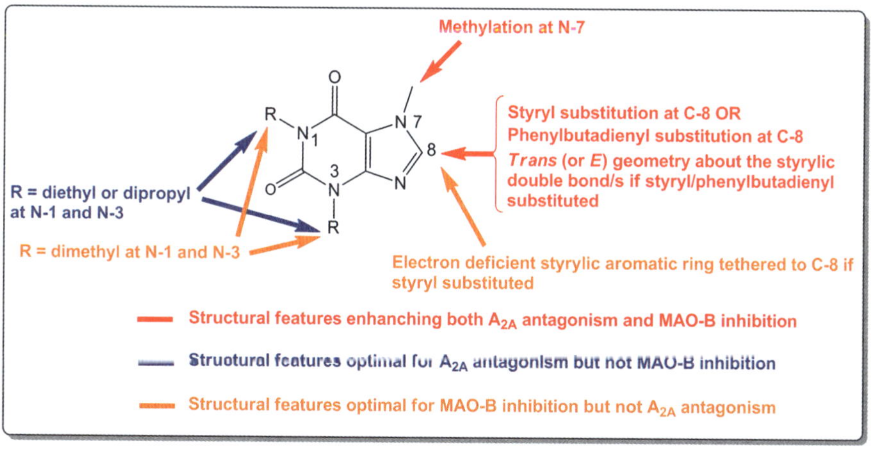

Figure 18.4 Structural requirements of alkylxanthine analogs in order to exhibit activity as A_{2A} antagonists exclusively (blue text and arrows), MAO-B inhibitors exclusively (orange text and arrows), or as both A_{2A} antagonists and MAO-B inhibitors (red text and arrows).

human recombinant MAO-B.[94–96] The active site of the enzyme consists of an entrance connected to the substrate cavity. Relatively large inhibitors, such as the reversible inhibitor 1,4-diphenyl-2-butene, exhibit a dual binding mode that involves traversing both the entrance and substrate cavities.[95] *Trans*-8-Styrylcaffeines probably exhibit a similar mode of binding with the xanthinyl core located in the substrate cavity of the active site while the styryl substituent extends into the entrance cavity.[89,97] Without the styryl side chain, caffeine itself is expected to bind to either the substrate or the entrance cavity leaving the remaining cavity unoccupied. Therefore the dual binding mode of the styryl side chain possibly enhances the interactions between the inhibitor and the active site amino acid residues and hence the binding affinity.[84] The finding that caffeine is a weak MAO-B inhibitor is clearly in support of this hypothesis.

*18.3.2.3 (*E,E*)-8-(4-Phenylbutadien-1-yl) Alkylxanthines*

In a 2008 study it was shown that substitution with a phenylbutadienyl (rather than styryl) side chain at C-8 of the xanthinyl core (Figure 18.5) may result in enhanced efficacy in promoting binding between the active site of MAO-B and caffeine.[92] For example, (*E,E*)-8-[4-(3-chlorophenyl)butadien-1-yl]caffeine (**9**) ($K_i = 42.1$ nM) was approximately 1.9 times more potent as an MAO-B inhibitor than **6** ($K_i = 80.6$ nM) while (*E,E*)-8-(4-phenylbutadien-1-yl)caffeine (**10**) ($K_i = 148.6$ nM), was almost 20 times more potent than the corresponding *trans*-8-styrylcaffeine (**8**) ($K_i = 2864$ nM).[89] The finding that (*E,E*)-8-(4-phenylbutadien-1-yl)caffeine analogs retain potent A_{2A} antagonism suggests that they also may be promising lead compounds for the development of dual-target-directed drugs. However, the observation that 1,3-diethyl substitution (which blunts MAO-inhibitory properties) is required for potent A_{2A}

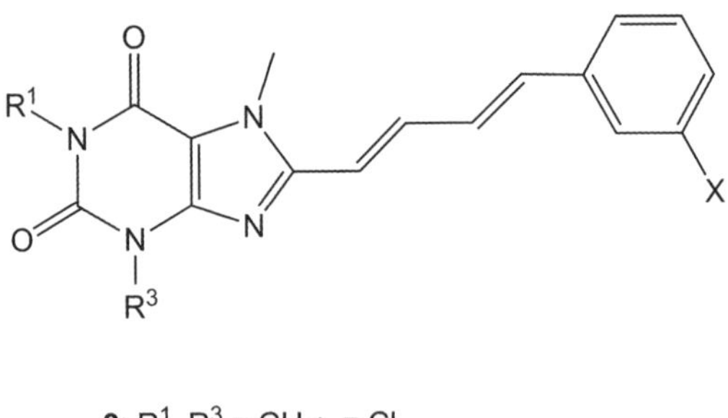

9: R^1, R^3 = CH_3; = Cl

10: R^1, R^3 = CH_3; = H

11: R^1, R^3 = C_2H_5; = H

Figure 18.5 Chemical structures of (*E,E*)-8-(4-phenylbutadien-1-yl)alkylxanthine analogs **9–11**.

antagonism also applies to the (*E*,*E*)-8-(4-phenylbutadien-1-yl) analogs (compare structure **10** with **11**).

Since A_{2A} antagonists and MAO-B inhibitors potentiate the motor restorative effects of levodopa by acting at different targets, the combination of these two activities in a single molecule may be particularly advantageous as an adjunct therapy to levodopa. The involvement of both the A_{2A} receptor and MAO-B in neuroprotection suggests that dual-target-directed drugs also may exhibit enhanced neuroprotective properties. While a number of methylxanthines have been shown to act as A_{2A} antagonists and MAO-B inhibitors, optimizing the structures for dual action remains a challenge since modifications that lead to enhanced A_{2A} antagonism frequently have an opposing effect on MAO-B inhibition and *vice versa*.

18.4 Discovery of Ladostigil, a Multi-Target Drug Candidate for AD and PD

18.4.1 Ladostigil, an MTDL by Design

The multimodal drug, ladostigil {TV3326; [(*N*-propargyl-(3*R*)-aminoindan-5-yl)ethyl methyl carbamate] (**12**, Figure 18.6) was designed to possess neuroprotective and neurorestorative activities by targeting various pathogenic mechanisms of AD. The underlying principle in the design of ladostigil was to amalgamate the neuroprotective propargyl moiety of the anti-parkinsonian drug and irreversible selective monoamine oxidase (MAO)-B inhibitor, rasagiline [*N*-propargyl-1-(*R*)-aminoindan] (**13**, Figure 18.6) to the carbamate ChE inhibitory moiety of the anti-AD drug, rivastigmine {(*S*)-3-[1-(dimethylamino)-ethyl]phenethylmethyl carbamate} (**14**, Figure 18.6). The resulting molecule, ladostigil, is a novel dual AChE-BuChE and brain-selective MAO-A and -B inhibitor intended for the treatment of dementia co-morbid with extrapyramidal disorders (parkinsonism) and depression (presently in phase IIb clinical studies).

18.4.1.1 Ladostigil's Building Blocks: (1) Rasagiline: an Irreversible Selective MAO-B Inhibitor

Rasagiline (**13**, Figure 18.6) is a second generation, selective, and irreversible inhibitor of MAO-B, indicated as monotherapy or adjunct to levodopa therapy for patients with early and late PD.[98–101] In preclinical studies, rasagiline was shown to possess broad neuroprotective/neurorestorative activities against a variety of neurotoxins *in vivo* and in neuronal cell culture models of neurodegeneration.[102] *In vivo* studies include its neuroprotective effect in the MPTP mouse[103] and monkey[104] models, in focal ischemia in rat,[105] in stroke-prone spontaneously hypertensive rats,[106] and in a traumatic brain injury mouse model.[107] Notably, rasagiline was also found to possess *in vivo* neurorestorative activity in substantia nigra pars compacta (SNpC) neurons, when given

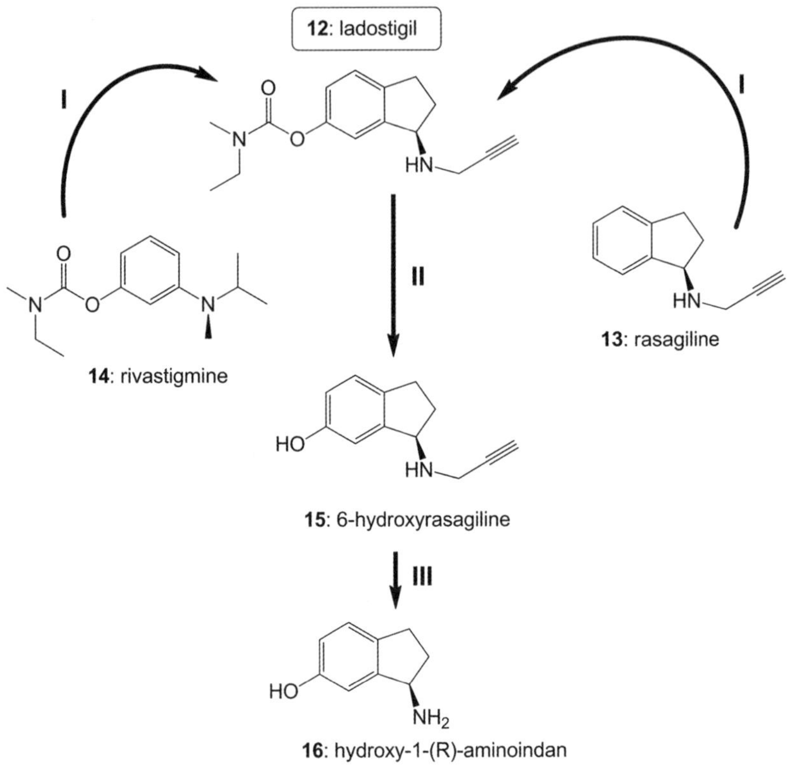

Figure 18.6 The MTDL design strategy towards ladostigil and its active metabolites: (I) the propargyl moiety of the irreversible selective MAO-B inhibitor, rasagiline (*N*-propargyl-1-(*R*)-aminoindan) is amalgameted with the carbamate ChE inhibitory moiety of rivastigmine {(*S*)-3-[1-(dimethyl-amino)ethyl] phenethylmethyl carbamate}; (II) and (III) hydrolysis of ladostigil's carbamate moiety by pseudo-inhibition of ChE yields 6-hydroxyrasagiline [6-hydroxy-*N*-propargyl-1(*R*)-aminoindan], and further gives rise to ladostigil's major metabolite, hydroxy-1-(R)-aminoindan.

post-MPTP,[103] amongst others. Structure–activity relationship studies that provided evidence that the *N*-propargyl moiety of rasagiline promotes neuronal survival via distinct neuroprotective/neurorescue pathways highlighted the importance of this moiety for the latter activities of rasagiline.[108,109]

18.4.1.2 Ladostigil's Building Blocks: (2) Rivastigmine: a Pseudo Cholinesterase Inhibitor

Rivastigmine (**14**, Figure 18.6) is an established ChEI that inhibits both AChE and BuChE and which is indicated for the treatment of mild to moderate dementia associated with AD and PD.[110,111] Early *in vitro* studies demonstrated that rivastigmine possessed a concentration-dependent neuroprotective effect against cell death induced by β-amyloid (Aβ) and okadaic acid.[112] This effect was obtained at a concentration that differs from its IC_{50} of AChE inhibition, suggesting that the neuroprotective effect of rivastigmine might be through a mechanism unrelated to ChE inhibition.[112] Recently, it was shown that rivastigmine favorably modulated SH-SY5Y cellular response to heat shock stimulus injury, indicating that one of the drug's putative neuroprotective mechanisms might be mediated through the heat shock response.[113]

18.4.2 Basic Characteristics and Neuropharmacology of Ladostigil

Figure 18.7 provides a graphic overview of the neuropharmacological mechanisms associated with ladostigil's multiple actions: shown are proteins and genes involved in the pathological features of AD, and that are targeted by ladostigil to elicit its neuroprotective activity. Shown are (1) ChE inhibitory activity, (2) brain-selective irreversible MAO-A and -B inhibition, (3) regulation of APP processing, (4) regulation of cell survival signaling pathways, (5) neurotrophic factors, and (6) protective activity against ROS.

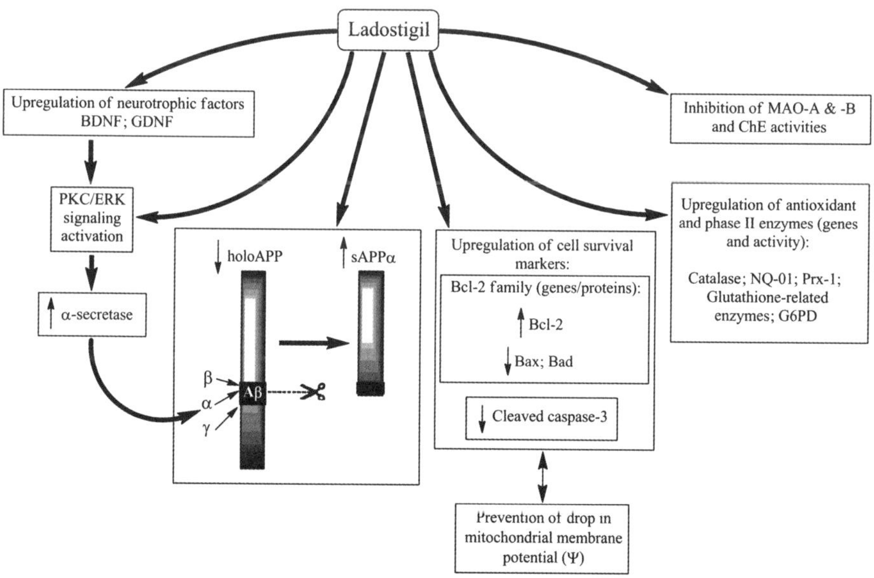

Figure 18.7 Proteins and genes involved in the pathological features of AD that are targets for the neuroprotective activity of ladostigil: ChE inhibitory activity, brain-selective irreversible MAO-A and -B inhibition, regulation of APP processing, regulation of cell survival signaling pathways, neurotrophic factors, and protective activity against ROS.

18.4.2.1 ChE Inhibition

The role of ACh in the activity of striatal projection neurons is still unclear, but mounting evidence suggest that cholinergic transmission, via muscarinic receptors, ensured the correct processing mechanisms of cortical inputs forward from the cortex to projection cells.[114,115] Ladostigil (12–26 mg/kg) caused a dose-related antagonism of scopolamine-induced deficits in reference and working memory and memory consolidation in rats[116] by sufficiently increasing the level of ACh in the synapse to compete with scopolamine for the muscarinic receptors subserving memory.[116,117] Notably, symptoms arising from excess cholinergic stimulation in the periphery (e.g. salivation, diarrhea, and muscle weakness) or the brain (e.g. respiratory depression or hypothermia) were only observed at doses in excess of 139 mg/kg.[118] This wide therapeutic index is likely due to the relatively small rate of increase in ChE inhibition with increasing drug doses that was shown to occur with ladostigil over the range of 17–69 mg/kg.[116,118] Ladostigil inhibits both AChE and BuChE *in vitro*,[117,118] with AChE being inhibited at levels 100× that of BuChE.[117,118]

18.4.2.2 MAO Inhibition

Previous studies have shown that brain MAO-A and -B are inhibited only after chronic but not acute treatment with ladostigil.[116,119,120] This is attributed to the formation of active MAO inhibitory metabolites as a consequence of hydrolysis of the carbamate moiety in ladostigil by ChE, which yields the 6-hydroxyrasagiline (**15**, Figure 18.6) metabolite[119–121] with high affinity towards MAO-A ($K_i = 1.9\,\mu M$) and -B ($K_i = 17\,\mu M$). Adequate concentrations of 6-hydroxyrasagiline and other metabolites of ladostigil may significantly inhibit MAO-A and -B to increase levels of striatal dopamine, 5-HT, and NE, thereby accounting for the respective anti-parkinsonian and antidepressant activities of ladostigil shown in preclinical models.[119–121] MAO inhibition by ladostigil, similar to other propargylamines such as clorgyline, selegiline, and rasagiline, was shown to be irreversible and slow to recover.[122]

18.4.2.3 Neuroprotective Effects of Ladostigil

In preclinical studies, ladostigil was shown to have a broad neuroprotective activity against a variety of neurotoxins in *in vivo* and neuronal cell cultures models of neurodegeneration (summarized in Table 18.2). Notably, ladostigil's *S*-enantiomer, TV3279, also possesses neuroprotective activity, observed in various animal and cell culture models of neurodegeneration (e.g. antagonizing the spatial memory impairment induced by scopolamine in rats,[116] reducing edema, motor function, and memory deficits after closed head injury in mice,[116,118] and protecting against ischemia-induced cytotoxicity in PC12 cells,[118] indicating that the neuroprotective activity of ladostigil is independent of MAO inhibition. It is likely that several different mechanisms are in operation (Table 18.2) and the accumulated data provide evidence that the

Table 18.2 The neuroprotective activities of ladostigil in *in vivo* and *in vitro* models of neurodegeneration.

Neuroprotective effects	*Experimental system*
Protection against MPTP-induced dopaminergic neurotoxicity	Mouse
Prevention of STZ-induced gliosis, oxidative/nitrative stress and memory deficits	Rat
Prevention of age-related glial activation and spatial memory deficits	Aged rat
Prevention of closed head injury-induced cerebral edema, and impaired motor and memory function	Rat
Antagonism of scopolamine-induced impairments in spatial memory	Rat
Improvement of cognitive performance	Aged monkey
Inhibition of behavioral hyperactivity syndromes	Rat
Prevention of sodium azide-induced ChAT immunoreactivity reduction	Rat
Protection against oxygen-glucose deprivation	NGF differentiated PC-12 cells
Activation of ERK1/2	Partially differentiated PC-12 cells
Activation of PKC	PC-12 and SH-SY5Y cells; SK-N-SH cells; Mouse hippocampus
Prevention of caspase-3 activation	SK-N-SH cells; SH-SY5Y cells
Regulation of Bcl-2 protein family	SK-N-SH cells
Prevention of reduction in mitochondrial membrane potential (ψ)	SK-SY5Y-cells
Induction of neurotrophic factors, GDNF and BDNF	SK-N-SH cells
Up-regulation of antioxidant activity and mRNA expression of antioxidant enzymes	SH-SY5Y and SK-N-SH cells; Aged rat
Regulation of APP processing	PC-12 and SH-SY5Y cells; SK-N-SH cells; Mouse hippocampus
Down-regulation of the familial AD-linked presenilin-1 gene expression	Aged rat

neuroprotective effects of ladostigil might be associated with some intrinsic pharmacological action of its propargyl moiety, similar to that observed in other propargylamine-containing drugs.[109,123–125]

18.4.2.4 Protective Activity against Oxidative Stress

Accumulative studies have suggested that oxidative stress in AD is mediated by Aβ-generated ROS.[126,127] One of the main ROS is H_2O_2 produced during the redox process and is considered as a messenger in intracellular signaling cascades, including cellular metabolism and proliferation.[128,129] Several lines of evidence suggested that AChE and BuChE activation might be involved in apoptosis associated with H_2O_2.[130,131] These links between cholinergic signals and oxidative stress provide an additional therapeutic target for AChEIs in AD. In line with these findings, ladostigil (1 and 5 μM) was found to exhibit a significant neuroprotective effect against H_2O_2-induced damage in SH-SY5Y cells.[132] In addition, MAO, which is located in the outer mitochondrial membrane, can produce a significant amount of ROS, in particular H_2O_2.[133] Consistent with this, ladostigil was found to increase antioxidant enzyme activities in brain dopaminergic system in rats[134] and at 1 μM exerted antioxidant activity through both direct scavenging effect on free radicals overproduced in H_2O_2-treated neuronal cells, as well as indirectly by stimulating the expression and activity of cellular antioxidant enzymes, catalase and glutathione reductase, suggesting their involvement in the cytoprotective effect of ladostigil.[132] Ladostigil also was shown to reverse the effect of aging on mRNA expression levels of various genes associated with metabolism and oxidation processes in the hippocampus of old rats, including glutathione peroxidase precursor (GSHPX-P), glutathione S-transferase, glutathione synthetase, thioredoxin peroxidase (TDPX1), and glucose-6-phosphate dehydrogenase (G6PD).[135] *In vivo* chronic administration of ladostigil (1 mg/kg) for a week before and after ICV injection of STZ significantly reduced microglia activation, loss of astrocytes and prevented the increase in nitrotyrosine, a marker of nitrative-oxidative stress.[136] These and other data suggest that the prevention of memory deficits by ladostigil may also in part be a result of its antioxidant activity.[136]

18.4.2.5 Regulation of APP Processing

An essential neuroprotective effect, shared by ladostigil, rasagiline, and other propargylamine-derivative compounds, relates to their ability to regulate the processing of APP by the non-amyloidogenic α-secretase pathway.[137] In this context, ladostigil (*in vivo*, 52 mg/kg/day, i.p. 14 days; *in vitro*, 1–10 μM) was reported to markedly suppress holo-APP protein levels and elevate sAPPα release,[108,138,139] indicating that this drug can be of clinical value towards accelerating the non-amyloidogenic APP processing and reduce Aβ generation. The stimulation of sAPPα release induced by ladostigil was blocked by a hydroxamic acid-based metalloprotease inhibitor, indicating an α-secretase metalloprotease mediation in this effect.[138]

*18.4.2.6 Neuroprotective Activity of Hydroxy-1-(*R*)-aminoindan*

Hydroxy-1-(*R*)-aminoindan (**16**, Figure 18.6) is the major metabolite of ladostigil[121] and markedly reduces the levels of the apoptosis-associated phosphorylated protein, H2A.X (Ser139), decreases the cleavage of caspase 9 and caspase 3, while increasing the anti-apoptotic proteins, Bcl-2 and Bcl-xl.[140] These findings suggest that hydroxyl-1-(*R*)-aminoindan may contribute to the overall neuroprotective activity of its parent compound, ladostigil.

18.5 Conclusions

The treatment of neurodegenerative diseases poses perplexing challenges. These challenges are partly due to the complex pathology involved in the etiology of these diseases. The utility of an emerging approach that purports to use a single polypharmacological drug molecule (i.e. one that acts on more than one drug target in the etiological pathway of the disease), may offer new hope in the treatment of many neurodegenerative diseases. This approach has led us and others to exploit a new paradigm in medicinal chemistry, the 'multi-target directed ligand' (MTDL) design strategy. Such a transition from a 'magic bullet' to a 'magic shotgun' approach may be key in designing future treatment regimens for neurodegenerative diseases and associated cognitive decline. In our opinion, drug discovery endeavors will have to shift focus from the design of selective agents that target only one pathophysiological pathway to the design of agents generated through an MTDL design strategy: drugs that operate through manifold mechanisms and are intended to target the very complexity of the disease state.

Acknowledgements

This work was supported by a grant from MIUR (FIRB RBNE03FH5Y) and the University of Bologna to MLB and CM; MBY gratefully acknowledges the support of Teva Pharmaceutical Co. (Netanya, Israel), Varinel Co. (USA), and Rappaport Family Research Institute, Technion - Israel Institute of Technology; CVDS acknowledges support from NEOMED.

References

1. R. L. Nussbaum and C. E. Ellis, *N. Engl. J. Med.*, 2003, **348**, 1356–1364.
2. R. Scatena, G. E. Martorana, P. Bottoni, G. Botta, P. Pastore and B. Giardina, *Expert Opin. Invest. Drugs*, 2007, **16**, 59 72.
3. C. J. Van der Schyf, *Expert Rev. Clin. Pharmacol.*, 2011, **4**, 293–298.
4. R. T. Bartus, R. L. Dean, 3rd, B. Beer and A. S. Lippa, *Science*, 1982, **217**, 408–414.
5. C. J. Van der Schyf and W. J. Geldenhuys, *Int. Rev. Neurobiol.*, 2011, **100**, 107–125.
6. K. A. Jellinger, *J. Neural Transmission*, 2003, 101–144.
7. C. J. Van der Schyf, W. J. Geldenhuys and M. B. Youdim, *J. Neurochem.*, 2006, **99**, 1033–1048.
8. C. J. Van der Schyf, S. Mandel, W. J. Geldenhuys, T. Amit, Y. Avramovich, H. Zheng, M. Fridkin, S. Gal, O. Weinreb, O. Bar-Am, Y. Sagi and M. B. Youdim, *Curr. Alzheimer Res.*, 2007, **4**, 522–536.
9. C. J. Van der Schyf and M. B. Youdim, *Neurotherapeutics*, 2009, **6**, 1–3.
10. M. L. Bolognesi, R. Matera, A. Minarini, M. Rosini and C. Melchiorre, *Curr. Opin. Chem. Biol.*, 2009, **13**, 303–308.
11. A. Cavalli, M. L. Bolognesi, A. Minarini, M. Rosini, V. Tumiatti, M. Recanatini and C. Melchiorre, *J. Med. Chem.*, 2008, **51**, 347–372.
12. U. Klotz, *Eur. J. Clin. Pharmacol.*, 2008, **64**, 225–226.

13. M. B. Youdim and J. J. Buccafusco, *Trends Pharmacol. Sci.*, 2005, **26**, 27–35.
14. M. B. Youdim and J. J. Buccafusco, *J. Neural Transmission*, 2005, **112**, 519–537.
15. M. L. Bolognesi, M. Rosini, V. Andrisano, M. Bartolini, A. Minarini, V. Tumiatti and C. Melchiorre, *Curr. Pharm. Des.*, 2009, **15**, 601–613.
16. R. Morphy and Z. Rankovic, *Curr. Pharm. Des.*, 2009, **15**, 587–600.
17. R. Morphy and Z. Rankovic, *J. Med. Chem.*, 2005, **48**, 6523–6543.
18. R. Morphy, C. Kay and Z. Rankovic, *Drug Discovery Today*, 2004, **9**, 641–651.
19. R. Morphy and Z. Rankovic, *J. Med. Chem.*, 2006, **49**, 4961–4970.
20. C. Melchiorre, M. L. Bolognesi, A. Minarini, M. Rosini and V. Tumiatti, *J. Med. Chem.*, 2010, **53**, 5906–5914.
21. C. Melchiorre, M. S. Yong, B. G. Benfey and B. Belleau, *J. Med. Chem.*, 1978, **21**, 1126–1132.
22. S. Palea, M. Corsi, J. M. Rimland and D. G. Trist, *Br. J. Pharmacol.*, 1995, **115**, 3–10.
23. B. Belleau, B. G. Benfey, T. J. Benfey and C. Melchiorre, *Br. J. Pharmacol.*, 1982, **75**, 573–578.
24. C. Melchiorre, A. Cassinelli and W. Quaglia, *J. Med. Chem.*, 1987, **30**, 201–204.
25. C. Melchiorre, *Med. Res. Rev.*, 1990, **10**, 327–349.
26. M. Rosini, M. G. Bixel, G. Marucci, R. Budriesi, M. Krauss, M. L. Bolognesi, A. Minarini, V. Tumiatti, F. Hucho and C. Melchiorre, *J. Med. Chem.*, 2002, **45**, 1860–1878.
27. A. Chahdi, L. Daeffler, J. P. Gies and Y. Landry, *Fundam. Clin. Pharmacol.*, 1998, **12**, 121–132.
28. P. D. Dobson and D. B. Kell, *Nat. Rev.*, 2008, **7**, 205–220.
29. R. A. Casero, Jr. and P. M. Woster, *J. Med. Chem.*, 2009, **52**, 4551–4573.
30. G. Estrada, E. Villegas and G. Corzo, *Nat. Prod. Rep.*, 2007, **24**, 145–161.
31. J. Li, K. M. Doyle and T. Tatlisumak, *Curr. Med. Chem.*, 2007, **14**, 1807–1813.
32. M. D. Senanayake, H. Amunugama, T. D. Boncher, R. A. Casero and P. M. Woster, *Essays Biochem.*, 2009, **46**, 77–94.
33. J. Massoulie, *Neuro-Signals*, 2002, **11**, 130–143.
34. N. C. Inestrosa, A. Alvarez, C. A. Perez, R. D. Moreno, M. Vicente, C. Linker, O. I. Casanueva, C. Soto and J. Garrido, *Neuron*, 1996, **16**, 881–891.
35. Z. Z. Chong, F. Li and K. Maiese, *Prog. Neurobiol.*, 2005, **75**, 207–246.
36. T. L. Wadsworth, J. A. Bishop, A. S. Pappu, R. L. Woltjer and J. F. Quinn, *J. Alzheimers Dis.*, 2008, **14**, 225–234.
37. K. Ono, K. Hasegawa, H. Naiki and M. Yamada, *Biochem. Biophys. Res. Commun.*, 2005, **330**, 111–116.
38. M. L. Bolognesi, R. Banzi, M. Bartolini, A. Cavalli, A. Tarozzi, V. Andrisano, A. Minarini, M. Rosini, V. Tumiatti, C. Bergamini, R. Fato,

G. Lenaz, P. Hrelia, A. Cattaneo, M. Recanatini and C. Melchiorre, *J. Med. Chem.*, 2007, **50**, 4882–4897.

39. M. L. Bolognesi, M. Bartolini, M. Rosini, V. Andrisano and C. Melchiorre, *Bioorg. Med. Chem. Lett.*, 2009, **19**, 4312–4315.
40. F. Mancini, M. L. Bolognesi, C. Melchiorre, A. Cavalli and V. Andrisano, *J. Sep. Sci.*, 2007, **30**, 2935–2942.
41. F. Mancini, M. L. Bolognesi, C. Melchiorre and V. Andrisano, *J. Pharm. Biomed. Anal.*, 2009, **50**, 164–170.
42. M. Bartolini, C. Bertucci, V. Cavrini and V. Andrisano, *Biochem. Pharmacol.*, 2003, **65**, 407–416.
43. A. Cavalli, M. L. Bolognesi, S. Capsoni, V. Andrisano, M. Bartolini, E. Margotti, A. Cattaneo, M. Recanatini and C. Melchiorre, *Angew. Chem., Int. Ed. Engl.*, 2007, **46**, 3689–3692.
44. A. Alvarez, C. Opazo, R. Alarcon, J. Garrido and N. C. Inestrosa, *J. Mol. Biol.*, 1997, **272**, 348–361.
45. M. L. Bolognesi, V. Andrisano, M. Bartolini, R. Banzi and C. Melchiorre, *J. Med. Chem.*, 2005, **48**, 24–27.
46. A. Frydman-Marom, M. Rechter, I. Shefler, Y. Bram, D. E. Shalev and E. Gazit, *Angew. Chem., Int. Ed. Engl.*, 2009, **48**, 1981–1986.
47. A. A. Reinke and J. E. Gestwicki, *Chem. Biol. Drug Des.*, 2007, **70**, 206–215.
48. D. P. Flaherty, S. M. Walsh, T. Kiyota, Y. Dong, T. Ikezu and J. L. Vennerstrom, *J. Med. Chem.*, 2007, **50**, 4986–4992.
49. M. L. Bolognesi, M. Bartolini, A. Tarozzi, F. Morroni, F. Lizzi, A. Milelli, A. Minarini, M. Rosini, P. Hrelia, V. Andrisano and C. Melchiorre, *Bioorg. Med. Chem. Lett.*, 2011, **21**, 2655–2658.
50. L. L. Blazer and R. R. Neubig, *Neuropsychopharmacology*, 2009, **34**, 126–141.
51. M. Bartolini, C. Bertucci, M. L. Bolognesi, A. Cavalli, C. Melchiorre and V. Andrisano, *ChemBioChem*, 2007, **8**, 2152–2161.
52. A. Mordente, G. E. Martorana, G. Minotti and B. Giardina, *Chem. Res. Toxicol.*, 1998, **11**, 54–63.
53. A. K. Raina, D. J. Templeton, J. C. Deak, G. Perry and M. A. Smith, *Redox Rep.*, 1999, **4**, 23–27.
54. P. I. Moreira, S. L. Siedlak, G. Aliev, X. Zhu, A. D. Cash, M. A. Smith and G. Perry, *J. Neural Transmission*, 2005, **112**, 921–932.
55. K. S. SantaCruz, E. Yazlovitskaya, J. Collins, J. Johnson and C. DeCarli, *Neurobiol. Aging*, 2004, **25**, 63–69.
56. M. L. Bolognesi, A. Cavalli, C. Bergamini, R. Fato, G. Lenaz, M. Rosini, M. Bartolini, V. Andrisano and C. Melchiorre, *J. Med. Chem.*, 2009, **52**, 7883–7886.
57. Y. H. Kang and J. M. Pezzuto, *Methods Enzymol.*, 2004, **382**, 380–414.
58. S. Capsoni, G. Ugolini, A. Comparini, F. Ruberti, N. Berardi and A. Cattaneo, *Proc. Natl. Acad. Sci. USA*, 2000, **97**, 6826–6831.
59. S. Capsoni and A. Cattaneo, *Cell. Mol. Neurobiol.*, 2006, **26**, 619–633.

60. S. Capsoni, S. Giannotta and A. Cattaneo, *Mol. Cell. Neurosci.*, 2002, **21**, 15–28.
61. V. Andrisano, M. Bartolini, M. L. Bolognesi, A. Cavalli, C. Melchiorre and M. Recanatini, *PCT Int. Appl. Pat.*, 2003, WO2003087035.
62. M. Lew, *Pharmacotherapy*, 2007, **27**, 155S–160S.
63. M. L. Bolognesi, A. Cavalli, L. Valgimigli, M. Bartolini, M. Rosini, V. Andrisano, M. Recanatini and C. Melchiorre, *J. Med. Chem.*, 2007, **50**, 6446–6449.
64. H. H. Fernandez and J. J. Chen, *Pharmacotherapy*, 2007, **27**, 174S–185S.
65. M. B. Youdim, G. G. Collins, M. Sandler, A. B. Bevan Jones, C. M. Pare and W. J. Nicholson, *Nature*, 1972, **236**, 225–228.
66. G. G. Collins, M. Sandler, E. D. Williams and M. B. Youdim, *Nature*, 1970, **225**, 817–820.
67. D. A. Di Monte, L. E. DeLanney, I. Irwin, J. E. Royland, P. Chan, M. W. Jakowec and J. W. Langston, *Brain Res.*, 1996, **738**, 53–59.
68. J. P. Finberg, J. Wang, K. Bankiewicz, J. Harvey-White, I. J. Kopin and D. S. Goldstein, *J. Neural Transmission Suppl.*, 1998, **52**, 279–285.
69. M. B. Youdim and Y. S. Bakhle, *Br. J. Pharmacol.*, 2006, **147**, S287–S296.
70. A. Nicotra, F. Pierucci, H. Parvez and O. Senatori, *Neurotoxicology*, 2004, **25**, 155–165.
71. J. S. Fowler, N. D. Volkow, G. J. Wang, J. Logan, N. Pappas, C. Shea and R. MacGregor, *Neurobiol. Aging*, 1997, **18**, 431–435.
72. B. Karolewicz, V. Klimek, H. Zhu, K. Szebeni, E. Nail, C. A. Stockmeier, L. Johnson and G. A. Ordway, *Brain Res.*, 2005, **1043**, 57–64.
73. C. J. Fowler, A. Wiberg, L. Oreland, J. Marcusson and B. Winblad, *J. Neural Transmission*, 1980, **49**, 1–20.
74. K. Xu, E. Bastia and M. Schwarzschild, *Pharmacol. Ther.*, 2005, **105**, 267–310.
75. A. Pinna, J. Wardas, N. Simola and M. Morelli, *Life Sci.*, 2005, **77**, 3259–3267.
76. M. Morelli, T. Di Paolo, J. Wardas, F. Calon, D. Xiao and M. A. Schwarzschild, *Prog. Neurobiol.*, 2007, **83**, 293–309.
77. S. Ferré, G. von Euler, B. Johansson, B. B. Fredholm and K. Fuxe, *Proc. Natl. Acad. Sci. USA*, 1991, **88**, 7238–7241.
78. S. Ferré, W. T. O'Connor, K. Fuxe and U. Ungerstedt, *J. Neurosci.*, 1993, **13**, 5402–5406.
79. S. Ferré, B. B. Fredholm, M. Morelli, P. Popoli and K. Fuxe, *Trends Neurosci.*, 1997, **20**, 482–487.
80. T. Kanda, M. J. Jackson, L. A. Smith, R. K. Pearce, J. Nakamura, H. Kase, Y. Kuwana and P. Jenner, *Ann. Neurol.*, 1998, **43**, 507–513.
81. J. F. Chen, R. Moratalla, F. Impagnatiello, D. K. Grandy, B. Cuellar, M. Rubinstein, M. A. Beilstein, E. Hackett, J. S. Fink, M. J. Low, E. Ongini and M. A. Schwarzschild, *Proc. Natl. Acad. Sci. USA*, 2001, **98**, 1970–1975.
82. M. A. Schwarzschild, L. Agnati, K. Fuxe, J. F. Chen and M. Morelli, *Trends Neurosci.*, 2006, **29**, 647–654.

83. F. Bibbiani, J. D. Oh, J. P. Petzer, N. Castagnoli, Jr., J. F. Chen, M. A. Schwarzschild and T. N. Chase, *Exp. Neurol.*, 2003, **184**, 285–94.
84. J. P. Petzer, N. Castagnoli, Jr., M. A. Schwarzschild, J. F. Chen and C. J. Van der Schyf, *Neurotherapeutics*, 2009, **6**, 141–151.
85. K. A. Jacobson, C. Gallo-Rodriguez, N. Melman, B. Fischer, M. Maillard, A. van Bergen, P. J. van Galen and Y. Karton, *J. Med. Chem.*, 1993, **36**, 1333–1342.
86. C. E. Müller, U. Geis, J. Hipp, U. Schobert, W. Frobenius, M. Pawłowski, F. Suzuki and J. Sandoval-Ramírez, *J. Med. Chem.*, 1997, **40**, 4396–4405.
87. J. F. Chen, S. Steyn, R. Staal, J. P. Petzer, K. Xu, C. J. Van Der Schyf, K. Castagnoli, P. K. Sonsalla, N. Castagnoli, Jr. and M. A. Schwarzschild, *J. Biol. Chem.*, 2002, **277**, 36040–36044.
88. J. P. Petzer, S. Steyn, K. P. Castagnoli, J. F. Chen, M. A. Schwarzschild, C. J. Van der Schyf and N. Castagnoli, Jr., *Bioorg. Med. Chem.*, 2003, **11**, 1299–1310.
89. N. Vlok, S. F. Malan, N. Castagnoli, Jr., J. J. Bergh and J. P. Petzer, *Bioorg. Med. Chem.*, 2006, **14**, 3512–2351.
90. D. Van den Berg, K. R. Zoellner, M. O. Ogunrombi, S. F. Malan, G. Terre'Blanche, N. Castagnoli, Jr., J. J. Bergh and J. P. Petzer, *Bioorg. Med. Chem.*, 2007, **15**, 3692–3702.
91. N. Castagnoli, Jr., J. P. Petzer, S. Steyn, K. Castagnoli, J. F. Chen, M. A. Schwarzschild and C. J. Van der Schyf, *Neurology*, 2003, **61**, S62–S68.
92. J. Pretorius, S. F. Malan, N. Castagnoli, Jr., J. J. Bergh and J. P. Petzer, *Bioorg. Med. Chem.*, 2008, **16**, 8676–8684.
93. H. Kase, S. Aoyama, M. Ichimura, K. Ikeda, A. Ishii, T. Kanda, K. Koga, N. Koike, M. Kurokawa, Y. Kuwana, A. Mori, J. Nakamura, H. Nonaka, M. Ochi, M. Saki, J. Shimada, T. Shindou, S. Shiozaki, F. Suzuki, M. Takeda, K. Yanagawa, P. J. Richardson, P. Jenner, P. Bedard, E. Borrelli, R. A. Hauser and T. N. Chase, *Neurology*, 2003, **61**, S97–100.
94. C. Binda, P. Newton-Vinson, F. Hubálek, D. E. Edmondson and A. Mattevi, *Nat. Struct. Biol.*, 2002, **9**, 22–26.
95. C. Binda, M. Li, F. Hubálek, N. Restelli, D. E. Edmondson and A. Mattevi, *Proc. Natl. Acad. Sci. USA*, 2003, **100**, 9750–9755.
96. F. Hubálek, C. Binda, A. Khalil, M. Li, A. Mattevi, N. Castagnoli, Jr. and D. E. Edmondson, *J. Biol. Chem.*, 2005, **280**, 15761–15766.
97. C. Binda, J. Wang, L. Pisani, C. Caccia, A. Carotti, P. Salvati, D. E. Edmondson and A. Mattevi, *J. Med. Chem.*, 2007, **50**, 5848–5852.
98. Parkinson Study Group, *Arch. Neurol.*, 2002, **59**, 1937–1943.
99. Parkinson Study Group, *Arch. Neurol.*, 2004, **61**, 561–566.
100. Parkinson Study Group, *Arch. Neurol.*, 2005, **62**, 241–248.
101. C. W. Olanow, O. Rascol, R. Hauser, P. D. Feigin, J. Jankovic, A. Lang, W. Langston, E. Melamed, W. Poewe, F. Stocchi, E. Tolosa and ADAGIO Study Investigators, *N. Engl. J. Med.*, 2009, **361**, 1268–1278.
102. O. Weinreb, T. Amit, O. Bar-Am and M. B. Youdim, *Prog. Neurobiol.*, 2010, **92**, 330–344.

103. Y. Sagi, S. Mandel, T. Amit and M. B. Youdim, *Neurobiol. Dis.*, 2007, **25**, 35–44.
104. R. E. Heikkila, R. C. Duvoisin, J. P. Finberg and M. B. Youdim, *Eur. J. Pharmacol.*, 1985, **116**, 313–317.
105. Z. Speiser, A. Mayk, S. Eliash and S. Cohen, *J. Neural Transmission*, 1999, **106**, 593–606.
106. S. Eliash, Z. Speiser and S. Cohen, *J. Neural Transmission*, 2001, **108**, 909–923.
107. W. Huang, Y. Chen, E. Shohami and M. Weinstock, *Eur. J. Pharmacol.*, 1999, **366**, 127–135.
108. O. Bar-Am, M. Yogev-Falach, T. Amit and Y. Sagi, *J. Neurochem.*, 2004, **89**, 1119–1125.
109. O. Bar-Am, O. Weinreb, T. Amit and M. B. Youdim, *FASEB. J.*, 2005, **19**, 1899–1901.
110. M. Weinstock, M. Razin, M. Chorev and A. Enz, *J. Neural Transmission*, 1994, **43**, 219–225.
111. J. C. Machado, *Evid.-Based Ment. Health*, 2009, **12**, 113.
112. E. Arias, S. Gallego-Sandin, M. Villarroya, A. G. Garcia and M. G. Lopez, *J. Pharmacol. Exp. Ther.*, 2005, **315**, 1346–1353.
113. X. Zhou, A. R. Patel, F. Perez and D. A. Jurivich, *Transl. Res.*, 2009, **153**, 132–141.
114. P. N. Izzo and J. P. Bolam, *J. Comp. Neurol.*, 1988, **269**, 219–234.
115. P. Calabresi, D. Centonze and G. Bernardi, *Trends Neurosci.*, 2000, **23**, S57–S63.
116. M. Weinstock, C. Bejar, R. H. Wang, T. Poltyrev, A. Gross, J. Finberg and M. B. Youdim, *J. Neural Transmission*, 2000, **60** (Suppl.), S157–S170.
117. M. Weinstock, E. Gorodetsky, T. Poltyrev, A. Gross, Y. Sagi and M. B. Youdim, *Prog. Neuropsychopharmacol. Biol. Psychiatry*, 2003, **27**, 555–561.
118. M. Weinstock, T. Goren and M. B. Youdim, *Drug Develop. Res.*, 2000, **50**, 216–222.
119. J. Sterling, Y. Herzig, M. B. Youdim and M. Weinstock, Israel Chemical Co., 2000.
120. F. Hubalek, C. Binda, M. Li, Y. Herzig, J. Sterling, M. B. Youdim, A. Mattevi and D. E. Edmondson, *J. Med. Chem.*, 2004, **47**, 1760–1766.
121. J. Sterling, Y. Herzig, T. Goren, N. Finkelstein, D. Lerner, W. Goldenberg, I. Miskolczi, S. Molnar, F. Rantal, T. Tamas, G. Toth, A. Zagyva, A. Zekany, J. Finberg, G. Lavian, A. Gross, R. Friedman, M. Razin, W. Huang, B. Krais, M. Chorev, M. B. Youdim and M. Weinstock, *J. Med. Chem.*, 2002, **45**, 5260–5279.
122. Y. Sagi, M. Weinstock and M. B. Youdim, *J. Neurochem.*, 2003, **2**, 290–297.
123. Y. Akao, W. Maruyama, S. Shimizu, H. Yi, Y. Nakagawa, M. Shamoto-Nagai, M. B. Youdim, Y. Tsujimoto and M. Naoi, *J. Neurochem.*, 2002, **82**, 913–923.

124. W. Maruyama, Y. Akao, M. B. Youdim and M. Naoi, *J. Neural Transmission*, 2000, Suppl., 171–186.
125. W. Maruyama, M. Naoi and M. B. H. Youdim, in *Neurotoxic Factors in Parkinson's Disease and Related Disorders*, ed. A. Storch and M. A. Collins, New York, Kluwer Academic/Plenum Publishers, 2000, p. 321.
126. M. Mamelak, *Neurobiol. Aging*, 2007, **28**, 1340–1360.
127. A. Nunomura, G. Perry, G. Aliev, K. Hirai, A. Takeda, E. K. Balraj, P. K. Jones, H. Ghanbari, T. Wataya, S. Shimohama, S. Chiba, C. S. Atwood, R. B. Petersen and M. A. Smith, *J. Neuropathol. Exp. Neurol.*, 2001, **60**, 759–767.
128. S. G. Rhee, K. H. Kim, H. Z. Chae, M. B. Yim, K. Uchida, L. E. Netto and E. R. Stadtman, *Ann. N.Y. Acad. Sci.*, 1994, **738**, 86–92.
129. J. R. Stone and S. Yang, *Antioxid. Redox Signal.*, 2006, **8**, 243–270.
130. K. U. Schallreuter and S. Elwary, *Life Sci.*, 2007, **80**, 2221–2226.
131. X. Q. Xiao, N. T. Lee, P. R. Carlier, Y. Pang and Y. F. Han, *Neurosci. Lett.*, 2000, **290**, 197–200.
132. O. Bar-Am, O. Weinreb, T. Amit and M. B. Youdim, *J. Mol. Neurosci.*, 2009, **37**, 135–145.
133. M. B. Youdim, D. Edmondson and K. F. Tipton, *Nature Rev.*, 2006, **7**, 295–309.
134. M. C. Carrillo, C. Minami, K. Kitani, W. Maruyama, K. Ohashi, T. Yamamoto, M. Naoi, S. Kanai and M. B. Youdim, *Life Sci.*, 2000, **67**, 577–585.
135. O. Weinreb, O. Bar-Am, T. Amit, N. Drigues, Y. Sagi and M. B. Youdim, *Chem. Biol. Interact.*, 2008, **175**, 318–326.
136. S. Shoham, C. Bejar, E. Kovalev, D. Schorer-Apelbaum and M. Weinstock, *Neuropharmacology*, 2007, **52**, 836–843.
137. K. S. Vetrivel and G. Thinakaran, *Neurology*, 2006, **66**, S69–S73.
138. M. Yogev-Falach, T. Amit, O. Bar-Am, Y. Sagi, M. Weinstock and M. B. Youdim, *FASEB J.*, 2002, **16**, 1674–1676.
139. M. Yogev-Falach, O. Bar-Am, T. Amit, O. Weinreb and M. B. Youdim, *FASEB J.*, 2006, **20**, 2177–2179.
140. O. Bar-Am, T. Amit and M. B. Youdim, *J. Neurochem.*, 2007, **103**, 500–508.

CHAPTER 19

Designing Drugs with Dual Activity: Novel Dual Angiotensin II and Endothelin Receptor Antagonists

NATESAN MURUGESAN

Department of Medicinal Chemistry, Bristol-Myers Squibb Research and Development, Princeton, NJ 08543-5400, USA
Email: natesan.murugesan@bms.com

19.1 Introduction

Hypertension is a highly prevalent condition affecting approximately 65 million individuals in the United States, based on a preliminary report from the National Health and Nutrition Examination Survey (NHANES) 2005–2006.[1] Hypertension is a major risk factor for human morbidity and mortality through its effects on heart, brain, and kidneys. While many drugs are available for treating the disease, approximately one-third of the hypertensive population is still not adequately treated.[2] To achieve the target blood pressure (defined as systolic blood pressure (BP) of <140 mm Hg and diastolic BP of <90 mm Hg) recommended by the latest guidelines, multiple anti-hypertensive drugs, including angiotensin II receptor blockers (ARB), ACE inhibitors, beta-blockers, calcium channel blockers, and diuretics are commonly needed in most patients, often in combination.[3] However, there is still a significant medical

RSC Drug Discovery Series No. 21
Designing Multi-Target Drugs
Edited by J. Richard Morphy and C. John Harris

Published by the Royal Society of Chemistry, www.rsc.org

need for newer therapeutic drugs that effectively reduce BP across a wide variety of patients.

Human essential hypertension involves two powerful and well characterized endogenous vasoconstrictors angiotensin II (Ang II) and endothelin-1 (ET-1).[4] The endogenous polypeptide Ang II plays a major pathophysiological role in human hypertension, and it is now well known that antagonism of the action of Ang II comprises one of the most effective therapies for the treatment of hypertension and congestive heart failure.[5] Endothelin-1 is one of the most potent vasoconstrictors known, being 10 times more potent than angiotensin II on a molar basis.[6] Since its discovery in 1988, knowledge of ET-1 and its actions has been pursued vigorously which has resulted in over 18 000 publications to date. These studies have revealed that endothelin is an important physiological mediator, regulating blood pressure through effects on blood vessels, heart, and kidneys, and that the ET system can be overactive in disease states such as hypertension, pulmonary hypertension, heart failure, and renal disease.[7] This myriad of pathologies which are associated with the ET system has resulted in extensive research efforts to discover both selective and nonselective antagonists of its receptors. The advances in this area have led to a number of ET antagonists either approved for clinical use or under development.[8]

A considerable body of preclinical evidence has shown that simultaneous antagonism of both the renin–angiotensin system and the endothelin system can produce a greater reduction in blood pressure and added cardiovascular benefit than antagonizing either system alone.[9] In fact, AT_1 receptor blockade alone leads to an increase in endothelin-1 production. Thus, it is expected that dual AT_1 and ET_A receptor antagonism in humans could be more effective than current standard therapies for treating hypertension and other related disorders. In this chapter, we will describe the design and development of single molecules which show combined potent antagonism of both the AT_1 and ET_A receptors. A lead drug candidate in this class of dual action receptor antagonists (DARAs) has shown exciting preclinical as well as clinical data and presents a potentially unique opportunity as an anti-hypertensive drug.

19.2 Angiotensin II Receptor Antagonists

The renin–angiotensin system (RAS) is an important regulator of blood pressure and fluid-electrolyte homeostasis.[4c,5a,10] The RAS has been implicated in pathogenesis of hypertension, congestive heart failure, and chronic renal failure. Inhibitors of the RAS, including angiotensin-converting enzyme (ACE) inhibitors, angiotensin II receptor blockers (ARBs), and direct renin inhibitors are commonly used in the treatment of hypertension. The RAS involves a proteolytic cascade in which the aspartyl protease renin catalyzes the conversion of angiotensinogen into a biologically inactive substance, angiotensin I. Angiotensin-converting enzyme (ACE) then converts Ang I to the physiologically active octapeptide angiotensin II (Ang II), which causes potent

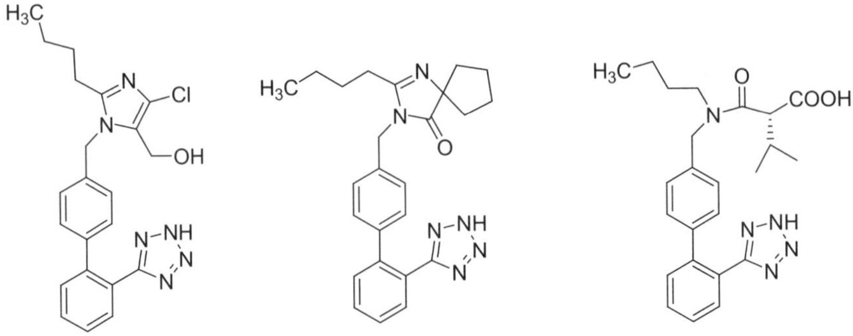

Losartan, 1 Irbesartan, 2 Valsartan, 3

Figure 19.1

vasoconstriction, increased aldosterone secretion, and enhanced sympathetic activation. The actions of angiotensin II are mediated through two specific G-protein coupled receptor subtypes named AT_1 and AT_2.

Angiotensin II receptor blockers represent a class of effective and well tolerated orally active anti-hypertensive drugs.[4c] Losartan (**1**) was the first drug of this class to reach the market, followed by irbesartan (**2**), valsartan (**3**), and others (Figure 19.1). ARBs selectively displace angiotensin II from the AT_1 receptor and produce their blood pressure lowering effects by antagonizing angiotensin II-induced vasoconstriction, aldosterone release, catecholamine release, arginine vasopressin release, water intake, and hypertrophic response. Monotherapy with ARBs in mild-to-moderate hypertension controls blood pressure in 40 to 50% of these patients; however, these drugs are often combined with thiazide diuretics because blocking the action of angiotensin II alone is frequently not sufficient to control hypertension.[11]

19.3 The ET System

The endothelins (ET-1, ET-2, and ET-3) constitute a family of 21-amino acid peptides, and are produced by the endothelium of blood vessels and by many other tissues.[4a,b] ET-1 is the major cardiovascular isoform, and is an extremely powerful vasoconstrictor. A large number of studies have elucidated that endothelins play many important roles in addition to vasoconstriction, including modulation of vascular tone, cell proliferation, and hormone production. The biosynthesis of ET-1 starts with an inactive prepropeptide that is cleaved by a furin-like convertase to yield 'big ET-1,' a 38-amino acid propeptide, with approximately 1% of the biological activity of the mature peptide. Cleavage of 'big ET-1' at the Trp-Val bond by endothelin-converting enzyme (ECE) releases the biologically active 21-amino acid peptide.[12]

It is generally believed that the ET peptides are not stored in secretory granules, but rather are produced in response to specific stimuli. The synthesis of these separate but related peptides is regulated at multiple levels. Transcription of the preproET-1 gene occurs in response to stimuli such as hypoxia, shear stress, and ischemia. Angiotensin II and TGF-α also stimulate ET-1 production. The half-life of the mRNA is approximately 15–20 minutes and the plasma half-life of ET-1 is approximately 4–7 minutes; thus, ET-1 levels can be rapidly adjusted as required.[12b]

19.3.1 Endothelin Receptors

The downstream effects of the ET peptides are mediated by two distinct receptor subtypes, endothelin-A (ET_A) and endothelin-B (ET_B), both of which belong to the rhodopsin-like G-protein coupled receptor superfamily.[12b,13] ET_A receptors display higher affinity for ET-1 over ET-2 and ET-3 while ET_B receptors bind all three ET isopeptides with equal affinity. Endothelin receptors are expressed in a variety of vascular as well as non-vascular human tissues, and possess different function depending on species and location. ET_A receptors are expressed abundantly on the vascular smooth muscle cells where they mediate vasoconstriction and cell proliferation. Stimulation of ET_A receptors in the myocardium yields positive inotropy and may also modulate the expression of genes leading to hypertrophy. The major site of ET_B receptor expression is on the vascular endothelium; activation of these receptors causes release of nitric oxide and prostacyclin which elicit vasodilation. A small population of ET_B receptors is also expressed on some vascular and non-vascular smooth muscle cells, where their activation leads to contraction. Clearance of ET-1 from the circulation depends on pulmonary ET_B receptors.[14]

19.3.2 Endothelin Receptor Antagonists

This myriad of pathologies which are associated with the endothelins has resulted in extensive research efforts to discover both selective and non-selective antagonists of their receptors.[15] The advances in this area have led to a number of ET antagonists which have either been approved for clinical use or are in clinical development.[16] Currently, the mixed antagonist, bosentan, and the ET_A selective antagonists, sitaxentan and ambrisentan, have been approved for the clinical treatment of pulmonary arterial hypertension.[8] The use of such compounds for the broader therapy areas of hypertension,[17] heart failure,[7b] cancer,[18] and nephropathy[19] is still being evaluated in late-stage clinical trials.

A series of biphenylsulfonamides has been reported by workers at Bristol-Myers Squibb as potent and selective ET_A antagonists.[20] In this medicinal chemistry effort, it was shown that compounds with a heterocyclic ring such as an oxazole (**4**, Figure 19.2) at the 4′-position of the biphenyl moiety showed

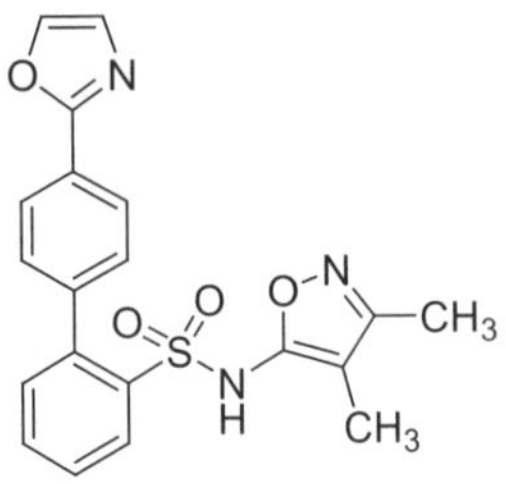

4

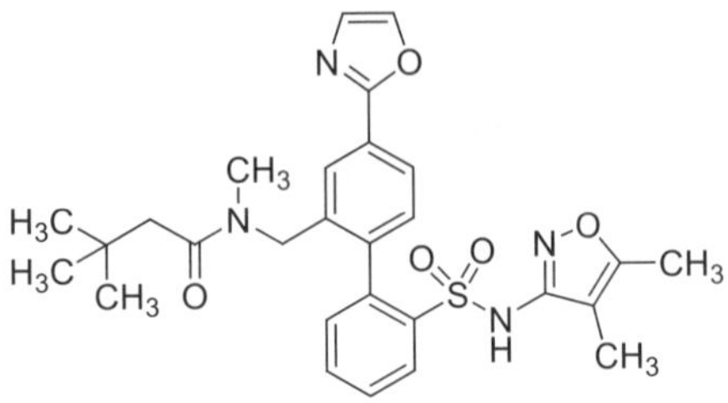

5

Figure 19.2

remarkable enhancement in potency and metabolic stability versus the 4′-unsubstituted biphenyl. This work led to the selection of BMS-193884 (**4**: $ET_A K_i = 1.4$ nM; $ET_B K_i = 18\,700$ nM) as a clinical development compound. In a phase I single ascending dose tolerance study, BMS-193884 was well tolerated up to a dose of 600 mg administered orally and showed a dose-dependent increase in plasma drug concentration. BMS-193884 also showed promising hemodynamic effects in a phase II clinical trial for congestive heart failure.[21]

Additional SAR studies at the 2′-position of **4** led to the identification of several analogs with improved binding affinity as well as selectivity for the ET_A receptor. Following the discovery that a 3-amino-isoxazole group displayed significantly improved metabolic stability in comparison to its 5-regioisomer, the 3-amino-isoxazole group was combined with the optimal 2′-neopentyl *N*-methyl amide substituent, leading to the second generation analog, **5** (BMS-207940). Compound **5** is an extremely potent ($ET_A K_i = 10$ pM) and selective (80 000-fold for ET_A versus ET_B) antagonist. The bioavailability of **5** was 100% in rats and the systemic clearance and volume of distribution were higher than that of **4**. In rats, intravenous **5** blocked big ET pressor responses with a 30-fold greater potency than **5**. After oral dosing at 3 μmol/kg, **5** displayed enhanced pharmacodynamic duration relative to **4**.[22]

19.4 Dual Angiotensin II and Endothelin Receptor Antagonists

19.4.1 Scientific Rationale

Angiotensin II augments the production and action of endothelin, and endothelin increases the production and function of angiotensin II, thus creating a positive feedback loop.[9a] Numerous preclinical studies have shown that simultaneous antagonism of both AT_1 and ET_A receptors may result in greater efficacy versus single receptor antagonists in diseases where either ET_A and/or AT_1 have been implicated.[9b] The superior therapeutic benefit of dual AT_1 and

ET_A receptor antagonism was demonstrated in a canine model of renovascular hypertension. The combination of an AT_1 receptor antagonist (losartan) with a mixed ET_A/ET_B antagonist (bosentan) produced a 40 mm Hg reduction in mean blood pressure, compared to a 20 mm Hg decrease with losartan alone.[9b] In another study, using a rat model of hypertension and heart failure, the combination of losartan with an ET_A receptor antagonist (LU-135252) worked synergistically to return blood pressure, heart weight, and mortality levels to those of the non-hypertensive controls, effectively reversing the heart failure in the rats.[9a] The greater physiological benefit of dual AT_1 and ET_A receptor antagonism has been demonstrated in several other animal models of hypertension such as in DOCA-salt rats, in spontaneously hypertensive rats (SHRs), and in diabetic rats.[23] Thus, it is hypothesized that concomitant blockade of AT_1 and ET_A receptors in humans should be more effective than current standard therapies for treating hypertension and other related disorders. In addition to causing hypertension, the angiotensin and endothelin endocrine/paracrine pathways also promote cell proliferation, inflammation, and fibrosis at the tissue level and vascular level. Therefore, it is also anticipated that DARA compounds would have additional benefit in treating/preventing cardiovascular and renal disease by reducing tissue remodeling and potential end organ damage via inhibition of the angiotensin–endothelin pathways. A report that described molecular recognition similarities within the AT_1 and ET_A receptors[24] lends further additional impetus to the discovery of dual action receptor antagonists (DARAs).

19.4.2 Rationale of DARA Drug Design

The ET_A receptor antagonist **4** shares the same biphenyl core as a large number of AT_1 receptor antagonists, including irbesartan (**2**) (Figure 19.3). In addition, ET_A receptor antagonists and AT_1 receptor antagonists share the same requirement of an acidic group at the 2-position of the biphenyl moiety (a

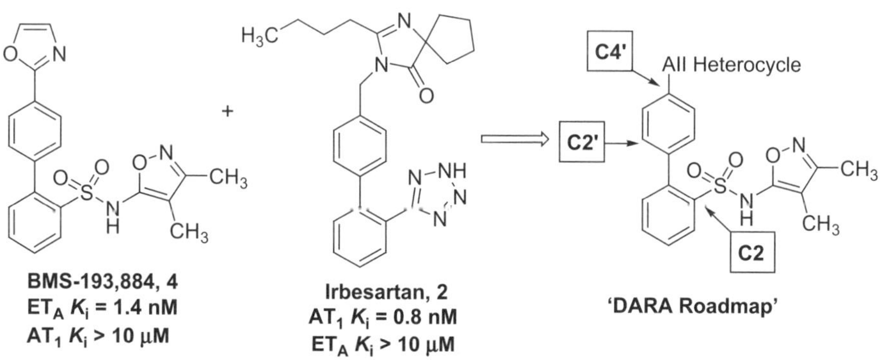

Figure 19.3

tetrazole for irbesartan and an oxazolylsulfonamide for compound **4**). Merging the AT_1 and ET_A receptor antagonist chemotypes for the preparation of a DARA requires a number of significant structural balancing acts. Our 'road-map' (Figure 19.3) for using the template of BMS-193884 (**4**) as an advanced scaffold for the preparation of a DARA required the following conditions.

a) The isoxazole sulfonamide located at C2 of the biphenyl core was demonstrated to be crucial to ET_A receptor activity,[20b] and in any biphenyl DARA it would be a requirement. The AT_1 receptor must tolerate the specific heteroaryl sulfonamide required for ET_A antagonism.
b) The 'AII-heterocycle' has been defined as the heterocyclic/heteroaromatic group located at the 4′-position on the biphenyl core of the DARA molecules (Figure 19.3). A large number of variations of this group have been used in known AT_1 receptor antagonists, and this group imparts the bulk of the AT_1 receptor affinity to the biphenyl scaffolds in the program. The ET_A receptor would need to be able to tolerate larger and more diverse groups than oxazole at C4′ of the biphenyl to accommodate heterocycles which would confer AT_1 receptor affinity.
c) SAR generated in our efforts to find potent ET_A receptor antagonists suggested that substitution at the C2′-position of the biphenyl core of an ET_A antagonist often led to an increase in ET_A receptor affinity. This approach improved the ET_A receptor affinity by >100-fold in some cases and was seen as a key option in the design of an ET_A/AT_1 dual receptor antagonist. Thus, the strategy of the DARA program was to garner as much AT_1 receptor affinity as possible using a potent 'AII heterocycle' at the 4′-position of the biphenyl, and then enhance ET_A receptor affinity with a potent 2′-substituent.

Following the strategy described above, incorporation of the irbesartan imidazolidinone 'AII heterocycle' at the 4′-position of the biphenyl moiety of BMS-193884 produced the biphenyl sulfonamide **6**. Compound **6** was only approximately 6-fold less potent than irbesartan at the AT_1 receptor (Scheme 19.1).[25] Unfortunately, its ET_A receptor affinity was diminished compared to **4** (approximately 40-fold less potent than **4**). This data clearly suggested that the imidazolidinone AII heterocycle at C4′ of the biphenyl was tolerated by the ET_A receptor but with a loss of potency compared with the simpler oxazole, and that the AT_1 receptor accommodated the isoxazolylsulfonamide moiety required for ET_A antagonism.

Our previous studies also demonstrated that certain substituents located at the C2′-position of the biphenyl moiety could be used to increase ET_A receptor affinity. In these studies, cyclic amides such as pyrrolidinones were especially useful for improving ET_A receptor activity. In an effort to improve the potency of **6**, a pyrrolidinone side chain was appended at the 2′-position of **6** leading to **7**. Compound **7** had potent affinity for both AT_1 ($IC_{50} = 6\,nM$) and ET_A ($IC_{50} = 10\,nM$) receptors (Scheme 19.1 and Table 19.2). Thus, AT_1 receptor potency was maintained as we added the necessary structural features to

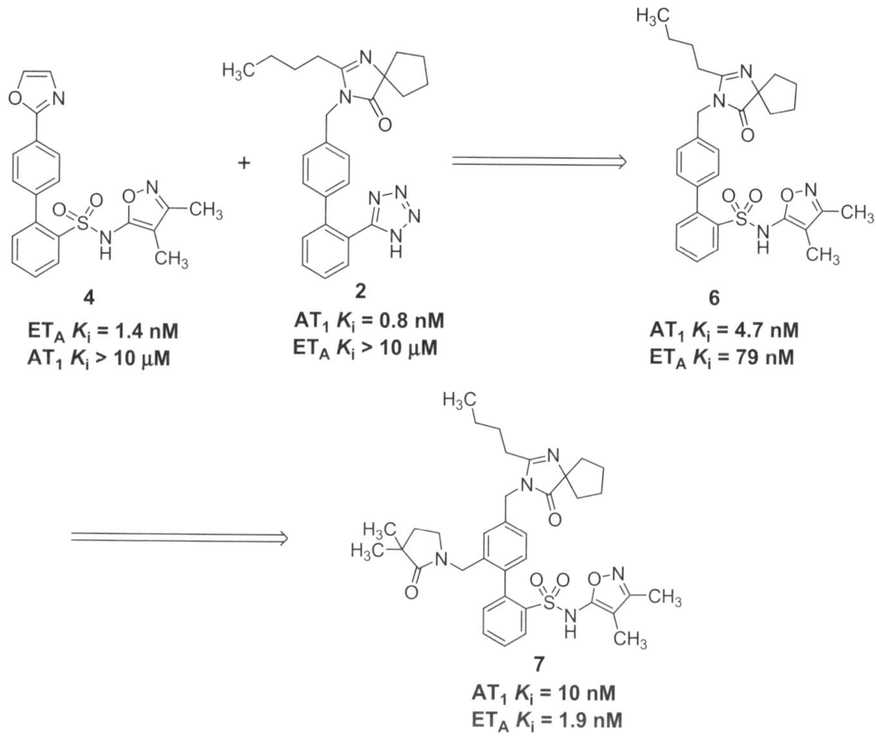

Scheme 19.1

enhance ET_A receptor affinity. Compound **7** showed no activity against either the AT_2 or ET_B receptor subtypes (IC_{50} values of $>10\,\mu M$).

To study the *in vivo* hypertensive (pressor) effects of Ang II and ET, these peptides can be injected directly into rats, causing substantial transient increases in blood pressure (BP). These effects can be blocked by pretreating the rats with either AT_1 or ET_A receptor antagonists, respectively. While irbesartan **2** was active only in blocking the pressor effect of Ang II, and **4** was active only in blocking bET-1 pressor effect, DARA **7** was uniquely effective as an inhibitor in both Ang II and bET-1 pressor studies when administered orally, indicating that dual antagonism of ET_A and AT_1 receptors could be achieved simultaneously *in vivo*. The oral bioavailability of **7** in rats at $10\,\mu g/kg$ was good (%F = 38) with excellent oral exposure ($C_{max} = 3.1\,\mu M$) and a reasonable elimination profile ($T_{1/2} = 5.5$ hours). However, in subsequent studies, **7** had low oral bioavailability (%F < 10) in dogs and monkeys. Given the prediction of high permeability from the Caco-2 cell assay (150 nm/s at pH 5.5), first pass metabolism of compound **7** was suspected as the cause of its poor oral bioavailability in higher species. These results prompted additional exploration of the SAR around **7** with the goal of finding new DARA analogs with improved pharmacokinetic properties (i.e., improved oral bioavailability in dogs and monkeys) and improved dual receptor potency.

Due to the stringent structural requirements at the 4′-imidazolinone as well as at the isoxazolyl sulfonamide moieties, attention was initially focused at optimizing the 2′-position of the molecule. A large number of 2′-substituted analogs of **7** were evaluated. Fortunately, a number of small lipophilic groups such as propyl, methoxymethyl, and ethoxymethyl were tolerated by both receptors and, in some cases, even enhanced the potency compared to compound **7**. In addition, a possible site of metabolism to be addressed could be the 5-aminoisoxazole ring, since it was shown in earlier studies that this moiety was susceptible to presystemic metabolism, resulting from enzymatic cleavage of the 5-isoxazole ring by bacteria in the rat GI tract.[22] Since we had also shown that analogs containing the 3-isoxazole regioisomer appeared to be resistant to this type of degradation of the isoxazole ring, 3-isoxazole DARA analogs were also explored.

On the basis of the above analysis, analogous combinations of the 3-isoxazole moiety with an optimal 2′-ethoxymethyl substituent afforded **8** (BMS-346567). Analog **8** (Figure 19.4) showed excellent *in vitro* and *in vivo* potency against both AT_1 and ET_A receptors (Table 19.2). Significantly, **8** ($hAT_1 K_i = 0.8$ nM and $hET_A K_i = 9.3$ nM) was 12-fold more potent than our earlier lead **7** at the human AT_1 receptor, but was 5-fold less active than **7** for the human ET_A receptor.[26]

The pharmacokinetic profile of **8** is summarized in Table 19.1. Compound **8** showed good oral bioavailability in rats, dogs, and monkeys, averaging 40%, 86%, and 21% F, respectively. The PK profile was substantially improved compared to **7**, which showed less than 10% oral bioavailability in dogs and monkeys.

In conscious normotensive rats, compound **8** dose-dependently antagonized the Ang II induced pressor response with an ED_{50} value of 0.8 μmol/kg i.v. and 3.6 μmol/kg p.o. As an AT_1 receptor antagonist, it was found to be substantially more potent and longer acting than irbesartan in this model. As reported previously, selective ET antagonists such as BMS-193884 were inactive in this Ang II-induced pressor model. Compound **8** was also more

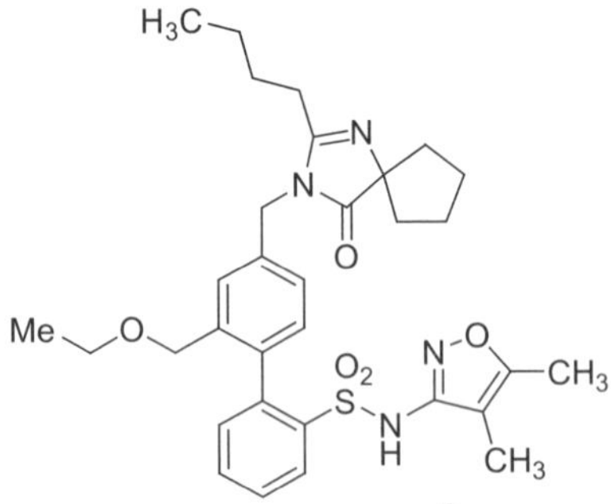

8

Figure 19.4

Table 19.1 SAR of 2′-substituted *N*-3-isoxazolyl biphenylsulfonamide derivatives.

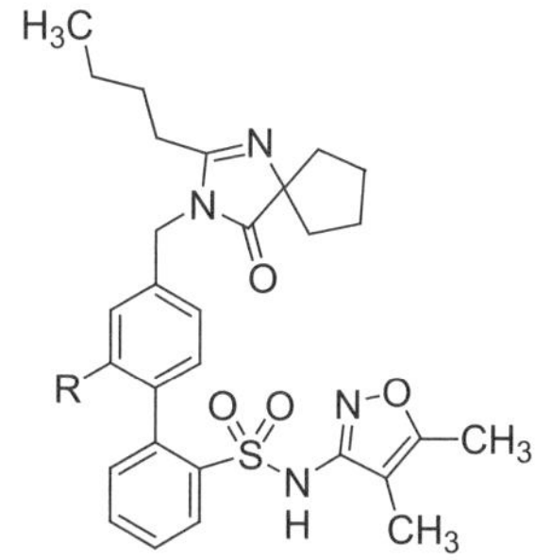

Compound	*R*	*AT_1 K_i (nM)* Rat	human	*hET_A K_i (nM)*	*Oral AII Pressor (30 μM/kg) (AUC units)*	*Oral big ET Pressor (30 μM/kg) (AUC units)*
1	—	0.8 ± 0.2	1.1 ± 0.3	$>10^4$	13,500	inactive
2	—	$>10^4$	$>10^4$	1.4 ± 0.2	inactive	9,700
7	—	7.9	10	1.9	5,800	7,900
8	$-CH_2OCH_2CH_3$	0.4	0.8	9.3	18,600	15,800

Table 19.2 Pharmacokinetic properties of **8**.[a]

Species, dose in mg/kg	*Rat, 6*	*Dog, 6.8*	*Cyno, 6*
Cl (mL/kg/min)	1.9 ± 0.6	1.3 ± 0.5	13.2 ± 0.8
Vss (L/kg)	0.5 ± 0.05	0.2 ± 0.03	1.3 ± 0.5
Plasma $T_{1/2}$ (hr)	4.1 ± 0.6	2.4 ± 1.1	13.6 ± 1.1
Mean residence time (hr)	5.1 ± 1.1	2.3 ± 0.5	1.6 ± 0.5
Cmax (μg/ml)	12 ± 3.1	11 ± 3	1.7 ± 0.9
T_{max} (hr)	0.6 ± 0.4	0.7 ± 0.3	0.5 ± 0.2
Oral bioavailability (%)	40	86	21

[a]All data are shown as mean ± SD ($n = 3$ for all studies except monkey oral leg which was $n = 5$).

efficacious and longer acting as a ET_A receptor antagonist than BMS-193884 in the big ET-1-induced pressor model. It should be noted that irbesartan and other ARBs are inactive in the big ET pressor model. Thus, compound **8** truly demonstrated potent DARA activity *in vitro* and *in vivo*. Compound **8** was also more potent and longer acting than **7** in these two pressor models as measured by the AUC values (Table 19.2).

A number of other known ‘AII heterocycles’ from the literature have been evaluated at the 4′-position of the biphenylsulfonamide DARAs. Although the irbesartan ‘AII heterocycle’ has consistently provided the most potent DARA compounds, a number of analogs containing other ‘AII heterocycles’ also showed comparable potency.[27] Notable among them are the imidazole amide derivative **11**, the substituted pyridine **13** and the cycloheptenoimidazole **14** (Table 19.3).[27,28]

Table 19.3 Substitution of the 4′-position of biphenylsulfonamides with different AT heterocycles.

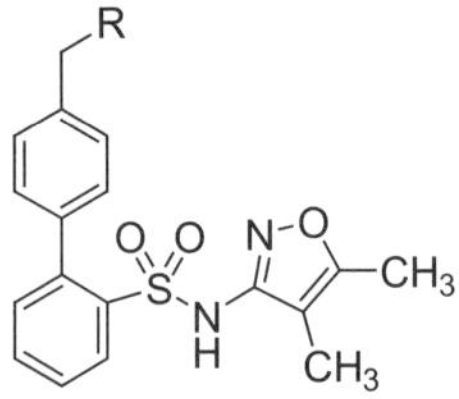

Compound	*R*	AT_1 $K_i(nM)$	ET_A $K_i(nM)$
1	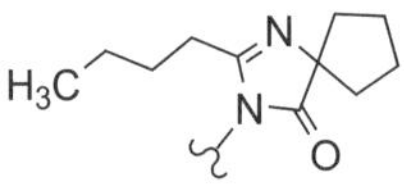	4.7	79
9	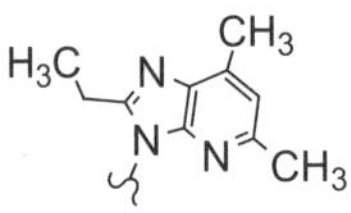	1.8	16000
10	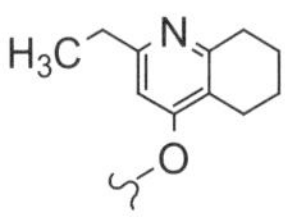	11	590
11	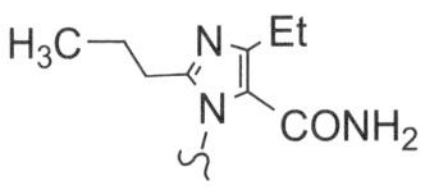	7	55
12	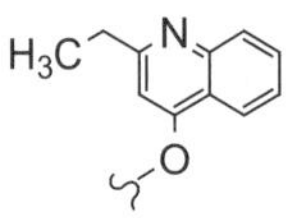	81	1000
13	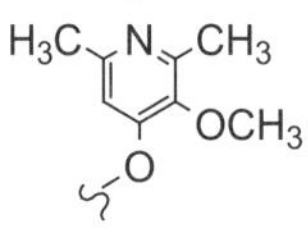	4	119
14	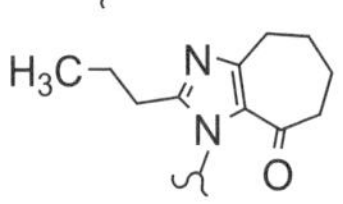	8	105
15	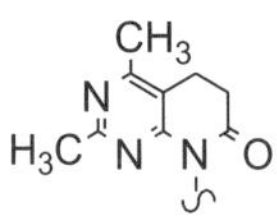	20	74
16	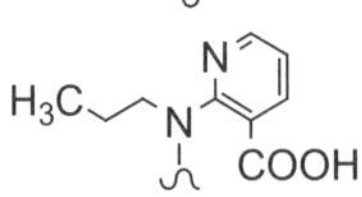	48	750
17	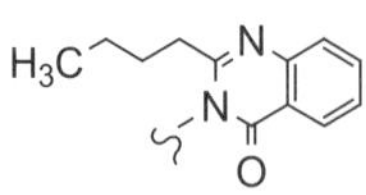	73	950
18	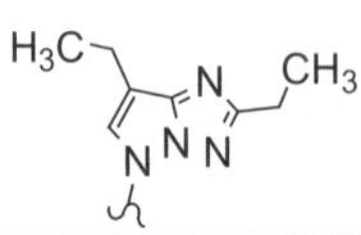	77	240

Analogs were then prepared which match these 'AII heterocycles' with optimal 2′-substituents. Inclusion of the 2′-ethoxymethyl substituent on the biphenyl core of these compounds, in particular, enhanced both AT_1 and ET_A receptor binding affinity. Thus, optimized compounds **19–21** showed potent AT_1 and ET_A binding affinity (Table 19.4) and all three analogs are orally active in the rat models of angiotensin-mediated as well as endothelin-mediated hypertension.[27]

Analog **22**, containing the 4′-valine amide moiety from the AT_1 antagonist valsartan, displayed potent DARA activity even in the absence of a 2′-substituent (Table 19.5).[29] However, this compound was subject to rapid metabolic clearance and exhibited low systemic exposure and marked inter-animal variability in oral exposure in cynomolgus monkeys. Subsequent *in vitro*

Table 19.4 SAR of 2′-substituted *N*-5-isoxazolyl biphenylsulfonamide derivatives.

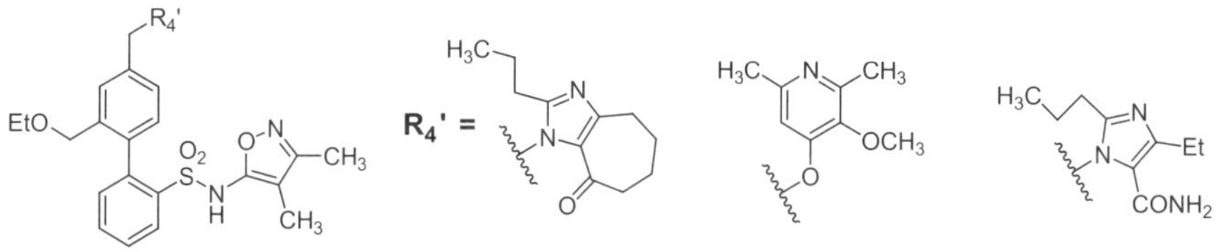

Compound #	*19*	*20*	*21*
AT_1 IC_{50} (nM)	5	3	3
ET_A IC_{50} (nM)	20	13	20

Table 19.5 Reduction of site-specific CYP3A-mediated metabolism by substitution of the isoxazole ring.

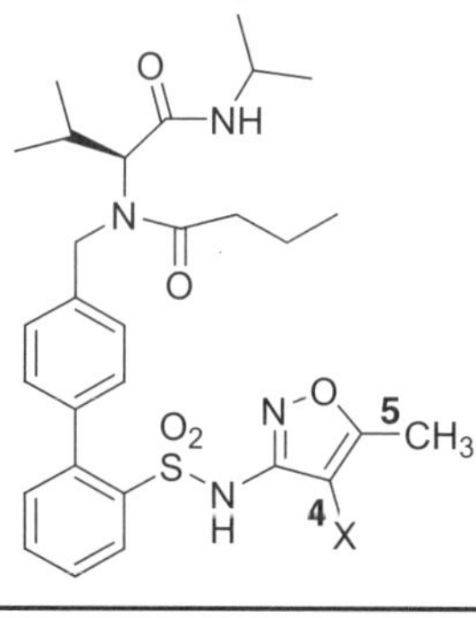

Compound	*X*	*AT_1 IC_{50} (nM)*	*ET_A IC_{50} (nM)*	*CYP3A4*
22	CH_3	5 nM	2 nM	45 μM (BFC) 15 μM (BzRes)
23	Cl	1.5 nM	1.5 nM	55 μM (BFC) 20 μM (BzRes)
24	F	2.0 nM	4.0 nM	>100 μM (BFC) 50 μM (BzRes)

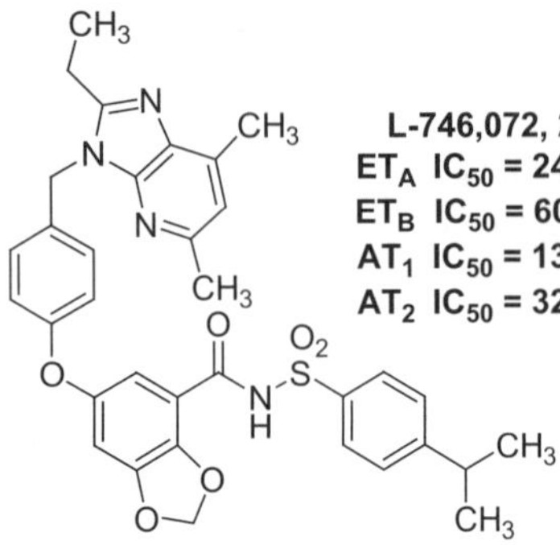

Figure 19.5

metabolism studies indicated that compound **22** was a substrate solely for cytochrome P450 CYP3A4, and was not metabolized by other human P450 enzymes. Analysis of key metabolites of **22** led to the identification of the 5-methyl group on the isoxazole as a metabolic soft spot. Hydroxylation of the isoxazole 5-methyl group was found to be the primary metabolite. As part of a strategy to block or slow down metabolism by modifying the methyl groups, it was found that replacement of the 4-methyl group on the isoxazole ring with halogens (Cl or F) not only improved overall metabolic stability but also decreased CYP3A-mediated hydroxylation of the isoxazole 5-methyl group (Table 19.5). As exemplified by the fluoro analog **24**, this structural modification resulted in an increase in the systemic exposure relative to previous analogs and a dramatic reduction in inter-animal variability in oral exposure in the monkeys after p.o. administration.[29] In addition, **24** was a substrate for metabolism by both CYP2C9 and CYP3A4 enzymes, thus avoiding the reliance on a single P 450 enzyme for metabolic clearance. Thus, this effort enabled the modulation of site-specific CYP3A4 metabolism and provided DARA analogs with improved overall metabolic profiles.[29,30]

A moderately potent dual AT_1 and ET_A receptor antagonist (L-746072, **25**) (Figure 19.5) was previously reported by Greenlee and co-workers at Merck.[31] However, this compound was non-selective for the AT_2 or the ET_B receptors and oral activity in this series of compounds had not been demonstrated. A number of biaryl isoxazolylsulfonamide DARA analogs related to **8** have also been reported in the patent literature.[32]

19.5 *In Vivo* Activity of DARA

Based on its substantially improved *in vitro* and *in vivo* profile versus **7**, **8** (BMS-346567) was chosen for additional evaluation. Compound **8** showed good oral bioavailability in rats, dogs, and monkeys, averaging 40%, 86%, and 21% F, respectively (Table 19.1). The PK profile was substantially improved relative to **7**, which showed less than 10% oral bioavailability in dogs and monkeys.

Spontaneously hypertensive rats (SHRs) are believed to model human essential hypertension. A number of important classes of anti-hypertensive drugs, such as calcium channel blockers, β-blockers, ACE inhibitors, and AT_1 receptor antagonists, have shown efficacy in this model, as well as in treating human essential hypertension. Using telemetrized SHRs, the ability of compound **8** to lower BP was compared to irbesartan in SHRs at increasing doses, both compounds starting at 10 μmol/kg/day (days 15–21) to 30 μmol/kg/day (days 22–28) and finally to 100 μmol/kg/day (days 29–35) over a 36-day period (Figure 19.6). Compound **8** showed a significant lowering of blood pressure at the lowest dose tested (10 μmol/kg/day), but this effect was not distinguishable from the effects of irbesartan at the same dose (Figure 19.6). At 30 μmol/kg/day, **8** was more effective compared to controls and the same dose of irbesartan (Figure 19.6). This trend continued throughout the remainder of the study. In fact, at the 100 μmol/kg/day dose, **8** reduced the blood pressure from 170 to less than 100 mm Hg during the course of the drug's pharmacodynamic duration. This result was significantly more efficacious than the effect of irbesartan at the same dose. The clear added efficacy of **8** in reducing blood pressure at the higher doses when compared to irbesartan was most likely attributable to the ET_A receptor antagonism activity of **8**, which is lacking in irbesartan. In fact, on the last day of the study at the 100 μmol/kg/day dose of **8**, the average BP of the SHRs treated with **8** was reduced by almost 70 mm Hg to 90 mm Hg for about 4 hours, almost doubling the BP reduction achieved by irbesartan at

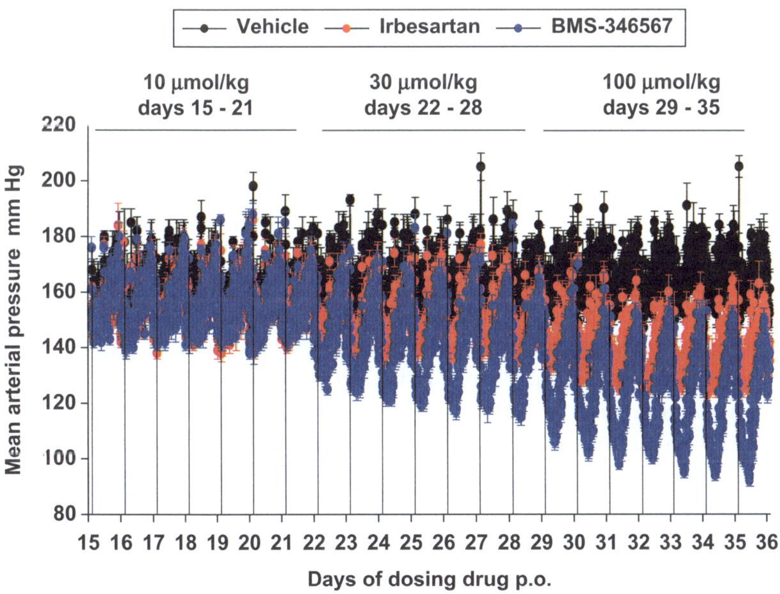

Figure 19.6 Effect of dosing irbesartan and **8** (BMS-346567) on mean arterial pressure in telemeterized SHR at 10, 30, and 100 μmol/kg/day.

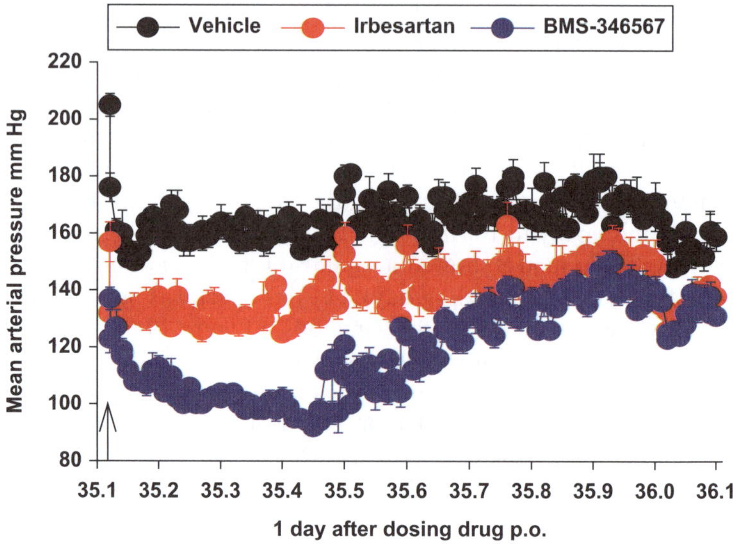

Figure 19.7 Effect of dosing irbesartan and **8** (BMS-346567) on mean arterial pressure in telemeterized SHR on the 7th day of dosing at 100 μmol/kg (day 35).

100 μmol/kg. Compound **8** at 100 μmol/kg/day essentially converted the SHR into normotensive rats (Figure 19.7) during the course of its pharmacodynamic action.[33]

19.6 Clinical Studies of DARA

Pharmacopeia (a subsidiary of Ligand Pharmaceuticals), licensed BMS-346567 from Bristol-Myers Squibb and advanced this compound as PS-433540, for the potential treatment of hypertension and diabetic nephropathy. In phase I clinical trials in normal subjects, PS433540 was well tolerated and demonstrated a pharmacokinetic profile consistent with once-daily oral administration. PS433540 fully blocked an increase in blood pressure in an angiotensin challenge test at both the 250 mg or 500 mg dose levels.

In a phase II clinical trial in hypertensive patients ($n = 170$), PS433540 given at either 200 mg or 500 mg once-daily produced a statistically significantly greater reduction in blood pressure compared to placebo. The drug was well tolerated by patients. Most of the adverse events reported were mild or moderate in severity and included headaches and minor musculoskeletal and respiratory complaints.[34]

In a 261-patient, randomized, double-blind, phase IIa trial, PS433540 was found to be safe and well tolerated and demonstrated statistically significant greater reductions in blood pressure than placebo. Patients were randomized to once-daily doses of PS433540 at 200 mg, 400 mg, or 800 mg doses, placebo, or irbesartan at a 300 mg dose for 12 weeks. PS433540 at 200 mg, 400 mg, and

800 mg reduced systolic blood pressure by 13, 14, and 23 mm Hg, respectively. For diastolic blood pressure, the reductions with PS433540 were 7, 9, and 14 mm Hg respectively. All doses reduced blood pressure statistically significantly greater than placebo. The 800 mg daily dose of PS433540 produced a statistically significant reduction in systolic and diastolic pressure when compared to irbesartan. The lower dose levels of PS433540 did not give anti-hypertensive effects that were statistically significantly from irbesartan.[35] Although efficacy was demonstrated in the phase IIa study with BMS-346567, this compound is not under active clinical development.

19.7 Summary

In conclusion, by rational design we have combined the structural features of AT_1 receptor antagonists such as irbesartan with those of BMS-193884, an ET_A-selective endothelin receptor antagonist, into single molecules. The resultant analogs show potent dual AT_1 and ET_A receptor antagonist activity. Additional optimization resulted in the discovery of BMS-346567 as a highly potent and orally bioavailable dual antagonist which blocked blood pressure elevations caused by intravenous infusion of Ang II or big ET-1 to a greater extent and with longer duration than either AT_1 or ET_A receptor antagonists alone. BMS-346567 clearly demonstrated superior efficacy versus irbesartan (a selective AT_1 receptor antagonist) in the normal SHR model of hypertension, demonstrating the synergy of AT_1 and ET_A receptor blockade in a single molecule. In phase I and phase II clinical trials, BMS-346567 (PS433560) was found to be safe and well tolerated, and demonstrated statistically significant greater reductions in blood pressure than placebo in hypertensive patients. The 800 mg oral dose of PS433540 produced a statistically significantly greater reduction in blood pressure than irbesartan (which was tested at its highest approved dose of 300 mg). Overall, these preclinical and clinical data demonstrate that the DARA mechanism has the potential to be a significant new addition to the armamentarium of anti-hypertensive drugs, and further holds the promise of being cardioprotective agents.

Acknowledgements

I would like to thank Drs John Macor, William R. Ewing, and Peter Cheng for their critical review of the manuscript and constructive comments.

References

1. J. A. Cutler, P. D. Sorlie, M. Wolz, T. Thom, L. E. Fields and E. J. Roccella, *Hypertension*, 2008, **52**, 818.
2. H. P. Gavras, *Am. J. Hypertension*, 2001, **14**, 231S.

3. (a) B. Waeber and H. R. Brunner, *J. Hypertension*, 2001, **19**, S9; (b) M. Moser, *J. Hypertension*, 2002, **20**, S3.
4. (a) M. Yanagisawa, H. Kurihara, S. Kimura, Y. Tomobe, M. Kobayashi, Y. Mitsui, Y. Yazaki, K. Goto and T. Masaki, *Nature (London)*, 1988, **332**, 411; (b) A. Inoue, M. Yanagisawa, S. Kimura, Y. Kasuya, T. Miyauchi, K. Goto and T. Masaki, *Proc. Natl. Acad. Sci. USA*, 1989, **86**, 2863; (c) R. R. Wexler, W. J. Greenlee, J. D. Irvin, M. R. Goldberg, K. Prendergast, R. D. Smith and P. B. Timmermans, *J. Med. Chem.*, 1996, **39**, 625.
5. (a) G. K. Aulakh, R. K. Sodhi and M. Singh, *Life Sci.*, 2007, **81**, 615; (b) Z. Cao, M. E. Cooper, L. L. Wu, A. J. Cox, K. Jandeleit-Dahm, D. J. Kelly and R. E. Gilbert, *Hypertension*, 2000, **36**, 561; (c) A. M. Salam, *Expert Opin. Invest. Drugs*, 2003, **12**, 879.
6. B. J. Epstein and S. Anderson, *Expert Rev. Cardiovasc. Ther.*, 2009, **7**, 675.
7. (a) R. C. A. Tostes and M. N. Muscara, *Curr. Drug Targets: Cardiovasc. Haematol. Disord.*, 2005, **5**, 287; (b) G. Ertl and J. Bauersachs, *Drugs*, 2004, **64**, 1029.
8. S. G. Raja, *Cardiovasc. Ther.*, 2010, **28**, e65.
9. (a) J. Bohlender, S. Gerbaulet, J. Kramer, M. Gross, M. Kirchengast and R. Dietz, *Hypertension*, 2000, **35**, 992; (b) P.-E. Massart, D. G. Hodeige, H. Van Mechelen, A. A. Charlier, J.-M. Ketelslegers, G. R. Heyndrickx and J. E. Donckier, *J. Hypertension*, 1998, **16**, 835.
10. (a) G. B. Appel and A. S. Appel, *Prog. Cardiovasc. Dis.*, 2004, **47**, 105; (b) P. R. Conlin, *Int. J. Clin. Practice*, 2005, **59**, 214.
11. (a) P. Meier, M. Maillard and M. Burnier, *Curr. Drug Targets: Cardiovasc. Haematol. Disord.*, 2005, **5**, 15; (b) A. B. Ribeiro, *Curr. Med. Res. Opin.*, 2006, **22**, 1; (c) R. E. Schmieder, *Am. J. Hypertension*, 2005, **18**, 720.
12. (a) Y. Kawanabe and S. M. Nauli, *Cell. Mol. Life Sci.*, 2011, **68**, 195; (b) A.-K. Khimji and D. C. Rockey, *Cell. Signalling*, 2010, **22**, 1615.
13. S. W. Watts, *Am. J. Physiol.*, 2010, **298**, R254.
14. (a) G. P. Rossi, M. Cavallin, G. G. Nussdorfer and A. C. Pessina, *J. Cardiovasc. Pharmacol.*, 2001, **38**, S49; (b) P. H. Sugden, *J. Mol. Cell. Cardiol.*, 2003, **35**, 871; (c) R. M. Touyz and E. L. Schiffrin, *Can. J. Physiol. Pharmacol.*, 2003, **81**, 533.
15. J.-D. Aubert and L. Juillerat-Jeanneret, *Expert Opin. Ther. Targets*, 2009, **13**, 1069.
16. (a) M. Kirchengast and M. Luz, *J. Cardiovasc. Pharmacol.*, 2005, **45**, 182; (b) M. J. Palmer, *Prog. Med. Chem.*, 2009, **47**, 203; (c) C. Wu, *Expert Opin. Ther. Patents*, 2006, **16**, 1337; (d) J. Iqbal, R. Sanghi and S. K. Das, *Mini-Rev. Med. Chem.*, 2005, **5**, 381.
17. D. E. Kohan, *Curr. Opin. Nephrol. Hypertension*, 2010, **19**, 134.
18. A. Bagnato and L. Rosano, *Int. J. Biochem. Cell Biol.*, 2008, **40**, 1443.
19. M. Barton, *Biochim. Biophys. Acta, Mol. Basis Dis.*, 2010, **1802**, 1203.
20. (a) N. Murugesan, Z. Gu, P. D. Stein, S. Spergel, S. Bisaha, E. C. K. Liu, R. Zhang, M. L. Webb, S. Moreland and J. C. Barrish, *Bioorg. Med. Chem. Lett.*, 2002, **12**, 517; (b) N. Murugesan, Z. Gu, P. D. Stein, S. Spergel,

A. Mathur, L. Leith, E. C. K. Liu, R. Zhang, E. Bird, T. Waldron, A. Marino, R. A. Morrison, M. L. Webb, S. Moreland and J. C. Barrish, *J. Med. Chem.*, 2000, **43**, 3111.
21. M. Hulpke-Wette and R. Buchhorn, *Curr. Opin. Invest. Drugs (Pharma-Press Ltd.)*, 2002, **3**, 1057.
22. N. Murugesan, Z. Gu, S. Spergel, M. Young, P. Chen, A. Mathur, L. Leith, M. Hermsmeier, E. C. K. Liu, R. Zhang, E. Bird, T. Waldron, A. Marino, B. Koplowitz, W. G. Humphreys, S. Chong, R. A. Morrison, M. L. Webb, S. Moreland, N. Trippodo and J. C. Barrish, *J. Med. Chem.*, 2003, **46**, 125.
23. (a) S. M. Gardiner, J. E. March, P. A. Kemp, J. J. Mullins and T. Bennett, *Br. J. Pharmacol.*, 1995, **116**, 2237; (b) S. Dhein, S. Hochreuther, C. A. D. Spring, K. Bollig, C. Hufnagel and M. Raschack, *J. Pharmacol. Exp. Ther.*, 2000, **293**, 351; (c) T. Ikeda, H. Ohta, M. Okada, N. Kawai, R. Nakao, P. K. S. Siegl, T. Kobayashi, T. Miyauchi and M. Nishikibe, *J. Cardiovasc. Pharmacol.*, 2000, **36**, S337.
24. D. Dascal, V. Nirula, K. Lawus, S.-E. Yoo, T. F. Walsh and K. Sandberg, *FEBS Lett.*, 1998, **423**, 15.
25. N. Murugesan, J. E. Tellew, Z. Gu, B. L. Kunst, L. Fadnis, L. A. Cornelius, R. A. F. Baska, Y. Yang, S. M. Beyer, H. Monshizadegan, K. E. Dickinson, B. Panchal, M. T. Valentine, S. Chong, R. A. Morrison, K. E. Carlson, J. R. Powell, S. Moreland, J. C. Barrish, M. C. Kowala and J. E. Macor, *J. Med. Chem.*, 2002, **45**, 3829.
26. N. Murugesan, Z. Gu, L. Fadnis, J. E. Tellew, R. A. F. Baska, Y. Yang, S. M. Beyer, H. Monshizadegan, K. E. Dickinson, M. T. Valentine, W. G. Humphreys, S.-J. Lan, W. R. Ewing, K. E. Carlson, M. C. Kowala, R. Zahler and J. E. Macor, *J. Med. Chem.*, 2005, **48**, 171.
27. J. E. Tellew, R. A. F. Baska, S. M. Beyer, K. E. Carlson, L. A. Cornelius, L. Fadnis, Z. Gu, B. L. Kunst, M. C. Kowala, H. Monshizadegan, N. Murugesan, C. S. Ryan, M. T. Valentine, Y. Yang and J. E. Macor, *Bioorg. Med. Chem. Lett.*, 2003, **13**, 1093.
28. Z. Gu, L. Fadnis, J. E. Tellew, B. Kane, L. Cornelius, R. A. Baska, S. Beyer, H. Monshizadegan, M. Valentine, S. Chong, R. A. Morrison, K. Carlson, W. R. Ewing, M. Kowala, J. E. Macor and N. Murugesan, *Abstracts of Papers*, 226th ACS National Meeting, New York, NY, United States, September 7–11, 2003, MEDI.
29. W. R. Ewing, N. Murugesan, Z. Gu, L. Fadnis, Y. Zhu, G. Yu, X. Wu, L. Cornelius, H. Mason, R. Zahler, D. Zhang, S. Authur, R. A. Baska, H. Monshizadegan, C. Ryan, K. Carlson, M. Kowala, C. D'Arienzo, H. Zhang, W. G. Humphreys and J. E. Macor, *Abstracts of Papers*, 232nd ACS National Meeting, San Francisco, CA, United States, September 10–14, 2006, MEDI.
30. (a) H. Zhang, D. Zhang, W. Li, M. Yao, C. D'Arienzo, Y.-X. Li, W. R. Ewing, Z. Gu, Y. Zhu, N. Murugesan, W.-C. Shyu and W. G. Humphreys, *Drug Metab. Dispos.*, 2007, **35**, 795; (b) W. R. Ewing, Y. Zhu, N. Murugesan, Z. Gu, L. Fadnis, X. Wu, L. A. Cornelius, J. E. Macor, R. Zahler, B. Kunst, D. Zhang, S. Auther, R. Baska, H. Monshizadegan,

C. Ryan, K. Carlson, W. G. Humphreys, H. Zhang, G. Yu, H. Mason, J. Tellew, M. Kowala and C. D'Arienzo, *Abstracts*, 38th Middle Atlantic Regional Meeting of the American Chemical Society, Hershey, PA, United States, June 4–7, 2006, MRM.

31. T. F. Walsh, K. J. Fitch, D. L. Williams, Jr., K. L. Murphy, N. A. Nolan, D. J. Pettibone, R. S. L. Chang and S. S. O'Malley, B. V. Clineschmidt, *et al.*, *Bioorg. Med. Chem. Lett.*, 1995, **5**, 1155.
32. (a) A. Cole, P. A. Jokiel, M.-R. Brescia, L. Qin, I. Henderson, (Ligand Pharmaceuticals Inc., USA). Application: WO 2009158309, 2009; (b) A. G. Cole, P. A. Jokiel, M.-R. Brescia, L. Qin, I. Henderson, (Ligand Pharmaceuticals, Inc., USA). Application: WO 2009155448, 2009.
33. M. C. Kowala, N. Murugesan, J. Tellew, K. Carlson, H. Monshizadegan, C. Ryan, Z. Gu, B. Kane, L. Fadnis, R. A. Baska, S. Beyer, S. Arthur, K. Dickinson, D. Zhang, M. Perrone, P. Ferrer, M. Giancarli, J. Baumann, E. Bird, B. Panchal, Y. Yang, N. Trippodo, J. Barrish and J. E. Macor, *J. Pharmacol. Exp. Ther.*, 2004, **309**, 275.
34. J. M. Teutel, *et al.*, American Society of Hypertension 2008 Annual Meeting, May 16, 2008, New Orleans, LA.
35. Ligand Pharmaceuticals, *Press Release*, February 2, 2009.

CHAPTER 20

Ethyl Urea Inhibitors of the Bacterial Type II Topoisomerases DNA Gyrase (GyrB) and Topoisomerase IV (ParE)

STEPHEN P. EAST,[a,*] LLOYD G. CZAPLEWSKI[b] AND DAVID J. HAYDON[b]

[a] Evotec (UK) Ltd, 114 Milton Park, Abingdon, Oxfordshire, OX14 4SA, UK; [b] Biota Europe Limited, Begbroke Business & Science Park, Sandy Lane, Yarnton, Oxfordshire, OX5 1PF, UK
*Email: Stephen.East@evotec.com

20.1 Introduction

In the broadest sense, antibacterial chemotherapy might be described as the archetypal multi-target approach to drug discovery and development. The use of empiric antibacterial drugs to treat a single indication that is probably caused by one or more bacterial species, and the desire for broad-spectrum antibacterials to treat multiple indications, places high demands on the specificity and spectrum of activity of a new antibacterial compound. A successful new broad-spectrum antibacterial must be promiscuous enough to inhibit perhaps 30 or more bacterial species and numerous sub species and

RSC Drug Discovery Series No. 21
Designing Multi-Target Drugs
Edited by J. Richard Morphy and C. John Harris

Published by the Royal Society of Chemistry, www.rsc.org

strains and yet, at the same time, be selective enough not to inhibit mammalian cells. This represents a daunting challenge even in the simplest case of a compound inhibiting a single target in each bacterial species, but when a multi-targeting antibacterial compound is considered the demand escalates.[1]

A dual-targeting antibacterial may require a compound to inhibit 60 plus related but different and sometimes quite divergent bacterial targets but not the closest human orthologues. Despite this seemingly impossible challenge, many successful antibacterials manage to do this rather well. The huge advantage that dual-targeting antibacterials offer is their very low level of spontaneous resistance development. Unless there are very special circumstances, target-based spontaneous resistance to a single-targeting drug is likely to occur at a frequency around 5×10^{-8} to 1×10^{-9} at best. A compound targeting two such independent essential targets may have a spontaneous resistance frequency of up to 10^{-16} (the product of two single target frequencies of 10^{-8} each) leading to a long effective clinical lifetime for the product before significant levels of spontaneous resistance develop in the population.

There are several approaches being pursued by academia and the pharmaceutical industry to identify new multi-target inhibitors to combat bacterial resistance: (i) combination therapies; (ii) hybrid inhibitors; and (iii) single pharmacophore multi-target inhibitors.[2,3]

In the combination therapy approach different drugs acting on different biological targets are co-administered so that if resistance develops against one of the drugs the other will still elicit its antibacterial effect. This principle is sound but the execution to identify the most appropriate combination therapies is often challenging; however, there are examples in the marketplace that include Bactrium®, Synercid® and Klacid HP7®. An alternative combination approach is to combine a drug with an inhibitor of the protein responsible for resistance to the drug: for example, Augmentin® combines the penicillin amoxicillin with a beta-lactamase inhibitor clavulanic acid.[4]

In the hybrid approach the small molecule drug is designed such that it combines pharmacophoric features of two or more established antibacterials that interact with different targets. The pharmacophores can be linked in the form of a pro-drug, whereby a metabolic event releases the antibacterial drugs *in vivo*. Alternatively, the pharmacophores can be covalently linked but they are still able to act on the different molecular targets. The latter approach is an exciting strategy but optimisation of the pharmacophores and the pharmacokinetic profiles is difficult. Theravance Inc. has advanced a hybrid bifunctional compound that is the marriage of a glycopeptide antibiotic with a cephalosporin antibiotic compound into clinical trials (TD-1792).[5] Cumbre have gone one step further with a tri-targeting rifamycin-quinolone hybrid antibiotic (CBR-2092) designed to treat Gram-positive infections. The rifamycin SV, an RNA polymerase inhibitor, and 4H-4-oxo-quinolizine, a potent and balanced inhibitor of DNA gyrase and topoiosomerase IV, are joined with a chiral linking group. The three inhibitory activities are maintained in the hybrid leading to exceptionally low levels of spontaneous resistance development at $<10^{-12}$ for *Staphylococcus aureus*.[6]

In the third multi-target strategy to evade existing bacterial resistance mechanisms, one pharmacophore acts on multiple targets that are independently essential. Ideally, the inhibitors should have similar potency against each target such that if resistance to one target develops the drug is still active on the other and hence is still an effective treatment. This approach is the subject of this chapter and, in particular, the focus will be on the ATP-binding domains of the bacterial type II topoisomerases DNA gyrase and topoisomerase IV.

20.2 Function of Bacterial Type II Topoisomerases

The type II bacterial topoisomerases DNA gyrase and topoisomerase IV are enzymes that are highly conserved across all bacterial species. They are responsible for maintaining DNA topology and do so by a mechanism involving double-strand breakage, transfer of a DNA duplex through the break followed by ligation of the cleaved DNA strand. These functions are essential for DNA replication and transcription and hence bacterial cell survival.[7] Both enzymes form heterotetrameric complexes and this is a feature that distinguishes the prokaryotic type II topoisomerases from their related eukaryotic counterparts, which only form heterodimeric complexes. This difference in structure presents an opportunity to selectively inhibit life cycle processes in bacterial cells and potentially avoid any issues, in terms of mechanism of action, with toxicity in mammalian systems. The bacterial type II topoisomerases are therefore considered ideal antibacterial targets.

DNA gyrase consists of two GyrA and two GyrB subunits that together form an A_2B_2 complex. Topoisomerase IV is comprised of two ParC and two ParE subunits that together form a C_2E_2 complex. Although there are differences between DNA gyrase and topoisomerase IV in terms of specific structure and function, both targets catalyse DNA double-stranded breaks and subsequent ligation and they both bind and hydrolyse ATP to provide the energy to perform these tasks. Consequently the catalytic and ATP-binding sites in DNA gyrase and topoisomerase IV provide suitable opportunities for the identification of inhibitors containing a single pharmacophore that target both enzymes. Indeed the type II bacterial topoisomerases are well-validated targets for antibacterial chemotherapy as exemplified by the quinolone/fluoroquinolone class of antibacterials which target the catalytic site. Within this class of antibacterials dual inhibitors such as gemifloxacin and moxifloxacin (Figure 20.1) have been shown to act on both the GyrA and ParC subunits. Developing dual GyrA/ParC inhibitors is an exciting and active area of research but it falls outside of the scope of this chapter and more information on this topic can be found elsewhere.[8]

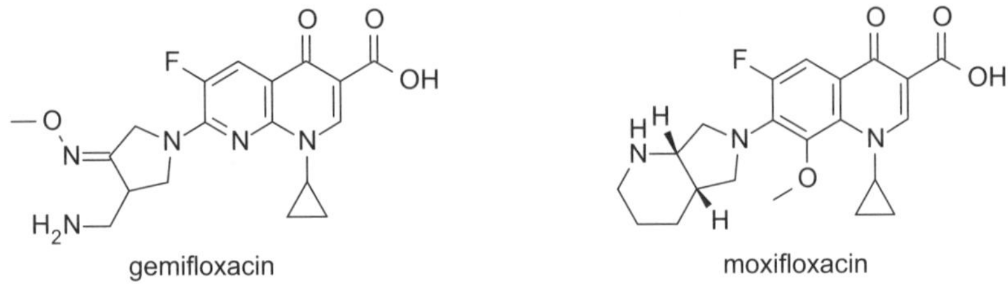

Figure 20.1 Dual-targeting GyrA/ParC fluoroquinoline antibacterials.

Inhibition of the ATP domains of DNA gyrase and topoisomerase has been the subject of an intense research effort but currently there are no compounds in the clinic. Nature has been the source of some inhibitors such as the aminocoumarins (e.g. novobiocin) and the cyclothialidines which compete with ATP binding to GyrB and to a lesser extent ParE. These chemical classes have provided significant interest from a structural and mechanistic perspective, and indeed novobiocin was approved for clinical use; however, it was subsequently withdrawn. A gap in the marketplace for antibacterials that are able to act on both GyrB and ParE inhibitors has therefore arisen. The advantage of targeting both enzymes is that these types of inhibitors would be expected to have lower spontaneous resistance frequencies. Additionally, inhibitors of the ATPase domains would also be a novel mechanism of action against the bacterial type II topoisomerases and hence they would be expected to overcome the resistance that has emerged in the GyrA/ParC antibacterials.

20.3 Structural Features of Ligand–Protein Complexes of GyrB and ParE

The emergence of structural information on GyrB and ParE has been instrumental in stimulating the search for new synthetic chemotypes that inhibit these enzymes.[9] From crystal structures of the hydrolytically stable ATP analogue ADPNP bound to *Escherichia coli* GyrB (N-terminal 24 kDa and 43 kDa fragments) together with GyrB structures containing novobiocin and cyclothialidine, Boehm *et al.*[10] observed that although these ligands adopt slightly different but overlapping binding modes, a network of hydrogen bonds involving Asp73 and a localised conserved water molecule (Figure 20.2) were present in all of the structures. Using this information, Boehm constructed a pharmacophore and performed a virtual screen to select low molecular weight fragments that could form the key hydrogen bonding interactions. This filtered subset of compounds was screened against GyrB and the most interesting chemotypes were confirmed using an array of biophysical techniques (analytical ultracentrifugation, surface plasmon resonance, nuclear magnetic resonance (NMR), X-ray). Subsequent structure-informed modification of one of the chemotypes led to the indazole (**1**) and suggested that in addition to the

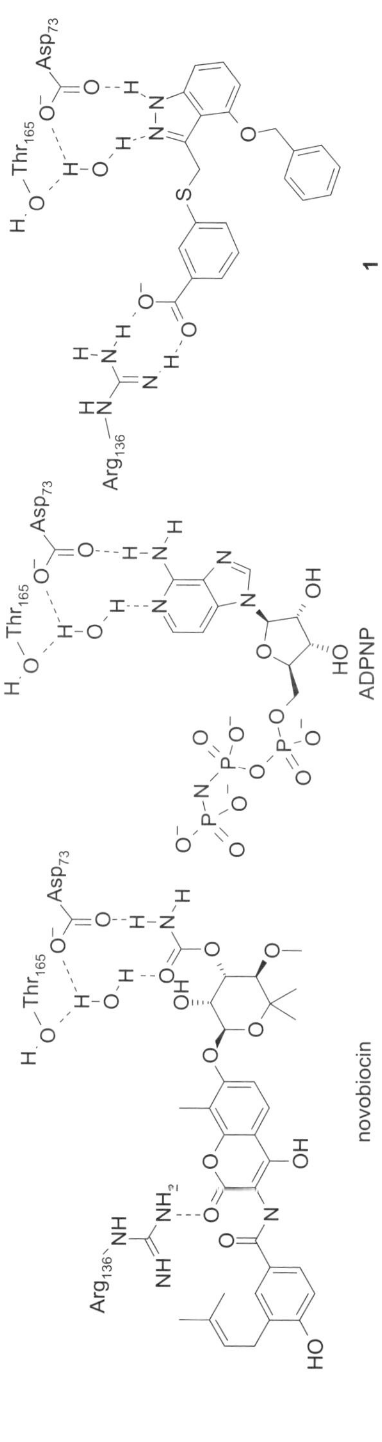

Figure 20.2 Key hydrogen-bonding interactions in *E. coli* GyrB structures complexed with novobiocin, ADPNP and **1**.

Asp73 H-bonding network, a stacking interaction with the Glu50-Arg76 salt bridge (not shown) and an additional interaction with Arg136 (observed in the novobiocin structure) were also important for ligand binding to GyrB.

Although this study by Boehm and co-workers did not provide information on how potent the indazole was as an inhibitor of ParE it is a key paper in the development of novel synthetic ATP competitive inhibitors of the type II bacterial topoisomerases and it also represents one of the first applications of fragment-based drug discovery using an array of techniques.[11]

Since the publication of the first GyrB structural information, data has also started to emerge on ParE. Bellon *et al.*[12] solved the structures of *E. coli* ParE (43 kDa N-terminal fragment) complexed with ADPNP and *E. coli* ParE (24 kDa N-terminal fragment) complexed with novobiocin and they aligned and compared the active sites with the corresponding related GyrB proteins (Table 20.1). They observed that novobiocin was ~5-fold less potent as a ParE inhibitor (*E. coli* IC_{50} 210 nM) than as a GyrB inhibitor (*E. coli* IC_{50} 46 nM) and they attributed this difference in potency to the single residue difference (GyrB Ile78 → ParE Met74) based on mutation studies.

Additional structural data has been published in the PDB by Vertex Pharmaceuticals that include *S. aureus* and *E. coli* GyrB crystallised with new synthetic inhibitors (PDB accession codes: 3G7B, 3G75 and 3G7E).[13] An *E. coli* ParE structure (PDB accession code: 3FV5) has also been deposited by Wei and co-workers with the benzimidazole ethyl urea **2** (Figure 20.3).[14] This last structure is of particular relevance as it represents the first crystal structure from the ethyl urea series of GyrB/ParE inhibitors to be published. The ethyl urea class of compounds are arguably the most well reported synthetic type II bacterial topoisomerase inhibitors that target the ATP-binding site and these are the subject of this chapter.

Table 20.1 A comparison of the key amino acid residues in the ligand binding sites of GyrB and ParE.

Key residues in ADPNP Binding Site (*E. coli*)		Key residues in novobiocin binding site (*E. coli*)	
43 kDa GyrB	43 kDa ParE	24 kDa GyrB	24 kDa ParE
Tyr5	Tyr5		
Asn42	Asn38		
Glu50	Glu46	Glu50	Glu46
Asp73	Asp69	Asp73	Asp69
Ile78	Met74	Ile78	Met74
		Gly81	Asp77
		Ile94	Ile90
Lys103	Lys99		
Tyr109	Tyr105		
		Arg136	Arg132
Thr165	Thr163	Thr165	Thr163
Gln335	Gln332		
Lys337	Lys334		

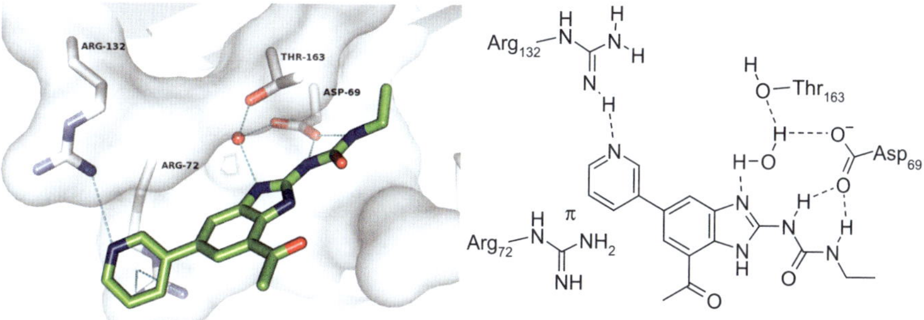

2

Figure 20.3 Key interactions of benzimidazole ethyl urea ligand **2** crystallised with *E. coli* ParE.

20.4 Ethyl Urea Inhibitors of GyrB and ParE

20.4.1 Benzimidazoles

The pioneering work on the dual-targeting GyrB/ParE inhibitors containing an ethyl urea moiety was performed by Vertex Pharmaceuticals. Following a high-throughput screen on GyrB using a focused library of ~30 000 compounds, the benzimidazole carbamate **3** surfaced as a modest inhibitor of both *S. aureus* (K_i 2 μM) and *E. coli* (K_i 20 μM) GyrB (Figure 20.4).[15] Compound **3** did not show inhibitory activity of *E. coli* ParE up to 60 μM and it was also inactive in antibacterial assays on *S. aureus*, *Streptococcus pneumoniae* and *Haemophilus influenzae* with no minimum inhibitory concentration (MIC) recorded up to 16 μg/ml.

Using the structural information on novobiocin crystallised into *E. coli* GyrB, Charifson *et al.* made two significant changes to compound **3** to improve the binding to GyrB. The carbamate oxygen was replaced with a nitrogen to improve the hydrogen bonding network of the ligand with the Asp73 residue and the second modification was to replace the 5-(4-hydroxy)styryl substituent on the benzimidazole core with a 3-pyridyl in order to enable an interaction with Arg136 (similar to the interaction of the coumarin ring in novobiocin). These two alterations to the structure of **3** provided the compound **4** that showed a dramatic increase in potency of *S. aureus* GyrB (K_i 130 nM) and an even more impressive increase in inhibitory potency of *E. coli* GyrB (K_i 81 nM). By improving the binding to the GyrB enzyme, Vertex were able to demonstrate antibacterial activity against *S. aureus* (MIC 16 μg/ml), *S. pneumoniae* (MIC 2 μg/ml) and *H. influenzae* (MIC 16 μg/ml). The compound was also shown to be a weak inhibitor of ParE (*E. coli* K_i 2.3 μM).

Analysis of the crystallographic data suggested that the 6- and 7-positions of the benzimidazole ring would be two appropriate sites at which to introduce substituents for further SAR exploration. In particular the 7-position appeared

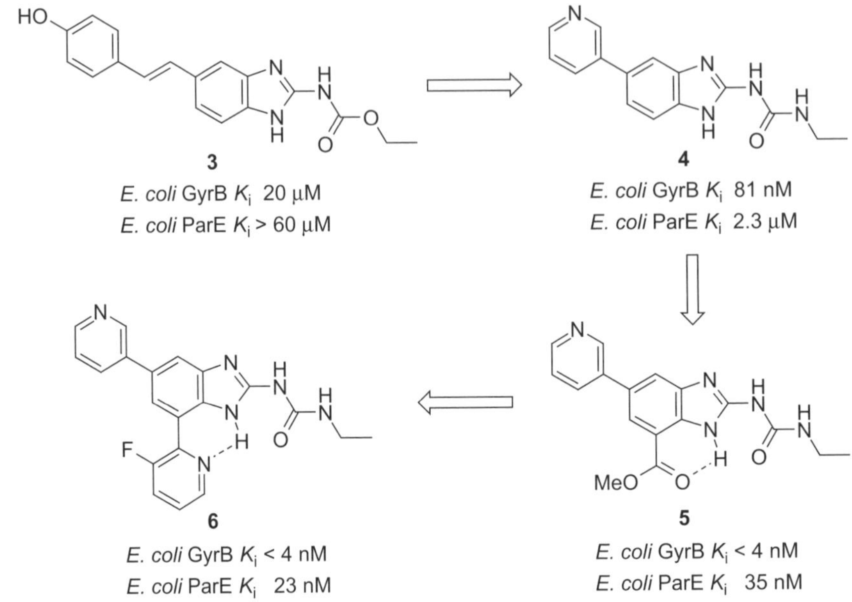

Figure 20.4 Evolution of the Vertex benzimidazole ethyl urea dual GyrB/ParE inhibitors.

to provide a suitable exit vector from the benzimidazole core that would access the space occupied by the ribose unit of ATP. The methyl ester derivative **5** was prepared and this compound was >15-fold more potent an inhibitor of GyrB (*S. aureus* K_i 8 nM, *E. coli* K_i <4 nM) compared with **4**. What was particularly striking about this derivative was that the increase in inhibitory potency of *E. coli* ParE was even more significant at >65× (K_i 35 nM). On the basis of the activity of further analogues with substituents in the 7-position (e.g. 1-pyrazole, 2-pyridyl – structures not shown) Charifson proposed that the exceptional and perhaps unexpected increase in inhibitory potency of ParE was the consequence of the ligands ability to adopt a coplanar conformation via a possible intramolecular hydrogen bonding with the benzimidazole NH. From analysis of co-crystal structures of benzimidazole complexes with ParE compared with GyrB it was postulated that the Met74 residue in ParE (the corresponding residue in GyrB is Ile78) results in narrowing of the ATP-binding pocket that the ligand occupies. The consequence of this is that flatter structures are able to interact more favourably with the wall of the protein in ParE. This was in line with the original differences noted by Bellon *et al.*[12] based on the novobiocin structures of GyrB and ParE. With the release of the co-crystal structure of **2** in *E. coli* ParE this certainly appears to be a plausible theory.

Of the 7-substituted compounds described in the SAR analysis conducted by Vertex, the compound **6** (VRT-752586), which contains a fluoropyridine in the 7-position, is one of the best characterised. This compound is a very potent inhibitor of GyrB (*S. aureus* K_i 14 nM; *E. coli* K_i < 4 nM) and ParE (*S. aureus* K_i <6 nM; *E. coli* K_i 23 nM).[16] In comparison with other compounds in this series it was commented that the relative ratio of the K_i values between the two targets is both compound structure and bacterial species specific. A counter-screen to check for activity against human topoisomerase II activity revealed that **6** showed no activity up to 25 μM, supporting the hypothesis that selectivity over related mammalian enzymes is possible.

MICs were measured for compound **6** against a large panel of resistant and susceptible aerobic, anaerobic and atypical bacteria. Of noteworthy importance, the MIC_{90} values were very low for Gram-positive organisms, e.g. *S. aureus* = 0.12 μg/ml (number of isolates/*n* = 54), *S. pneumoniae* = 0.03 μg/ml (*n* = 64), *Enterococcus faecalis* = 0.06 μg/ml (*n* = 35) and *Streptococcus pyogenes* = 0.12 μg/ml (*n* = 22). Some activity was shown against Gram-negative organisms, e.g. *H. influenzae* (MIC values = 0.5–>16 μg/ml, *n* = 36) and *Moraxella catarrhalis* (MIC values = 0.03–0.25 μg/ml, *n* = 23) suggesting that GyrB/ParE inhibitors might be useful for the treatment of respiratory tract infections. Interestingly there was no antibacterial activity against the wild type strain of *E. coli* (MIC >64 μg/ml) but both the *E. coli* permeability (*impA*) mutant and the efflux (*tolC*) mutant did show an MIC of ≤0.5 μg/ml suggesting that **6** has both limited permeability and is subject to efflux in *E. coli*.

To confirm that **6** was acting on GyrB/ParE in bacterial cells Grossman *et al.*[17] measured plasmid supercoiling in *E. coli* (using the *tolC* mutant) and they observed that the steady state levels were reduced in a dose-dependent manner. For a related compound in the benzimidazole series it was also reported that DNA synthesis and to a lesser extent RNA and protein synthesis were inhibited in *S. aureus*, thus confirming the mode of action on GyrB/ParE.

Through mutation studies it was revealed that both the structure of the compound and the species can play a role in the location of the primary mutant *in vitro*. For example in *S. aureus* the primary target for **6** based on the observed genetic mutations was GyrB whereas in *S. pneumoniae* the first mutants were observed in the gene that encodes ParE. Further experiments to identify the consequence of mutants in *E. faecalis* showed that for single mutations in either GyrB or ParE the effect on the MIC for **6** was minimal but when the doubly mutated bacterial strain was treated with **6** the MIC was significantly higher (>10× increase). What Grossman *et al.*[17] also observed was that spontaneous resistance frequencies at 2×, 4× and 8× the MIC of **6** are $<5.2\times10^{-10}$ but that these steadily increase with single mutations in either *gyrB* or *parE* and then more significantly to 1.2×10^{-8} in the doubly mutated strain of *E. faecalis* at 8× the MIC. Mani *et al.*[16] showed that in a more comprehensive analysis of spontaneous resistance frequencies in a wider panel of species at multiples of the MIC values, **6** completely suppressed the formation of mutants at 8× the MIC for all Gram-positive organisms tested. Resistance frequencies in *H. influenzae* were detectable suggesting that **6** may not inhibit both targets in this species. The conclusion from these experiments is that dual-targeting GyrB/ParE inhibitors show a beneficial effect at reducing the occurrence of resistance *in vitro*; however, they also showed that the primary target (i.e. GyrB or ParE) varies depending on the species. This phenomenon has also been observed in the GyrA/ParC dual inhibitors.[18]

The bactericidal activity of **6** was assessed *in vitro* using four bacterial strains (*S. aureus*, *S. pneumoniae*, *E. faecalis* and *H. influenzae*).[16] In all four cases the compounds demonstrated a 3-log unit reduction in colony-forming units in time kill experiments at 4× and 8× or 10× the MIC, which confirmed the possibility of GyrB/ParE dual inhibitors having a cidal activity. Charifson *et al.* presented the *in vivo* data for **6** and showed that it is efficacious in a *S. aureus* rat thigh infection model and it also shows a dose-dependent reduction of colony-forming units in a *S. pneumoniae* lung infection model.[15]

20.4.2 Triazolopyridines

Biota and Evotec[19] explored alternative [6,5]-fused bicyclic ring systems containing the ethyl urea moiety of **4** and prioritised those that maintained the key interactions with the Asp73 residue and the conserved water when docked into the ATP-binding site of 1EI1.[20] The 3-pyridyl was also fixed in an effort to maintain the interaction proposed by Vertex with the Arg136 residue, although in the docking model and in the structure deposited by Wei *et al.*[14] it appears more likely a π-interaction. This resulted in three compounds **7**–**9** (Figure 20.5) that inhibit *E. coli* GyrB with IC_{50} values < 10 μM. The imidazopyridine **7** and the triazolopyridine **8** also show some antibacterial activity, albeit weak, against *S. aureus* and *E. faecalis*. Both compounds were active against an efflux pump mutant strain of *E. coli* but they were ineffective against the wild-type strain, in line with observations noted by Vertex, i.e. **7** and **8** might also display limited permeability or greater susceptibility to efflux mechanisms in Gram-negative bacteria.

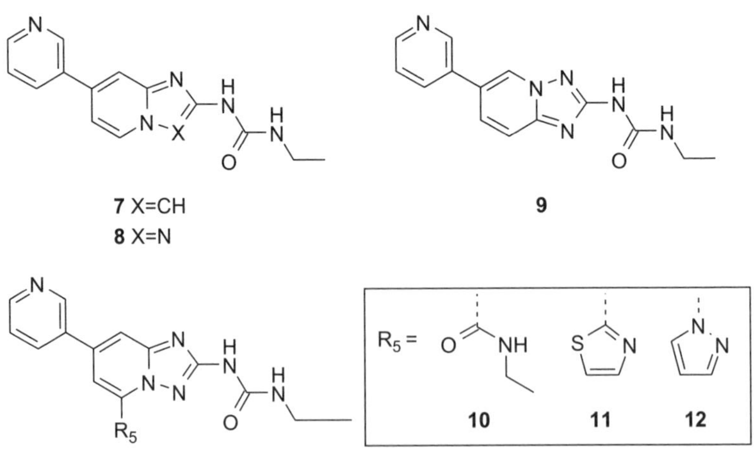

Figure 20.5 Biota/Evotec triazolopyridine ethyl ureas.

The triazolopyridine **8** is also a weak inhibitor of *E. coli* ParE (IC_{50} 145 μM) and it was selected for further elaboration to improve the potency. Based on the postulated binding mode and in comparison with the substitution pattern in the Vertex benzimidazoles, the 5-position was targeted for modification. This study led to three compounds, **10–12** with *E. coli* GyrB IC_{50} values <100 nM which represented a >50-fold improvement in the inhibitory activity of GyrB. The compounds also inhibit *E. coli* ParE although the improvement in potency against this target is less pronounced (>4×). Overall the increase in potency resulted in improved antibacterial activity. The compounds were more potent against Gram-positive pathogens (e.g. **12** MICs: *S. aureus* 2–4 μg/ml, *S. pneumoniae* 2 μg/ml, *E. faecalis* 2–4 μg/ml) although some weak activity against respiratory tract Gram-negative bacteria (e.g. **12** MICs: *H. influenzae* 16 μg/ml, *M. catarrhalis* 1 μg/ml) was noted. Pfizer has also synthesised triazolopyridine ethyl ureas as DNA gyrase inhibitors and reported compounds including **10** in a patent disclosure.[21] Antibacterial activity of **10** was measured in the Gram negative organism *Neisseria gonorrhoeae* and the observed MIC was 1 μg/ml.

Spontaneous resistance frequencies for **10–12** were estimated as $<1.8 \times 10^{-9}$ in *S. aureus* at multiples of 2×, 4× and 8× the MIC values, which is in line with the dual-targeting antibacterial activity in the benzimidazole series. Time kill experiments also in a *S. aureus* strain indicated a cidal mode of action with a >3-fold drop in colony-forming units after 15 hours;[22] again this data is consistent with compounds from the Vertex benzimidazole ethyl ureas. Compounds from this chemical series were active via intraperitoneal administration in a lethal murine *S. aureus* infection model and one compound offered complete protection to a group of 10 mice when dosed at 100 mg/kg at 1 h and 6 h post inoculation. Mortality was monitored for seven days, with all mice surviving the experiment. The structure of the compound active in this model is not disclosed.

20.4.3 Imidazopyridines

Starr *et al.*[23] described how the imidazopyridine scaffold in **7** could be adapted to include a pyrimidine (e.g. **13–16**) in the 5-position thus accessing the same region in GyrB and ParE as the fluoropyridine moiety in **6**. They evaluated the substitution pattern on the ethyl urea and concluded that whilst small substituents other than the ethyl group were tolerated in this position (e.g. cyclopropyl, *t*-butyl) they all showed a drop of potency against *S. pneumoniae* GyrB, *S. pneumoniae* ParE and in the antibacterial assays of at least 4-fold, suggesting that the ethyl moiety was optimal for this template. The SAR of a variety of 5-position pyrimidines was presented and modifications to the pyrimidine that retained co-planarity with the imidazopyridine scaffold were observed to be more potent in the enzymatic assay. This conclusion correlates with the proposal by Charifson *et al.*[15]

Certain modifications to the pyridyl group at the 7-position were also reported and an extensive body of data for the inhibition of GyrB and ParE

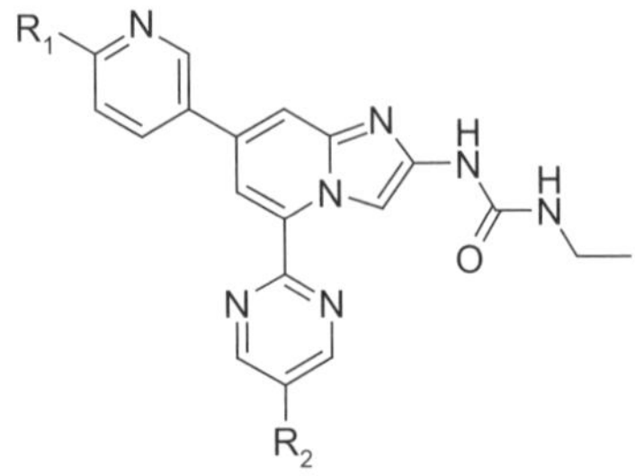

13 R_1 = H, R_2 = H
14 R_1 = H, R_2 = Me
15 R_1 = Me, R_2 = H
16 R_1 = OMe, R_2 = Me

Figure 20.6 Pfizer imidazopyridine ethyl ureas.

was discussed. Overall, the compounds in this chemical series were at least 4-fold more potent against GyrB *in vitro* compared with ParE; however, it is not clear if the differences in enzyme activity observed *in vitro* translate into reduced dual targeting in the bacteria. The two best compounds in terms of their dual enzyme inhibitory activity *in vitro* were **15** (IC_{50} *S. pneumoniae* GyrB 44 nM; IC_{50} *S. pneumoniae* ParE 119 nM) and **16** (IC_{50} *S. pneumoniae* GyrB 117 nM; IC_{50} *S. pneumoniae* ParE 147 nM) (Figure 20.6).

In the antibacterial assays, the imidazopyridines also showed good activity against Gram-positive organisms (e.g. **13** MIC values: *S. aureus* 0.5 μg/ml, *S. pyogenes* 0.25 μg/ml, *S. pneumoniae* 0.125 μg/ml). Of noteworthy importance is the activity against fluoroquinoline-resistant *S. pneumoniae* organisms (e.g. **13** MIC 0.25 μg/ml) underpinning the potential for GyrB/ParE inhibitors as an opportunity to overcome the resistance mechanisms associated with GyrA/ParC inhibitor class. Starr *et al.*[23] also reported the *in vivo* activity of compounds **13** and **14** in mouse *S. pyogenes* sepsis and *S. pneumoniae* lung infection models with encouraging PD_{50} values of ~20 mg/kg.

20.4.4 Benzothiazoles/Thiazolopyridines

In a series of patent disclosures, Biota has reported benzothiazole ethyl ureas and thiazolopyridine ethyl ureas (Figure 20.7). The first patent application describes examples that include compounds such as **17** where the benzothiazole scaffold is decorated with heteroaryl substituents in the 5- and 7-positions.[24] Compound **17** is reported to have an IC_{50} of <750 nM in the *E. coli* GyrB assay but no ParE data is provided. Extensive MIC data illustrated that these compounds are typically active against Gram-positive pathogens. Perhaps the most significant data presented is against multiple strains of *S. aureus* and *S. pneumoniae* where the antibacterial activity is <0.25 μg/ml and included in the *S. pneumoniae* panel were two fluoroquinoline-resistant organisms. *In vivo* data for **17** is also presented in the disclosure and in a murine septicaemia model 100% of the mice survived for 7 days (the duration of experiment) when dosed intraperitoneally with one 30 mg/kg dose, 1 hour after a lethal dose of *S. aureus* (survival rate of vehicle control = 0%).

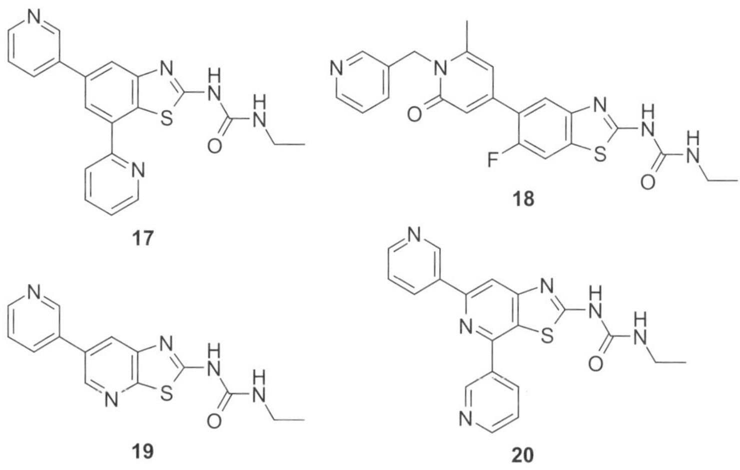

Figure 20.7 Biota benzothiazole and thiazolopyridine ethyl ureas.

A subsequent patent application from Biota presented benzothiazole compounds such as **18** where the substitution pattern is predominantly 5,6-disubstituted with small substituents such as F, MeO and Me being described in the 6-position.[25] The MIC_{90} values for **18** against clinical isolates are impressive and include but are not limited to the Gram-positive pathogens: *S. aureus* = 0.06 μg/ml ($n = 10$), *S. pneumoniae* = 0.015 μg/ml ($n = 10$), *Staphylococcus epidermidis* = 0.03 μg/ml ($n = 10$) and *E. faecalis* = 0.25 μg/ml ($n = 10$). Additionally, there are a few examples of thiazolopyridines presented (e.g. **19** and **20**[26] and the IC_{50} values of compounds in these disclosures are <750 nM in the *E. coli* GyrB assay. Compound **20** is reported to have MIC values of 0.25–1 μg/ml against *E. faecalis*, *S. pneumoniae* and *M. catarrhalis*.

Additional experiments performed by Biota to determine spontaneous resistant frequencies at 8× the MIC demonstrated a very low propensity for the development of mutant strains *in vitro* for members of the benzothiazole/thiazolopyridine series. The species tested included *S. aureus* (resistant frequency: $<6.9\times10^{-10}$), *S. epidermis* ($<1.7\times10^{-9}$), *E. faecalis* ($<5.3\times10^{-10}$), *H. influenzae* ($<7.6\times10^{-10}$) and *S. pyogenes* ($<1.6\times10^{-11}$). In *S. pneumoniae* the resistance frequencies were measured as $\sim7.7\times10^{-9}$ and the MIC values of the isolated mutants increase by ~4-fold on treatment with the same compound but the compound still showed good antibacterial activity with an MIC of 0.03 μg/ml. Another compound from this series is also active in the *S. aureus* septicaemia mouse model with a calculated ED_{50} of 1.73 mg/kg and it is also active in *S. aureus* thigh infection model in neutropenic mice.[27] The structures are not available.

Two additional patent applications from AstraZeneca[28] and Ranbaxy[29] have described thiazolopyridine and benzothiazole ureas respectively.

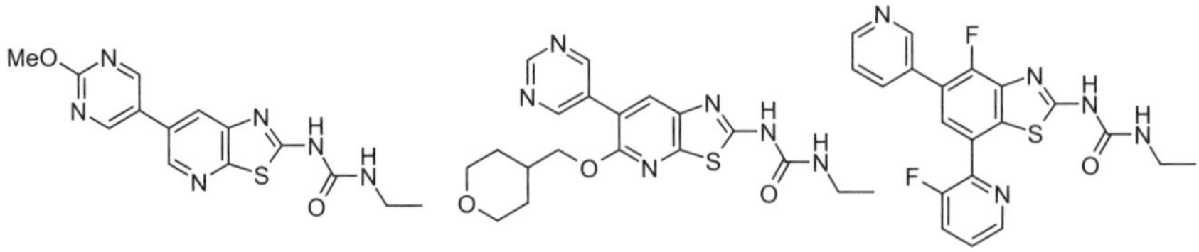

Figure 20.8 AstraZeneca thiazolopyridines and Ranbaxy benzothiazole ethyl ureas.

Representative examples from the AstraZeneca compounds are related to **19** and show good activity on *Mycobacterium smegmatis* GyrB (**21** IC_{50} 140 nM; **22** IC_{50} <3 nM) and antibacterial activity on *S. pneumoniae* (**21** MIC 0.5 µg/ml; **22** MIC 0.0625 µg/ml) (Figure 20.8). There are also a few oxazolopyridines reported in the same disclosure but they are significantly less active in the GyrB and antibacterial assays. The Ranbaxy patent application is more closely related to analogues of compound **17**. All of the examples appear to have either a 2-pyridyl or the fluoropyridyl substituent in the 7-position (as seen in the benzimidazole **6**) and a fluorine in the 4-position (e.g. **23**) with the 5-position being the major point of diversity. No specific data for any of the compounds are reported but the compounds have been tested in DNA supercoiling assays for GyrB inhibition and a DNA relaxation assay for ParE inhibition. Under both assay conditions there are compounds reported to be active in the sub-micromolar range (IC_{50} values 30–550 nM) and are active in antibacterial assays against Gram-positive pathogens.

20.4.5 Imidazopyridazines

Adding a twist to the efforts on structural classes containing [6,5]-bicyclic heteroaryl ethyl ureas, AstraZeneca has reported a series of imidazopyridazines where the ethyl urea substituent is appended to the 6-membered ring instead of the 5-membered ring (Figure 20.9).[30] Compound **24** is reported to be active on *S. pneumoniae* ParE (IC_{50} 872 nM) and the related example **25** displays an IC_{50} of 508 nM (also on ParE). It was not reported whether these compounds are similarly potent inhibitors of GyrB. The MIC of **25** is 100 µg/ml against a strain of *S. pneumoniae.*

20.4.6 Pyridines

Additional disclosures from AstraZeneca have reported a series of pyridines (e.g. **26**) containing the ethyl urea moiety (Figure 20.9).[31,32] Compounds in this patent have been tested in *E. coli* GyrB and ParE assays. The only data that is provided is on *S. aureus* GyrB and the majority of the compounds appear to have IC_{50} values ≤ 1 µM. The MIC against *S. aureus* for compound **26** is

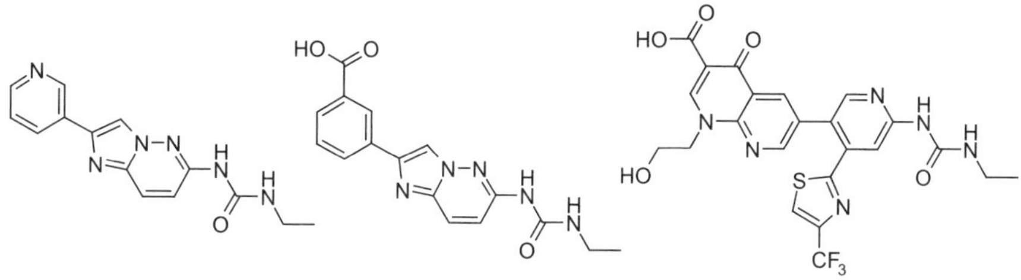

24 **25** **26**

Figure 20.9 AstraZeneca imidazopyridazine and pyridine ethyl ureas.

reported as 0.55 μM (~0.3 μg/ml). What is intriguing about this series in terms of its structure is that it appears to be a hybrid of the ethyl urea GyrB/ParE inhibitors together with the quinolone class of antibiotics.

20.5 Summary and Outlook

Bacterial DNA gyrase and topoisomerase IV have proved their value as targets for antibacterial therapy in the form of the quinolones/fluoroquinolones. Members of this class of antibiotics have also demonstrated that they can target both essential enzymes (GyrA and ParC subunits respectively) and the perceived benefit of this is that dual-acting compounds are likely to help slow the emergence of resistance in a clinical setting. Nevertheless, bacterial resistance is likely to be a continual threat to world health and new opportunities for the development of antibacterials that help combat this threat are required. The ATP-binding domains of DNA gyrase (GyrB) and topoisomerase IV (ParE) have been identified as an alternative opportunity to find dual-targeting inhibitors that affect the function of the bacterial type II topoisomerases.

Interest in the ATP-binding site of these targets has been fuelled by structural data on both GyrB and ParE and the similarity of their binding sites. Application of this structural data has resulted in the emergence of a synthetic class of GyrB/ParE inhibitors referred to as the ethyl ureas, that were first discovered by Vertex Pharmaceuticals but have since been exploited by several other groups (Table 20.2). Based on the preclinical data available to date, the ethyl ureas are emerging as one of the most interesting series of single pharmacophore dual acting inhibitors of GyrB and ParE. Ethyl ureas are typically active against Gram-positive organisms and exhibit very low spontaneous resistance frequencies, which is an indication of their potential dual-targeting ability. They have also been shown to have a cidal mode of action and there are now several examples of compounds which show *in vivo* efficacy in rodent models of infection. The challenge now is to translate the preclinical effectiveness of this class of compounds into a clinically relevant compound.

Table 20.2 Ethyl urea chemical series with representative structures and selected data.

Representative Structure	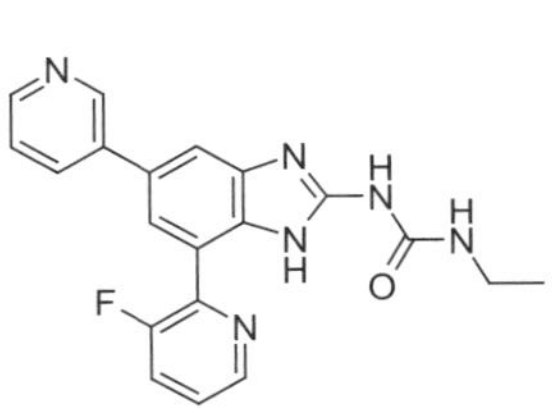	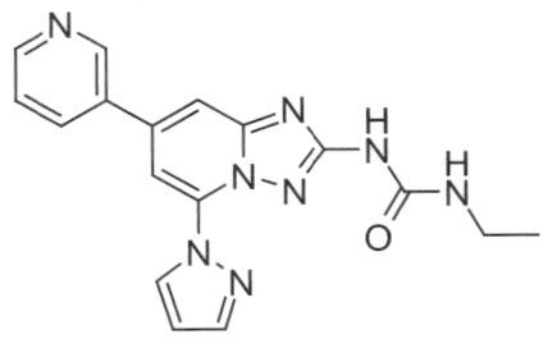	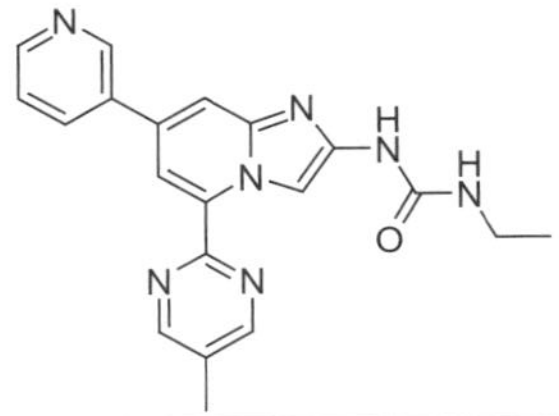	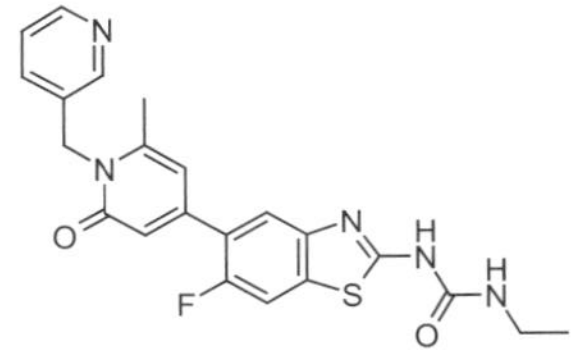
Organisation	Vertex	Biota/Evotec	Pfizer	Biota
GyrB K_i/IC_{50}	*S. aureus* 14 nM *E. coli* <4 nM	*E. coli* 54 nM	*S. pneumoniae* 53 nM	*E. coli* <750 nM
ParE K_i/IC_{50}	*E. coli* 23 nM	*E. coli* 25 000 nM	*S. pneumoniae* 250 nM	not available
MICs	*S. aureus* 0.03 μg/ml *S. pneumoniae* 0.03 μg/ml *H. influenzae* 1 μg/ml	*S. aureus* 2–4 μg/ml *S. pneumoniae* 2 μg/ml *E. faecalis* 2–4 μg/ml *E. coli* >128 μg/ml	*S. aureus* 0.5 μg/ml *S. pneumoniae* 0.125 μg/ml *S. pyogenes* 0.5 μg/ml	*S. aureus* ≤0.12 μg/ml *S. pneumoniae* ≤0.03 μg/ml *S. pyogenes* ≤0.5 μg/ml *E. faecalis* ≤0.5 μg/ml
In vivo efficacy	√	√	√	√
Source	Primary literature	Primary literature	Primary literature	Patent
Representative Structure	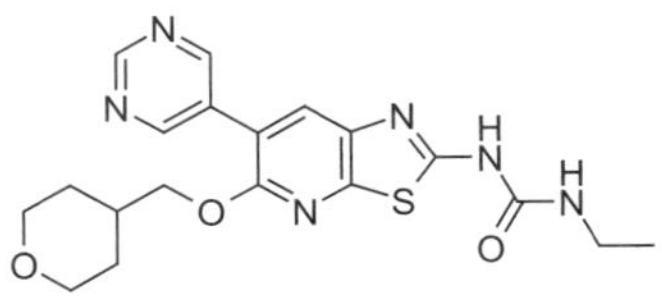	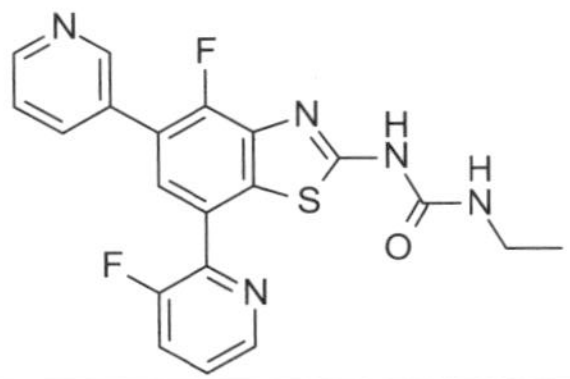	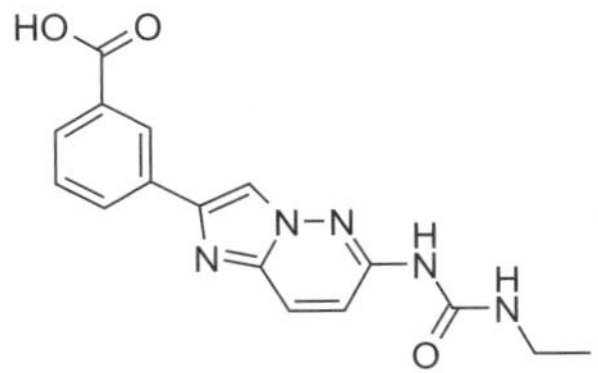	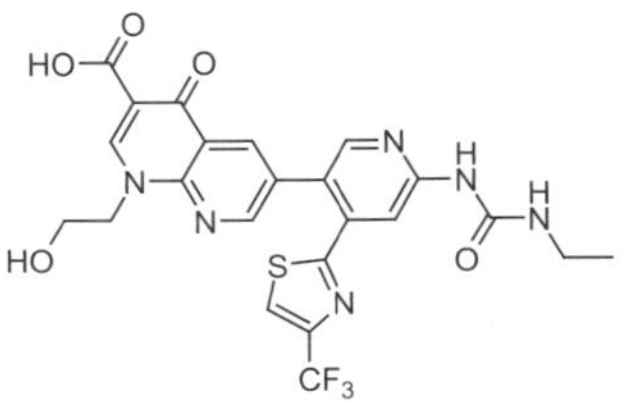
Organisation	AstraZeneca	Ranbaxy	AstraZeneca	AstraZeneca
GyrB K_i/IC_{50}	*M. smegmatis* <3 nM	Supercoiling <550 nM	not available	*S. aureus* 93%@1000 nM
ParE K_i/IC_{50}	not available	Supercoiling <550 nM	*S. pneumoniae* 508 nM	*E. coli* <200 000 nM
MICs	*S. pneumoniae* 0.0625 μg/ml	*S. aureus* <2 μg/ml *S. pneumoniae* <0.5 μg/ml *S. pyogenes* <2 μg/ml	*S. pneumoniae* 100 μg/ml	*S. aureus* 0.3 μg/ml
In vivo efficacy	not available	not available	not available	not available
Source	Patent	Patent	Patent	Patent

References

1. L. L. Silver, *Clin. Microbiol. Rev.*, 2011, **24**, 71.
2. J. B. Bremner, J. I. Ambrus and S. Samosorn, *Curr. Med. Chem.*, 2007, **14**, 1459.
3. L. L. Silver, *Nat. Rev. Drug Discovery*, 2007, **6**, 41.
4. R. N. Brogden, A. Carmine, R. C. Heel, P. A. Morley, T. M. Speight and G. S. Avery, *Drugs*, 1981, **22**, 337.
5. D. D. Long, J. B. Aggen, J. Chinn, S. K. Choi, B. G. Christensen, P. R. Fatheree, D. Green, S. S. Hegde, J. K. Judice, K. Kaniga, K. M. Krause, M. Leadbetter, M. S. Linsell, D. G. Marquess, E. J. Moran, M. B. Nodwell, J. L. Pace, S. G. Trapp and S. D. Turner, *J. Antibiot.*, 2008, **61**, 603.
6. G. T. Robertson, E. J. Bonventre, T. B. Doyle, Q. Du, L. Duncan, T. W. Morris, E. D. Roche, D. Yan and A. S. Lynch, *Antimicrob. Agents Chemother.*, 2008, **52**, 2324.
7. J. C. Wang, *Annu. Rev. Biochem.*, 1996, **65**, 635.
8. K. Drlica, H. Hiasa, R. Kerns, M. Malik, A. Mustaev and X. Zhao, *Curr. Top. Med. Chem.*, 2009, **9**, 981.
9. M. Oblak, M. Kotnik and T. Solmajer, *Curr. Med. Chem.*, 2007, **14**, 2033.
10. H.-J. Boehm, M. Boehringer, D. Bur, H. Gmuender, W. Huber, W. Klaus, D. Kostrewa, H. Kuehne, T. Luebbers, N. Meunier-Keller and F. Mueller, *J. Med. Chem.*, 2000, **43**, 2664.
11. M. Whittaker, *Drug Discovery Today*, 2009, **14**, 623.
12. S. Bellon, J. D. Parsons, Y. Wei, K. Hayakawa, L. L. Swenson, P. S. Charifson, J. A. Lippke, R. Aldape and C. H. Gross, *Antimicrob. Agents Chemother.*, 2004, **48**, 1856.
13. S. M. Ronkin, M. Badia, S. Bellon, A.-L. Grillot, C. H. Gross, T. H. Grossman, N. Mani, J. D. Parsons, D. Stamos, M. Trudeau, Y. Wei and P. S. Charifson, *Bioorg. Med. Chem. Lett.*, 2010, **20**, 2828.
14. Y. Wei and A. Letiran, PDB accession code: 3FV5.
15. P. S. Charifson, A.-L. Grillot, T. H. Grossman, J. D. Parsons, M. Badia, S. Bellon, D. D. Deininger, J. E. Drumm, C. H. Gross, A. LeTiran, Y. Liao, N. Mani, D. P. Nicolau, E. Perola, S. Ronkin, D. Shannon, L. L. Swenson, Q. Tang, P. R. Tessier, S.-K. Tian, M. Trudeau, T. Wang, Y. Wei, H. Zhang and D. Stamos, *J. Med. Chem.*, 2008, **51**, 5243.
16. N. Mani, C. H. Gross, J. D. Parsons, B. Hanzelka, U. Müh, S. Mullin, Y. Liao, A.-L. Grillot, D. Stamos, P. S. Charifson and T. H. Grossman, *Antimicrob. Agents Chemother.*, 2006, **50**, 1228.
17. T. H. Grossman, D. J. Bartels, S. Mullin, C. H. Gross, J. D. Parsons, Y. Liao, A.-L. Grillot, D. Stamos, E. R. Olson, P. S. Charifson and N. Mani, *Antimicrob. Agents Chemother.*, 2007, **51**, 657.
18. P. G. Higgins, A. C. Fluit and F. J. Schmitz, *Curr. Drug Targets*, 2003, **4**, 181.
19. S. P. East, C. B. White, O. Barker, S. Barker, J. Bennett, D. Brown, E. A. Boyd, C. Brennan, C. Chowdhury, I. Collins, E. Convers-Reignier, B. W. Dymock, R. Fletcher, D. J. Haydon, M. Gardiner, S. Hatcher, P. Ingram,

P. Lancett, P. Mortenson, K. Papadopoulos, C. Smee, H. B. Thomaides-Brears, H. Tye, J. Workman and L. G. Czaplewski, *Bioorg. Med. Chem. Lett.*, 2009, **19**, 894.
20. L. Brino, A. Urzhumtsev, M. Mousli, C. Bronner, A. Mitschler, P. Oudet and D. Moras, *J. Biol. Chem.*, 2000, **275**, 9468.
21. D. C. D. Butler, H. Chen, V. R. Hegde, C. Limberakis, R. M. Rasne, R. J. Sciotti and J. T. Starr, *Int. Pat. Appl.*, WO 2006038116, 2006.
22. D. Haydon, R. Halsey, J. Bennett and L. Czaplewski, unpublished results.
23. J. T. Starr, R. J. Sciotti, D. L. Hanna, M. D. Huband, L. M. Mullins, H. Cai, J. W. Gage, M. Lockard, M. R. Rauckhorst, R. M. Owen, M. S. Lall, M. Tomilo, H. Chen, S. P. McCurdy and M. R. Barbachyn, *Bioorg. Med. Chem. Lett.*, 2009, **19**, 5302.
24. D. J. Haydon, L. G. Czaplewski, N. J. Palmer, D. R. Mitchell, J. F. Atherall, C. R. Steele and T. Ladduwahetty, *Int. Pat. Appl.*, WO 2007148093, 2007.
25. D. J. Haydon and L. G. Czaplewski, *Int. Pat. Appl.*, WO 2009074812, 2009.
26. D. J. Haydon and L. G. Czaplewski, *Int. Pat. Appl.*, WO 2009074810, 2009.
27. D. Haydon, S. Barker, J. Bennett, D. Brown, I. Collins, P. Lancett, C. Smee, H. Thomaides and L. Czaplewski, unpublished results.
28. S. R. Ghorpade, M. G. Kale, D. C. Mckinney, S. H. Peer Mohamed and A. K. V. Raichurkar, *Int. Pat. Appl.*, WO 2009147431, 2009.
29. J. A. Sattigeri, N. Kumar, A. Yadav, L. Sharma, I. A. Cliffe, S. B. Varughese, S. R. Shabbir, V. S. Raj, D. J. Upadhyay and P. K. Bhatnagar, *Int. Pat. Appl.*, WO 2009156966, 2009.
30. J. Dumas and B. Sherer, *Int. Pat. Appl.*, WO 2009027733, 2009.
31. P. Hill, J. I. Manchester, B. Sherer and A. L. Choy, *Int. Pat. Appl.*, WO 2009147433, 2009.
32. N. Bifulco, A. L. Choy, O. Quiroga and B. Sherer, *Int. Pat. Appl.*, WO 2010136817, 2010.

Epilogue

The sister fields of multi-target drug discovery and systems biology are moving increasingly quickly and it is in the nature of publishing books such as this that reported work gets out of date equally quickly. The editors are acutely aware of this issue and take this opportunity to add a few comments on very recent developments and views.

Very recently, Gerald Maggiora has published a very provocative perspective,[1] where he argues that biological networks, like other complex networks, possess emergent properties.[2] Thus, attempting to predict the output of such networks from its component parts (e.g. reductionist assays/models) is essentially impossible unless one is in possession of an all-embracing theory for the entire system. In practice, he considers that comprehensive mathematical modelling (e.g. including the use of fuzzy logic and probabilistic approaches) will be needed in order to gain sufficient understanding of biological systems to de-risk future drug discovery. He recognises that this will require a paradigm shift for biologists, biochemists and, especially in the context of this book, medicinal chemists.

Such a paradigm shift will be uncomfortable and, no doubt, slow to achieve. We feel that much of the technology necessary to generate the data needed for such modelling already exists. The power (and 30-year legacy) of analytical 'one-target' approaches allied to a willingness to embrace the complexity of drug development using phenotypic rather than reductionist read-outs should lead to better and, ultimately, less risky approaches to the drugs needed to treat the complex diseases that society now faces. We feel very strongly that medicinal chemists need to be in the vanguard of this 'new' way of thinking about drug discovery, deploying their traditional mix of pragmatism, pattern-recognition expertise and analytical capabilities to this unavoidable challenge.

Given concerns expressed in this book over the physiological irrelevance of many modern assay technologies, it is perhaps unsurprising that we are already

RSC Drug Discovery Series No. 21
Designing Multi-Target Drugs
Edited by J. Richard Morphy and C. John Harris

Published by the Royal Society of Chemistry, www.rsc.org

witnessing a resurgence of interest in phenotypic approaches. Alongside classical *in vivo* pharmacology, the need for new disease-relevant phenotypic assays, such as cheaper and higher throughout cell-based systems, is well appreciated as is the desirability of more robust technologies for subsequently deconvoluting mechanism of action.

The MTDD field is already in transition from one based largely upon serendipity to a new era where more rational target selection and more designed medicinal chemistry becomes possible. Critically, however, serendipitous discoveries are likely to remain a key source of new agents for the foreseeable future given our currently deficient knowledge of biological complexity. We must stay alert for such opportunities and be willing to exploit them as and when they arise. There is no better example of this need to balance the 'designed' and the 'serendipitous' than recent developments in the multi-kinase inhibitor (MKI) field. Until recently, MKIs were discovered serendipitously by screening compounds from selective inhibitor programmes, as exemplified by the first generation dual PI3K/mTOR inhibitors described in Chapter 13. Now a gradual transition has begun to a more rational and prospective approach whereby profiles are predefined based on pathway knowledge and medicinal chemists seek to balance desired activities, eliminate undesired activities and optimise pharmacokinetics.[3] Encouraging though this trend is, it remains unclear to what extent MKIs (and indeed other multi-target agents) with finely tuned profiles can truly be rationally designed. Serendipitous discoveries from large-scale panel screening will likely remain important.

Of the challenges of lead optimisation outlined in Chapter 10, medicinal chemists as a group are becoming much more focussed on optimising physicochemical properties such as lipophilicity to avoid non-specific promiscuity and the associated toxicity that has dogged many previous MTDs in the clinic. One area that needs more attention is the optimal activity ratio which has been ignored in most MTDD projects. An interesting computational approach with the potential to address both the target selection and the optimal balance of activity issues has been reported recently.[4] A combinatorial library of targeted inhibitors was optimised under the control of a genetic algorithm resulting in the automated selection of the best inhibitor combinations to inhibit the release of specific cell mediators *and* information about the ratio producing synergy. Also, in this respect, the large number of MTDs currently in clinical development should provide invaluable feedback to drive future preclinical efforts. An increased focus on translational biomarkers that relate clinical efficacy to target engagement and modulation is essential.

It is becoming ever clearer that tackling complex diseases by addressing multiple targets makes perfect sense and the only remaining debate of consequence is between single agents with multi-target activity versus combining agents with individual activities. It is our view that arguing that one approach will always be better, and therefore ignoring the alternative approach, is misguided. Multi-target drugs (MTDs) and combinations have distinctive advantages and disadvantages. Whichever is preferred is highly situation dependent from the technical feasibility of a particular MTDD target combination right through to the needs of individual patients. Indeed a hybrid approach using combinations of MTDs may even be necessary in some cases.

Since MTDD projects can be inherently challenging, there are two key questions that must be addressed in order that the field fulfils its clear potential. First, how can we improve our capabilities to identify and validate those disease-associated target combinations that will provide efficacy without side effects? New approaches for identifying synergistic combinations are needed and are emerging.[4,5] Second, how do we assess the medicinal chemistry feasibility of those disease-relevant combinations, not only in terms of generating initial hit matter, but importantly that can support efficient optimisation towards a candidate drug? A large number of computational methods for assessing feasibility have already emerged,[6] but more predictive polypharmacology methods are still needed, especially with respect to less obvious combinations where the targets are not obviously related but for which it is still possible to discover drug-like ligands.

Although it is impossible to predict future directions and breakthroughs in MTDD and systems biology research, one thing is clear: research in these areas will continue to build momentum, and over time will provide a solid foundation for addressing unmet clinical needs.

J. Richard Morphy, C. John Harris
31st December 2011

References

1. G. M. Maggiora, *J. Comput. Aided Drug Design*, 2011, **25**, 699–708.
2. *Emergent Properties*: unexpected outputs that stem from interaction between the components of a complex system and its environment; see M. H. V. Van Regenmortel, *EMBO Report*, 2004, **5**, 1016–1020.
3. D. P. Sutherlin, L. Bao, M. Berry, G. Castanedo, I. Chuckowree, J. Dotson, A. Folks, L. Friedman, R. Goldsmith, J. Gunzner, T. Heffron, J. Lesnick, C. Lewis, S. Mathieu, J. Murray, J. Nonomiya, J. Pang, N. Pegg, W. W. Prior, L. Rouge, L. Salphati, D. Sampath, Q. Tian, V. Tsui, N. C. Wan, S. Wang, B. Wei, C. Wiesmann, P. Wu, B. Y. Zhu and A. Olivero. *J. Med. Chem.*, 2011, **54**(2), 7579.
4. B. G. Small, B. W. McColl, R. Allmendinger, J. Pahle, N. J. Rothwell, J. Knowles, P. Mendes, D. Brough and D. B. Kell, *Nat. Chem. Biol.*, 2011, **7**, 902.
5. S. J. Dixon and B. R. Stockwell, *Nat. Chem. Biol.*, 2010, **6**, 318.
6. B. Severyn, R. Liehr, A. Wolicki, K. Nguyen, E. M. Hudak, M. Ferrer, J. Caldwell, J. Hermes, J. Li and Matthew Tudor, *ACS Chem. Biol.*, 2011, 1391.
7. A. Koutsoukas, B. Simms, J. Kirchmair, P. J. Bond, A. V. Whitmore, S. Zimmer, M. P. Young, J. L. Jenkins, M. Glick, R. C. Glen and A. Bender, *J. Proteomics*, 2011, **74**(12), 2554.

Subject Index